Silicon Epitaxy

SEMICONDUCTORS
AND SEMIMETALS
Volume 72

Semiconductors and Semimetals

A Treatise

Edited by R. K. Willardson
CONSULTING PHYSICIST
12722 EAST 23RD AVENUE
SPOKANE, WA 99216-0327

Eicke R. Weber
DEPARTMENT OF MATERIALS
SCIENCE AND MINERAL
ENGINEERING
UNIVERSITY OF CALIFORNIA
AT BERKELEY
BERKELEY, CA 94720

Silicon Epitaxy

SEMICONDUCTORS AND SEMIMETALS

Volume 72

Volume Editors

DANILO CRIPPA

LPE EPITAXIAL TECHNOLOGY
BOLLATE, ITALY

DANIEL L. RODE

DEPARTMENT OF ELECTRICAL ENGINEERING
WASHINGTON UNIVERSITY
ST. LOUIS, MISSOURI

MAURIZIO MASI

DIPARTIMENTO DI CHIMICA
FISICA APPLICATA
POLITECNICO DI MILANO
MILANO, ITALY

ACADEMIC PRESS
A Harcourt Science and Technology Company

San Diego   San Francisco   New York   Boston
London   Sydney   Tokyo

This book is printed on acid-free paper.

Copyright © 2001 by ACADEMIC PRESS

All Rights Reserved.
No part of this publication may be reproduced or transmitted in any form or by any means, electronic or mechanical, including photocopy, recording, or any information storage and retrieval system, without permission in writing from the Publisher.

The appearance of the code at the bottom of the first page of a chapter in this book indicates the Publisher's consent that copies of the chapter may be made for personal or internal use of specific clients. This consent is given on the condition, however, that the copier pay the stated per copy fee through the Copyright Clearance Center, Inc. (222 Rosewood Drive, Danvers, Massachusetts 01923), for copying beyond that permitted by Sections 107 or 108 of the U.S. Copyright Law. This consent does not extend to other kinds of copying, such as copying for general distribution, for advertising or promotional purposes, for creating new collective works, or for resale. Copy fees for pre-2001 chapters are as shown on the title pages. If no fee code appears on the title page, the copy fee is the same as for current chapters.
0080-8784/01 $35.00

Explicit permission from Academic Press is not required to reproduce a maximum of two figures or tables from an Academic Press chapter in another scientific or research publication provided that the material has not been credited to another source and that full credit to the Academic Press chapter is given.

Academic Press
A Harcourt Science and Technology Company
525 B Street, Suite 1900, San Diego, California 92101-4495, USA
http://www.academicpress.com

Academic Press
Harcourt Place, 32 Jamestown Road, London NW1 7BY, UK
http://www.academicpress.com

International Standard Book Number: 0-12-752181-X

International Standard Serial Number: 0080-8784

PRINTED IN THE UNITED STATES OF AMERICA
01 02 03 04 05 EB 9 8 7 6 5 4 3 2 1

To my sweet daughter Camilla

I hope you will see all the marvelous things described in this book come through.

Your dad Danilo

Contents

PREFACE xv
LIST OF CONTRIBUTORS xix

Chapter 1   CVD Technologies for Silicon: A Quick Survey 1

A. M. Rinaldi and D. Crippa

    ABBREVIATIONS USED 2
  I. INTRODUCTION 3
      1. *Chapter Layout* 3
      2. *An Historical Perspective* 4
      3. *Classification Paths for CVD Technologies* 5
 II. BULK POLYCRYSTALLINE SILICON PROCESSES 7
      1. *Silicon Polycrystal Reactors* 7
      2. *Process and Technology Comparison* 9
III. FILM DEPOSITION: REACTORS 10
      1. *Principles of Reactor Design* 10
      2. *Atmospheric-Pressure Reactors* 17
      3. *Low-Pressure Reactors* 24
      4. *Rapid Thermal Process Reactors* 27
      5. *Plasma Reactors* 28
IV. FILM DEPOSITION: PRODUCTS AND CHEMISTRIES 37
      1. *Properties of Films and Materials* 37
      2. *Common Reaction Chemistries* 38
      3. *Plasma Peculiarities* 41
 V. AEROSOL CVD 41
      1. *Basics on Optical Fibers* 41
      2. *Aerosols and Principles of Reactors* 43
      3. *Inside Deposition Reactors* 43
      4. *Outside Deposition Reactors* 47
      5. *Basics of Chemistry for Optical Fibers* 49
VI. CONCLUSIONS 49
    REFERENCES 50

Chapter 2 Epitaxial Growth Theory: Vapor-Phase and Surface Chemistry ... 51

C. Cavallotti and M. Masi

I. Introduction ... 51
II. Development of a Detailed Kinetic Scheme ... 52
III. Gas-Phase Kinetics for Silanes and Chlorosilanes ... 56
   1. *Silane's Gas-Phase Reactivity* ... 56
   2. *Chlorosilane's Gas-Phase Reactivity* ... 61
IV. Surface Kinetics for Silanes and Chlorosilanes ... 71
   1. *Silane's Surface Reactivity* ... 71
   2. *Chlorosilane's Surface Reactivity* ... 74
V. Gas-Phase Precursors to Deposition and Overall Kinetic Scheme ... 80
VI. Overall Kinetic Scheme and Concluding Remarks ... 84
   References ... 86

Chapter 3 Epitaxial Growth Facilities, Equipment, and Supplies ... 89

V. Pozzetti

   Abbreviations Used ... 90
I. Introduction ... 90
II. The Epitaxial Reactor ... 91
   1. *Radiant Heating* ... 94
   2. *Induction Heating* ... 95
   3. *Single-Wafer Reactors* ... 96
   4. *Susceptors* ... 96
   5. *Quartzware* ... 98
   6. *Process Pressure: Atmospheric and Reduced Pressure* ... 98
III. Facilities ... 100
   1. *General Facilities Requirements* ... 102
   2. *Power Supply* ... 102
   3. *Cooling Water* ... 103
   4. *The Clean Room* ... 103
IV. Process Gas and Delivery ... 104
   1. *Process Plumbing* ... 104
   2. *Process Gases* ... 104
   3. *Gas Purity* ... 105
   4. *Gas Filtration* ... 107
   5. *Delivery Systems* ... 108
V. Exhaust Treatment ... 108
   1. *Wet Scrubbers* ... 110
   2. *Burn Systems* ... 110
   3. *Thermal Decomposition* ... 110
VI. Equipment for Power Epitaxy ... 111
   1. *High-Thickness Deposition Capability* ... 111
   2. *Susceptor Design and the Slip Problem* ... 113
   3. *Run-to-Run Repeatability* ... 113
   4. *Robust Equipment Requirements* ... 119
   5. *High-Resistivity Control* ... 119
   6. *Temperature—Uniformity and Profiling* ... 121
   7. *Reaction Chamber Design* ... 121

      8. *Gas Panel* 122
      9. *Gas Injection and Flow Patterns* 123
  VII. ADVANCED APPLICATIONS 123
 VIII. CONCLUSION 124
      REFERENCES 125

Chapter 4   Epitaxial Growth Techniques: Low-Temperature Epitaxy   127

J. Murota

      ABBREVIATIONS USED 127
    I. INTRODUCTION 128
   II. CVD MACHINES 128
  III. SURFACE TREATMENT 133
  IV. EPITAXIAL GROWTH MECHANISMS 135
      1. *Adsorption and Reaction in Undoped $Si_{1-x}Ge_x$ Epitaxial Growth on a (100) Surface* 135
      2. *In Situ Doping of B and P in $Si_{1-x}Ge_x$ Epitaxial Growth* . . . 138
   V. HIGH-QUALITY $Si/Si_{1-x}Ge_x/Si$ HETEROSTRUCTURE GROWTH AT HIGH Ge FRACTIONS 143
  VI. CONCLUSIONS 147
      REFERENCES 148

Chapter 5   Epitaxial Growth Techniques: Molecular Beam Epitaxy   151

Y. Shiraki

      ABBREVIATIONS USED 151
    I. INTRODUCTION 152
   II. MBE MACHINES 153
      1. *Solid-Source MBE Machines* 153
      2. *Gas-Source MBE Machines* 154
  III. SURFACE TREATMENT 155
  IV. GROWTH MECHANISMS 157
      1. *Solid-Source MBE* 157
      2. *Gas-Source MBE* 158
   V. DOPING 162
      1. *Solid-Source MBE* 162
      2. *Gas-Source MBE* 163
  VI. GROWTH OF SiGe(C) ALLOYS 166
      1. *Solid-Source MBE* 166
      2. *Formation and Control of SiGe Alloys by Gas-Source MBE* . . . 166
 VII. FORMATION OF Si/Ge HETEROSTRUCTURES 169
      1. *Band Structures of Si/Ge Heterostructures* 169
      2. *The Critical Thickness* 170
      3. *Surface Segregation* 172
      4. *Surfactant-Mediated Growth* 173
VIII. GROWTH OF SiGe BUFFER LAYERS 174
  IX. SELECTIVE GROWTH 175
   X. FORMATION OF SUPERLATTICES, QUANTUM WIRES, AND QUANTUM DOTS 177
  XI. CONCLUSION 181
      REFERENCES 182

Chapter 6 Epitaxial Growth Modeling 185
M. Masi and S. Kommu

    NOMENCLATURE 185
  I. INTRODUCTION 187
      1. *Physical and Chemical Aspects of Silicon Epitaxy* 189
      2. *The Multiscale Modeling Concept* 190
      3. *Conditions for Epitaxial Growth: The Terrace Step Kink Mechanism* 192
  II. DETAILED MODELING OF EPITAXIAL REACTORS 193
      1. *Conservation Equations* 193
      2. *Computational Issues* 196
      3. *Estimation of Involved Parameters* 197
      4. *Examples of Simulations* 198
  III. REDUCED-ORDER MODELS FOR EPITAXIAL SILICON DEPOSITION 209
      1. *Model Equations* 209
      2. *Boundary Layer Relationships for Mass and Heat Transport Parameters* 210
      3. *Examples of Simulations* 214
  IV. ATOMISTIC ASPECTS: CONTROL OF CRYSTAL MORPHOLOGY 215
      1. *Modeling of Epitaxial Silicon Growth on the Terrace Scale* 215
      2. *Linking the Atomic Scale with the Reactor Scale* 217
      3. *Stability of Epitaxial Growth* 218
  V. SUMMARY 220
    REFERENCES 220

Chapter 7 Epitaxial Layer Characterization and Metrology 225
V.-M. Airaksinen

    ABBREVIATIONS USED 226
  I. INTRODUCTION 226
  II. ON SAMPLING AND ACCURACY 229
      1. *Sampling and Cost* 229
      2. *Precision and Accuracy* 229
      3. *Measurement Pattern, Mapping, and Variation* 230
  III. DOPING CONTROL 235
      1. *General* 235
      2. *The Four-Point Probe* 239
      3. *Mercury Probe CV* 241
      4. *Spreading Resistance Profilometry* 243
      5. *Noncontact Resistivity Measurements: Air-Gap CV, Air-Gap SPV CV,*
        and Air-Gap SPV 245
      6. *Secondary-Ion Mass Spectrometry* 247
  IV. THICKNESS MEASUREMENTS 248
      1. *Standard FTIR* 248
      2. *Model-Based FTIR* 251
      3. *Thickness Measurements on n/N and p/P Structures* 252
      4. *Resistivity and Thickness Measurements on Multilayer Structures* 253
  V. CONTAMINATION AND SURFACE QUALITY 255
      1. *General* 255
      2. *Surface Inspection: Particles, Defects, and Microroughness* 256
      3. *Metal and Molecular Contamination* 261
  VI. FUTURE DEVELOPMENTS AND CONCLUSIONS 270

		1. Measurements on and Characterization Requirements for 300-mm Wafers	270
		2. Should We Monitor the Wafers or the Process?	271
	REFERENCES		272

Chapter 8 Epitaxy for Discrete and Power Devices 277
G. Beretta

	NOMENCLATURE		277
I.	INTRODUCTION		278
	1.	Epitaxy at High Voltages	279
	2.	Epitaxy at High Currents	279
II.	GENERAL CONSIDERATIONS		280
	1.	Breakdown Voltage in Epitaxial Junctions	280
	2.	Dopant Redistribution in the Epitaxial Process	282
	3.	Resistivity and Carrier Density	283
III.	SPECIFIC EPITAXY PROCESSES FOR POWER AND DISCRETE DEVICES		283
	1.	Bipolar High-Voltage Transistors	283
	2.	Bipolar Medium-Voltage Mesa Transistors	285
	3.	Small-Signal Bipolar Devices	285
	4.	RF Power Devices	287
	5.	Medium- and High-Voltage Power-MOS	287
	6.	Low-Voltage Power-MOS	288
	7.	Insulated Gate Bipolar Transistors	288
	8.	Mixed Epitaxy Devices	289
	9.	Diodes and Thyristors	290
IV.	DEFECTS AND PROBLEMS		290
	1.	Epi-Crown	290
	2.	Faceting	291
	3.	Back-Side Defects	291
V.	ACCESSORY CONSIDERATIONS		292

Chapter 9 Epitaxy on Patterned Wafers 295
S. Acerboni

I.	INTRODUCTION		295
	1.	What Is a Pattern on an Epitaxial Wafer?	296
II.	DEVICE REQUIREMENTS AND PROCESS COMPLEXITY		300
	1.	Bipolar, Bipolar/C-MOS, and Mixed Technologies	300
	2.	General Comparison Between Bare and Patterned Epitaxial Wafers	303
	3.	Operations and Epitaxial Equipment	306
III.	SURFACE PREPARATION: PRE-EPITAXIAL CLEANING		306
IV.	GEOMETRICAL PATTERN INTEGRITY		308
	1.	Pattern Shift	309
	2.	Pattern Distortion and Washout	312
	3.	Evaluation of Geometrical Pattern Integrity	313
V.	DOPING PATTERN INTEGRITY		316
	1.	The Autodoping Problem	316
	2.	Doping Profile in The Presence of Autodoping	318
	3.	Autodoping Characterization	321
	4.	Buried Layer Parameters Affecting Autodoping	322
	5.	Recipe Parameters Affecting Autodoping	324

	VI. Crystal Defectivity	327
	1. Thermal Treatments Before Epitaxial Growth	329
	2. Lattice Damage and Recovery After Implantation	332
	3. Dislocations, Slip-Lines, Epitaxial Stacking Faults, and Others	335
	VII. Conclusion: The Best Epitaxial Recipe?	343
	Bibliography	343

Chapter 10   Si-Based Alloys: SiGe and SiGe:C 345

D. J. Meyer

I. Introduction	345
II. Applications of Si Alloys	349
1. Bipolar Applications of $Si_{1-x}Ge_x$	349
2. CMOS Applications of $Si_{1-x}Ge_x$	353
3. Optoelectronics: Superlattices for Photodetectors and Emissive Devices	358
4. Alternative Substrate Material: Pseudo-Substrates for GaAs Solar Cells	358
5. Other Si Alloys	360
6. The Role of Carbon in $Si_{1-x}Ge_x{:}C$ Alloys	362
III. Low-Temperature Surface Preparation	362
1. Native Oxide Removal by H_2 Bake	363
2. Native Oxide Removal by a Liquid-Phase HF Process	364
3. Native Oxide Removal by Vapor-Phase HF Processes	365
4. Other Potential Vapor-Phase Native Oxide Removal Processes	368
IV. Process Chemistry for Si Alloy Deposition	370
1. Influences of Temperature, Pressure, and Gas Composition	371
2. Nonselective Mixed Poly/Epi Deposition	376
3. Selective Deposition of $Si_{1-x}Ge_x$	378
4. Inclusion Of C in $Si_{1-x}Ge_x$	378
5. Oxygen Control in $Si_{1-x}Ge_x$ Alloys	382
V. Metrology of $Si_{1-x}Ge_x$ Layers	386
1. SIMS	386
2. Auger	387
3. Ellipsometry	388
4. X-ray Diffraction	388
5. The Four-Point Probe	388
VI. Production Robust $Si_{1-x}Ge_x$ Processing	389
1. Composing a High-Throughput Process	389
VII. Summary	393
References	394

Chapter 11   Silicon Epitaxy: New Applications 397

D. Dutartre

I. Introduction	397
II. Equipment: State of the Art	398
1. Molecular Beam Epitaxy	399
2. Chemical Beam Epitaxy	400
3. Plasma-Enhanced Chemical Vapor Deposition	401
4. Ultrahigh-Vacuum Chemical Vapor Deposition	402
5. Rapid Thermal Chemical Vapor Deposition	404

III. Epitaxy Processes 406
   1. *New Challenges* 406
   2. *Growth Regimes* 407
   3. *Process Details* 415
IV. New Applications 422
   1. *Low-Cost Epitaxial Wafers* 422
   2. *SOI Wafers* 424
   3. *Bases of RF Bipolar Transistors* 426
   4. *Elevated Sources and Drains in CMOS* 435
   5. *Channel Engineering in CMOS* 438
   6. *New CMOS Architectures* 446
   7. *Other Applications* 451
V. Conclusion 453
   References 455

Index 459
Contents of Volumes in This Series 471

Epitaxial Si 4 μm thick grown by the $SiCl_4/H_2$ CVD process on a precisely oriented (111) silicon substrate as viewed in Nomarski contrast near the edge of the wafer.[1] The tri-pyramids measure about 400 μm across. Low-temperature growth at 900°C for 1 minute probably caused vapor-phase nucleation of silicon-bearing clusters which decorated the substrate and led to enhanced growth at the apexes of the tri-pyramids. Initial low-temperature growth was followed by growth at 1130°C for 3 minutes. The tri-pyramid facets are caused by long-range monatomic step-step attractive forces which are balanced by surface energy forces, yielding facets which lie 0.45° off the (111) crystal orientation. Notice the exceptionally smooth and stable growth at the wafer edge when the substrate orientation reaches the precise critical orientation.[2]

[1] Epitaxy by C.L. Paulnack and K.E. Benson, Bell Labs, unpublished.
[2] D.L. Rode, W.R. Wagner and N.E. Schumaker, Appl. Phys. Lett. **30,** 75 (1977).

Preface

Forty years after it appeared, it is still inspiring to read H. C. Theurer's classic 1961 paper (1) introducing the important technique of epitaxial growth of silicon (chemical-vapor deposition or CVD). This is the silicon-tetrachloride/hydrogen process. Theurer's work set the stage for the preparation of the first epitaxial mesa transistors and for planar silicon technology in general. For over two decades, this was the dominant approach to silicon epitaxy.

Even in those early days, the critical nature of the silicon-substrate interface was understood regarding crystalline order, oxide and carbonaceous contamination, and the vapor-phase nucleation of polymers such as $Si_{10}Cl_{20}H_2$. Nowadays, we have access to greatly improved high-purity apparatus as well as control and analytical instrumentation which permit the use of various chlorosilanes in silicon-epitaxy processes. Early on, trichlorosilane was used because of its ready availability in high-purity form by reaction of silicon-tetrachloride with silicon. More recently, reduced oxide contamination of the silicon substrate due to hydrogen passivation, and due to greatly improved equipment, has lead to the preferential use of dichlorosilane because it allows epitaxial deposition at lower temperatures, thus alleviating pattern distortion and thermal budgets. Indeed, with the advent of high-purity apparatus with respect to oxidation of the silicon surface, it has become feasible to dispense with chlorine altogether and to use silane directly. With the introduction of molecular-beam epitaxy during the last decade, researchers now can prepare silicon and related hetero-epitaxial structures on the scale of monoatomic layer thicknesses under conditions which are far from many of the thermodynamic-equilibrium considerations which pertain to CVD.

Nevertheless, CVD processes at various pressures still hold the advantages of large throughput and low cost. Wafer batches consisting of several dozen wafers are not uncommon for smaller wafer sizes. However, as wafer sizes have increased into the 200 to 300 mm range, single-wafer CVD equipment has become much more important along with greater requirements for planarity and thermal uniformity as well as doping and thickness uniformity—all while achieving low defect levels.

While epitaxy has been almost exclusively the workhorse for bipolar technology for many years, over the last two decades it has come to also serve the CMOS logic, microprocessor, and memory industries very well, especially as MOSFET dimensions have moved into the submicron regime and as BiCMOS requirements increase, as discussed recently by Samoilov *et al.* (2). Thin gate-oxides, latch-up problems, and short-gate effects are all favorably addressed by epitaxial technologies, especially reduced defect levels due to low-temperature epitaxial growth. Thanks to an aggressive competitive environment, the price premium for epitaxy on large wafers is a mere 20 to 30%, and it has been estimated by Royalty that by 2005 nearly half of all 200 mm wafers will be epitaxial wafers (3). It may be interesting to note that silicon's cousin, gallium-arsenide, has undergone a similar evolution during the 1990's, from predominantly direct-implant into semi-insulating wafers for FETs in the 1980's, to heteroepitaxy today along with enormous performance improvements for the wireless and high-speed fiber-optics industries.

So, unlike the early thinking of the 1970's, which hoped that eventually (or ideally) we would evolve toward essentially one epitaxial growth technique and essentially one semiconductor, quite the opposite has occurred. We now have access to quite a range of CVD and epitaxy techniques, as well as heteroepitaxy and a variety of Group-IV and III-V semiconductor alloys such as SiGe/Si and SiGeC/Si, to optimally address a wide variety of applications. It does seem that expanding market diversity is driving technology resources, and there is no end in sight. To date, silicon epitaxy has been tethered to the lattice parameter of the bulk silicon substrate upon which it is grown. However, even that fundamental limitation has been broken in the III-V semiconductor area with the introduction of metamorphic technology. We look forward to its expanded use in the silicon-germanium area too.

One feels obliged, with the ready availability of other fine books on the subject at hand such as Baliga's (4) *Epitaxial Silicon Technology* and Wolf and Tauber's (5) *Silicon Processing*, to explain the need for the present tome. Our goal, set at the beginning of this four-year effort, was to bring together authors, each a leading expert in the field, who not only could bring readers up to date on the latest epitaxial technology for silicon and its related alloys, but who also could present their work in a way that is palatable both to those who practice epitaxy and to those who may be entering the field.

Needless to say, this has taken considerable effort and understanding on all sides, and it is our pleasure to acknowledge the good nature and hard efforts of all the authors represented here. They have been most generous with their time and with their knowledge, some of which has not appeared publicly heretofore. Considering the passage of some 15 years since Baliga's important work, we thought it appropriate to again publish a book dedicated entirely to silicon epitaxy, an important part of the presently 170 billion dollar industry. For those who wish to delve into other related silicon processes, Wolf and Tauber have given us a great resource. In the present work, we focus exclusively on epitaxy.

It is a great pleasure to acknowledge input from our colleagues during the more than four years since January 28, 1997 when we first met to set out the plan for this book. First of all, we are grateful to Jon Rossi, Jon Deluca, Greg Wilson, and Graham R. Fisher (MEMC) who provided the initial enthusiasm and much further advice. For technical discussions, we benefited from talks with David B. Fraser (Intel), James M. Early and Bernard T. Murphy (Bell Labs), H. Ming Liaw (Motorola), Don W. Shaw (Texas Instruments), Paul J. Shaver (PerkinElmer Opto), and Barbara Abraham-Shrauner (Washington University). Students at Washington University, Lieutenant Colonel Li-Hsin Ouyang, Tun Zulkifli, Ashraf Wahba, Chang Hee Lee, and Steve Puntigan were most helpful with reviews and comments.

We wish you interesting and rewarding reading.

DANILO CRIPPA
DANIEL L. RODE
MAURIZIO MASI

1. Theurer, H. C., *J. Electrochem. Soc.* **108,** 649 (1961).
2. Samoilov, Arkadii V., Du Bois, D., Comita, P. B., and Carlson, D., *Semicond. Int.,* 73 (2000).
3. Royalty, R., *Solid State Technol.,* 60 (2001).
4. Baliga, B. J., Ed., *Epitaxial Silicon Technology,* Academic Press (1986).
5. Wolf, S., and Tauber, R. N., *Silicon Processing for the VLSI Era,* Lattice Press (2000).

List of Contributors

Numbers in parentheses indicate the pages on which the authors' contributions begin.

S. ACERBONI (295), *ST Microelectronics, CFM-AGI Department, Agrate Brianza, Italy, email: sara.acerboni@st.com*

V.-M. AIRAKSINEN (225), *Okmetic Oyj R&D Department, Vantaa, Finland, email: veli-matti.airaksinen@okmetic.com*

G. BERETTA (277), *ST Microelectronics, DSG Epitaxy Catania Department, Catania, Italy, email: giorgio.beretta@st.com*

C. CAVALLOTTI (51), *Dipartimento di Chimica Fisica Applicata, Politecnico di Milano, Milano, Italy, email: carlo@chfi.polimi.it*

D. CRIPPA (1), *MEMC Electronic Materials, Epitaxial and CVD Department, Operations Technology Division, Novara, Italy, email: danilo.crippa@lpe-epi.com*

D. DUTARTRE (397), *ST Microelectronics, Central R&D, Crolles, France, email: didier.dutarte@st.com*

S. KOMMU (185), *MEMC Electronic Materials Inc., EPI Technology Group, St. Peters, Missouri, email: skommu@memc.com*

M. MASI (51, 185), *Dipartimento di Chimica Fisica Applicata, Politecnico di Milano, Milano, Italy, email: maurizio.masi@polimil.it*

D. J. MEYER (345), *ATMI, Phoenix, Arizona, email: dmeyer@atmi.com*

J. MUROTA (127), *Research Institute of Electrical Communication, Laboratory for Electronic Intelligent Systems, Tohoku University, Sendai, Japan, email: murota@riec.tahoku.ac.jp*

V. POZZETTI (89), *Consultant, 20047 Brugherio, Italy, email: vipoz@yahoo.com*

LIST OF CONTRIBUTORS

A. M. RINALDI (1), *MEMC Electronic Materials, Epitaxial and CVD Department, Operations Technology Division, Novara, Italy, email: arinaldi@memc.it*

Y. SHIRAKI (151), *Department of Applied Physics, Graduate School of Engineering, The University of Tokyo, Tokyo 113-8656, Japan, email: shiraki@p.rcast.n-tokyo.ac.jp*

CHAPTER 1

CVD Technologies for Silicon: A Quick Survey

A. M. Rinaldi and D. Crippa[1]

MEMC ELECTRONIC MATERIALS
EPITAXIAL AND CVD DEPARTMENT
OPERATIONS TECHNOLOGY DIVISION
NOVARA, ITALY

ABBREVIATIONS USED	2
I. INTRODUCTION	3
1. *Chapter Layout*	3
2. *A Historical Perspective*	4
3. *Classification Paths for CVD Technologies*	5
II. BULK POLYCRYSTALLINE SILICON PROCESSES	7
1. *Silicon Polycrystal Reactors*	7
2. *Process and Technology Comparison*	9
III. FILM DEPOSITION: REACTORS	10
1. *Principles of Reactor Design*	10
2. *Atmospheric-Pressure Reactors*	17
3. *Low-Pressure Reactors*	24
4. *Rapid Thermal Process Reactors*	27
5. *Plasma Reactors*	28
IV. FILM DEPOSITION: PRODUCTS AND CHEMISTRIES	37
1. *Properties of Films and Materials*	37
2. *Common Reaction Chemistries*	38
3. *Plasma Peculiarities*	41
V. AEROSOL CVD	41
1. *Basics on Optical Fibers*	41
2. *Aerosols and Principles of Reactors*	43
3. *Inside Deposition Reactors*	43
4. *Outside Deposition Reactors*	47
5. *Basics of Chemistry for Optical Fibers*	49
VI. CONCLUSIONS	49
REFERENCES	50

[1]Present address: LPE Epitaxial Technology, Via Falzarego 8, 20021 Bollate, Italy.

Abbreviations Used

AP-CVD	Atmospheric pressure CVD
BPSG	Borophosphosilicate glass
BSG	Borosilicate glass
CPU	Central processing unit
CVD	Chemical vapor deposition
CVI	Chemical vapor infiltration
CZ	Czochralsky
DCS	Dichlorosilane (SiH_2Cl_2)
DECR	Distributed electron cyclotron resonance
ECR	Electron cyclotron resonance
FBR	Fluidized bed reactor
FSG	Fluorine-doped silicon glass
FTES	Fluorotriethoxysilane
HDP	High-density plasma
HSG	Hemispherically grained polysilicon
ILD	Interlevel dielectric
IMD	Intermetal dielectric
IVPO	Inside vapor-phase oxidation
LOCOS	Local oxidation of silicon
LP-CVD	Low-pressure CVD
LTE	Low-temperature epitaxy
MBTC	Model-based temperature control
MCVD	Modified CVD
MDES	Methyldimethoxysilane
MMP	Microwave multipolar plasma
MOS	Metal–oxide semiconductor
MPU	Microprocessor unit
OMCTS	Octamethylcyclotetrasiloxane
ONO	Oxide–nitride–oxide
OVD	Outside vapor deposition
PA-CVD	Plasma-assisted CVD
P-CVD	Plasma CVD
PE-CVD	Plasma-enhanced CVD
PH-CVD	Photoassisted CVD
PI-CVD	Plasma impulse CVD
PMCVD	Plasma-modified CVD
PSG	Phosphosilicate glass
RF	Radiofrequency
RIE	Reactive ion etching
RP-CVD	Reduced-pressure CVD
RTP	Rapid thermal process
sccm	Standard cubic centimeter per minute

SIPOS	Semiinsulating polysilicon
slm	Standard liter per minute
SMIF	Standard mechanical interface
SP-CVD	Surface plasma CVD
STI	Shallow trench isolation
TCS	Trichlorosilane ($SiHCl_3$)
TEOS	Tetraethyl orthosilicate
TET	Silicon tetrachloride ($SiHCl_4$)
TFT	Thin-film transistor
TMCTS	Tetramethyl cyclotranssiloxane
TMOS	Tetramethyl orthosilicate
TMS	Tetramethylsilane
TRIES	Triethoxysilane
UHV-CVD	Ultrahigh-vacuum CVD
UPS	Uninterruptible power supply
VAD	Vapor axial deposition
WJ	Watkins Johnson

I. Introduction

1. Chapter Layout

Epitaxy has been a primary application of CVD. Because of this fact, and to favor a didactic approach to a relevant part of this book's subject, this first chapter is devoted to a survey of CVD technologies as they are applied in the realm of silicon. These reasons also make clearer the characteristics of these pages. The contents are quite widespread and commonly known, but it is handy to have them together on starting this journey through silicon epitaxy. There are also peculiarities that we deemed worth including in these pages.

Focusing on silicon technology, it is important, in our opinion, but not very common in other reviews, to spend some time introducing bulk processes for the production of high-purity silicon. Although they are very different from all other processes considered here (of course, apart from basic chlorosilane chemistry), they fully share the CVD definition and are the practical starting point for a great part of the microelectronic industry. It is also interesting to consider that electronic-grade silicon is by far the most abundant material (exceeding 20,000 metric tons per year) synthesized by CVD.

Another interesting digression will involve optical fiber manufacturing, taking a glance there, too, at silicon compounds and CVD.

Within the limit of squeezing a sufficiently complete review of these topics into a few pages, the topic list and the layout structure are a way to transmit a personal view of and industrial experience in the field. To the same purpose, some practical considerations that sometimes emerge exemplify gaps in research and distances between development and large-scale manufacturing still to be filled.

It is a difficult task to lay out a bibliography consistent with this selection of topics. By no means are the references cited either exhaustive or fully representative; on the contrary, they are just the best-known to the authors and are intended as a simple starting point for the reader. It is in this spirit that we cite here some textbooks that can serve as general references for the whole chapter (Cerofolini and Meda, 1989, 1991; Chang and Sze, 1996; Hitchmann and Jensen, 1993; Jayant Baliga, 1986; Sherman, 1987; Sze, 1983; Vossen and Kern, 1991; Wolf and Tauber, 1986). We encourage the interested reader to have a look at these books and the references therein.

In this rapidly evolving field, review books quickly lose their utility as a view of the state of the art in the sector. To preserve this view, the proceedings of periodical international meetings on CVD *in toto* are invaluable, for instance, those of the Electrochemical Society, the Material Research Society, and EuroCVD.

2. A Historical Perspective

Without going into details, it is instructive to note some relevant milestones of CVD evolution over time.

It is commonly agreed upon to place the birth of CVD application some thousand years ago, in the prehistoric paintings drawn with soot on cave walls. It is curious enough that the first conscious and industrial applications, in the last decade of the 18th century, still used carbon-based deposition.

After that unsteady beginning, CVD expanded to more and more elements, notably metals and silicon, in the first years of the 19th century. The initial connection with silicon was the factor that boosted the exponential growth of CVD just after World War II; the same impressive growth of the microelectronic industry was transferred to one of its founding production tools.

It is curious to consider rare CVD applications such as the "arsenic mirror," used to identify arsenic poisoning: by heating a sample, metallorganics are formed and their consequent CVD of arsenic on an exposed surface is easily detected from the mirroring properties of the metallic film. This test was also described by the famous writer Agatha Christie.

Silicon-based electronics expanded rapidly into many branches: discrete devices and integrated circuits, power applications, CPU and MPU, and memories. In this very expansion are the seeds for other CVD applications. First are the varied panorama of new semiconductors with high-speed and telecom applications and further spinoffs in the optoelectronic industry. Still connected to the microelectronic sector is the development of integrated sensors and micromachining, magnetic coatings, superconductive films, and exotic structures for quantum devices. We see in parallel, benefiting from (relatively) low-cost deployment of technologies consolidated in the electronics area, an increasing number of applications to coatings for industrial and consumer areas, including coatings for cutting tools and erosion protection, by deposition of refractory metals, carbides and silicides, borides and nitrides, ceramics, and diamond-like materials.

To complete the big picture, we must also mention chemical vapor infiltration (CVI), which can be regarded in principle as CVD applied to the complex microscopic feature surface of porous or fibrous materials. Here, too, the founding principles and many materials are the same, but CVI has now evolved its own ambit of essential applications and techniques.

Closely connected to the applications just listed are several areas of support and study: detailed characterization of reaction kinetics, numerical simulations, research on new precursors, improvement and development of reactors and related technologies, and metrological, monitoring, and characterization techniques.

It is undoubtedly clear, from the overview of these activities, that CVD is widely used in industry and research. Whether it is "well-known," however, is uncertain. There is surely plenty of experience, but full-fledged theoretical descriptions and systematic approaches to many issues are still lacking. A striking example is the pyrolysis of silane, resulting in the deposition of a (more or less crystalline) form of silicon. This is undoubtedly one of the simplest silicon-compound molecules, and a reaction used and studied for more than 50 years, but

- there are still unknowns concerning the relevant reaction steps and constants,
- there is no ready-to-use simulation allowing a process engineer to develop a new process (i.e., a simulation reliable over nonstandard process conditions),
- there are no tools to predict film properties from process parameters, and
- there are still wide gaps in understanding and characterizing the structure and morphology of the deposited film.

3. CLASSIFICATION PATHS FOR CVD TECHNOLOGIES

A good starting point, which we have delayed until now, is some reasoning on the definition of CVD. What the acronym directly signifies is Deposition (of a solid material) from a Vapor phase by means of Chemical reactions.

About the *deposition* concept, we remark that it can refer to films as well as bulk materials. Present applications fully span this very broad scope, covering a range of thicknesses of eight orders of magnitude, from a single atomic layer to several centimeters. Moreover, the term should strictly exclude from the CVD family processes such as thermal oxidation, in which where the film is *grown* using the substrate material itself.

Working from a *vapor* phase is intended in a broad sense since, depending on the types of precursors and the process temperatures and pressures, different components of the gaseous mixture may behave like vapors while others behave almost like ideal gases. A further extension is required for plasmas.

Finally, stress is put also on the role of *chemical* reactions as the *origin* of the deposited material. Thus a clear distinction is made from physical methods of deposition (sputtering, sublimation, and molecular beam epitaxy) and modification of a preexisting material (diffusion, implantation, and annealing). Since the objective is to deposit a solid material, heterogeneous reactions will of course be the most relevant ones, and generally (but not always) homogeneous reactions will be undesirable, because they are commonly a source of particulate contamination of the deposited material. We do not forget that some of these processes (oxidation, diffusion, and annealing) are closely intertwined with CVD processes, often sharing the same reactors and technologies, many times being subsequent steps in the production line.

Considering the impressive variety of applications, even restricted to those that are silicon related, it is a natural consequence that there exist many methods of classification and approaches to studying the subject. It is instructive to understand how different schemes may be created in the same process of apprehending the subject.

The most straightforward approach starts by considering all the subsets related to physical parameters. We may group applications according to process temperature, pressure, and chemicals. This scheme leads to a quite fragmented and detailed subdivision.

Another classification is based on film structure (from amorphous to monocrystalline) and film electrical conductivity (from conductor to dielectric) and/or other physical properties. This subdivision is clearly geared toward the final application of the different materials.

For still another approach we can consider that the great majority (if not all) of the reactions involved need to be activated by some external energy source. A very common method of reaction activation is the thermal one, with heat being generated and supplied to the desired volume in different ways. It happens that the principal heating systems (at least in the silicon-related CVD subset we are considering) are quite well matched with the pressure regime. The latter is roughly split into high pressure, atmospheric pressure (AP), reduced pressure (RP), low pressure (LP), and ultrahigh vacuum (UHV). Another diffused activation energy source uses plasma instead of thermal activation. Although these reactors work almost in the LP range, they are better dealt with separately, focusing on the transduction of electromagnetic power to chemical reactions.

We have thus provided an interpretation of a quite common classification scheme. This line of thought should favor a pragmatic approach for the engineer, who has to look at CVD from the viewpoint of the machine and of the product obtained. We hope that it will be insightful even for the researcher.

Accordingly, we start with a self-contained section on bulk processes (Section II), move on to the hardware used for film processes (Section III), and then skim through the product families and their related chemistries (Section IV). Another self-contained section follows, devoted to aerosol CVD (Section V), because of its special application to optical fibers.

II. Bulk Polycrystalline Silicon Processes

1. SILICON POLYCRYSTAL REACTORS

a. Siemens Reactors

The Siemens reactor is constructed around thermal hydrogen reduction of trichlorosilane (TCS). Thermal power, in huge quantities given the reactor dimensions, is provided by means of resistive heating. In the early days of CVD, a very simple method for activating a reaction thermally was to place a metal wire heated by the Joule effect in the gas mixture. This technique is still widely used in the laboratory and research because it is simple and effective and allows a quick study of the process and product before expensive apparatus is built or modified. For this process, while the principle is the same, the stringent purity requirements for the product make the choice for the filament material silicon itself. Slim rods (0.5 cm per side) of electronic-grade polysilicon are positioned in the reactor as shown in Fig. 1. The U-shaped configuration provides the benefits of easy machining and assembly of the rods and mechanical stability against the relevant weight at the end of the process, having both electrical connections close to each other. Another advantage of using silicon as starting slim-rods is that it does not have to be removed from the final product, either when the rods are crushed into nuggets (for Czochralsky crucible feed) or for direct floating-zone monocrystal growth. Alternating current is fed to the rods for the whole process duration (typically 3 to 5 days) but increasing with time. The effect of a reduction in resistance with increasing cross section is largely offset by the higher mass conversion over the larger rods' surfaces. Considering the high currents and power needed for long times, especially at the end of the process (more than a thousand amperes and a few hundred kilowatts, respectively), power control is a critical element of this reactor system.

The reaction chamber was initially built as a quartz bell, with dimensions of up to almost 1 m in diameter and 2 m in height, over a basement of stainless steel with a gas inlet and outlet and a ceramic feedthrough for each electrical connection. The bell was cooled with forced airflow. Cooling of the walls is required to prevent deposition on them; for the considered reaction, the upper temperature limit for the inner side of the process chamber is some 700°C.

The discovery that the use of stainless steel did not contaminate the polysilicon allowed the construction of bigger and more rugged vessels (more than 1 m in diameter and almost 3 m in height), with more efficient water-cooling. Another relevant advantage of the new process chamber was the possibility of operating at pressures higher than atmospheric, with improvements in growth rate and process efficiency.

Advances in process and reactor design are related mostly to greater efficiency in energy utilization (now roughly 100 kWh/kg) and increased dimensions of the final rods, to offset the long times needed between each run to unload the product and

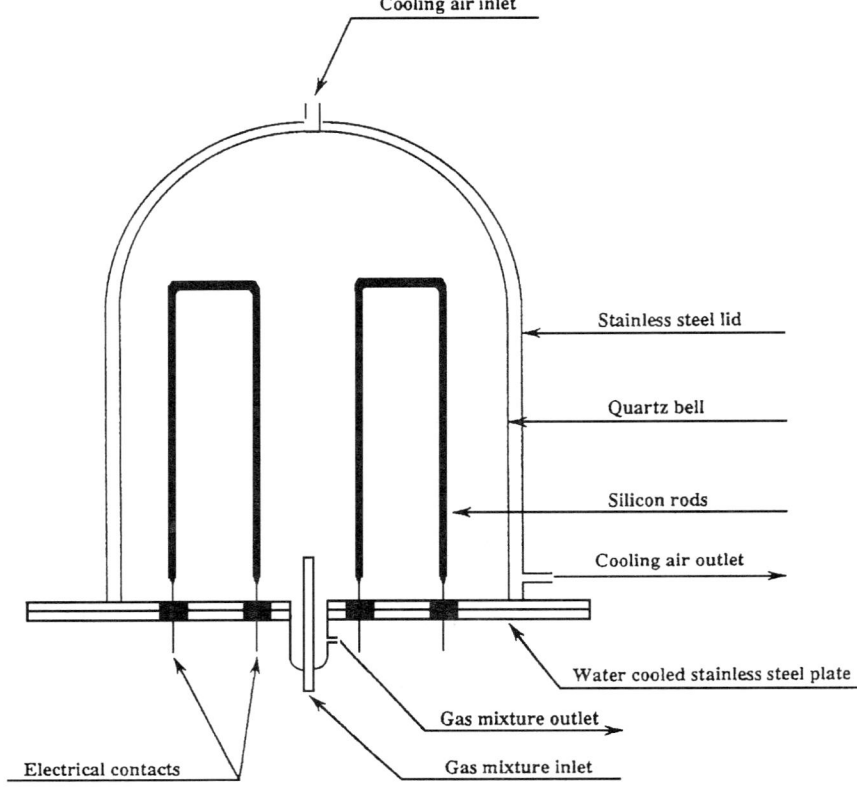

FIG. 1. Simplified cross section of a Siemens reactor for polycrystalline silicon deposition.

restart a new run; also, practical factors related to the weights of the rods (which can approach 100 kg for each U-couple) are relevant in industrial applications to avoid breakage of the rods during processing or unloading. Rods are now of impressive dimensions: easily exceeding 15 cm in diameter (depending on their total number) and 170 cm in length. Reactors are now designed to produce 12 or more rods in U-shaped connections.

Closely related to the reactor and the process of decomposition to silicon are the auxiliary processes and facilities for generation and recycling of gases. Efficiency of recycling (of the huge quantities of gas involved) is a factor comparable to the decomposition process in determining the profitability of the production system. We recall now, because of the connection with the "Siemens process" as a whole, that TCS for the process is generated in a fluidized bed reactor (FBR) from metallurgical-grade silicon and HCl.

b. *Komatsu and Union Carbide Reactors*

The Komatsu reactor is based on the same concept as the Siemens reactor, but is designed around silane decomposition. Its structure is very similar to that of the Siemens reactor, and the differences are dictated by the higher reactivity of silane. Gas distribution is more critical and sensitive to depletion effects, so that process gases (silane strongly diluted in hydrogen) are introduced into the reactor by means of multiple cooled injectors. Similarly, reactor-wall cooling is more critical and the wall temperature has to be below 100°C to have ample margin for avoiding spurious reactions.

Process activation is achieved in exactly the same way, but typically with only four slim rods making two U-connections. The grown rods are smaller than those of Siemens reactors, but they are still about 10 cm in diameter. Silane generation, like TCS generation for the Siemens process, is an integral part of the production cycle and is done by means of a FBR reacting Mg_2Si and NH_4Cl.

The Union Carbide (second-process) reactor is just a modified Komatsu reactor that can hold up to 12 rods. Here, too, the design of the injection is of utmost importance, as is cooling. For these reactors, also, the average power efficiency is roughly 100 kWh/kg. For the Union Carbide (second-process) reactor, the choice for silane generation was via hydrogenation of TET and silicon in a high-pressure FBR and by successive conversion and distillation to TCS, then to dichlorosilane (DCS), and eventually to silane.

2. PROCESS AND TECHNOLOGY COMPARISON

The main parameters of the Siemens process, as a consequence of the TCS precursor choice, are the temperature and precursor dilution. The process temperature is chosen to be between 1000 and 1200°C, while TCS is diluted approximately 1:10 (molar) in hydrogen; the dilution ratio is exactly determined according to the process temperature, for best efficiency of conversion to silicon. Some quantity of TET is added to the TCS. This is because the presence of TET hydrogenation products helps move the equilibrium of TCS toward decomposition; details on concurrent reactions and their interactions are given in Chapter 2, on the chemistry of hydrides and chlorosilanes. It is important to note that there are generally strong variations in parameter settings over each batch in process. The reason, as already noted for power usage, comes from the change in the active growth surface available. It starts at $0.1\ m^2$ and increases during the tens of hours of each run up to about $2\ m^2$. While the process could indeed proceed with fixed settings, the quality of the product (even the crystalline structure of it was found to influence Czochralsky pulling) and the profitability of the process are related to uniform growth and efficient decomposition of the reactants, to reduce the cost of the recycling steps. For these purposes, as the power is increased during the run, so is the gas flow, up to 1500 L/min.

Following a similar path, the Komatsu process takes place in the temperature range between 800 and 850°C. Pure pyrolysis of silane is not possible because of fast homogeneous reactions so that, even if there is no chlorine to bind, hydrogen dilution is mandatory; the determined dilution ratio is now about 1:50 in hydrogen.

The Union Carbide process, besides having the previously listed differences from the Komatsu process, is characterized by a slightly higher process temperature, toward 900°C, and hence a still higher silane dilution, up to and over 1:100 (molar).

Practice has confirmed that the silane processes are characterized by a much higher production cost than the original Siemens process, which is an essential factor considering that the Komatsu and Union Carbide processes were developed to avoid the royalty burden of using the Siemens. Another disadvantage of these processes is the lower specific deposition rate. There are some advantages, however, which are interesting. This polysilicon is the best one since product purity is directly related to gas purity, and silane, with its low boiling point, can be purified to a higher degree than chlorosilanes. Other practical factors are the absence of corrosive compounds and the easy recycling of by-product hydrogen.

We note here, for the sake of completeness, that the study of other technologies for bulk polysilicon production has moved in two other directions. One of them is the use of DCS as the precursor (Dow Corning pilot experiments) in Siemens-like reactors, trying to balance the good and bad properties of TCS and silane with its intermediate properties; although interesting, it presented problems on the industrial scale because of instability. The other direction is based, using various precursors, on FBRs, also for decomposition to silicon (Texas Instruments second process, Ethyl Corporation process, Union Carbide third process). The product, in the form of spherules (pellets) in the (sub)millimeter range, is known as light-poly, as opposed to chunk-poly (the nuggets obtained from crushing the rods), and is also used for Czochralsky pulling, allowing the possibility of continuous feed to the crucible.

III. Film Deposition: Reactors

1. PRINCIPLES OF REACTOR DESIGN

a. Some Concepts Behind Reactors

We now venture a tentative explanation of the preceding observation on the matter-of-fact relationship between pressure and temperature regimes in most reactors. Information on the reaction activation energy and growth rate is synthetically but well described in an Arrhenius plot, which has the general structure shown in Fig. 2. Similar plots are drawn for many deposition reactions, with accurate experimentation at different temperatures but at a given pressure. The plot

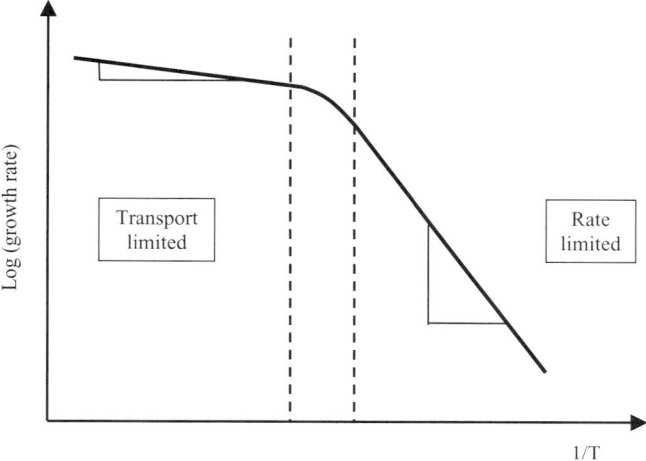

Fig. 2. Typical Arrhenius plot; logarithm of the growth rate as a function of the inverse of the absolute temperature T (K).

has three clearly distinct regions, as marked in Fig. 2. We do not consider here the complexities of the transition region (the "knee") or deviations (not plotted) possible at the extreme left.

The region at the right has a manifest slope, which is a direct measure of the reaction activation energy. Here the temperature, even if in the lower range, has strong effects on the growth rate. Recalling that differences in growth rate translate into nonuniformity (of thickness and other properties) of the deposited layer, the requirement for a performing temperature control when operating in this region is evident. This region is commonly referred to as surface rate limited or kinetically controlled.

On the left side, at higher temperatures, the growth rate shows relatively modest variation with temperature and the relevant factors are related to species motion (large-scale convection and small-scale diffusion across the boundary layer). Therefore this is designated the mass transport-limited or diffusion-controlled region.

Purely theoretical reasoning would suggest that a reactor operating in the kinetic region should be designed with the focus on the temperature generation and control system, leaving relaxed requirements for other parts. On the other hand, similar reasoning would imply that the greatest attention should be given to achieving a uniform gas distribution in transport-limited operating reactors. The obvious way to overcome this limitation is to increase the diffusivity by working at lower pressures.

How do we explain the fact that AP reactors are coupled with transport-limited regimes and LP reactors with kinetically limited regimes? We have to mesh various perspectives.

If we go back to the beginning of CVD application for silicon, the developmental approaches were at AP. Throughput was not an issue in getting things working in the first place, and uniformity requirements were quite different from those of today. Moreover, a constraint on high-temperature reactors is that they have to be cold-wall; otherwise the quartz bell would crack quickly due to deposition on its inner walls. In those early stages, the heat source matched to cold-wall operation was radiofrequency induction (and lamps shortly thereafter). Therefore, these facts determined high-temperature reactors to be AP ones. Technically, the simplest and most precisely controllable and repeatable way to operate isothermally is by the use of resistive heating wrapped around a furnace tube. This fact brings with it two constraints: furnace operation will be limited by the relatively long times of thermal stabilization (inherent in thermal masses and resistive heating slowness) and by the negative effect on isothermal conditions of heat transport by gases at AP. The first constraint is clearly an issue of productivity and can be circumvented only by increasing the number of wafers processed together. This solution works in the same direction as the second issue, since wafers can be packed more closely only by increasing diffusion, hence going to lower pressures, which caused low-temperature reactors to work at LPs. These elements together and the tradition that "you don't change the winning team" explain our present categories of reactors.

What happened later, however, introduced a slow drift toward indistinct borderlines. With the ready availability at reduced cost of low-vacuum technology and under the continuous push for better performance, cold-wall AP systems are becoming RP ones. This fact redirects attention to the crucial issue of diffusivity and, at the same time, requires improvement in temperature control (as soon as temperature variations became as detrimental as flow inhomogeneities).

Low-pressure hot-wall reactors supposedly had already achieved the best performance they could (accurate temperature control and higher diffusivity). However, an Arrhenius plot may be deceptive, since it focuses on temperature and leads to forgetting about all the other factors relevant to determining the operating regime: pressures, flow rates, deposition-exposed surfaces, characteristic dimensions (wafer spacing), and so on. It follows that LP reactors (you can bet on it for all *production line* reactors) have been pushed up to their process performance limits so that they no longer operate in the truly kinetic region (and are thus far from standard simulation approaches).

Definition of pressure regimes is a tough subject. Different authors give different (although similar, of course) ranges. The underlying problem is that there are no fundamental parameters, either physical–chemical or technical, that can mark transitions bordering the pressure regions. Other issues may arise from possible dependencies of pressure on measuring techniques (at very low pressures) and reactor design (for uniformity in reaction region and measuring points). It happens, too, that there are different borders depending on the technique: plasma, for instance, is generated at pressures overlapping the LP and UHV regimes of conventional CVD.

These thoughts lead us to list very rough indications, mainly for readers who do not have previous knowledge of the subject.

1. Atmospheric pressure (AP): The process pressure is different from the meteorological pressure, because of safety issues and exhaust extraction; settings are within inches of water (as traditionally measured) from the ambient pressure value.
2. Reduced pressure (RP): Here pressure is measured in Torr and spans from AP down to a few Torr.
3. Low pressure (LP): Usually measured in milliTorr (mTorr), LP extends from the previous border down to a few tens of mTorr.
4. Ultrahigh vacuum (UHV): UHV is still lower pressures over several decades.

b. *Energy Sources for Reaction Activation*

Activation energy sources for CVD must come down, at the end, to the range of some electron volts that is characteristic of chemical reactions. The possibility of supplying energy in this range by indirect means, through some converting medium, has broadened the possible sources, as shown in the following short overview.

We should better define the terms cold-wall and hot-wall reactors; they are applicable to thermally operated reactors.

1. In *cold-wall reactors,* even if they operate at very high temperatures (up to 1300°C), heat is supplied mainly to the wafers and/or the susceptor. In this way it is possible (with the help of some air or water circulation) to keep the quartz bell relatively cold and to reduce greatly flaky film deposition on the quartz inner surface.
2. *Hot-wall reactors* operate at midrange temperatures (typically from 400° to 900°C) but the process chamber is also heated. The furnace tube is thus an isothermal ambient, where the temperature can be controlled (directly, by means of thermocouples inserted in the process tube) with an unsurpassed degree of accuracy and stability, today both on the order of a few tenths of a degree Celsius. The drawback is the deposition of film on the inner surface of the process tube. Consequences are the possibility of particle contamination (but only at higher deposited thicknesses, since the adherence is strong), initial "break-in" when the tube is only partially coated, mechanical stress in front of the temperature ramps (because of expansion coefficients), and the necessity for more frequent cleaning (or changing) of the process tube.

We should also consider that there are at least two ways of supplying thermal energy.

1. *Resistance heating* generates thermal energy by the Joule effect. The resistor structure depends strongly on the specific application. In hot-wall systems resistive wires are wound around the quartz process tube and supported by ceramic casing. While former technology was based on series-parallel arrays of thin wire resistors, now a single wire (a few millimeters in diameter)

spirals around the tube, providing improved ruggedness. The heating element is subdivided into zones, which are separately controlled to allow compensation for external differences (such as losses at the ends) and utilization of temperature ramps along the furnace. The number of independent zones was previously three; now it is typically five and, in more recent products, up to seven. Sometimes when used in cold-wall systems, heating resistors are embedded in the susceptor. This design feature is also chosen for the low-temperature substrate heating required in plasma reactors.

2. With *radiofrequency (RF) induction* it is the susceptor itself (and partly the wafers too) that is used as the resistor, traversed by the eddy currents induced by the RF field of a surrounding coil (powered at frequencies of tens of kilohertz).

Sources of "light" (generically indicating radiation from infrared to ultraviolet) are the core of other heating techniques.

3. *High-intensity lamps,* with an emission spectrum matching the absorption of the susceptor and/or substrate, are probably the most common energy source in cold-wall reactors and make possible the concept and operation of RTP reactors.
4. Power *lasers* tuned to the substrate operate on the same principle; however, they have the clear advantage of heating just the beam-exposed surface. For special, small-scale applications, this is ideal for direct patterning (by pulsed rastering of the surface), even of a room-temperature substrate, without the need for flowing gases since the reaction is localized on microscopic areas.
5. In *photo-assisted CVD* (photoCVD or, less commonly, PHCVD), the target of radiation absorption is now one or more reactant gases, so that gas-phase reactions and their products are now at the center of the deposition mechanism.

Use of *plasma* is intended both as a chemistry enhancement and as an activation energy source. Electromagnetic field energy (in various ways, as we will see later in this section) is transferred primarily to electrons. The plasma is far from equilibrium, there being a difference of some (tens of) thousands of kelvins between the electron and the ions temperatures; electrons are thus the intermediate transducer to carry the energy over to precursor molecules in order to sustain chemical reactions.

There are indeed still other methods to induce the chemical reactions for a CVD process but their description would lead into laboratory and experimental techniques.

c. The Complete Reactor

For several years review books and some literature have strongly stressed the relevance of looking at the reactor globally, including in the study all supporting systems and facilities. For proper and steady functioning of a reactor, which rewards both product quality and productivity as required by the present competition in industry, all subsystems have an influence on and are influenced by the process. It is for this very reason that the process engineer has special responsibility in supervising and connecting all these linked blocks and should have sufficient knowledge about each one. It is neither proper nor possible to address these topics in such a short review as this chapter, but we recall at least the main related issues, as learned from experience. Chapter 3 is devoted largely to specializing these general concepts to some epitaxial reactor technologies. In the following sections on reactor typologies we limit the description to functioning principles and schematics of the process chamber.

The word *facilities* is usually broad-spanning and includes all that stands around and is connected to the machine itself. Supplies of process and auxiliary gases (or fluids) have to be properly dimensioned, filtered, and provided with safety and maintenance sectioning and purging. The electrical power supply is an issue not just of passing a sufficient quantity as in the specification of power consumption but, according to process type, of providing separate uninterruptible power supplies (UPS) to control electronics, valves, or even pumps. Cooling water and air are simpler issues but no less relevant to guaranteeing safety and equipment uptime. Exhaust of process gases is very critical. Within the definition of exhaust system are included pumping or extraction apparatus which directly influence process pressure regulation. Even more, exhaust is critical for safety, for maintenance sectioning and purging of the lines, and for disposal of hazardous and toxic gases or other residues. It is unfortunate that severe accidents have occurred in laboratories and industrial plants because of careless design of exhaust and lack of exchange of information on by-products between process and facilities engineers. Ambient class is evidently an essential element in particle performance of products, but it should be viewed not as a static specification, but along the whole cycle of machine utilization (including maintenance). Special care should be given to pressure transients during operation of wall-mounted systems (i.e., when the operating front side of the reactor is flush with the clean-room wall and the body of the system is located in the lower-class gray room).

In the *machine body* itself, there are several subsystems around the process chamber: increasingly complex control electronics and PC interfaces, fluid handling areas (gas panels), power sections for energy input to the process, wafer storage, load-locks, and robots for wafer handling. Performance and productivity are dominant parameters in commercial reactor design; this in turn pushes for the highest automation and clustering of units (either for paralleling of a slow process or for better management of different steps in sequence, more often now including surface pretreatment). As an example of the strict connection of process with productivity, consider when a process step in a clustered sequence becomes just slightly longer, only a few seconds, than the following one: this may halve the throughput! At the time of writing, the dominant wafer diameters in the silicon industry are 150 and 200 mm, the latter being a mature technology but still increasing in its production share; 300-mm wafers also, although more slowly than predicted, are increasing in prevalence. This is different from the situation for other semiconductors, where 150-mm wafers are (if they exist at all) the leading edge of technology. This simple geometrical factor has profound consequences for reactor and process design and poses performance challenges (think of parameter uniformity over a wafer). This factor is surely one of those driving the trend to short processes combined within single-wafer clustered systems.

The *susceptor* provides mechanical support for the wafer(s) and quite often, at the same time, direct or indirect heating. We may roughly divide susceptors into three classes.

1. In cold-wall reactors, whatever the heating system, SiC-coated graphite is used. These materials may be produced with high degrees of chemical purity, are mechanically stable at high temperatures and can be machined with high precision, and SiC is chemically superb at resisting corrosive environments.
2. In hot-wall reactors, the common terminology is that of boats or carriers. They serve only as mechanical support (and gas flow shielding in a few special cases). The most common material is quartz, but SiC carriers are slowly gaining acceptance because of the lower cost of ownership.
3. In plasma systems (parallel plates), the susceptor coincides with the lower electrode, thus coupling the wafer to RF power; more generally the susceptor is also the means of thermal heating of the wafer. In these systems, both because of the low temperatures involved and for electrical reasons, the susceptor is made of a passivated metal, commonly aluminum.

The *process chamber* is another important element.

1. Clear-fused quartz is the choice for cold-wall reactors, especially because of its transmittivity over the required light frequencies.
2. For hot-wall reactors, quartz still dominates, but there are some applications for silicon carbide tubes; however, SiC use is less common for process tubes than for carriers. This occurs both because of the much higher unit costs of the pieces and for technical reasons: its higher thermal conductivity poses problems when the process makes use of temperature ramps.
3. Plasma chambers (excluding some tubular configurations, not so common nowadays) have metal walls like the wafer support, made of passivated aluminum or stainless steel.

Good management (consisting of cleaning and maintenance) of both the process chamber and the susceptor is essential in achieving adequate performance in terms of process quality (particle contamination) and machine downtime. This too is the responsibility of the process engineer, by means of proper choice of materials (when possible) and by tuning the process, paying attention to film deposition on auxiliary materials.

Experimental laboratory reactors fall into three categories: they may be built for a very special purpose; more commonly, they are tentative precursors of future commercial reactors or anticipate the evolution and improvement of present commercial reactors. It may be rewarding to follow up these applications also, since it may help determine the vision of industrial trends.

2. ATMOSPHERIC-PRESSURE REACTORS

a. *Common Considerations*

Apart from the continuous type, which has its own niche in this family, all other reactors in this group are also classified as cold-wall reactors. The ability to operate cold-wall reactors (differently from hot-wall ones) in high-temperature ranges has made them the elective choice for epitaxial deposition, because of the strong effect of temperature on atomic rearrangement in a crystalline matrix. Throughput or efficiency considerations favor a different use in only a few special cases of high-thickness polycrystalline films. With relatively simple modifications, they can operate at reduced pressure to improve film uniformity; as film specification requirements become tighter, this alternative is gaining more popularity. A common operating characteristic of all atmospheric reactors is very high dilution of precursors, to reduce undesired gas-phase reactions; the resulting precursor partial pressures are not very different from those used in LPCVD. Diluting gases may be inert ones (nitrogen is the most common, argon and helium following at some distance because of their higher costs) or part of the reaction species [clearly exemplified by the use of hydrogen in epitaxy from (cloro)silanes].

Since the primary application for cold-wall reactors is epitaxy, they are given adequate attention later in the book.

b. *Continuous-Type Reactors*

Key elements of a continuous-type reactor are

- a system to separate the process region from the ambient atmosphere, which does not interfere with the transport mechanism;
- a heating system which can handle variations imposed by moving parts and still provide adequate uniformity in the deposition region; and
- methods for injecting and exhausting process gases in a highly uniform and laminar (hence repeatable) way.

What became a *de facto* standard for this type of system is the belt reactor manufactured by Watkins Johnson (WJ; Scotts Valley, CA). We can understand, from the schematic in Fig. 3 and the detail of the gas injection system in Fig. 4, how the highlighted elements are implemented in that reactor family.

Full separation from the ambient atmosphere of the process region is achieved by means of multiple nitrogen curtains (i.e., a high-flow "blade" of nitrogen with adequate nearby exhaust). All relevant parts of the process assembly are made of

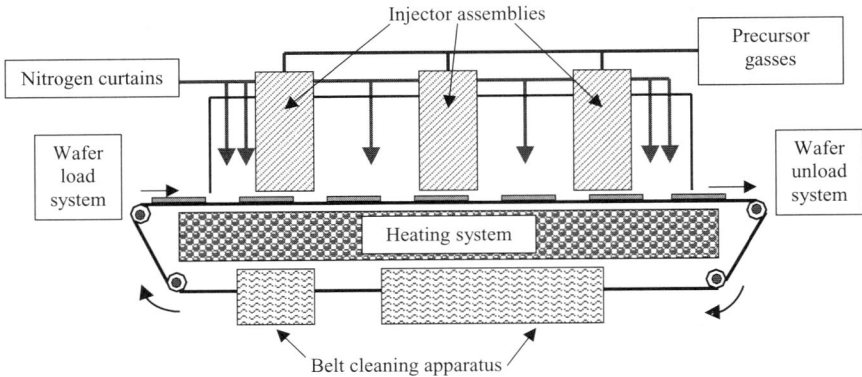

FIG. 3. A highly simplified schematic of a WJ-family reactor.

some steel alloy, selected for a low level of metal contamination and for thermal and mechanical stability. Since wafers are transported over a (woven metal) belt in continuous movement, the system is very well suited for high automation. Heat is supplied by resistors under the full extension of the process assembly. To control the multiple factors affecting the temperature (dispersion toward ambient, transport by belt and wafers, removal by impinging gas jets), resistor banks are subdivided into many units (as many as 20–30) with independent control. The increased

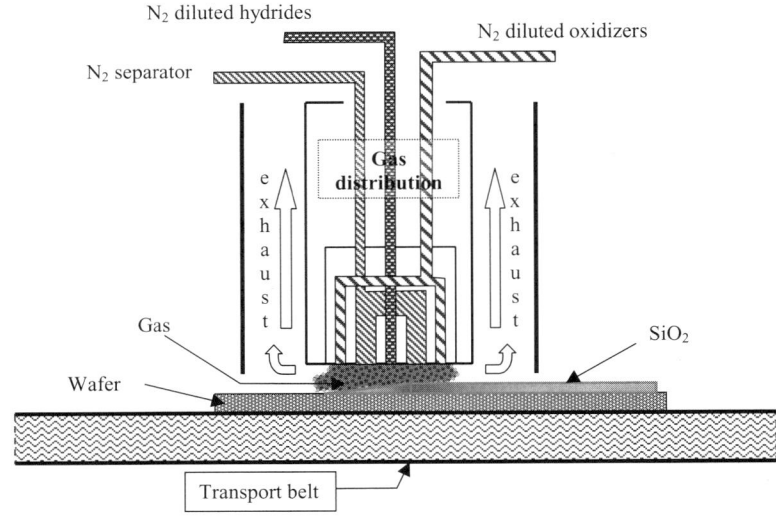

FIG. 4. Detailed structure for working principle description of the WJ gas injection system.

number of separately controllable heating zones was one of the factors marking the progress between reactor generations. Practical construction constraints and limits on metal contamination diffusion make this reactor type suitable for processing in the low-temperature region (up to 500°C), depositing various types of amorphous films. While in the wafer industry films are undoped and doped silicon oxides, this type of reactor has also found widespread use with other chemicals in the flat panel display industry. Since wafers have a steady longitudinal motion (on the order of inches per minute), uniform deposition may be achieved by having reactant gases injected in a sharp jet crossing the whole belt perpendicular to motion.

What determined the success of WJ was the design of the injectors:

- Reactant species [hydrides (silane and dopants) and oxygen, for instance] mix just a few millimeters above the wafer surface.
- Deposition takes place in a region a few millimeters wide along the motion.
- However, the useful width can be in excess of 12 in. (the latest model can in fact process 300-mm wafers).
- Three (or four) injectors in series find a place in the deposition assembly, so that there is the option of a higher uniformity and throughput (processing the same film) or of reducing manufacturing steps by depositing different films in one pass.
- Injectors can be easily removed for cleaning and maintenance.

Among process parameters, while temperature plays a relevant role as expected, pressure and gas flows are of utmost importance. Key to a uniform film is a very homogeneous laminar flow of process gases impinging on the wafer. Much of the available literature states that the process works at a pressure slightly higher than atmospheric. While that may not even be true under some operating conditions, it focuses attention on the wrong point, i.e., an absolute value for pressure. In fact, the so-called process pressure (ca. 1 in. of water) is a *relative* pressure (between the process region and the exhaust plenum, over calibrated orifices). It causes exhaust gas removal, and it is influential because it determines the fluid-dynamics regime of the reactants. Although these reactors have been in wide use for several years, there is still work ongoing to understand this process better, as evident in some recent works (Masi *et al.*, 1997; Nieto *et al.*, 1999).

It is important to mention the high quantity of gas (N_2, on the order of several hundreds of liters per minute) consumed by these reactors even in standby mode. The reasons for this fact are related strictly to the application of the principles presented here on an industrial scale and merit more detail as an example.

A typical process recipe for plain oxide from silane and oxygen at 400°C is as follows:

1. Use 50 sccm silane and 2 slm oxygen (from a typical O_2/SiH_4 ratio of 20 : 1) for each injector supply.

2. To maintain a stable laminar flow, the injector must be fed with fixed fluxes of gas for each part (skirt) of the jet impinging on the wafer; if the total of process gases is 5 slm, the diluting nitrogen is 4.95 slm for silane and that 3 slm for oxygen.
3. As illustrated in Fig. 4, the silane jet is separated from the split oxygen jet by a pure nitrogen injection, which amounts to 14 slm.
4. Given the to-be-deposited material properties (i.e., chosen deposition temperature, O_2/SiH_4 ratio, pressure), most other process features (such as layer uniformity and gas-phase particulate contamination) are controlled to a large extent by the correct settings of all main and some auxiliary gas flows; the latter, which are integral parts of the recipe, include nitrogen flow from some shields within the exhaust inlet (a few liters per minute here too, to reduce strongly oxide powder deposition in this region) and from the chamber floor beneath the belt (this, too, to reduce oxide deposition on both the floor and the belt).

Let us conclude this discussion of the WJ reactor with two remarks. The first is about the fast pace of microelectronic technology, which, because of its ever more demanding specifications, is already close to surpassing the capability of this reactor, even though it was, just a short time ago, major advance over its competing systems. The second remark is about the point we should have made that setting up and runing a specific process on a specific reactor (even when it is a "very well-known and common process") are still tough work, and collecting expertise and tips from the production line would produce a book as long as the present one.

c. Barrel Reactors

These reactors take their curious name from the shape of the susceptor, a hollow polygonal prism slanted gently inward at the upper side, which does look like the top half of a barrel, as shown in Fig. 5. Some authors prefer the name *cylinder reactors*.

The main purpose of the slant is so that the wafers can stand by gravity in shallow pockets machined on the sides plus some fluid dynamics compensation for depletion. The susceptor is positioned inside a quartz bell jar, the process environment. Process gases are injected at some spots from the top and exhausted from the bottom; to attain reasonable uniformity, the susceptor is slowly rotated during the process. Heating can be by means of either banks of high-intensity lamps or RF induction. Implemented design variations concern susceptor insertion into the bell jar, which can be either from the top or from the bottom (as illustrated). Quite diffused representatives are the Applied Materials 7700 and 7800 series (Applied Materials, Inc., Santa Clara, CA; www.appliedmaterials.com).

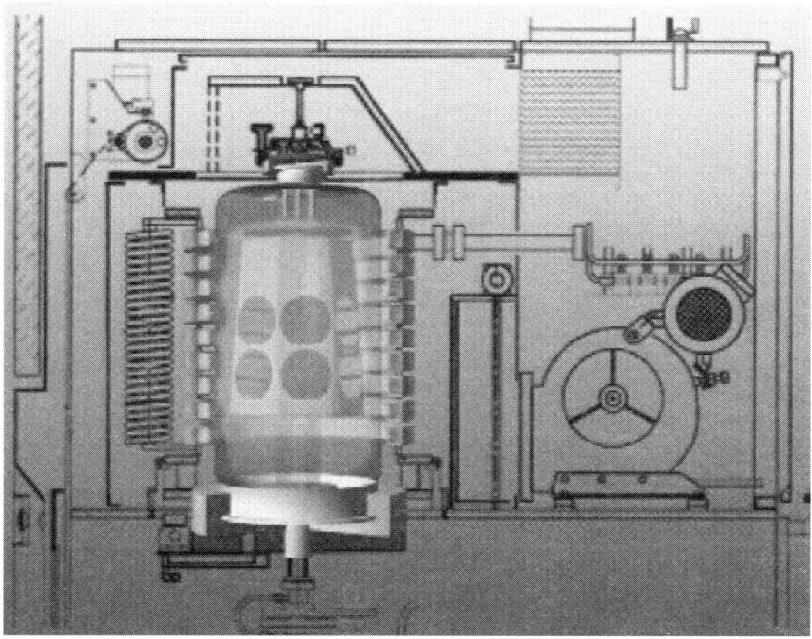

FIG. 5. Barrel reactor ensemble with a view of the susceptor heating and external bell air-cooling (courtesy of LPE).

d. Pancake Reactors

In this type of reactor, the flat susceptor is indeed like a pancake with a hole in the middle. From the center hole there is the gas inlet, with exhaust taking place downward at the disk rim; the process chamber is completed by a dome-shaped quartz upper enclosure. Depending on the relative size of the susceptor and wafers, the latter find place in shallow pockets arranged in one or more rings. Heat is delivered typically by an RF coil located under the susceptor but a configuration with lamps from the upper side is also used. See Fig. 6 for a simplified schematic. While this configuration is ideally axisymmetric, inhomogeneities in the gas extraction system easily arise. These and the others coming from heating are then averaged out by susceptor rotation. All design aspects of this reactor make it well suited for operation on small-diameter wafers. An intrinsic flaw of this configuration is that gases, in their radial outward motion, have a higher physical density when they are fresh, and spread to a lower density when they are already reacting, thus increasing the effect of depletion. Arguably, the most common representative of this type of reactor is the Gemini (Lam Research Corporation, Fremont, CA; www.lamrc.com).

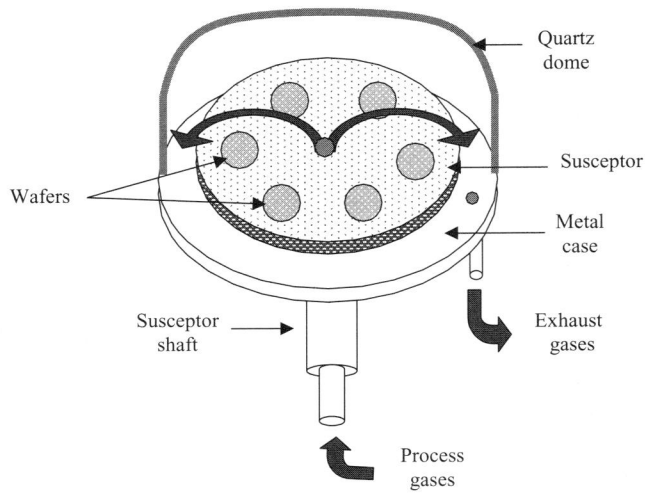

FIG. 6. The most common structure of a pancake reactor.

e. Cellular Reactors

The design flaw we pointed out for pancake reactors is corrected for cellular reactors: here gases are injected from the periphery and exhausted from the center. However, what makes this reactor design peculiar is that the wafers do not just lie flat. The top of the doughnut susceptor has on it vertical radial disks (like teeth) with slanted sides (most like a big conical gearwheel), each one with shallow pockets on both sides where the wafers sit. Resistance heating is placed directly under the susceptor; the design may include supplementary heating from the top using resistors or lamps, to compensate for the inevitable losses.

The Tetron (Lam Research Corporation) reactor (an explicative sketch of the susceptor is shown in Fig. 7) is, to our best knowledge, the only industrywide representative of this concept. Although this reactor is unsurpassed in load size and probably productivity (think of one run with 50 wafers of 150 mm), it has significant drawbacks for the same reasons. Because of the great number of wafers, long loading and unloading operations may adversely affect particle performance; the system may be automated but not with efficiency; there are of course limits on process tuning; and in the case of failure, losing one run has a very high cost.

f. Vertical-Flow Reactors

In the naming of these reactors, stress is placed on gas motion, which is in the vertical direction. It is really a way to encompass a family of intrinsically similar solutions:

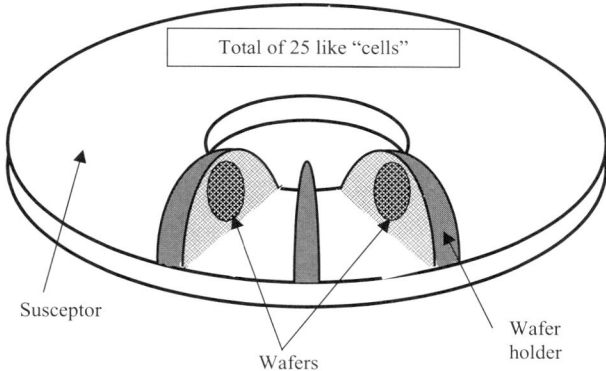

FIG. 7. Simplified sketch of the susceptor used in cellular reactors; the basic element is the doubly tapered (upward and radially inward) "tooth," with pockets on each side to hold the wafers.

1. The process chamber is delimited by a dome-shaped quartz upper enclosure.
2. Wafers sit in pockets on the flat circular susceptor, which is rotating; the shape of the susceptor gave origin to names such as planar and satellite reactor.
3. Gases enter from the top, either by direct injection or through a dispersing showerhead (Fig. 8).
4. Exhaust is between the susceptor's periphery and the quartz dome.
5. Heating is by an RF coil placed under the susceptor.

For the sake of clarity, we report that other authors consider the vertical and the pancake reactors to be in the same family and interchange their names. There is

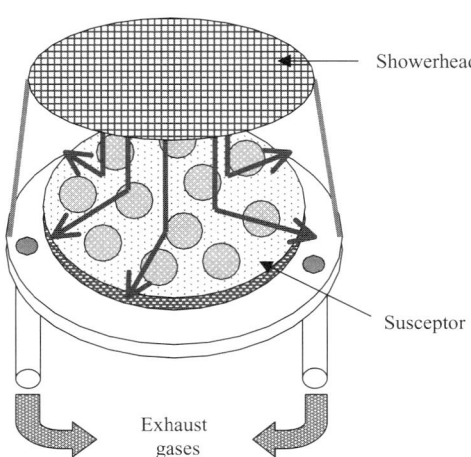

FIG. 8. Schematic of a vertical reactor, with gases flowing from a showerhead.

some reason for this; with a sufficiently high dome and inlet flow, even in a pancake reactor gases may reach the wafer surfaces in an almost-vertical direction instead of the usual horizontal one.

g. *Horizontal Single-Wafer Reactors*

Early horizontal reactors were tubular reactors housing flat susceptors; the susceptors were tilted upward at the exhaust side to compensate for depletion and the wafers lay on the upper sides of the susceptors. This starting design had its primary applications in the laboratory, for process study on small-diameter wafers. It was soon realized that depletion could best be corrected by working on a single wafer and making it rotate. What seemed to be a loss of productivity was offset by good engineering and improvement in film parametric performance. It is curious and instructive to recall that as recently as the early 1990s the future of single-wafer horizontal reactors was in strong doubt; today, they are the standard for 200-mm wafers and have gained some use also for 150-mm wafers. From the schematic in Fig. 9, the functioning of the reactor should be quite evident.

Design efforts are focused on controlling the gas flow, possibly having a laminar condition over the wafer; wafer rotation helps homogenization of both thermal and chemical nonuniformities. Very diffused implementations of this reactor in silicon epitaxy are the ASM Epsilon (ASM Epitaxy, Tempe, AZ; www.asm.com) and the Applied Materials Centura (Applied Materials, Inc.).

3. LOW-PRESSURE REACTORS

a. *The Horizontal (Tubular) Furnace*

Figure 10 is a schematic of a horizontal furnace LPCVD reactor. Like the AP reactor, the adjective horizontal indicates the main geometrical dimension and the

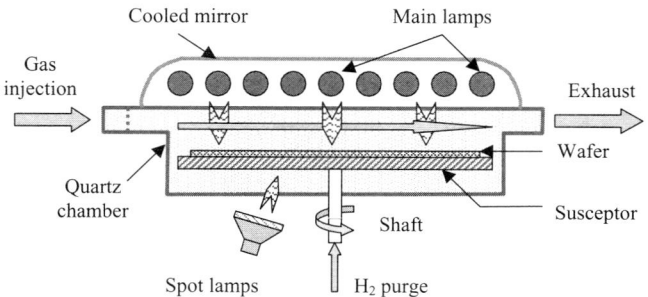

FIG. 9. Typical cross section of a horizontal single-wafer reactor, with the most relevant process elements; many design features may differ according to the manufacturer, to maintain cleanliness and implement wafer handling, gas distribution, and heating optimization.

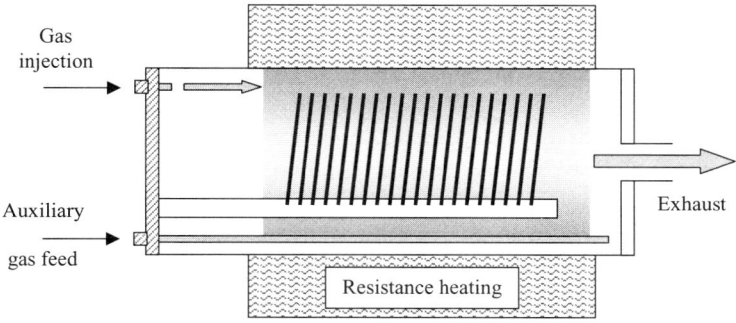

FIG. 10. Main elements of a horizontal furnace low-pressure reactor.

direction of gas flow in the process chamber. The difference is that the flow is perpendicular to several stacked wafers (instead of parallel to their surfaces) and species transport to the wafer surface (i.e., between wafers now) is carried on by diffusion across the wafer scale and not only across a thin boundary layer. This change of scale is of course permitted by the LP regime.

Although the LP reactor was introduced later (in the early 1960s) than the atmospheric one, its structure and functioning were immediately to the point, and in fact they have remained unchanged over these years. What marked the steady evolution of this reactor design was improvements in auxiliary items.

1. Temperature control: More independent heating zones with programmable controllers, the introduction of paddle thermocouples (i.e., thermocouples permanently inside the tube for continuous autocalibration), and improved accuracy of control loops.
2. Gas flow control: From the first flux-meters and rotameters to multifeatured (soft-start, ramp opening, programmable, digital) mass flow controllers.
3. Pressure control: From simple rotary pumps to multistage rotary pumps, dry (oil-free vanes) pumps, roots boosters, compensated capacitive sensors, and autotuning PID controllers.
4. More automated wafer handling: From full manual operation with boats sliding onto the tube's inner surface (thus originating particles) to almost fully automatic systems with mass wafer transfer (between cassettes and boats) and cantilever loading.
5. Improved system management: With various process recipes available and automatic process cycle execution,
6. Extension of process types: From simple silane pyrolysis and oxidation to the most complex chemistries and products.

The issue of precursor depletion progressing toward the exhaust end of the tube is addressed by different and partly coexisting solutions.

1. Precursors may be diluted.

2. Spacing between wafers may be varied depending on the position along the tube.
3. Different single- and multipoint injection manifolds may be placed inside the tube.
4. The temperature may ramp, increasing toward the tube end.
5. Wafer boats may be caged (being surrounded by slit shells to modify gas diffusion).

Adoption of one or a combination of these solutions is, as often happens, not just a technical solution but a delicate compromise among process, product, and productivity specifications and requirements. (See MRL Industries, Sonora, CA; www.mrlind.com.)

b. *The Vertical (Tubular) Furnace*

There are, theoretically, no differences in process and performance between horizontal and vertical furnaces, since process settings can be exactly the same and gravity has a negligible effect on low-pressure gas motion. All information in the preceding section applies here as well.

However, the superiority of vertical configurations is clearly marked by experience. It is determined by technical and pragmatic factors.

1. The vertical configuration is exceptionally well suited to compact design, complete automation (cassette-to-cassette operation), clean-room wall-mounting, standard mechanical interface (SMIF) pod integration, and machine clustering.
2. Because of the aforementioned advantages, all technological improvements in machine constituents [machine management and control systems, special features such as a fast temperature ramp and model-based temperature control (MBTC), faster robots, and more] are designed for and implemented in vertical furnaces.
3. There are still some features [for instance, the different-diameter liner (Fig. 11, right)] that are practical in a vertical and not in an horizontal design.

The most common configurations are those shown in Fig. 11, where the relevant aspect is of gases flowing upward in one case (right) and downward in the other case (left). This allows the former scheme to have inlet and outlet of the gases from the flange and the quartz tube, and the heating element to be simpler and closed at the top, but with more complex inner quartzware parts. There are also configurations in which the carrier is standing on a fixed flange and the complete assembly of tube and heating element is lowered onto it for processing and then raised for load/unload operations.

More details on industrial implementation and special features are available from the suppliers, including Hitachi Kokusai Electric (Tokyo; www.h-kokusai.com),

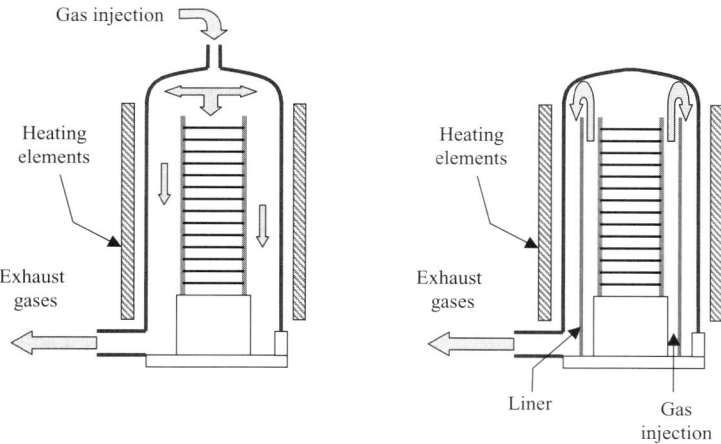

FIG. 11. Two possible variations of the basic configuration of a vertical LP furnace.

ASM Europe (Bilthoven, The Netherlands; www.asm.com), Semitool (Kalispel, MT; www.semitool.com), and Tokyo Electron Limited (Austin, TX; www.telusa.com).

A published process simulation carried out in a vertical furnace with large-diameter wafers (200 mm) is reported by Rinaldi *et al.* (1999).

c. Ultrahigh-Vacuum Reactors

What makes a UHV reactor design different from a LP one is the higher complexity of the vacuum/pumping system and the presence of high-flux turbomolecular pumps and a load-lock chamber. Turbo pumps are essential to reach an ultimate vacuum better than 10^{-5} Torr in the first outgassing phases of the process and to maintain a pressure lower than a few milliTorr with process gases flowing. The complexity of the load-lock system forces a process chamber of the horizontal type to be relatively small, accommodating just one carrier of almost-standard size. With vertical structures, the better automation allows maintenance of the same structure and size as the corresponding LP reactor, thus increasing the load size with respect to the horizontal one. The better surface desorption possible with this reactor structure and the high levels of diffusivity in the molecular regime are the main factors that make the system effective for low-temperature epitaxy (LTE) and good conformal step coverage.

4. RAPID THERMAL PROCESS REACTORS

As can be immediately seen from comparison of Fig. 12 and Fig. 9, there are few and small differences between horizontal AP and RTP reactors. The latter were

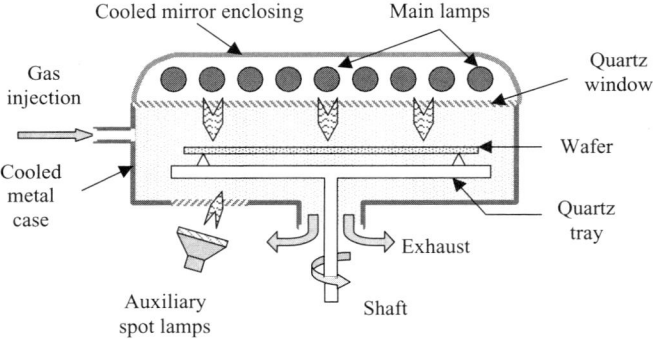

FIG. 12. Main constituent elements of a rapid thermal process CVD reactor.

developed first for thermal treatment only (which is an AP process), and the goal of rapid processing (almost no temperature ramps and a very low thermal budget) was attainable far from thermal equilibrium. This was achieved by lamp heating with the light spectrum in the wafer absorption region and by not using a susceptor [as in Steag-AST (now Mattson) systems].

Because of these thermal features, the reactor was (relatively) easily modified for CVD operation at AP or RP, providing the opportunity to flow gases at room temperature and switch the deposition reaction by turning the lamps on and off.

5. PLASMA REACTORS

a. *Basics of Plasma Use in CVD*

The first use of plasma in CVD was reported in 1962. It is not surprising, however, when we examine the complexity of this processing technique that the first industrial applications came about a dozen years later. The central idea of plasma-enhanced CVD [(PECVD), or plasma-assisted CVD (PACVD)] is to input only a fraction of the energy required by the reaction thermally (susceptor resistive heating) and convey the remainder by means of electron energy in the plasma. It is very important to understand that these plasmas (which are named quite intuitively *cold plasmas*) are not in thermal equilibrium, because energy is supplied mainly to electrons, and transfer to ions (by means of collisions) is very slow due to their different masses. As a side note, to understand the distinction better, we recall that the simplest way to generate an equilibrium plasma is by direct heating to a very high temperature, which then becomes the temperature of the plasma itself and of its components.

The reader will already understand that all plasma reactors can be classified as cold-wall ones, since only the wafer and susceptor are heated. Since only a fraction of the energy is supplied in this new form, this fact explains the choice of the term *enhanced* CVD. With this observation in mind it may be expected, as is really the case, that every type of film deposited with thermal CVD can also be deposited with PECVD, but at a much lower substrate temperature (typically less than 400°C). Epitaxial layers may indeed be obtained, even if not of a high quality because of the low temperature; the layer crystalline quality is, however, greatly improved by remote plasma, as we shall see later.

The lower substrate temperature is probably the greatest advantage of PECVD and may also force its use as more and more layers are added in device fabrication, with the latter layers having a lower thermal budget (i.e., being strongly limited in maximum temperature and step duration so that they do not damage already present structures). The second advantage of PECVD is the great flexibility in film properties. This will be better understood when we see more process parameters come into play and an increase in the complexity of the chemistry. The third advantage is that PECVD reactors (apart from very early tubular structures) are typically single-wafer and structured for complete automation and cluster integration. There are modern multiwafer machines, but they are carefully engineered to parallelize two or four wafers in compact and automatic structures, improving the throughput and allowing more flexibility when sequencing different steps. [A commercial unit is from Mattson (Fremont, CA; www.mattson.com).]

All these listed goods have their costs.

1. The complexity of the chemistry and the physics involved is challenging, so that processes are hardly understood, and modeling and simulations are very rough.
2. Because of the complexity of the chemistry and the abundance of species, there are higher levels of (typically unwanted and difficult to control) impurities in the films.
3. The presence of higher-energy particles in the plasma causes some degree of sputtering of the process chamber (hence incorporation of contaminants) and of the processed wafer itself, with possible physical damage to the existing layer(s).

Together with the standard CVD process parameters, there are more parameters of the system to control. To substrate temperature, pressure, and gas flows, we now add the electromagnetic input power (from a few hundred watts to some 30 kW), its frequency [from 0 (dc) to some 10 GHz], the power partition between electrodes, and the magnetic field intensities in some configurations.

Strongly influencing the process are the reaction chamber geometry and the conductivity of its constituent parts, because they determine the electric field generating and superimposed on the plasma. It can be demonstrated that, at fixed process parameters, changing the surface ratio of the main electrodes could drive the process from sputter ablation to sputter deposition.

Directly deriving from the process parameters, even if not always easily determinable, are the physical parameters of the system:

1. the electronic density, equaling (globally) the ionic density;
2. the average electron energy, corresponding to the electronic–gas temperature, of several thousand kelvins;
3. the average ion energy, usually (but not always) neglected in the plasma balance, because it is just above the gas temperature;
4. the plasma frequency, proportional to the square root of the electronic density, which determines the transparency of the plasma to an applied frequency;
5. various electrical potentials (plasma, floating, and sheath, related to different surfaces);
6. the ionization fraction (ratio of ion density to total plasma density), which is typically less than 0.1 (i.e., plasmas are weakly ionized) but can approach unity in special cases such as ECR as we will see.

There are two other differences with respect to standard CVD that are worth pointing out. In a CVD reactor, with a given geometry, gas fluxes, and a tuned pressure control system, the flow pattern is easily there, stationary (even if probably not known). On the other hand, plasma stability and extension in the chamber are less easily accomplished tasks. Also, in PECVD some special *in situ* diagnostic tools are required to access the new properties of plasma which have to be monitored to guarantee process consistency.

Before moving on to the subject of this section, let us recall other common applications of plasmas in the microelectronics industry, which may use reactor configurations similar to or even totally different from PECVD:

- physical sputtering and (reactive) sputter deposition,
- plasma etching and reactive ion etching (RIE),
- ashing (i.e., removal of photoresist), and
- ion plating.

b. Plasma Generation Techniques

Standard techniques for plasma generation in a gas chamber at sufficiently low pressures are the following:

(a) plasma directly *excited* between powered electrodes,
(b) plasma *induced* through nonconducting walls,
(c) plasma induced by *microwaves*, and
(d) plasma induced by (thermionic) electron bombardment.

Excluding the last one, which is used for generation of ion beams, we now give an introductory description of these techniques.

The most intuitive method (and historically the first one) for generating plasma (from gas at a low pressure) is by applying a sufficiently high tension (either dc or ac) between two conducting plates, as shown in Fig. 13. The electric field

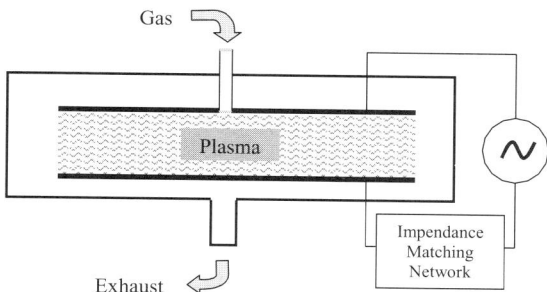

FIG. 13. Schematic of plasma generation by direct excitation through powered electrodes.

established between the plates extracts and accelerates electrons from the gas molecules, starting an avalanche reaction. Gas discharges (and the plasma therein) have quite complex conducting properties (i.e., current–voltage relation). It is up to the power source and auxiliary circuitry (an impedance matching "box" for ac supply and current limiting for the dc case) to start conduction in the gas, reach a preset functioning point, and then maintain its stability. Alternating current supplies (quite often RF, since the typical frequencies for these applications are in the megahertz region) have distinct advantages over dc ones. In the RF case it is possible to achieve a higher plasma density and efficiency, and the existence (or deposition!) of dielectric layers on the electrodes does not constitute a problem.

We remind the reader that different electrical configurations (with correspondingly different deposition behaviors) can be exploited from the basic one in Fig. 13. They can be obtained just by changing the relative area of the plate electrodes or the distance-to-diameter ratio, grounding one of the electrodes and/or the metallic chamber, or even powering the electrodes with two distinct supplies. Another common enhancement is the use of multifrequency systems, where (typically) two frequencies are used (one in the hundreds of kilohertz range and the other in the tens of megahertz range), either at the same time or alternately. This trick, based on the great effect of frequency on many film properties, allows improved film parameter tuning.

A method that is not really dissimilar to the previous, since it starts the plasma from an electric field, is the induction of the generating field through nonconducting walls. For power frequencies lower than 1 MHz (but typically of a few hundred kilohertz), both inductive and capacitive coupling are effective, as simplified in Fig. 14. The plasma produced in these structures is suited to isotropic etching, so that reactors built following these schematics are used for ashing. In the case of a microwave (gigahertz-range) source, it is the direct exposure to the beam emitted by a suitable waveguide that creates the plasma; this configuration is exemplified in Fig. 15.

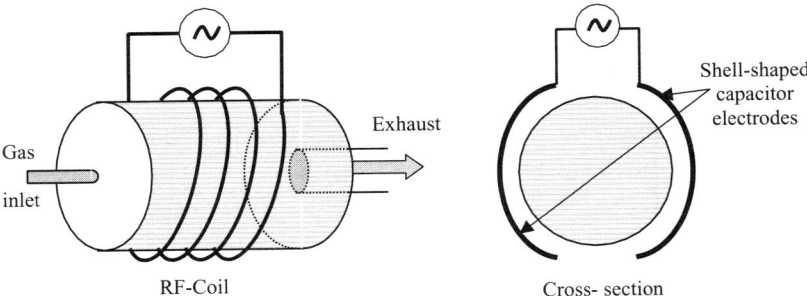

FIG. 14. Generation of plasma by inductively coupled induction (left) and capacitively coupled induction (right).

c. *Parallel-Plate Plasma Reactors*

Without delving into the details of plasma conductivity, we can rely on the enlightening similarity between plasmas and semiconductors to understand some of the basic electrical behavior of the former. In CVD plasmas, where only electrons are at higher temperatures, while ions are relatively just above ambient temperature, the current carriers are very similar to those in semiconductors, electrons to electrons and ions to holes respectively. Figure 16 is a plot of the direct and reverse

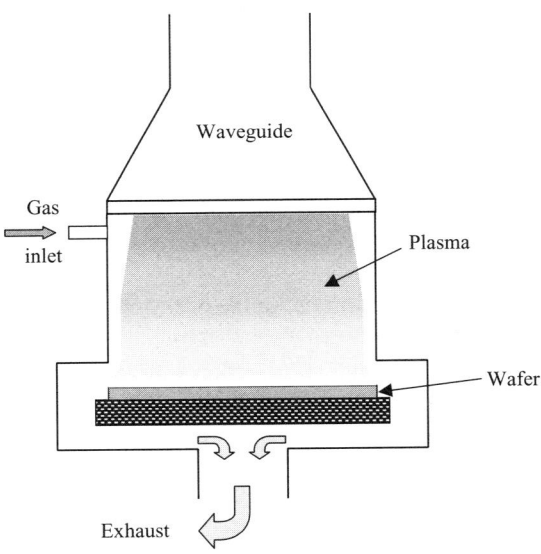

FIG. 15. A microwave source induces plasma in a nearby low-pressure chamber.

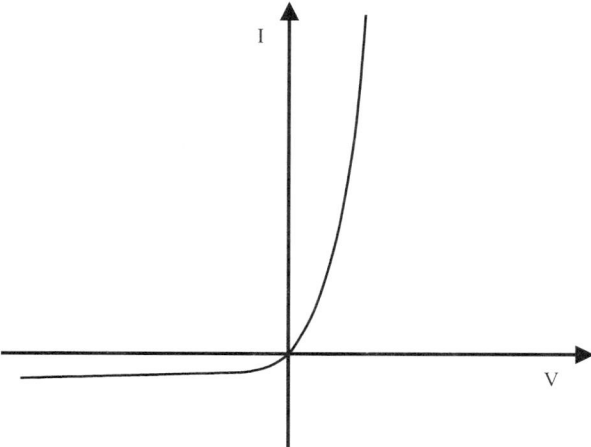

FIG. 16. The current–tension (I–V) characteristic of plasma with one dc-biased (by sheath potential) electrode.

conduction of a plasma (not dc-discharge generated) between parallel plates. The remarkable similarity to the I–V characteristic of a diode is just the result of an electron depletion region (sheath), which is formed on each surface in contact with the plasma. The latter phenomenon is caused by the greater electron mobility, hence all surfaces in contact with the plasma build up a negative potential. For these reasons, the terminology (RF)-diode plasma is often used for these electrical configurations.

To complete the sidestep in terminology, recall that triode glow discharges (where an additional grid electrode confines and screens the plasma) are characterized by having the sample not in the plasma region; for this reason they are an intermediate toward remote plasma structures.

In industrial systems electrodes are geometrically symmetrical plates, with a small spacing in between (of the order of centimeters; smaller for single-wafer reactors and larger for other configurations, as in Figs. 17a and 17b, respectively). Because of RF standards and FCC (or similar) regulations, most often the supply frequency is 13.56 MHz; experience and theory both prove that for this configuration there are no benefits from input powers higher than 5 kW.

In single-wafer designs (or, at most small batches of two to four wafers), process gases are delivered from a showerhead, an electrically conducting distribution grid which also serves as the upper electrode. For larger-batch processing (10 or more wafers), the reactor structure is very similar to that of the pancake reactor (cf. Fig. 6). We now have, instead of the quartz dome, the upper electrode in the form of a solid conducting plate. Both radial flow directions (outward and inward) are common. Resistances in the lower electrode accomplish heating of the wafer(s);

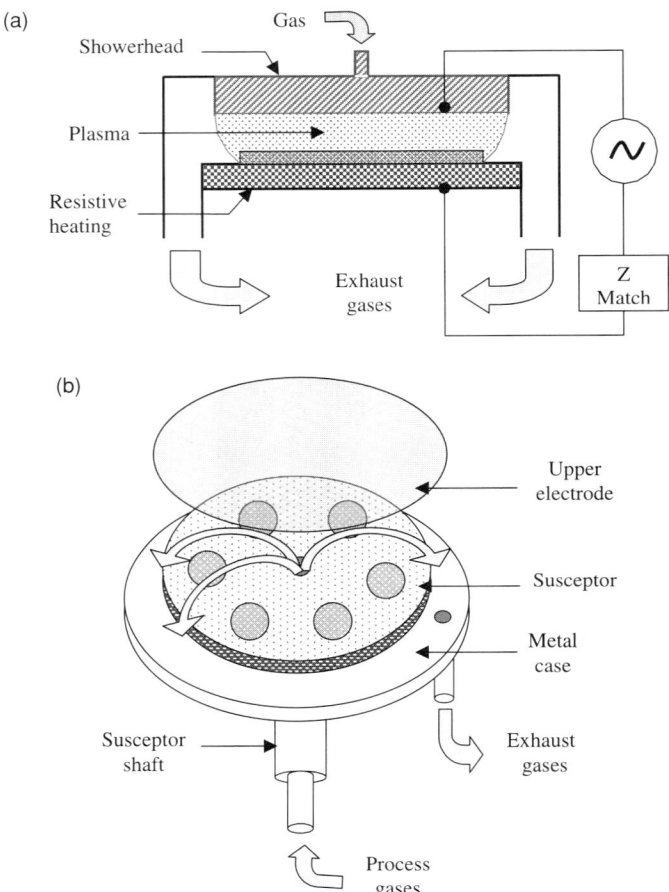

FIG. 17. (a) Schematic of the basic parallel-plate configuration of plasma reactors: compact structure for single-wafer processing. (b) Schematic of the basic parallel-plate configurations of plasma reactors: batch structure for smaller diameter wafers.

the chamber is typically aluminum or stainless steel, coated with a protective and insulating material film.

A variation of this scheme (magnetically enhanced plasma) is the addition of a magnetic field to confine electrons near an electrode, thus making possible plasma stability at higher pressures. For the sake of completeness we add that magnetron reactors, used for sputtering, are special cases of parallel-plate reactors with this magnetic field enhancement. Here a magnetic field is imposed so that electrons are forced into closed and dense trajectories parallel to the cathode electrode, where the material is positioned to be sputtered.

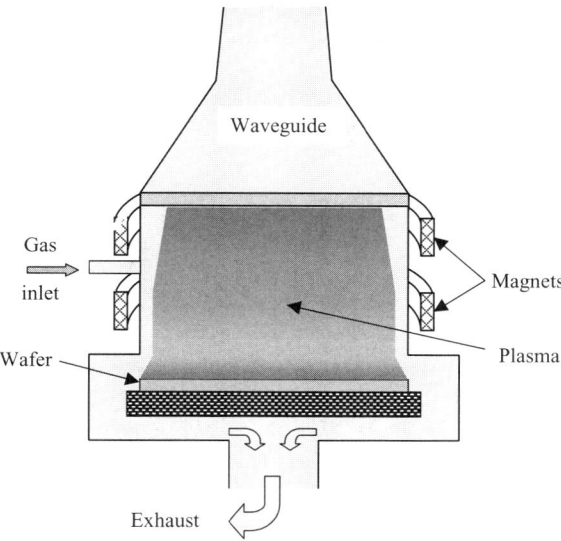

FIG. 18. Generation of high-density plasma by means of appropriate coupling of microwaves and a static magnetic field in an ECR reactor.

d. *Electron Cyclotron Resonance Reactors*

The principle of electron cyclotron resonance (ECR) plasma generation requires the introduction of magnetic fields; the explanation is best understood with the help of Fig. 18. Weakly ionized plasma may easily be generated by microwave induction in the region just below the waveguide horn. In the same region, a pair of Helmholtz coils imposes a uniform magnetic field (directed as the vertical axis); the effect of this field is to induce rotational motion in electron trajectories. If the characteristic electron (cyclotron) frequency of this motion is matched to the microwave frequency, the resonance condition greatly enhances power transfer to the plasma and allows the reaching of a high-density (ionization) plasma (HDP). With a common microwave source at 2.45 MHz (and a power of only a few hundred watts), the resonance condition is met with a magnetic field of 875 G.

Various enhancements are possible with different positioning of the injection points of auxiliary and process gases and by using an optional magnetic coil under the wafer. Other technical variations are based on multipolar rings of magnets and multiple antennas for microwave injection [microwave multipolar plasma (MMP), distributed electron cyclotron resonance (DECR)].

The advantage of these reactors lies in the high density, which gives good gap-filling. Although the substrate is placed far from the plasma generation region, because of the high plasma density and the very low pressure required, all excited species take part in the deposition, so that this configuration is not a remote plasma but some intermediate one.

e. Remote Plasma Reactors

Various levels of sputtering of all plasma-exposed surfaces and multiple and complex reaction pathways are real drawbacks of plasma processing. A solution is in the concept of remote PECVD (RePCVD or RPCVD; sometimes called indirect plasma). The idea is simple: separate the regions of generation and use of plasma, i.e., have one section of the reactor where sufficiently dense plasma is produced and let it flow into another section where it reacts on the wafer. Since the wafer is now at some distance downstream, in what is called afterglow plasma (the most unstable states are already decayed), it is possible to lower ion bombardment of the substrate substantially.

A further improvement comes from injecting only some of the reactants (or even the diluting gas only) into the plasma generation region and introducing the remainder in the afterglow region; depending on the relative injection positions, they may or not be lightly excited. This makes the reaction pathways simpler and, to some extent, selectable; the benefits are deposited films with possible exact stoichiometry, excellent control of dopants and impurities, and good-quality homo- and heteroepitaxial layers.

A schematic of a remote plasma implementation is shown in Fig. 19; the plasma may also be generated by microwave sources.

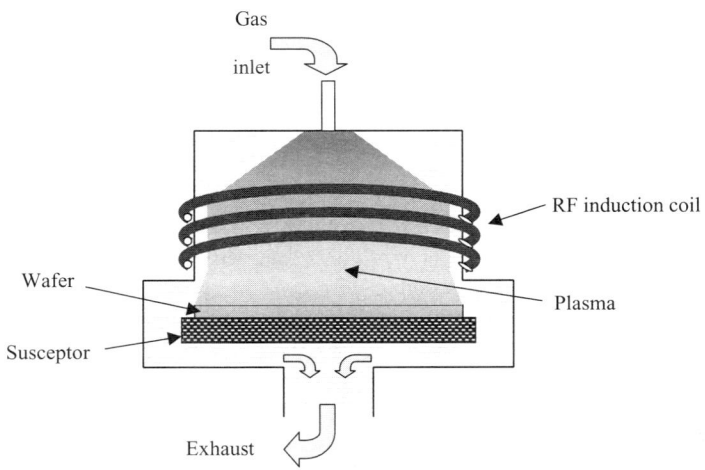

FIG. 19. Basic schematic of a remote plasma reactor.

IV. Film Deposition: Products and Chemistries

1. PROPERTIES OF FILMS AND MATERIALS

This section serves as a preliminary introduction to some of the silicon products of CVD systems in the microelectronics industry. As such, it does not strive for completeness; it is intended to give just an overview of the applications of the reactors described previously and of the most common chemistries involved. Some of these materials are covered in more detail in subsequent chapters.

Another cause of the breadth of content is the definition of film itself. In the present context, it could be best taken as complementary to bulk deposition, to designate "any layer thinner than or comparable to the substrate on which it is deposited." It may be dubious to distinguish between thin and thick film, when in the final application the ratio of width to height can favor one or the other dimension. As examples among current applications, consider epitaxial or polycrystalline films, spanning from a few tens of angstroms (in ULSI high-speed devices) to almost 100 μm (epi for power devices and poly for micromachining applications). Let us recall also that the lower dimension limit has already been reached with quantum device construction from stacks of monoatomic layers.

The field of film and material properties too is a bit involved. As a matter of fact, some properties (intrinsic stress, for instance) depend on many and small details of the chemical composition (with wanted and unwanted impurities) and physical structure, so that the relationships among parameters are not well known (if not just guessed by practice). While the complexity of dependencies may provide flexibility in film engineering and utilization, it also makes it difficult and lengthier to design processes for new products.

Let us recall some of those properties, relevant for manufacturing process flow or device construction and characteristics: etch rate, bulk and surface resistivity, dielectric constant and strength, impurity migration, refractive index, step coverage, thermal expansion and stability, intrinsic stress, and film adhesion. As for epitaxial silicon, each compound deserves a whole book, with space devoted to process details and comparisons and in which each compound's properties are studied and its dependencies and significance for applications described.

A field that is in rapid evolution, driven by safety considerations and the quest for lower process temperatures, is that of organic precursors. Because of the complexity of the chemistry (often in just the first stages of understanding) and the almost-chaotic growth of research, we cite reactions for TEOS only. A few silicon organic compounds that are referenced in the literature are TRIES (triethoxysilane), TMOS (tetramethylorthosilicate), OMCTS (octamethylcyclotetrasiloxane), TMCTS (tetramethylcyclotranssiloxane), MDES (methyldimethoxysilane), and TMS (tetramethylsilane). It is interesting, however, to note that organic precursors are more common for other semiconductors than for silicon, probably because of inertia in the legacy process and regarding equipment.

In the following section, after giving some descriptive information about each compound, we list as a quick reference overall stoichiometric reactions for the most used precursors (with reactants on the left and relevant reaction by-products on the right). We note, however, that quite often compounds exist in nonstoichiometric form; even the global reactions are then intended as "sensible" indications and do not exclude the presence (although at low levels) of more complex by-products.

2. Common Reaction Chemistries

a. Doped and Undoped Silicon

Apart from monocrystalline epitaxial layers, silicon is deposited in polycrystalline or amorphous form almost exclusively from silane pyrolysis. Doped polycrystalline silicon is quite often used for gate electrodes, in shallow junction devices and transistor structures such as thin-film transistors (TFT), resistors, and short-distance interconnects. A special application is in solar cell technology. Doped hemispherically grained (HSG) polysilicon is used as an electrode to enhance area factors in capacitive structures; its characteristic structure is obtained just by careful utilization of surface atom mobility. Hydrogenated amorphous silicon (αSi:H) is deposited at lower temperatures, with conditions (especially in plasma reactors) favoring hydrogen incorporation. By means of annealing it restructures in polysilicon, with improved electrical performance. Semiinsulating polysilicon (SIPOS) is polysilicon "doped" with oxygen (SiO_x) by the presence during deposition of low levels of nitrous oxide (N_2O); its high resistivity is the reason for using it as a carrier barrier in advanced structures.

b. Silicon Oxide (SiO_2) and Glasses

Many applications of silicon oxides are common in the microelectronics industry: as a sacrificial layer during a masking operation, a dielectric in capacitors and MOS structures, a gap-fill in shallow trench isolation (STI), a reflow layer, an interlevel dielectric, a planarization layer, and a protective coating. While some of the applications mentioned require pure silicon dioxide, many other oxide types would be better named glasses. This is because of their nonstoichiometric structure and the presence of different impurities, in the form of partially controlled reaction by-products and intentionally introduced dopants. Examples are borosilicate glass (BSG), phosphosilicate glass (PSG), and borophosphosilicate glass (BPSG), from doping oxide with boron, phosphorus, or both, and fluorine-doped silicon glass (FSG), a low-k (dielectric constant) material used as an intermetal dielectric (IMD).

(a) Silane and oxygen:
$$SiH_4 + O_2 \rightarrow SiO_2 + 2H_2 \qquad (1)$$
$$SiH_4 + 2O_2 \rightarrow SiO_2 + 2H_2O \qquad (2)$$

(b) Silane and nitrous oxide:
$$SiH_4 + 2N_2O \rightarrow SiO_2 + 2H_2 + 2N_2 \qquad (3)$$

(c) Dichlorosilane and nitrous oxide:
$$SiH_2Cl_2 + 2N_2O \rightarrow SiO_2 + 2HCl + 2N_2 \qquad (4)$$

(d) TEOS (tetraethoxysilane; formerly tetraethylorthosilicate) pyrolysis:
$$Si(OC_2H_5)_4 \rightarrow SiO_2 + 2H_2O + 4C_2H_4 \qquad (5)$$

(e) TEOS and oxygen:
$$Si(OC_2H_5)_4 + 12O_2 \rightarrow SiO_2 + 10H_2O + 8CO_2 \qquad (6)$$

(f) TEOS and ozone:
$$Si(OC_2H_5)_4 + 8O_3 \rightarrow SiO_2 + 10H_2O + 8CO_2 \qquad (7)$$

(g) FTES (fluorotriethoxysilane) and water:
$$FSi(OC_2H_5)_4 + 3H_2O \rightarrow FSiO_{3/2} + 3/2\,H_2O + 3C_2H_5OH. \qquad (8)$$

c. Silicon Nitride (Si_3N_4)

Silicon nitride has some uses similar to those of silicon dioxide: as a gate dielectric [alone or in oxide–nitride–oxide (ONO) structures], as an interlevel dielectric (ILD), as a masking layer during oxidation in local oxidation of silicon (LOCOS) processes, and in final passivation.

(a) Silane and ammonia:
$$3SiH_4 + 4NH_3 \rightarrow Si_3N_4 + 12H_2 \qquad (9)$$

(b) Dichlorosilane and ammonia:
$$3SiH_2Cl_2 + 4NH_3 \rightarrow Si_3N_4 + 6H_2 + 6HCl \qquad (10)$$
$$3SiH_2Cl_2 + 10NH_3 \rightarrow Si_3N_4 + 6H_2 + 6NH_4Cl. \qquad (11)$$

d. Silicon Oxinitrides (SiO_xN_y)

Besides some applications in common with nitride, such as final passivation, these have special applications as antireflective coatings for UV lithography, low-stress materials in micromachining, and small-scale optical devices.

(a) Silane, ammonia, and nitrous oxide:
$$SiH_4 + yNH_3 + xN_2O \rightarrow SiO_xN_y + xN_2 + (1.5y + 2)H_2 \qquad (12)$$

(b) Silane, ammonia, and carbon dioxide:
$$SiH_4 + yNH_3 + x/2\,CO_2 \rightarrow SiO_xN_y + x/2\,CO + (2 + 1.5y)H_2. \qquad (13)$$

e. Silicides

Silicides of refractory metals are used as gate electrodes (in conjunction with a polysilicon layer) and for interconnect construction.

(a) Tungsten silicide (WSi_2) from tungsten hexafluoride and silane:
$$WF_6 + 2SiH_4 \rightarrow WSi_2 + 6HF + H_2 \qquad (14)$$

(b) Tungsten hexafluoride and dichlorosilane:
$$WF_6 + 7/2\,SiH_2Cl_2 \rightarrow WSi_2 + 3/2\,SiF_4 + 7HCl \qquad (15)$$
$$WF_6 + 7/2\,SiH_2Cl_2 \rightarrow WSi_2 + 3/2\,SiCl_4 + 6HF + HCl \qquad (16)$$

Possible by-products are also SiF_xCl_y and Cl_2.

(c) Titanium silicide ($TiSi_2$) from titanium tetrachloride and silane:
$$TiCl_4 + 2SiH_4 \rightarrow TiSi_2 + 4HCl + 2H_2 \qquad (17)$$

(d) Tantalum silicide ($TaSi_2$) from tantalum pentachloride and silane:
$$TaCl_5 + 2SiH_4 \rightarrow TaSi_2 + 5HCl + 3/2\,H_2 \qquad (18)$$

(e) Tantalum pentachloride and dichlorosilane:
$$TaCl_5 + 2SiH_2Cl_2 + 5/2\,H_2 \rightarrow TaSi_2 + 9HCl \qquad (19)$$

f. SiGe and SiC

So-called *IV–IV* semiconductors are attracting a lot of interest and effort. SiGe heterostructures have boosted silicon into the high-frequency application domain, while SiC shows unique ruggedness because of its wide bandgap and high temperature stability:

$$SiH_4 + xGeH_4 \rightarrow SiGe_x + 2(1 + x)H_2 \qquad (20)$$
$$SiH_4 + CH_4 \rightarrow SiC + 4H_2 \qquad (21)$$
$$SiH_4 + 1/3\,C_3H_8 \rightarrow SiC + 20/3\,H_2 \qquad (22)$$
$$1/2\,Si_2H_6 + 1/2\,C_2H_2 \rightarrow SiC + 4H_2. \qquad (23)$$

3. PLASMA PECULIARITIES

Film deposition with plasma techniques, with attention to reaction and product characterization, is covered by an extensive literature. A common pitfall and limit of current studies is to correlate the chemistry and material properties with *process parameters* (such as power and frequency) and not *physical plasma parameters* (such as electron density and temperature). Our simple introduction to plasma reactor principles is sufficient to show how, much more than for thermal CVD, the process parameter effect is strongly dependent on the specific technology and hardware used. There are additional possibilities of process chemistries in plasma reactors, which are precluded to thermally activated CVD. Several mechanisms [excitation, dissociation, ionization, recombination, charge transfer, atom abstraction (exchange between molecules), deexcitation, sputtering] now involve not only active molecules and substrate, but also inert carriers, ions, and electrons. Another interesting fact is the lower (still comparing to thermal CVD) sensitivity of growth rate to *in situ* doping.

Plasma processes for silicon dioxide also use CO or CO_2 with SiH_4. FSG is commonly deposited from conventional plasma SiO_2 by doping with SiF_4 or CF_4. Hydrogen-rich, nonstoichiometric silicon nitride is obtained by simple reaction of silane and ammonia or silane and nitrogen.

V. Aerosol CVD

1. BASICS ON OPTICAL FIBERS

The advantages of using optical fibers in signal communication (all those of standard wire transmission plus electromagnetic noise immunity and far larger bandwidth) favor their wide use from short to very long transmission distances. The possibility now exploited by optoelectronics of tighter integration between the electrooptical transducers and electronic signal control and digitalization have brought the world of optical fiber into close touch with those of microelectronics, computers, and networks.

Manufacturing of optical fibers has in common with microelectronics just using some special CVD furnaces and precursors and being silicon-based. The differences are, however, so interesting that we now step aside for an introduction into the related technologies. Good rapid papers on these topics are those by Roba (1986) and Cognolato (1995).

Without going into the physics of optical fibers and their functioning, there are two basic and simple facts that allow us to understand the requirements and the processing of these materials. The purpose of an optical fiber (a sort of "special-glass thread") is to transmit "light signals," i.e., pulses of light radiation (in the visible or near-infrared region). Ideal requirements of any transmission are that

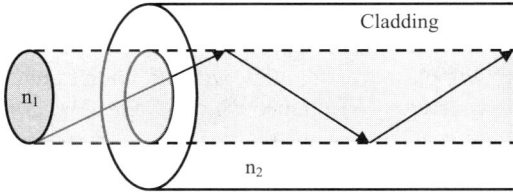

FIG. 20.   Propagation of a light signal along an optical fiber.

the signal is not attenuated (it does not lose power during its travel) and that the information is not corrupted (the shape of the signals is preserved). In the real world, the goal is to approximate these ideals as best as possible.

The first item, attenuation, is related very intuitively to the amount of generic imperfections (of mechanical type or chemical impurities) that may disturb the signal propagation. The resulting requirement is thus for high-purity materials and processes, homogeneity of materials, and close matching of relevant parameters (such as thermal expansion coefficients).

The second item, the amount of signal deformation, is physically related to the way the light propagates along the fiber: it continuously bounces off the "walls" of the fiber because of reflection from a medium with a lower refractive index, as in Fig. 20. The complete structure of the fiber is shown in section in Fig. 21. The core diameter is either 8 or 50 μm, depending on the type of transmission mode used. The cladding diameter is 125 μm. Since reflection from cladding of light rays traveling in the core is determined by refractive indices, the capability for accurate control is fundamental for transmission properties and performance. Final values of refractive indices are determined down to ten thousandths, with the difference between core and cladding typically 0.4 to 0.8%.

CVD can address both issues: CVD technology and processes (as learned from the microelectronics field) are especially suited to control the high intrinsic purity of materials and gradients of dopant incorporation. Note that the term dopant is a wide extension of the microelectronics concept, since these different compounds

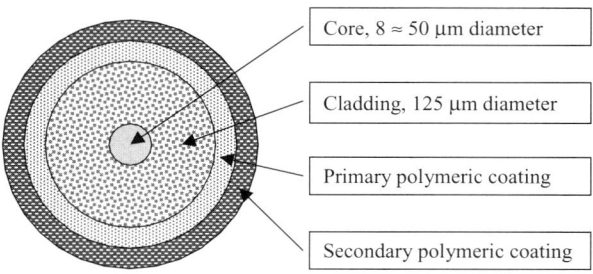

FIG. 21.   Cross section of an optical fiber.

can constitute a large percentage (up to 10–20%) of the material. It is fundamental for understanding the use and evolution of CVD in this area to know the main steps of fiber manufacturing.

It is relatively easy in glassworking, and also a consolidated practice, to obtain thinner rods from larger ones by drawing when the material is softened at a high temperature. It was learned that, with some control of the process, it is possible to preserve the structure of the initial rod and just scale it down. We can thus construct a gigantic optical fiber, with the desired diameter ratio of core to cladding and the spatial profile values of refractive index, then shrink it to the required diameter while making it longer.

The starting rod is called "the preform" and has a diameter ranging from 10 mm (as was more common in the early stages) to 100 mm (as is more common nowadays with OVPO techniques), with a length usually of less than 1 m. Such preforms can yield fibers several kilometers long, up to even 400 km, but typical values are in the range of a few tens of kilometers. At this point, the fiber, which is already functioning "optically," receives polymer coatings for wear and chemical protection (the primary one) and mechanical strength (the secondary one); the outside end diameter is about 1 mm.

2. AEROSOLS AND PRINCIPLES OF REACTORS

A few concepts are common to the various techniques used in this sector: solid material is generated as an *aerosol* and is deposited mainly by the *thermophoretic* effect; deposition takes places in many layers (with the capability for controlling the properties of each).

Aerosols are a suspension in a gas of liquid or solid particles, with dimensions ranging from molecular to 100 μm and considered over time scales such that their suspension characteristics are not strongly modified. In recent years, there have been dedicated studies of this state, which is growing in importance in industry as well as in fundamental sciences. A recent and extensive book is that by Kodas and Hampden-Smith (1999).

Thermophoresis is motion due to temperature gradients, accomplished toward the lower-temperature zone. Down to the very basic, a particle suspended in a fluid with a temperature gradient is pushed because of the imbalance in the net forces exerted by fluid molecules from the higher- and the lower-temperature sides. Studies on the thermophoretic effect applied to IVPO reactors are reported by Simpkins *et al.* (1979).

3. INSIDE DEPOSITION REACTORS

The point of *inside vapor-phase oxidation* (IVPO) is to use a fused silica tube both as the reactor tube and as the deposition substrate. At the end of the fiber processing, the silica tube will be the main fraction of the cladding. Common to inside deposition techniques is that, after deposition, the tube is processed in several passes with a burner on a glassworking lathe, at a high temperature (some 2000°C). This manufacturing step is called preform collapsing (squeezing the central empty space to obtain a full solid rod).

a. MCVD Reactors

Modified chemical vapor deposition (MCVD) was patented in 1974. Its use is quite widespread, because of its relative simplicity and good productivity, even if its chemical efficiency is poor (i.e., less than half the reactants flowing are effectively deposited in the product, while the remaining fraction is wasted). Figure 22 is a schematic illustrating the process.

A silica tube is mounted on a structure similar to a glassworking lathe with special arrangements for inlet and exhaust of gases. The tube itself rotates and an oxyhydrogen burner slides along the axis and is set to heat the tube to circa 1600°C. The process is at AP and gas flow is a few liters per minute; within the tube dimension this allows a laminar flow, critical for deposition homogeneity and uniformity. For each layer deposition, the burner moves slowly in the same direction as the flowing process gases, and at the end of the run, it is quickly moved to the inlet side to start another cycle. When gases reach the heated zone, they react quite rapidly (it is believed that the conversion is complete) and the solid phase is suspended in the flow, constituting an aerosol state, with a particle dimension on the order of micrometer. As the aerosol, carried by the flow, moves downstream from the reaction zone, it gets between walls at temperatures much lower than the gas itself. The temperature gradient is in fact now reversed with respect to the

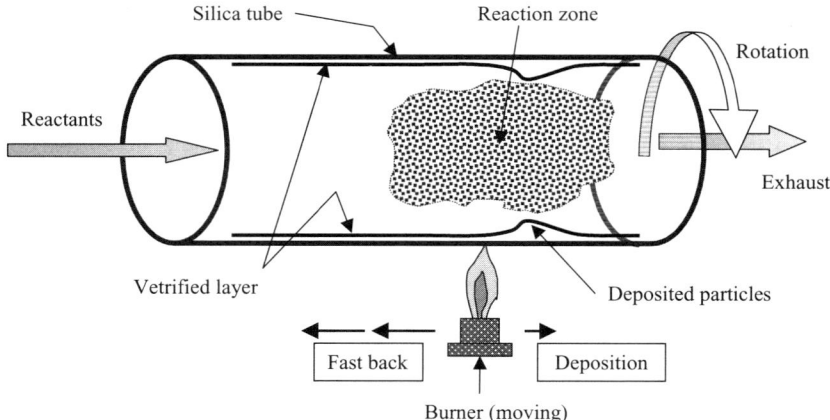

FIG. 22. Deposition-zone cross section of a modified CVD reactor.

deposition zone; the hottest points are now in the center and aerosol particles are deposited on the walls by the thermophoretic effect. After a short time, the heated zone is over the just-deposited material and the temperature is more than sufficient to make it coalesce and vitrify. Layers thus obtained have thicknesses ranging from 10 to 50 μm.

A variation of this technique is to operate at a pressure higher than the AP, with some modifications to the gas exhaust system. The advantage is that the pressure difference can oppose the surface tension forces of the heated zone and stop the trend of collapse and deformation already present at the process temperature.

b. *PCVD, PICVD, and SPCVD Reactors*

The plasma CVD (PCVD) technique is a quite different one, as illustrated in Fig. 23. A resonant cavity fed by a microwave source (typically 2.45 GHz) generates a plasma region in the low-pressure gas; operation is at approximately 10 Torr. It is now the plasma energy that starts the gas-phase reactions. As the resonant cavity slides along the tube, deposition takes place on the wall and is coalesced by the nearby coaxial furnace (placed on both sides of the resonator), which keeps the tube at a temperature ranging from 1000 to 1200°C. Deposition takes place with the cavity moving in both directions along the tube; the deposited layers are thinner than those obtained by MCVD (typically 1 μm each), but the efficiency in terms of reactant use is much higher.

Plasma impulse CVD and surface plasma CVD (PICVD and SPCVD, respectively) are variations of the plasma technique. Both have in common that there are no moving parts. The silica tube stands in a standard resistive heating furnace,

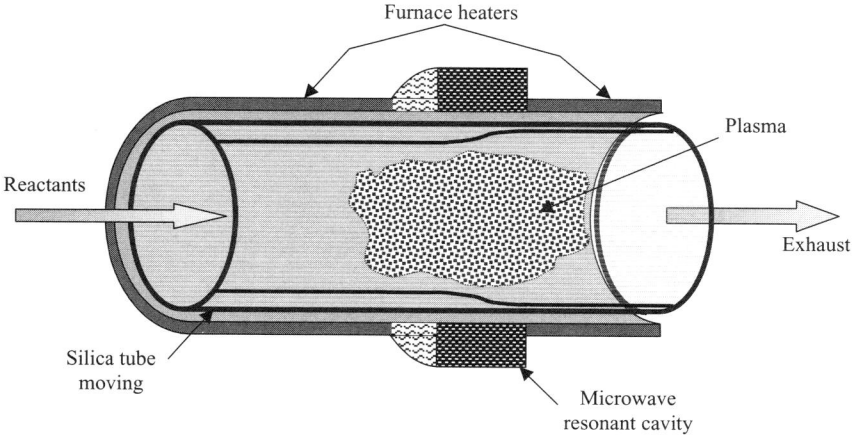

FIG. 23. Principle functioning of a plasma CVD reactor.

which serves for vitrification, while the activation energy for the reactions is supplied by a microwave plasma excited at 2.45 GHz.

In the first case, PICVD, a waveguide surrounds the process tube and the power is pulse-injected from the exhaust end. This arrangement generates a plasma wave propagating along the tube, and follows the reaction deposition. At that point, reacted gases are flushed with fresh ones and another cycle starts. Typical cycle frequencies are about 100 Hz. In the second case, SPCVD, microwaves are transmitted to the reacting gases by an illuminator (a waveguide terminating horn) that introduces power from the exhaust end of the tube. Reaction and deposition characteristics are similar to those of PICVD.

c. PMCVD Reactors

As exemplified in Fig. 24, plasma modified CVD (PMCVD) takes advantage of both MCVD and PCVD. This is accomplished by separating the functions of reaction activation, handled by plasma, vitrification, provided by a relatively low-temperature burner, and deposition. Flowing gases are preheated in the burner region while vitrification takes place. Next is the reaction region, where plasma is generated by an induction coil powered at a few megahertz. Just afterward the tube wall is cooled to enhance deposition by the thermophoretic effect. This combination gives quite good results in terms of both reactant efficiency and deposition rate.

One problem common to all plasma-based processes is that the initial tube has a larger diameter than in MCVD (to allow the plasma sufficient stability) and this can worsen the preform geometry during collapse.

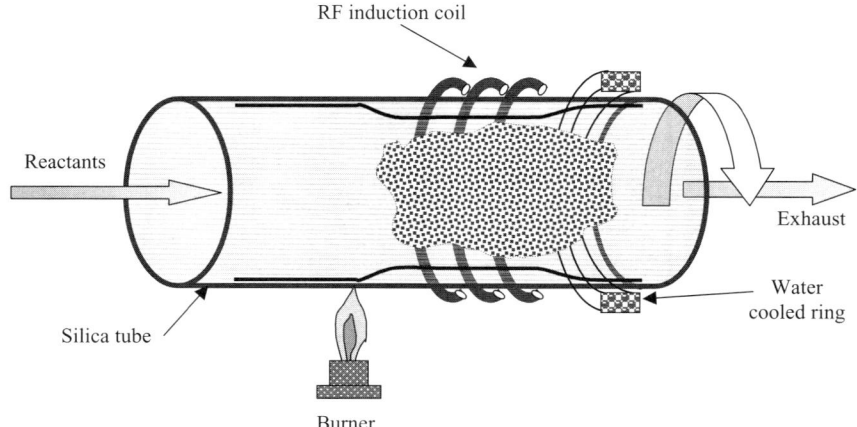

FIG. 24. Deposition-zone schematic for a plasma modified CVD reactor.

4. OUTSIDE DEPOSITION REACTORS

a. *OVD Reactors*

Outside vapor deposition (OVD) reactors were the first ones to be devised for CVD aerosols, in the early 1940s, and to be tentatively used for optical applications. Although their products cannot be used directly for optical fibers, because of insufficient purity, with some additional manufacturing steps they can yield material of a quality comparable to that from IVPO. Nowadays this is probably the most common technique used in large-scale manufacturing. The structure of an OVD reactor, not really changed from the first design, is shown in Fig. 25.

Precursors, oxygen, and a combustible (hydrogen, methane, or propane) are fed together into a burner. The aerosol particles, some tenths of a micrometer in dimension, deposit on the rod. Uniformity is accomplished by rotation of the rod and translation of the burner in both directions; the hardware is basically a glassworking lathe. The rod on which deposition initially starts was originally graphite, but now for optical applications the preferred material is alumina. When deposition of the layer structure is completed, the support rod is removed by simply dragging it out. Since the material is now homogeneous and completely colloidal, the final mechanical properties of the preform are better than those of IVPO origin. The other fundamental advantage is the porosity, which is a requirement for the subsequent purification process. The colloidal preform is flushed with chlorine and helium while moving in a furnace, exposed at temperatures increasing from 500 to 1500°C. Three effects are accomplished: chlorine removes free molecular hydrogen and OH groups (dehydration); helium (because of its small size and fast diffusion) forces other impurities to leave the structure; and at the highest temperature, the rod is consolidated and collapsed to its final preform dimension.

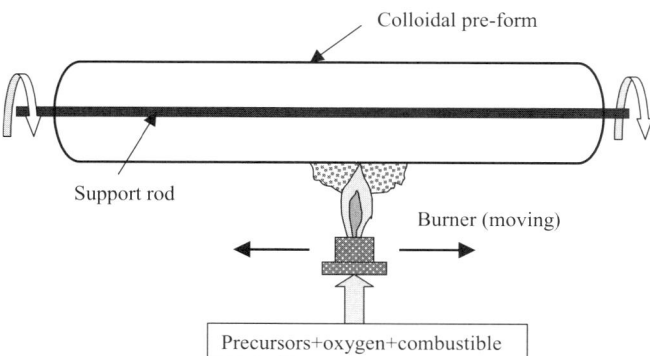

FIG. 25. Descriptive view of an outside vapor deposition reactor.

With this method the cladding, too, has to be deposited by layers, since it is not already present as in IVPO. This is not really an issue for process productivity since those layers can be deposited at much higher rates, given the lower material requirements.

In a slight variation of these processing steps, the initial rod can be a partial preform, previously consolidated and drawn thinner. In this way, there is no need to extract it and there is additional flexibility in controlling the refractive index profile.

b. VAD Reactors

Vapor axial deposition (VAD) is very similar in principle to OVD, since the aerosol for deposition is generated by burning reactants in the flame and is deposited on a rotating solid. In addition, the posttreatments (dehydration, purification, consolidation, and collapse) are the same. The main difference is that, as shown in Fig. 26, the steps are integrated in the same processing system and operated continuously. A simple piece of glass rod is used as a starting surface for the deposition; the rod is kept in rotation and slowly drawn upward along the axis as the preform grows. A central burner (with a complex inner gas distribution to

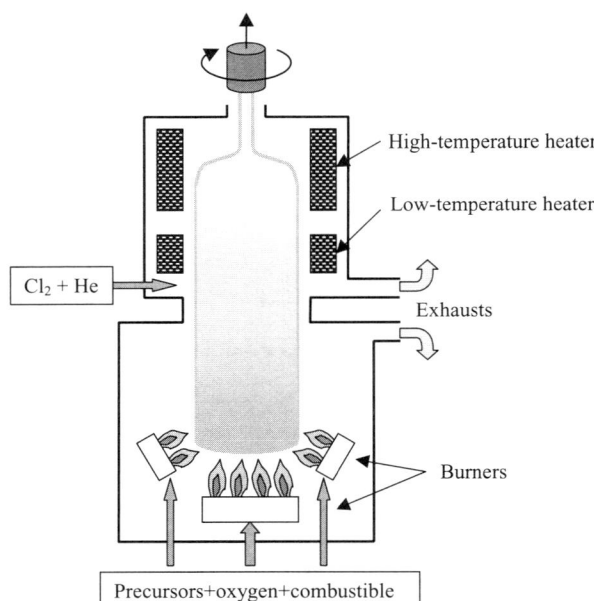

FIG. 26. Cross-sectional drawing showing the main elements of a vapor axial deposition continuous reactor.

control the dopant profile) and possibly lateral burners are located in fixed positions. Here the whole cross section of the preform is grown at the same time, which is why the dopant profile has to be controlled by gas distribution variations in the flame region. There are two heaters in the upper chamber of the system, the first one at a lower temperature for chlorine dehydration and the second at a higher temperature to enhance helium cleaning and allow preform consolidation.

Although the system is intrinsically designed for continuous operation, this possibility has not been fully exploited in practice, and, apart from a few special cases, preform sizes are similar to those produced with discontinuous methods.

5. BASICS OF CHEMISTRY FOR OPTICAL FIBERS

For silica, which is the basic component, the preferred deposition reaction is TCS oxidation. The main reason for this choice, beside stability considerations related to the reaction temperatures and methods used in these special applications, was to avoid the incorporation of molecular hydrogen and hydroxyl groups into the deposited structure. Their effects on fiber quality are indeed serious, especially because of the strong absorption in some wavebands.

Among the compounds that increase the refractive index are (in order of increasing effect with respect to concentration) P_2O_5, GeO_2, Al_2O_3, TiO_2, ZrO_2, and Nb_2O_3; those that decrease the refractive index include (in order of increasing effect with respect to concentration) B_2O_3 and F_2. For several reasons (including chemical stability, vitreous oxide form, and exclusion from the reactions of compounds damaging fiber performance), the preferred ones are GeO_2 and F_2. Germanium oxide is obtained, similarly to silica, from tetrachlorogermane oxidation, while fluorine has the following precursors: CCl_2F_5, CF_4, SiF_4, and SF_6.

Current methods of manufacturing silica-based fibers have already reached such high levels of material perfection that attenuation is almost down to the intrinsic material properties. Thus current research is focused mainly on other materials.

VI. Conclusions

This chapter has shown how CVD is a widely applied process throughout industry. Although many processes and applications have been described, a lot of other industrial applications exist, such as in polymer coatings (Lamendola, 1988; Lamendola and Tenuto, 1998; Valentini *et al.*, 1998), metallurgy, and construction. Despite its complexity, nowadays the most widely used technique is plasma CVD, in all its variations. The traditional processes are limited to low-added value products.

We hope that this "quick survey" on CVD has helped the reader to find his or her way in the messy world of silicon-based thin films.

Acknowledgments

We thank Andrea Cranna, whose thesis dissertation was used as the starting point for this chapter, and Massimo Rustioni (who is one of the last experts on bulk polysilicon for solar-cell applications) for his contribution to the description of bulk CVD in Section II. Special thanks go to our wives, Chiara and Mirella, and our children, Erica, Giacomo, and Camilla, for their patience and support.

References

Cerofolini, G. F., and L., Meda, *Physical Chemistry of, in and on Silicon,* Springer Series in Materials Science, Vol. 8, Springer-Verlag, Berlin, 1989.

Cerofolini, G. F., and L., Meda, in *Chemistry for Innovative Materials,* edited by G. F. Cerofolini, R. M. Mininni and P. Schwarz, p. 64, EniChem, Milano, 1991.

Chang, C. Y., and S. M., Sze, *ULSI Technology,* McGraw–Hill, New York, 1996.

Cognolato, L., *J. Phys. IV Coll. C5* **5**, 975 (1995).

Hitchmann, M. L., and K. F., Jensen, *Chemical Vapor Deposition,* Academic Press, London, 1993.

Jayant Baliga, B., *Epitaxial Silicon Technology,* Academic Press, Orlando, FL, 1986.

Kodas, T. T., and M., Hampden-Smith, *Aerosol Processing of Materials,* John Wiley & Sons, New York, 1999.

Lamendola, R., in Materials Research Society Fall Meeting Tutorial Program, Boston, 1988.

Lamendola, R., and G., Tenuto, *Proceedings, 41st Annual Technical Conference Society of Vacuum Coaters,* p. 458, Boston, Apr. 18–23, 1998.

Masi, M., S., Carrà, G., Vaccari, and D., Crippa, in *Chemical Vapor Deposition XIV,* edited by M. D. Allendorf and C. Bernard, p. 1167, Electrochemical Society, Pennington, NJ, 1997.

Nieto, J.-P., B., Caussat, J.-P., Couderc, S., Coletti, and L., Jeannerot, *J. Phys. IV France* **9**, 149 (1999).

Rinaldi, A. M., S., Carrà, M., Rampoldi, M. C., Martignoni, and M., Masi, *J. Phys. IV France* **9**, 189 (1999).

Roba, G., *Fisica e tecnologia, VII-3,* Editrice Compositori, Bologna, 1986.

Sherman, A., *Chemical Vapor Deposition for Microelectronics,* Noyes, Park Ridge, NJ, 1987.

Simpkins, P. G., S., Greenberg-Kosinski, and J. B., MacChesney, *J. Appl. Phys.* **50**, 5676 (1979).

Sze, S. M., *VLSI technology,* McGraw–Hill, New York, 1983.

Valentini, A., A., Convertino, M., Alvisi, R., Cingolani, R., Lamendola, and L., Tapfer, *Thin Solid Film* **1**(5), 162 (1998).

Vossen, J. L., and W., Kern, *Thin Film Processes II,* Academic Press, New York, 1991.

Wolf, S., and N., Tauber, *Silicon Processing for the VLSI Era: Process Technology,* Lattice Press, Sunset Beach, CA, 1986.

CHAPTER 2

Epitaxial Growth Theory: Vapor-Phase and Surface Chemistry

C. Cavallotti and M. Masi

DIPARTIMENTO DI CHIMICA FISICA APPLICATA
POLITECNICO DI MILANO
MILANO, ITALY

I.	INTRODUCTION	51
II.	DEVELOPMENT OF A DETAILED KINETIC SCHEME	52
III.	GAS-PHASE KINETICS FOR SILANES AND CHLOROSILANES	56
	1. *Silane's Gas-Phase Reactivity*	56
	2. *Chlorosilane's Gas-Phase Reactivity*	61
IV.	SURFACE KINETICS FOR SILANES AND CHLOROSILANES	71
	1. *Silane's Surface Reactivity*	71
	2. *Chlorosilane's Surface Reactivity*	74
V.	GAS-PHASE PRECURSORS TO DEPOSITION AND OVERALL KINETIC SCHEME	80
VI.	OVERALL KINETIC SCHEME AND CONCLUDING REMARKS	84
	REFERENCES	86

I. Introduction

The growth of silicon thin solid films through chemical vapor deposition (CVD) techniques can be regarded as the most important process of the microelectronic industry. As reported by Hitchman and Jensen, "CVD is a process whereby a thin solid film is synthesized from the gaseous phase by a chemical reaction" [41]. In this definition two points characterize CVD with respect to other growth methods: the use of a gas-phase precursor and the occurrence of a reaction, which can take place on the surface or in the gas phase. Among the gas-phase molecules containing silicon, those that are more convenient and therefore have been adopted to grow silicon films industrially are chlorosilanes and silanes. In this chapter the kinetics of the gas-phase and surface reactions active during the growth of silicon films from cholrosilanes and silanes precursors is investigated. The final aim is the understanding of the reaction pathway that determines the transition of silicon from the gas to the solid phase. This is necessary since, to control all the process features with the degree of sophistication required nowadays, most of the physical and chemical aspects nature have to be understood. As a corollary, the theoretical and experimental methods adopted in chemical kinetics are also discussed.

Identification of the elementary reactions and of their kinetic constants for a CVD system is usually a difficult task because of the occurrence of gas-phase and surface processes at the same time. From this point of view the deposition of silicon from gas-phase precursors can be considered as a particular case. In fact, because of its industrial relevance, it has been thoroughly investigated by both experimental and theoretical methods.

This chapter is organized as follows. In Section II the process usually followed to develop a kinetic mechanism is addressed. Here particular emphasis is placed on the individuation of the physical and chemical parameters necessary for the realization of a satisfactory and effective model. Experimental and theoretical methods nowadays adopted to determine kinetic constant values for gas-phase and surface processes are briefly described. In Section III gas-phase reactions of silanes and chlorosilanes are considered. In particular, they are grouped into thermally activated precursors decomposition, radical reactions, plasma-assisted precursor decomposition, and radical reactions. In this section thermodynamic parameters for the known SiH_xCl_y gas-phase species are also reported. In Section IV surface reactions are considered. In particular, the following processes are examined: precursor and radical adsorption, surface desorption, and HCl interaction with a silicon surface.

The chapter reports the details necessary to develop a kinetic scheme (Section V). Much information has been collected on the subject, of both an experimental and a theoretical nature. In many cases it is difficult to match all these data into an overall detailed kinetic scheme. An effort has been made to harmonize all the available information in an elementary mechanism that can be adopted for the growth of silicon films by means of the most important precursors under quite broad process conditions (e.g., from high to low pressure and from thermally activated to plasma-assisted depositions).

II. Development of a Detailed Kinetic Scheme

A detailed chemical mechanism describes a single overall chemical reaction, such as that determining the growth of a thin solid film, by means of an ensemble of irreducible elementary chemical reactions [75]. Kinetic constants of each elementary reaction can be determined either by experimental methods or via theoretical calculations. A kinetic constant can be estimated adopting traditional theories of kinetic process (e.g., the kinetic theory of gases and transition-state theory) based on data (vibrational frequencies, transition-state structures) that have been experimentally determined or calculated through quantum chemistry methods. The field of chemical kinetics has seen considerable progress in recent years, mainly for two reasons: first, the improvement of the theory of reaction rates, which can now be considered a mature field [5, 7, 31, 77], and second, the capability of estimating kinetic constants of gas-phase and surface elementary reactions by shock tube experiments and surface science techniques and the ever-increasing reliability of quantum chemical calculations. Based on such progress, a systematic method of realization of kinetic schemes can be proposed.

A generic procedure for realization of a detailed kinetic scheme requires as the first step the identification of all the chemical species (i.e., neutral species, radicals, and ions) that are likely to be present in the gas or on the surface. Subsequently, all the elementary reactions that can occur among these species must be considered and a kinetic constant must be determined. At this stage physical intuition is of great importance to cancel from the hypothesized events all those reactions that are unlikely to occur (e.g., a reactive event between two neutral species in a gas is possible only at very high temperatures because of the stability of the reagents). The next step is the determination of the kinetic constants for all the selected reactions. Many kinetic constant values for gas-phase reactions are readily available from the literature or can be found in kinetic databases now accessible on the World Wide Web.

However, when a system different from those already studied is examined, the kinetic constants for many possible processes are often not known. For silicon deposition, this is particularly true for new metal organic precursors that are under design to lower the deposition temperature. In these cases kinetic constant values can be either calculated or determined experimentally. The experimental evaluation of a kinetic constant is usually a time-consuming process and can be hampered by many problems in the realization of a reliable experimental setup. Moreover, it is not known a priori whether or not the considered reaction will have a relevant impact on the reactivity of the overall system or not. It is therefore a better approach to evaluate the kinetic constants of all the processes considered from a theoretical approach and then to identify the key reactions in the developed kinetic scheme from a sensitivity analysis and evaluate experimentally the kinetic constants of these processes.

A kinetic scheme for CVD usually consists of two subschemes, one dealing with the gas-phase chemistry and the other with the surface.

The rate of the mth gas-phase reaction can be expressed as

$$\tilde{r}_m^g = k_m \prod_{i=1}^{NC} C_i^{g_{im}} (1 - e^{\Delta G_m/RT}), \tag{1}$$

where k_m is the reaction rate constant and g_{im} is the reaction order with respect to the ith species, whose molar concentration is C_i. Because a detailed chemical mechanism should involve only elementary reactions, g_{im} assumes a value equal to the stoichiometric coefficient (i.e., $g_{im} = -\nu_{im}^g$) for the reactants involved in the considered reaction, zero being its value for all other species. The rate of a gas-phase reaction is always expressed as moles per cubic meter per second (mol · m^{-3} · s^{-1}) if the SI system is used; otherwise, other consistent units can be adopted. Equation (1) considers a reversible reaction, the exponential term on the right-hands side (r.h.s.) being the contribution of the reverse reaction. The free energy change associated with the mth reaction can be calculated as

$$\Delta G_m(T) = \Delta G_m^0(T) + RT \ln \prod_{i=1}^{NC} (C_i RT)^{\nu_{im}^g}, \qquad (2)$$

where the standard value ΔG_m^0 is calculated through the free energy of formation of each reactant involved in the reaction [70, 73].

The reactions that can occur at the growing surface are mainly of three types: adsorption, surface recombination, and desorption. A general formulation suitable for all three cases mentioned is

$$\tilde{r}_m^s = k_m^s \left(\prod_{i=1}^{NC} (C_i)^{s_{im}} \right) \prod_{j=1}^{NSITE} \prod_{i=1}^{NCA_j} (C_{ij}^s)^{p_{im}}, \qquad (3)$$

where C_{ij}^s is the concentration of the adsorbed ith species over the jth kind of surface site [expressed as moles per square meter (mol · m^{-2}) if SI units are used]. In particular, the surface concentration is often expressed as a function of the surface coverage θ_{ji} (i.e., $C_{ij}^s = \psi_j \theta_{ji}$), where ψ_j and θ_{ji} are the surface concentration of the jth kind of surface site and the coverage fraction of the ith species on the jth site, respectively. Finally, k_m^s and p_{im} are the surface reaction rate constant and the reaction order with respect to the ith adsorbed species, respectively. Equation (3) reflects that the rate of a surface reaction depends on the probability of a gas-phase species hitting a free surface site or interacting with one or more adsorbed species. The above formulation considers different types of surface sites on which the interaction can take place. This aspect is particularly important when the growth of semiconductor compounds is examined (e.g., the case of Si–Ge alloys). In any case, in SI units, the rate of a surface reaction is always expressed as moles per square meter per second (mol · m^{-2} · s^{-1}).

The adsorption rate can be expressed as the frequency of the collision of the molecules coming from the gas-phase on the surface, multiplied by both the sticking coefficient and the probability of finding an empty surface site. On the whole, the rate constant to be used together with Eq. (3) is

$$k_m^s = \gamma_m \sqrt{\frac{RT}{2\pi M_m}} e^{-E_m/RT} \left(\prod_{j=1}^{NSITE} \prod_{i=1}^{NCA_j} \psi_j^{-p_{im}} \right), \qquad (4)$$

where γ_m is the steric factor for adsorption of the mth species and E_m is the activation energy of the process. If $\gamma_m = 1$ and $E_m = 0$, Eq. (4) gives the maximum possible adsorption rate. Unfortunately, such a limit cannot be established for surface recombination and desorption reactions. In these cases, experimental information about surface bond frequency and strength are often used. If such information is not available, approximated values are usually adopted. Arrhenius preexponential values for surface recombination and desorption reactions are, in the SI system, 10^{13}–10^{14} s^{-1} and 10^{13}–10^{14} m^2 · mol^{-1} · s^{-1} for unimolecular and bimolecular reactions, respectively. The activation energy can be determined by adopting simplified estimates with standard thermochemical methods. Nowadays, quantum chemical calculations allows the direct estimation of the overall rate constant (e.g., activation energy and preexponential factor) [23, 39, 72].

The considerations reported above have to be slightly modified when plasma-activated reactions are present in the mechanism. Plasma enhanced CVD systems are nowadays frequently used also for epitaxial deposition at low pressures, but their largest use is still in the deposition of amorphous hydrogenated silicon. The most common adopted precursor is, in any case, silane. In fact, the highly energetic electrons are able to ionize and dissociate the adopted precursors at very high rates. Furthermore, they are able to promote reaction pathways unlikely to occur in pure thermally activated systems. The presence of free electrons in the systems adds a further degree of freedom to the problem of the definition of the detailed reaction mechanism. If the applied electric field that promotes the plasma, usually radio frequency (RF) powered, is "far" from the deposition surface, all the charged species can be considered to be confined between the two electrodes, and accordingly their interaction with the deposition surface can be neglected. In this case, the reactions activated by the electrons (i.e., the so-called electron impact reactions) contribute only to the reactions occurring in the gas phase. If one of the electrodes is also the substrate, ion-activated surface processes can take place. In this case the situation is more complex and is not addressed here because of its specificity.

Weakly ionized plasmas, such as those adopted for silicon deposition, are characterized by a very large number of elementary reactions, which involve a correspondingly large number of chemical species. Usually that scheme can be obtained by adding to the thermally activated mechanism all the species and the reactions originated by the electron impact reactions. Electron reactions can be subdivided into two categories: ionization reactions, which sustain the plasma, and neutral dissociation reactions, which are the most important source of radical species. The rate of the electron impact ionization and neutral dissociation reactions can be evaluated as a function of the electronic temperature as

$$k_m(T_e) = \int_0^\infty \sqrt{2\varepsilon/m_e} \cdot f(\varepsilon, T_e) \cdot \sigma_m(\varepsilon) d\varepsilon. \tag{5}$$

Equation (5) can be adopted provided that the electron energy distribution function $f(\varepsilon, T_e)$ and the reactive cross section $\sigma_m(\varepsilon)$ are known [34, 58, 60]. The electron energy distribution function can be obtained through the solution of the Boltzmann equation [33]. In fact, due to the absence of thermodynamic equilibrium in weakly ionized plasmas, the electron energy distribution function is not Maxwellian. An isotropic approximate solution of the mentioned Boltzmann equation yields the so-called Druyvesteyn distribution function, which can be adopted in the following form parameterized as a function of the electronic temperature

(T_e), evaluated in electorn volts (eV), and where $f(\varepsilon, T_e)$ has units of eV^{-1} [33]:

$$f(\varepsilon, T_e) = 0.566 \cdot \sqrt{\varepsilon} \cdot T_e^{-1.5} \exp\left(-0.244 \cdot (\varepsilon/T_e)^2\right). \quad (6)$$

Under glow discharge conditions, $f(\varepsilon, T_e)$ can reasonably be approximated by Eq. (6) because the electron inelastic collision frequency is some order of magnitude lower than the elastic one (e.g., electron elastic and inelastic collision frequencies are of the order of 10^9 and 10^4 Hz, respectively). Due to the complexity of the electron–molecule interactions, experimentally measured reactive cross sections for electron impact reactions have to be adopted. Theoretical formulations for $\sigma_m(\varepsilon)$ usually give rough estimations.

III. Gas-Phase Kinetics for Silanes and Chlorosilanes

1. Silane's Gas-Phase Reactivity

The deposition of silicon from silane has been the subject of much experimental and theoretical research. These works have investigated the gas-phase [42, 48, 49, 62–64] as well as the surface chemistry [25, 26, 49, 76]. Besides direct kinetic studies, different comprehensive models of the description of transport phenomena and gas-phase and surface kinetics have also been developed [11–13, 35, 46, 51, 54, 55, 59]. Thus, the deposition of silicon from pure silane and silane diluted in hydrogen represents one of the few systems where the construction of a detailed chemical mechanism was performed through a systematical analysis of all the chemical species and reactions that are likely to be present in the gas-phase and at the surface.

Thermochemical parameters for molecular and radical species that can be produced in the gas-phase of a reactor adopting silanes or chlorosilanes as precursors were taken from the compilation of Kee *et al.* [50] and are reported in Table I.

The gas-phase kinetic mechanism for silanes and disilanes has been the subject of intensive experimental and theoretical work. In the first proposal it consisted of about 120 elementary reactions [12]. Then, through sensitivity analysis and comparison of model predictions with experimental data, the total number of equations was substantially reduced. A gas-phase kinetic scheme that is known to reproduce experimental data with a good accuracy is that reported in Table II. It consists of a set of 10 reversible reactions and 8 species: SiH_4, SiH_2, Si, H_3SiSiH, H_2SiSiH_2, Si_2H_6, Si_3H_8, and H_2.

As can be observed, only a few of the gas-phase species that can be produced from the dissociation of silanes are considered in the kinetic scheme. SiH_3, Si_2H_5, and SiH, which are, however, likely to be produced in the gas phase, are not considered because of their instability with respect to silane species with an even number of hydrogen atoms (e.g., SiH_2, Si, SiH_4, and Si_2H_6). To exemplify this behavior the dissociation of SiH$_4$ can be considered: the loss of the first hydrogen atom requires 383 kJ/mol, while the energy required to remove a hydrogen from SiH$_3$ is 291 kJ/mol, more than 90 kJ/mol lower. With respect to these reaction steps the loss of a hydrogen molecule (SiH$_4$ → SiH$_2$ + H$_2$) requires 237 kJ/mol. It is therefore likely that the first step in the dissociation of SiH$_4$ will be the loss of a hydrogen molecule instead of a hydrogen atom. It is interesting to observe that the opposite reaction pathway is followed by methane, for which it is known from pyrolysis studies that the first step is the loss of a hydrogen atom to give CH$_3$. SiH$_3$ and SiH can be formed if plasma-activated systems are considered [54, 55, 58–60].

A gas-phase mechanism able to describe the gas-phase reactivity of silanes is reported in Table II. All reported reactions are considered reversible. According to the fragmentation pathway reported, silane decomposes following reaction G1 if the activation of the gas phase is thermal and reaction P1, P2, and P3 if the system is activated by plasma. If disilane is fed to the reactor, then reactions G2 and G3 are both possible, while if the precursor molecule is Si$_3$H$_8$ the two possible reaction pathways lead to the formation of either SiH$_4$ or of SiH$_2$. Two isomers are possible for Si$_2$H$_4$: HSiSiH$_3$ and H$_2$SiSiH$_2$. The equilibrium between these two species is given by reaction G6. Other reactions that are likely to occur in the gas-phase of a silane gas are G7–G10, with G9 and G10 responsible for the formation of Si atoms.

In many cases of technological interest the deposition of silicon is carried out in low-pressure reactors. In such cases the dependence of the kinetic constants on the gas pressure has to be considered. The kinetic constants reported in Table II are written according to Troe's formalism [34]. Troe's expression for the pressure dependence of the rate constant is an improvement of Lindeman's original theory of unimolecular reactions. The kinetic constant is expressed as a function of the rate constant in the low (k_0)- and the high-pressure limits (k_∞):

$$k = \frac{k_0 C_M}{(1 + k_0 C_M)/k_\infty} F. \qquad (7)$$

C_M is the third body concentration (e.g., the gas-phase density). The parameter F, in this framework, is given by the expression

$$\log F = \left[1 + \left[\frac{\log P_r + c}{n - 0.14(\log P_r + c)}\right]^2\right]^{-1} \log F_{\text{cent}}, \qquad (8)$$

where $c = -0.4 - 0.67 \log F_{\text{cent}}$, $n = 0.75 - 1.27 \log F_{\text{cent}}$, $P_r = k_0 C_M/K_\infty$, and $F_{\text{cent}} = (1-a)e^{-T/T^{***}} + ae^{-T/T^*} + e^{-T^{**}/T}$. Thus, the pressure dependence, besides the high- and low-pressure parameters, is obtained by means of four parameters, a, T^{***}, T^*, and T^{**}. In the above equations, the subscript m, referring to the mth reaction, has been omitted for the sake of simplicity.

In Table II the subscripts 0 and ∞ indicate the rate parameters for the low- and high-pressure limits, while the subscripts f and b refer to the forward and the backward reactions.

TABLE I

THERMOCHEMICAL DATA FOR SILANES AND CHLOROSILANES[a]

Species	MW	$S°$	$\Delta H^{\circ f}$	C_{Pa}	C_{Pb}	C_{Pc}	C_{Pd}	C_{Pe}
(a) Thermochemical parameters for Si–H molecules								
H_2	2.0	130.6	0	3.298	8.2×10^{-4}	-8.14×10^{-7}	-9.47×10^{-11}	4.13×10^{-13}
H	1.0	114.7	218,143	2.991	7.0×10^{-4}	-5.63×10^{-8}	-9.23×10^{-12}	1.58×10^{-15}
				2.50	0	0	0	0
				2.50	0	0	0	0
SiH_4	32.08	204.66	33,915	2.475	9.00×10^{-3}	2.18×10^{-6}	-2.68×10^{-9}	-6.62×10^{-13}
SiH_3	31.08	216.80	198,506	6.894	4.03×10^{-3}	-4.18×10^{-7}	-2.23×10^{-10}	4.38×10^{-14}
SiH_2	30.08	207.13	271,192	2.947	6.47×10^{-3}	5.99×10^{-7}	-2.22×10^{-9}	3.05×10^{-13}
SiH	29.08	184.98	383,780	5.016	3.73×10^{-3}	-3.61×10^{-7}	-3.73×10^{-10}	8.47×10^{-14}
Si	28.08	167.90	450,772	3.475	2.14×10^{-3}	7.67×10^{-7}	5.22×10^{-10}	-9.90×10^{-13}
Si_2H_6	62.16	270.06	79,930	4.142	2.15×10^{-3}	-2.19×10^{-7}	-2.07×10^{-10}	4.74×10^{-14}
Si_2H_5	61.16	288.15	233,132	3.836	-2.70×10^{-3}	6.85×10^{-6}	-5.42×10^{-9}	1.47×10^{-13}
$H_2Si_2H_2$	60.16	279.98	263,111	3.110	1.09×10^{-3}	2.90×10^{-8}	-2.74×10^{-10}	7.05×10^{-14}
HSi_2H_3	60.16	279.86	335,672	3.113	-2.33×10^{-3}	3.52×10^{-6}	-2.42×10^{-9}	6.39×10^{-13}
Si_2H_3	59.16	274.16	442,398	2.775	-6.21×10^{-4}	4.84×10^{-7}	-1.28×10^{-10}	1.13×10^{-14}
Si_2H_2	58.16	245.06	349,447	0.530	4.18×10^{-2}	-4.68×10^{-5}	3.18×10^{-8}	-9.48×10^{-12}
Si_3H_8	92.24	341.32	120,962	8.882	1.15×10^{-2}	-1.22×10^{-6}	-1.90×10^{-9}	5.54×10^{-13}
				1.579	3.55×10^{-2}	-4.27×10^{-5}	3.06×10^{-8}	9.36×10^{-12}
				8.451	9.28×10^{-3}	-1.09×10^{-6}	-1.44×10^{-9}	4.25×10^{-13}
				5.133	1.25×10^{-2}	-4.62×10^{-7}	-6.60×10^{-9}	2.86×10^{-12}
				8.987	5.40×10^{-3}	-5.21×10^{-7}	-5.31×10^{-10}	1.19×10^{-13}
				2.778	1.51×10^{-2}	-6.73×10^{-7}	-8.92×10^{-9}	4.11×10^{-12}
				7.697	5.66×10^{-3}	-5.21×10^{-7}	-5.62×10^{-10}	1.26×10^{-13}
				3.335	2.16×10^{-2}	-2.93×10^{-5}	2.29×10^{-8}	-7.27×10^{-12}
				7.258	5.12×10^{-3}	-7.63×10^{-7}	-6.66×10^{-10}	2.05×10^{-13}
				1.624	1.49×10^{-2}	-8.71×10^{-6}	-1.70×10^{-11}	1.10×10^{-12}
				5.778	4.07×10^{-3}	-4.26×10^{-7}	-7.92×10^{-10}	2.38×10^{-13}
				0.632	6.41×10^{-2}	-7.77×10^{-5}	5.49×10^{-8}	-1.65×10^{-11}
				13.423	1.56×10^{-2}	-1.94×10^{-6}	-2.39×10^{-9}	7.12×10^{-13}

(b) Thermochemical parameters for Si–H–cl molecules

Species	MW	S°	ΔH°	$C_{p,a}$	$C_{p,b}$	$C_{p,c}$	$C_{p,d}$	$C_{p,e}$
HCl	36.45	186.82	−92,365	3.338	1.26×10^{-3}	-3.67×10^{-6}	4.70×10^{-9}	-1.84×10^{-12}
				2.755	1.47×10^{-3}	-4.97×10^{-7}	8.11×10^{-11}	-5.07×10^{-15}
SiCl$_3$H	135.43	313.48	−496,369	2.883	3.31×10^{-2}	-5.17×10^{-5}	3.95×10^{-8}	-1.17×10^{-11}
				9.663	3.56×10^{-3}	-1.21×10^{-6}	-1.61×10^{-10}	5.64×10^{-14}
SiCl$_2$H$_2$	100.98	286.26	−311,806	1.100	3.26×10^{-2}	-4.69×10^{-5}	3.49×10^{-8}	-1.04×10^{-11}
				7.727	5.03×10^{-3}	-1.09×10^{-6}	4.42×10^{-10}	1.63×10^{-13}
SiClH$_3$	66.53	250.25	−134,779	0.506	2.70×10^{-2}	-3.30×10^{-5}	2.30×10^{-8}	-6.78×10^{-12}
				5.964	6.28×10^{-3}	-8.20×10^{-7}	-9.28×10^{-10}	2.80×10^{-13}
SiHCl$_2$	99.98	294.22	−143,572	2.368	2.40×10^{-2}	-3.72×10^{-5}	2.85×10^{-8}	-8.53×10^{-12}
				7.230	2.87×10^{-3}	-8.85×10^{-7}	-7.50×10^{-11}	5.75×10^{-14}
SiH$_2$Cl	65.53	260.72	32,659	1.700	1.96×10^{-2}	-2.62×10^{-5}	1.94×10^{-8}	-5.85×10^{-12}
				5.556	4.05×10^{-3}	-6.58×10^{-7}	-5.03×10^{-10}	1.60×10^{-13}
SiHCl	64.53	250.26	71,137	3.073	9.06×10^{-3}	-1.16×10^{-5}	8.31×10^{-9}	-2.48×10^{-12}
				4.900	1.98×10^{-3}	-3.63×10^{-7}	-2.28×10^{-10}	7.63×10^{-14}
SiCl$_4$	169.88	331.02	−662,970	5.252	3.12×10^{-2}	-5.25×10^{-5}	4.10×10^{-8}	-1.22×10^{-11}
				11.709	1.97×10^{-3}	-1.27×10^{-6}	3.90×10^{-10}	-4.76×10^{-14}
SiCl$_3$	134.43	315.99	−320,180	4.486	2.24×10^{-2}	-3.79×10^{-5}	2.98×10^{-8}	-8.86×10^{-12}
				9.098	1.41×10^{-3}	-9.32×10^{-7}	2.98×10^{-10}	-3.82×10^{-14}
SiCl$_2$	98.98	281.24	−168,652	3.827	1.31×10^{-2}	-2.22×10^{-5}	1.76×10^{-8}	-5.27×10^{-12}
				6.491	8.24×10^{-4}	-5.77×10^{-7}	1.97×10^{-10}	-2.71×10^{-14}
SiCl	63.53	241.90	158,604	3.096	5.74×10^{-3}	-9.74×10^{-6}	7.64×10^{-9}	-2.28×10^{-12}
				4.258	4.02×10^{-4}	-2.89×10^{-7}	1.01×10^{-10}	-1.41×10^{-14}
Cl$_2$	70.90	223.08	0	3.439	2.87×10^{-3}	-2.38×10^{-6}	2.89×10^{-10}	2.91×10^{-13}
				4.274	3.72×10^{-4}	-1.89×10^{-7}	5.34×10^{-11}	-5.06×10^{-15}
Cl	35.45	165.13	125,526	2.38	8.89×10^{-4}	4.07×10^{-7}	-2.17×10^{-9}	1.16×10^{-12}
				2.92	-3.60×10^{-4}	1.29×10^{-7}	-2.16×10^{-10}	1.38×10^{-15}

[a] All data taken from the compilation of Kee *et al.* [50]. Entropies and enthalpies are referred to standard conditions (i.e., 1 atm and 298 K) and are expressed as J/mol K and J/mol. Molecular weights are in kg/kmol. Specific heats can be calculated from the reported values as C_p (J/mol K) = $8.31 \, (C_{p,a} + C_{p,b} \cdot T + C_{p,c} \cdot T^2 + C_{p,d} \cdot T^3 + C_{p,e} \cdot T^4)$. The first-row values of $C_{p,x}$ are interpolated between 300 and 1000 K; the second-row values, between 1000 K and 2000 K.

TABLE II

SILANE AND DISILANE GAS-PHASE REACTION MECHANISMS[a]

	Reaction	$\log A$	α	E	A	T^{***}	T^*	T^{**}	Ref. No.(s.)
G1$_\infty$	$SiH_4 \leftrightarrow SiH_2 + H_2$	9.49	1.70	229,016					42
G1$_0$		23.72	−3.54	240,904	−0.4984	888.3	209.4	2760	42
G2$_\infty$	$Si_2H_6 \leftrightarrow SiH_4 + SiH_2$	10.26	1.70	210,150					42
G2$_0$		47.71	−10.37	234,558	-4.37×10^{-5}	438.5	2726	438.2	42
G3$_\infty$	$Si_2H_6 \leftrightarrow H_2 + HSiSiH_3$	9.96	1.80	226,869					42
G3$_0$		38.29	−7.77	247,070	−0.1224	793.3	2400	11.39	42
G4$_\infty$	$Si_3H_8 \leftrightarrow SiH_2 + Si_2H_6$	12.84	1.00	220,505					42
G4$_0$		63.24	−15.07	255,099	-3.5×10^{-5}	442.0	2412	128.3	42
G5$_\infty$	$Si_3H_8 \leftrightarrow SiH_4 + HSiSiH_3$	12.57	1.00	212,858					42
G5$_0$		76.64	−17.26	248,242	0.4157	365.3	3102	9.72	42
G6$_\infty$	$HSiSiH_3 \leftrightarrow H_2SiSiH_2$	13.40	−0.20	22,525					42
G6$_0$		27.04	−5.76	38,310	−0.4202	214.5	103	136.3	42
G7$_f$	$HSiSiH_3 + H_2 \leftrightarrow SiH_2 + SiH_4$	7.97	0.00	17,130					42
G7$_b$		4.97	1.10	24,238					42
G8$_f$	$HSiSiH_3 + SiH_4 \leftrightarrow Si_2H_6 + SiH_2$	8.24	0.40	37,250					42
G8$_b$		9.42	0.10	35,468					42
G9$_\infty$	$HSiSiH_3 \leftrightarrow Si + SiH_4$	13.15	0.54	240,896					42
G9$_0$		36.37	−7.42	255,166	0.5336	629.2	2190	625.5	42
G10	$Si + Si_2H_6 \leftrightarrow SiH_2 + HSiSiH_3$	9.11	0.00	52,744					42
P1	$SiH_4 + e \rightarrow SiH_2 + 2H + e$	7.71							54, 60
P2	$SiH_4 + e \rightarrow SiH_3 + H + e$	7.74							54, 60
P3	$SiH_4 + e \rightarrow SiH + H + H_2 + e$	7.82							54, 60
P4	$Si_2H_6 + e \rightarrow SiH_4 + SiH_2 + e$	8.76							54, 60
P5	$Si_2H_6 + e \rightarrow Si_2H_4 + H_2 + e$	8.15							54, 60
P6	$H_2 + e \rightarrow 2H + e$	6.43							54, 60
P7	$SiH_4 + H \rightarrow SiH_3 + H_2$	8.00		10,465					13

[a] Kinetic parameters given in terms of mol, m, s, J, and K. Kinetic constants are expressed as $k = AT^\alpha e^{-E/RT}$. Subscripts 0, ∞, f, and b refer to low-pressure and high-pressure limits and forward and backward reactions, respectively. Plasma reaction rates evaluated at $T_e = 2$ eV.

2. Chlorosilane's Gas-Phase Reactivity

Compared to silanes, we have much less knowledge of the gas-phase reactivity of chlorosilanes. This is due mainly to the higher level of complexity brought into the system by the presence of another atomic species, namely, chlorine.

Silicon is deposited from chlorosilane's precursors fed into the reactor diluted in a hydrogen carrier gas according to the following overall reaction system:

$$SiCl_4 + 2H_2 \rightarrow Si_s + 4HCl \tag{9}$$

$$SiHCl_3 + H_2 \rightarrow Si_s + 3HCl \tag{10}$$

$$SiH_2Cl_2 \rightarrow Si_s + 4HCl. \tag{11}$$

Hydrogen chloride, which is a known silicon etching agent, appears as a by-product of the growth process.

In the following, issues concerning the kinetics of the gas-phase and surface reactions occurring during the growth of Si are addressed. First, the geometric structures and thermochemical parameters of different gas-phase and surface species, as determined through theoretical calculations or experimental techniques, are considered. Next, the composition of the gas phase is examined as a function of the temperature under thermodynamic equilibrium conditions.

a. Structure and Energetics of Gas-Phase Chlorosilanes

The geometric structures of $SiHCl_3$, SiH_2Cl_2, $SiCl_4$, and SiH_4 are reported in Fig. 1. Since Si is sp^3 hybridized when coordinated with four atoms, it can be observed that all angles are almost-tetrahedral. All data reported in Fig. 1 were taken from the JANAF thermochemical tables [9].

Molecular weights, entropies at 298 K, enthalpies of formation at 298 K, and specific heats for the most important Si–H–Cl gas-phase species are reported in Table I. All data were taken from the Sandia compilation [50].

While the structure and thermochemical data of SiHCl stable molecules could be determined experimentally with a good degree of accuracy, the same could not be done for chlorosilane radicals, because of their intrinsic instability. The values of the entropy and enthalpy of formation of chlorosilane radicals can be evaluated by adopting quantum chemistry calculations. Quantum chemistry can nowadays be considered a mature field, and for many systems the accuracy with which thermochemical data can be predicted is near the so-called "chemical accuracy" limit, i.e., ± 5–10 kJ/mol from the real value.

Different a priori studies were conducted in the literature on the structure and energetics of SiHCl molecules. Here we report the results of the calculations Su and Schlegel [79] performed adopting a Hartree–Fock calculation followed by a Møller–Plesset correlation energy correction truncated at the second order

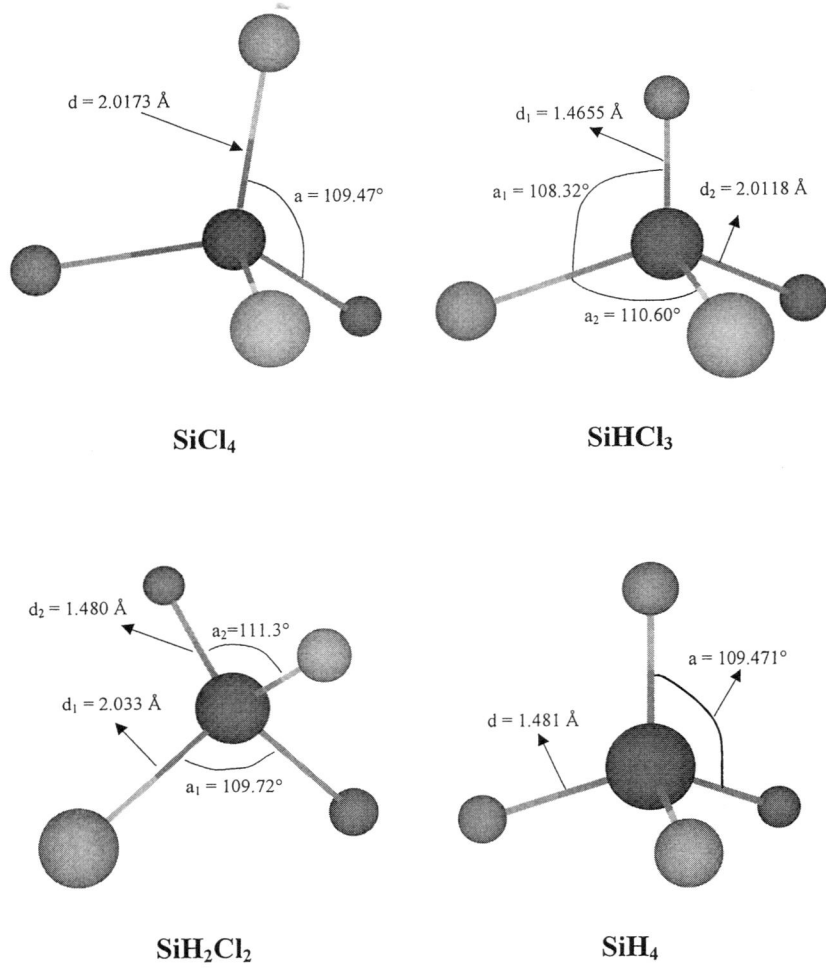

FIG. 1. Geometrical structures of the four most important precursors to epitaxial silicon deposition. The reported distances and angles were taken form the JANAF thermochemical tables [9].

(MP2). The basis set, i.e., the set of functions that describe the molecular orbital structure, that was adopted in the calculations was the Gaussian double-zeta basis set 6-31G with added diffuse and polarization functions. Following this method geometries and vibrational frequencies were calculated for the following chlorosilane species: SiH, SiCl, SiH_2, SiHCl, $SiCl_2$, SiH_3, SiH_2Cl, $SiHCl_2$, $SiCl_3$, SiH_4, SiH_3Cl, SiH_2Cl_2, $SiHCl_3$, and $SiCl_4$. A comparison between calculated and experimental vibrational frequencies showed good agreement in all cases. For instance, a summary of the experimental vibrational frequency values for the molecular species is reported in Table III. Vibrational frequencies calculated for $SiHCl_3$

TABLE III
VIBRATIONAL FREQUENCIES OF GAS-PHASE SiH_xCl_{4-x} SPECIES[a]

	Vibrational frequency (cm^{-1})		
SiCl$_4$	SiCl$_3$H	SiCl$_2$H$_2$	SiH$_4$
620 ± 1 (3)	2261 (1)	2237 (1)	2189.08 (3)
425 ± 1 (1)	811 (2)	2224 (1)	2185.7 (1)
220 ± 4 (3)	600 (2)	954 (1)	972.1 (2)
149 ± 2 (2)	499 (1)	876 (1)	913.28 (3)
	254 (1)	710 (1)	
	176 (2)	602 (1)	
		590 (1)	
		527 (1)	
		188 (1)	

[a] All data from the JANAF thermochemical tables [9]. Degeneracies are reported in parentheses.

were 180* (176*), 262 (254), 511 (499), 628* (600*), 852* (811*), and 2420 (2261), where the superscript asterisk refers to a doubly degenerate vibrational frequency and the experimental values are reported in parentheses. This kind of comparison between calculated and experimental values is important to assess the validity of the calculation method so that it can safely be extended to cases where no experimental values are available.

To obtain extremely accurate values of bond energies, and therefore of formation enthalpies, the total energy of the molecules was studied at a higher level of theory. In particular, the Møller–Plesset correlation energy correction to the self-consistent field energy was extended to the fourth order and diffuse and polarization functions were added to improve the description of the molecular orbital structure. To limit the systematic error implicit in each quantum chemistry calculation, isodesmic reactions were adopted to determine the values of the enthalpy of formation. An isodesmic reaction is a reaction where reagents and products have the same total number of lone pairs and similar bonds. The similarity between the chemical species appearing on the left and right sides of the reaction are meant to minimize the errors due to the approximate nature of the solution of the Shrödinger equation. It is interesting to note here that Curtiss et al. [16, 17] report that the mean absolute error of the enthalpies calculated adopting isodesmic reactions and a calculation method similar to that followed by Schlegel et al. is about 5.4 kJ/mol. The isodesmic reaction adopted to calculate enthalpies of formation for SiHCl species was

$$\frac{m-n}{m}SiH_m + \frac{n}{m}SiCl_m \rightarrow SiH_{m-n}Cl_n \qquad (12)$$

This reaction gave results in good agreement with experimental data and was chosen since the experimental heats of formation of SiH$_4$ and SiCl$_4$ are well known.

Calculated values are in good agreement with experimental values except for SiCl, whose formation enthalpy value is still a subject of discussion in the literature [44, 45]. Su and Schlegel suggested adopting a value of 152.8 ± 6.3 kJ/mol for SiCl until further experimental evidence is available.

The possibility of adopting density functional theory (DFT) to study SiHCl molecules is intriguing because of the relatively small computational effort it requires. If DFT predictions are sufficiently accurate, then this method can be adopted to study chlorosilane surface reactivity through a cluster approach. The performance of density functional methods for calculating SiHCl molecule's properties was assessed by Hay and co-workers. They found that good agreement with experimental data may be attained when the Becke hybrid three-parameter exchange functional is adopted in conjunction with the Lee–Young–Parr correlation energy functional. Calculated vibrational frequencies were found to be systematically 3–5% higher than experimental values. The mean absolute error for atomization energies was found to be 42 kJ/mol. Based on these results it can be concluded that DFT can be adopted to study the surface reactivity of the SiHCl system, provided that a comparison with experimental data is performed whenever possible.

b. Chlorosilane's Gas-Phase Decomposition Pathways

The gas-phase chemistry of chlorosilanes is less known than that of silanes. This is due mainly to the fact that the growth of silicon from chlorosilanes is usually conducted in the diffusion-controlled regime and at high temperatures. Under these conditions the composition of the gas phase is usually near thermodynamic equilibrium and the growth rate, which is the most important parameter to control to optimize the reactor performances, is controlled by the diffusion of the gas-phase precursors toward the deposition surface. It appears therefore that detailed knowledge of the gas-phase kinetics of chlorosilanes is not necessary to study the reactors used to deposit silicon epitaxial films. However, recently some discussion about which is the gas-phase precursor to the growth of silicon, i.e., which is the gas-phase chemical species that determines the growth of the film when chlorosilane gases are adopted, has arisen in the literature. In particular, the aim of the discussion is to understand if the nondissociated gas-phase chlorosilane molecule or the $SiCl_2$ species are responsible for the growth of the film. This is an important point since if $SiCl_2$ determines the growth of silicon, this would imply that the gas-phase behavior must be given great attention.

The thermal decomposition of $SiHCl_3$ has been studied through laser pyrolysis by Lavrushenko et al. [56]. The products of the decomposition reaction were HCl and a powder that probably derived from the polymerization of $SiCl_2$. Based on this observation it was assumed that the reaction proceeds as $SiHCl_3 \rightarrow SiCl_2 + HCl$, which is particularly thermodynamically favored with respect to other possible decomposition reactions. A theoretical analysis of the possible decomposition reactions for $SiHCl_3$ has been performed by Su and Schlegel. The three competing reactions are the loss of a chlorine atom to give $SiCl_3$ and H, the decomposition of $SiHCl_3$ into $SiCl_2$ and HCl, and the dissociation of $SiHCl_3$ into SiHCl and Cl_2. The enthalpy changes for the three reactions are 394, 475, and 235 kJ/mol, respectively. The third reaction therefore seems to be energetically favored over the other two. However, because of the rupture of two bonds and the formation of one, it is likely that the extraction of HCl from $SiHCl_3$ proceeds through a transition state and has a relevant activation energy. The activation energy for the reaction $SiHCl_3 \rightarrow SiCl_2 + HCl$ was calculated by Su and Schlegel using the G2 theory, which is known to give extremely accurate predictions of bond energies. The calculated transition state, characterized by a single imaginary frequency, has an activation energy of about 300 kJ/mol, 94 kJ/mol lower than that for the abstraction of atomic hydrogen, which proceeds without a transition state. The kinetic constant calculated adopting classical transition-state theory is $4.9 \times 10^{14} \cdot \exp(-308.5/RT)$ s^{-1}.

A different approach to the evaluation of the dissociation kinetic constant of $SiHCl_3$ was pursued by Narusawa [65, 66]: a coupled fluid-dynamic and kinetic model of the deposition of epitaxial silicon in a horizontal reactor. The physical and chemical parameters required to realize the model, such as the diffusion coefficients for all the involved gas-phase species, were found in the literature or calculated. The only data not available were the kinetic constants for the dissociation of $SiHCl_3$. Narusawa hypothesized that the gas-phase precursor to the growth of silicon was $SiCl_2$, which was the principal product of the dissociation of $SiHCl_3$. Since the kinetic constant for the decomposition of $SiHCl_3$ was the only parameter in the model, it was fitted over the experimental growth rate data. The kinetic constant so determined was $k = 10^{40.653} T^{-9.762} e^{-219330/RT}$, with the activation energy reported as kilojoules per mole. The values of the kinetic constants calculated by Su and Schlegel and by Narusawa can be compared at different temperatures: at 1000 K the kinetic constant of Narusawa has a value of 0.97 s^{-1}, versus the 0.048 s^{-1} that can be calculated from the a priori kinetic constant, while at 1300 K the kinetic constant values are 31.8 and 239.6 s^{-1}, respectively. As can be observed, the dissociation kinetic constant calculated by Narusawa appears to be more sensitive to a change in temperature than the other kinetic constant. That difference can be ascribed to various factors, one being the very different estimation procedure followed by the two authors and the other that the work of Narusawa is based on the hypothesis that the gas-phase growth precursor is $SiCl_2$ only, which might not be true. However, it must also be pointed out that the difference between the two kinetic constants is not very high, considering the uncertainties related to the evaluation of a kinetic constant through quantum chemistry and classic transition theory and the fitting procedure of Narusawa. It can therefore be concluded that this agreement between the two kinetic constants leaves open the possibility that $SiCl_2$ is the growth species.

The dissociation of SiCl$_4$ in the gas phase can proceed either through the formation of SiCl$_3$ and Cl or through the formation of SiCl$_2$ and Cl$_2$. A thermodynamic analysis of the enthalpy changes for the two reactions shows that the first channel is favored over the second for 33 kJ/mol. Moreover, the reaction that determines the formation of SiCl$_2$ and Cl$_2$ is likely to require an activation energy higher than the enthalpy change since in the reaction two bonds are broken and one is formed. In contrast, the transition state for the reaction yielding SiCl$_3$ and Cl proceeds through a simple fission transition state, in which case the activation energy is usually equivalent to the reaction enthalpy change. Since no kinetic constant values for this reaction were proposed in the literature, here we propose an estimation of the kinetic constant based on loose transition-state theory [31]. The required input parameters for this theory are the vibrational frequencies of SiCl$_4$, which were taken form the Janaf thermochemical tables, those of SiCl$_3$, and the geometrical structures of SiCl$_3$ and SiCl$_4$. A hypothesis on the structure of the transition state has to be devised to use loose transition-state theory. Here we have assumed that the transition-state structure is near that of the products and therefore we imagine the transition state as a molecule with a SiCl$_3$ part arranged as in the SiCl$_3$ radical and a chlorine atom 4.5 Å from the silicon atom and equidistant from the other three chlorine atoms. This hypothetical structure is consistent with the Hammond postulate, according to which the transition-state structure for an endothermic reaction is near that of the products, and vice versa for an exothermic reaction [38]. Accordingly the kinetic dissociation kinetic constant was calculated adopting the expression

$$k_{\text{diss}} = 4 \cdot \frac{k_b T}{h} \cdot \frac{Q^{\neq}_{\text{rot}} \cdot Q^{\neq}_{\text{vib}}}{Q^{\text{SiCl}_4}_{\text{rot}} \cdot Q^{\text{SiCl}_4}_{\text{vib}}} \cdot e^{-\frac{E}{RT}}, \tag{13}$$

where k_b and h are the Boltzmann and Plank kinetic constants, T is the temperature expressed in K, Q^j_t are the rotational and vibrational partition functions of the transition state ($\neq$) and of SiCl$_4$, and E is the activation energy of the dissociation process. The reaction path degeneracy for a reaction of this kind is 4, which is the first term that appears on the right side of Eq. (13). The rotational partition functions $Q^{\neq}_{\text{rot}}$ were calculated using the suite of programs UNIMOL [32]. The following kinetic constant value could thus be calculated:

$$k_{\text{diss}}(\text{SiCl}_4) = 4.8 \times 10^{15} \cdot e^{-56,000/T} \text{s}^{-1}. \tag{14}$$

An approach similar to that described above for SiHCl$_3$ was adopted by Narusawa [65] to determine the dissociation kinetic constant for SiCl$_4$. The decomposition reaction considered was SiCl$_4$ + H$_2$ → SiHCl$_3$ + HCl, which differs from that above studied in that H$_2$ appears as a reactive species. Indeed this reaction can be considered a lumped form of the following reaction pathway:

$$\text{SiCl}_4 \rightarrow \text{SiCl}_3 + \text{Cl} \tag{15}$$

$$\text{SiCl}_3 + \text{H}_2 \rightarrow \text{SiHCl}_3 + \text{H} \tag{16}$$

$$Cl + H_2 \rightarrow HCl + H \tag{17}$$

$$2H \rightarrow H_2 \tag{18}$$

$$SiCl_4 + H_2 \rightarrow SiHCl_3 + HCl. \tag{19}$$

All reaction steps, from (15) to (19), proceed fast, since they are reactions among radicals, except for reaction (15), which is the initiation reaction and is therefore likely to be the rate-determining step of all the overall process. According to rate-determining step theory, the kinetic constant of reaction (19) can be considered equal to that of reaction (15) and a comparison between the values of the two constants becomes possible. Following the fitting procedure above described Narusawa was able to determine the following dissociation kinetic constant for reaction (19): $k_{diss}(SiCl_4) = 10^{5.875} T^{-0.873} e^{-11574/T} m^3 \cdot mol^{-1} \cdot s^{-1}$. A comparison between the two kinetic constants at 1300 K shows that the value of the expression proposed by Narusawa, calculated as $k = k_{diss}[H_2]$, is 1.8×10^{-3} s^{-1}, while that calculated with Eq. (14) is 9.4×10^{-4} s^{-1}. The value of the kinetic constant for reaction (19) had to be multiplied for the gas-phase concentration (calculated at 1 bar and 1300 K) to be compared with the kinetic constant for the unimolecular dissociation reaction. The agreement between the values of the two calculated kinetic constants is fairly good, considering the many uncertainties and differences implicit in the evaluation process.

An approach similar to that described above for SiHCl$_3$ and SiCl$_4$ was adopted to study the decomposition pathway of SiH$_2$Cl$_2$. In this case some discussion is still present in the literature about which of the two reactions is the first step in the cleavage of the molecule. The two competing reactions are SiH$_2$Cl$_2$ → SiCl$_2$ + H$_2$ and SiH$_2$Cl$_2$ → SiHCl + HCl. The chemical reason for the uncertainty about which of the two reactions is faster lies in the fact that both SiCl$_2$ and HCl are very stable molecules. In fact the enthalpy changes for the two reactions are 324 kJ/mol for the reaction yielding SiCl$_2$ as product and 317.3 kJ/mol for the other. As can be observed, the difference is minimal. The kinetic constants calculated by Su and Schlegel for the two reactions were $k_1 = 8.3 \times 10^{13} \cdot \exp(-324/RT)$ and $k_2 = 6.9 \times 10^{14} \cdot \exp(-317/RT)$. At 1300 K the second reaction proceeds about 15 times faster than the other does. It would therefore seem that the fastest dissociation pathway leads to the formation of SiHCl and HCl. This is in contrast with experimental evidence, which showed that the only observable products of the dissociation of SiH$_2$Cl$_2$ were SiCl$_2$ and H$_2$. Kruppa et al. [53] in fact studied the dissociation of SiH$_2$Cl$_2$ by means of heterogeneous vacuum flash pyrolysis and found that the only observable products were SiCl$_2$ and HCl. The gas-phase composition was investigated adopting photoelectron and mass spectroscopy. Sausa and Ronn [74] adopted IR multiphoton dissociation to study the same reaction and found that in the gas phase, only SiCl$_2$ and H$_2$ were present. To solve this disagreement between experimental results and theoretical quantum chemistry calculations recently Wittbrodt and Schlegel [85] reexamined the thermal decomposition of dichlorosilane at a high level of theory. The calculations were performed adopting geometrical structures optimized with the MP2/6-311 + G(2df,2p) method. Successively energies were corrected, to calculate better the correlation energy value, adopting the QCISD(T)/6-311++G(2df,2p) and MP2/6-311++G(3df,3dp) methods, where very large basis sets were adopted to describe accurately the behavior of electrons. Transition-state theory was again adopted to determine the kinetic constant values. Despite the increase in the level of theory adopted, it was still found that the elimination of HCl proceeds from 14 to 120 times faster than that of H_2. Wittbrodt and Schlegel [85] suggest different possible reasons for this disagreement. One possibility is that HCl and SiHCl, which are very reactive species, can be consumed very rapidly under the experimental conditions examined. Another possibility is that transition-state theory is not refined enough to treat this kind of reactions. However, it is evident that the homogeneous decomposition pathway for SiH_2Cl_2 needs further investigation to be understood completely.

From what has been reported above it follows that the study of the gas-phase chemistry of chlorosilanes has focused mainly on the decomposition of the most important precursors: $SiHCl_3$, SiH_2Cl_2, and $SiCl_4$. Swihart and Carr [81] have theoretically explored the possibility that the radicals produced by the fragmentation of the precursor molecules can successively react through a radical chain mechanism. They hypothesized that radicals such as SiHCl might react with SiH_2Cl_2 to give disilane species, which, decomposing, might give the formation $SiCl_2$ and regenerate a SiHCl radical through a sort of radical chain mechanism. An example of the reaction mechanism they propose is as follows.

$$\text{Initiation:} \quad SiH_2Cl_2 \rightarrow SiHCl + HCl \quad (20)$$

$$\text{Propagation:} \quad SiHCl + SiH_2Cl_2 \rightarrow HCl_2SiSiH_2Cl \quad (21)$$

$$HCl_2SiSiH_2Cl \rightarrow HCl_2SiSiCl + H_2 \quad (22)$$

$$HCl_2SiSiCl \rightarrow Cl_2SiSiHCl \quad (23)$$

$$Cl_2SiSiHCl \rightarrow SiCl_2 + SiHCl \quad (24)$$

The kinetic constants for the above reaction set were calculated through quantum chemistry. Simulations performed adopting a kinetic scheme that comprised these reactions showed that they can determine a substantial increase in the rate of decomposition of SiH_2Cl_2. As a final remark it must be pointed out that even if the results obtained by Swihart and Carr are interesting, they were not compared to experimental data, which should be done to test the correctness of the proposed hypothesis.

In conclusion, while the decomposition pathways for $SiHCl_3$ and $SiCl_4$ can be considered definitive, there is still discussion about what is the mechanism of decomposition of SiH_2Cl_2. However, the data produced by experimental and theoretical work seem to be sufficient to realize a gas-phase kinetic scheme for chlorosilanes that might be adopted, in conjunction with a surface kinetic scheme, to assess the importance of gas-phase reactions during the growth of silicon epitaxial films from chlorosilane's precursors. In particular, two questions need to be addressed, the first being whether the gas phase is at thermodynamic equilibrium during the growth process and the second the identification of the gas-phase species that determines the growth of silicon.

c. Thermodynamic Equilibrium Calculations for Chlorosilanes

The study of the gas-phase composition of a chlorosilane gas at thermodynamic equilibrium is important to identify which are the most stable gas-phase species. Accordingly the composition of the gas phase at thermodynamic equilibrium at different temperatures and at two pressures (760 and 1 Torr) was computed for an overall composition similar to those adopted in industrial deposition reactors (97 mol% H_2 and mol3% $SiHCl_3$). The results of the calculations are reported in Figs. 2a and b. $SiHCl_3$ proved to be the thermodynamic species most stable at low gas-phase temperatures and high pressures, followed by SiH_2Cl_2 and $SiCl_4$. As the temperature increases the radicals $SiCl_2$ and HCl become the most abundant gas-phase species (except H_2). When the pressure is decreased the fragmentation of the chlorosilanes increases greatly and the chemical composition changes. Si atoms are now present in the gas phase at a relevant concentration at high temperatures, together with SiCl and $SiCl_2$. At low pressures an inversion in the stability of the SiH_xCl_{4-x} species with respect to that at high pressures is observed.

At 1 Torr and low gas-phase temperatures $SiCl_4$ is the most stable chlorosilane species (except for $SiCl_2$, which is produced to a great extent even at low temperatures when the pressure is low), followed by $SiHCl_3$ and SiH_2Cl_2. This result is in agreement with the thermodynamic calculations of Violette *et al.* [84], who studied the role of chlorine in selective Si epitaxy. Simulations were also performed for different initial gas-phase compositions (e.g., 3 mol% $SiCl_4$ in H_2), and results only slightly different from those reported in Figs. 2a and b were obtained. The temperature at which the $SiHCl_3$ mole fraction becomes lower than that of $SiCl_2$ is about 1000°C at a reactor pressure of 760 Torr, while $SiCl_2$ is the most abundant gas-phase Si species over the temperature range examined here at 1 Torr. This consideration is of particular importance for identifying the gas-phase precursor to the deposition of epitaxial silicon and it is addressed in detail in the following. HCl is also produced at a high concentration in the gas phase at high temperatures. Because of the etching effect of HCl, its presence at a high concentration can lead to local nonuniformity of the Si layer. Moreover, HCl has a much higher molecular weight than H_2 and is subject to a high thermal diffusion effect. It can therefore accumulate near the cold walls of the reactor and give rise to effects of local etching.

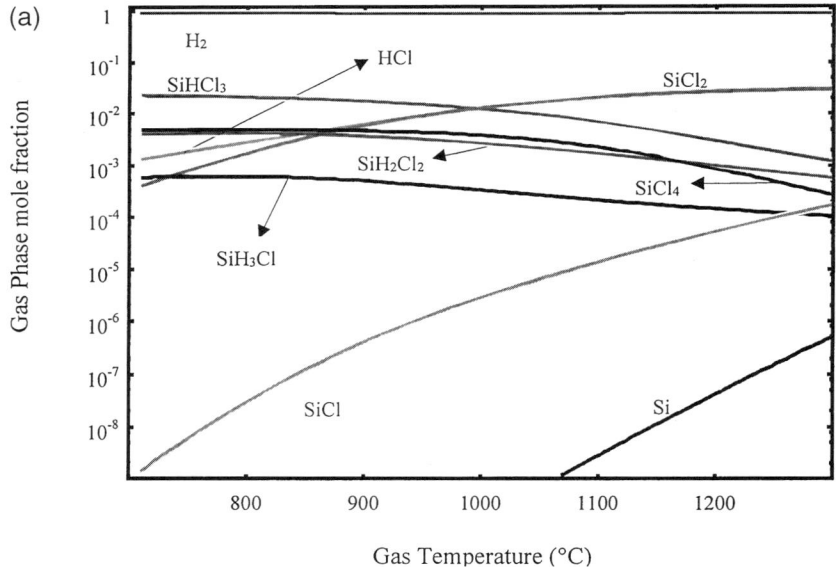

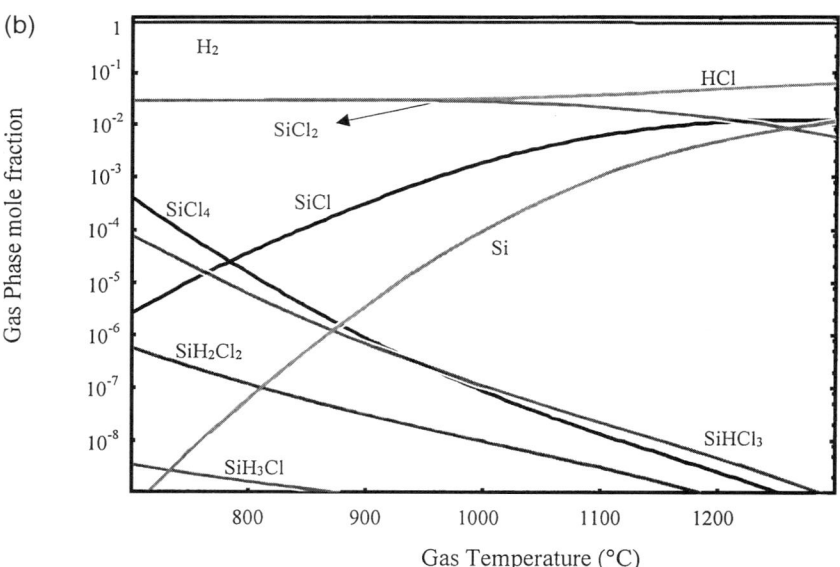

FIG. 2. Composition of the gas phase calculated at thermodynamic equilibrium as a function of the gas-phase temperature using the data reported in Table I. The total pressure is (a) 760 Torr and (b) 1 Torr.

IV. Surface Kinetics for Silanes and Chlorosilanes

1. SILANE'S SURFACE REACTIVITY

The determination of the surface reactivity of silanes gases require an understanding of the way in which not only SiH_4, but also H_2 interacts with a silicon epitaxial surface. In fact, hydrogen, which is usually fed into the growth reactors together with SiH_4, also adsorbs over silicon and can successively react with other adsorbed species. The adsorption of SiH_4 on silicon has been well studied in the literature both theoretically and experimentally, for two reasons: first, because of the importance of silicon in the microelectronics industry, and, second, because the epitaxial surface of silicon offers the possibility of studying a surface with sites of known structure and total number. In particular, the advantage of an epitaxial surface is that there are almost no preferential sites for adsorption, which make it easier to link macroscopic observations with elementary surface reactions.

In the following, experimental studies of the adsorption kinetics of silane on silicon are reviewed [8, 26–29, 76]. Basically these works were conducted at low pressures and used data on surface and gas-phase compositions to extract kinetic parameters.

Static secondary ion mass spectroscopy (SSIMS) and temperature-programmed desorption were adopted to determine the surface concentration of different hydrogenated species on Si(100) [28, 29]. Subsequently Gates and Kulkarni [26, 30] adopted the data collected in those experiments to extract kinetic parameters per a simple surface reaction scheme able to describe the growth of silicon from silane at low pressures. These studies were conducted at low pressures to limit the extent of gas-phase homogeneous decomposition reactions. Under such conditions the rate of the adsorption process is determined by the density of surface dangling bonds and by the concentration of silane in the gas phase according to the equation for the reaction rate,

$$R_{ads} = k_{ads} \cdot [SiH_4] \cdot [\sigma]^2, \qquad (25)$$

where k_{ads} is the adsorption kinetic constant and σ is the concentration of free surface sites. At low pressures no species are adsorbed on the surface and the number of surface sites can be considered equal to the density of silicon surface atoms [6.8×10^{18} at/m^2 on a Si(100) surface]. Under these conditions the adsorption kinetic constant can be calculated directly if the total adsorption rate is known. The adsorption kinetic scheme proposed by Gates is reported in Table IV. The main feature of this surface kinetic scheme is the dissociative adsorption of SiH_4 on two surface sites, followed by the fast decomposition of adsorbed SiH_3 in Si and three adsorbed hydrogen atoms. Subsequently hydrogen desorbs from the surface as molecular hydrogen. A kinetic analysis of the growth processes showed that two regimes are possible. At temperatures higher than 600°C and for a gas phase where only silane is present, the growth rate is determined by the rate of adsorption of SiH_4. All successive reaction steps are fast and do not limit the film deposition

TABLE IV

SURFACE REACTIONS FOR SILICON GROWTH FROM SILANE AND DISILANE[a]

	Reaction	log A	α	E	Ref. Nos.[b]
S1	$SiH_4 + 2\sigma \rightarrow SiH_3^* + H^*$	9.06	0.5	12,600	25–29
S2	$SiH_3^* + \sigma \rightarrow SiH_2^* + H^*$	13.64	—	113,000	25–29
S3	$2SiH_2^* \rightarrow 2SiH^* + H_2$	20.38	—	188,400	25–29
S4	$Si + \sigma \rightarrow Si^*$	5.77	0.5	0	c
S5	$SiH + \sigma \rightarrow SiH^*$	5.77	0.5	0	c
S6	$Si^* \rightarrow \sigma Si_{(s)} + \sigma$	20.90	—	184,200	25–29
S7	$SiH_2 + \sigma \rightarrow SiH_2^*$	5.76	0.5	0	c
S8	$H + \sigma \rightarrow H^*$	6.63	0.5	0	c
S9	$SiH_3 + \sigma \rightarrow SiH_3^*$	5.76	0.5	0	c
S10	$Si_2H_2 + 2\sigma \rightarrow 2SiH^*$	10.56	0.5	12,600	c
S11	$SiH^* \rightarrow Si_{(s)} + \frac{1}{2}H_2 + \sigma$	11.90	—	196,700	25–29
S12	$2H^* \rightarrow H_2 + 2\sigma$	18.11	—	196,700	25–29

[a] Kinetic parameters given in terms of mol, m, s, J, and K. Rate constants are written $k = AT^\alpha e^{-E/RT}$. Surface reactions are written as irreversible, while rates are in mol/m^2/s.
[b] c, estimated through collisional theory [7].

rate. The activation energy for the adsorption of silane was found to be 12 kJ/mol. At temperatures lower than 600°C the rate-determining step is the desorption of hydrogen from the surface. This reaction is in fact highly activated, and therefore at low surface temperatures it proceeds slowly. The physical interpretation of having hydrogen desorption as the rate-determining step is that under these conditions the surface is almost completely covered by adsorbed hydrogen atoms. To check these results the hydrogen coverage during the growth process was investigated experimentally [14, 76]. It is interesting that, if the deposition surface is Si(111), the desorption kinetics for hydrogen changes from double site to single site. In both cases, however, the overall growth kinetics remains the same.

In Table IV reactions S5, S7, S8, and S10 refer to the adsorption of a gas-phase radical on the growth surface. These reactions, which were not originally present in the kinetic scheme proposed by Gates, were added to take into consideration that if a gas-phase radical species hits the surface, then, because of its high reactivity, it is very likely that it sticks. Accordingly, the kinetic constants for the adsorption of H, SiH$_2$, Si$_2$H$_2$, and SiH$_3$ were estimated by adopting the collision theory [Eq. (4)] with a unary sticking coefficient.

If the deposition of silicon is conducted in a plasma-activated reactor, the electronic reactions also have to be considered. This can be done by introducing the electron impact reactions reported in Table II into the gas-phase kinetic scheme. The kinetic constants of these reactions are a function of the mean electronic temperature of the plasma, which can be either experimentally measured through a Langmuir probe or calculated by adopting a plasma model. Examples where the latter approach was successfully adopted to study the performance of a plasma reactor are reported by Kushner [54, 55] and by Masi [59, 60]. The reason to adopt a plasma reactor is that the gas-phase electronic reactions determine the formation in the gas phase of a relatively high concentration of radical species. Successively these species interact with the silicon deposition surface, which is usually placed in proximity to the plasma, and determine the growth of the film. The energy required to activate the system is consequently delivered directly to the gas phase, and not through heating of the deposition surface. With plasma devices, therefore, high growth rates can be attained working at low deposition temperatures, which may be needed in certain cases.

Plasmas adopted for deposition systems usually operate under glow discharge conditions, which means that the plasma is only partially ionized (the molar fraction of ions usually ranges between 10^{-5} and 10^{-8}). Because of large high difference in mobility between electrons and neutral or positively charged species, plasma systems operate far from thermodynamic equilibrium [33].

The parameters that characterize a plasma are the electronic temperature T_e and the elctronic concentration C_e. A simple approach to determining their values consists in adopting the ambipolar theory, which relates the electron energy and concentration to the power applied to the reactor [60, 33]. Once the electronic temperature is known, the kinetic constants for the electronic reactions can be determined by adopting Eqs. (8) and (9) and experimentally determined cross sections [10, 20, 69].

In conclusion, the rate constants necessary to characterize the gas-phase and surface chemistry active in a thermal or plasma CVD reactor are reported in Tables II and IV. The rate constants are written using Troe's formulation for the reactions in the gas phase and $k = AT^\alpha e^{-E/RT}$ for the surface reactions, while for electron impact reactions the values calculated at 2 eV are reported. The adopted units are consistent with the SI system and the surface reactions are calculated for a dangling bond density of 1.13×10^{-5} mol/m^2, which corresponds to a Si(100) surface.

To diminish the number of reactions that have to be considered in the kinetic scheme and to identify the growth pathway, a sensitivity analysis was performed on the mechanism reported in Tables II and IV [47, 71, 75, 83]. The adopted sensitivity parameter was $S_j = \partial \ln GR / \partial \ln k_j$, which allowed us to determine the sensitivity of the growth rate with respect to the variation of a generic kinetic constant included in the scheme. The reactions that were found to have a significant impact on the growth rate were G1, P1, P2, S1, S10, and S11. Accordingly a simple kinetic scheme able to represent the deposition kinetics is the following:

$$SiH_4 \rightarrow SiH_2 + H_2 \quad \text{(G1)}$$

$$SiH_4 + e \rightarrow SiH_2 + H_2 + e \quad \text{(P2)}$$

$$SiH_4 + 2\sigma \rightarrow SiH^* + H^* + H_2 \quad \text{(S1b)}$$

$$SiH_2 + 2\sigma \rightarrow SiH^* + H^* \quad \text{(S7b)}$$

$$SiH^* \rightarrow Si_{(s)} + 0.5H_2 + \sigma \quad \text{(S10)}$$

$$2H^* \rightarrow H_2 + 2\sigma. \quad \text{(S11)}$$

The main advantage of having a simplified kinetic scheme consists in the fact that it can be adopted in fluid dynamic simulations with a complicated geometry. With a decrease in the number of gas-phase species considered, there is a corresponding increase in the complexity of the computational grid that can be adopted without increasing the computational burden.

Finally, it can be observed that the surface reaction mechanism reported above is consistent with the findings of Holleman and Verweij [46].

2. Chlorosilane's Surface Reactivity

a. Structure and Energetics of Adsorbed Chlorosilanes

The geometrical structure and energetics of surface species present on the surface during the growth of epitaxial Si films has been the subject of many theoretical and experimental investigations. Digital electron-stimulated desorption ion angular distribution (ESDIAD) and high-resolution electron energy loss spectroscopy (HREELS) were adopted by Gao et al. [24] and by Bennett et al. [4] to determine the bonding configuration of chlorine on the Si surface. In particular, ESDIAD is a powerful method to measure the orientations of chemical bonds with a high level of accuracy ($\approx 1°$) [18, 57, 86]. Usually, these experiments are performed in an ultra high-vacuum (UHV) chamber equipped with Auger electron spectroscopy, an Ar sputtering gun, a digital ESDIAD apparatus, and a quadrupole mass spectrometer (QMS) for ion mass analysis. A scheme of the experimental apparatus is shown in Fig. 3. In ESDIAD the silicon surface is bombarded through an electron beam, which determines the evaporation of surface atoms as ions. In the experiment described here a 120-eV electron beam approximately 1 mm in diameter was adopted to study the surface configuration of chlorine atoms adsorbed over a Si(100) 2×1 surface. With respect to ESDIAD, HREELS allows the identification of surface species through vibrational characterization. The main result of the work of Gao et al. [24] is that chlorine can form a bond with a silicon surface mainly in two configurations. In the first case chlorine bonds with two symmetric reconstructured surface Si atoms ($SiCl_{surf}$), while the other possible configuration is bridge bonding between two surface Si atoms (Si_2Cl_{surf}). The more abundant surface species is the first one, while the second is characterized by a high cross section for ESD. Two surface vibrational frequencies were identified through HREELS: 550–600 cm^{-1} for $SiCl_{ads}$ and $\sim$295 cm^{-1} for the bridged Si_2Cl_{ads} compound. As the surface temperature was increased from 273 to 673 K the bridged surface species Si_2Cl_{ads} dissociates and gives the more stable $SiCl_{ads}$ species. Finally, it was found that the major desorption products at high temperatures are $SiCl_2$ and $SiCl_4$.

Successive experiments conducted by Sterrat et al. [78] further cleared the details of chlorine surface bonding at high chlorine surface coverages. In this work *Si2p* soft X-ray photoemission and photon-stimulated ion desorption (PSID) were adopted. PSID and photoemission data were recorded at room temperature. The

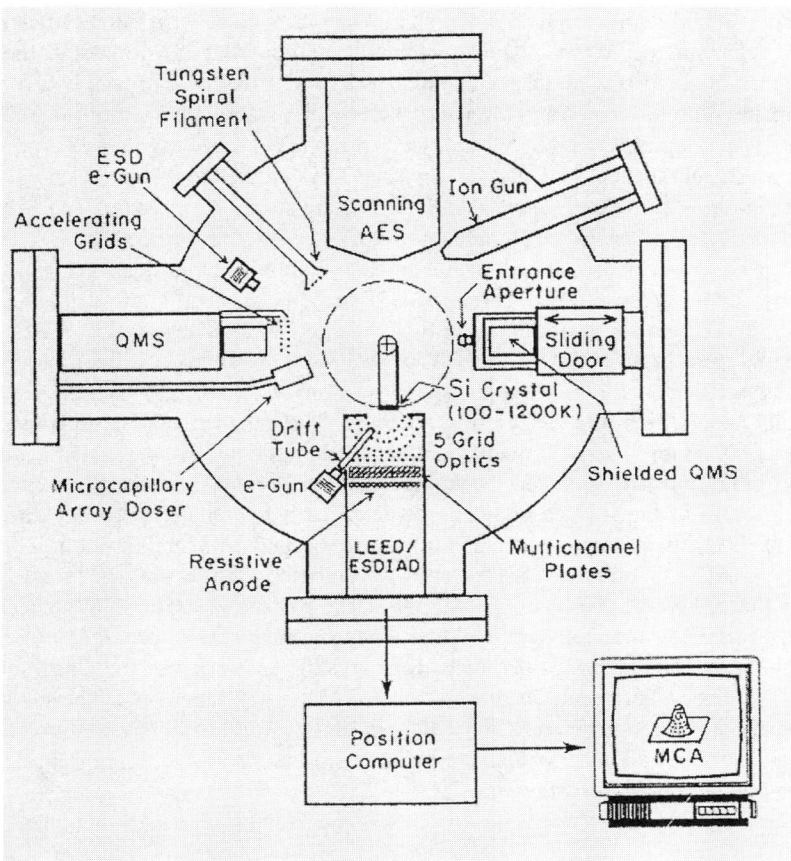

FIG. 3. A schematic diagram of the apparatus adopted by Yates and co-workers to study the adsorption of chlorine on a silicon surface [86].

results of these experiments indicated the presence of monochloride and dichloride species adsorbed on the Si surface at a ratio of 2.5:1. It is important to observe that Ref. 78 proposed that the off-normal bonding of chlorine observed by Gao *et al.* [24], who attributed this configuration to Cl bonded at both ends of a surface dimer structure, could also be explained by the formation of a $SiCl_2$ surface species.

b. *Adsorption and Desorption Kinetics*

The interaction of chlorosilane gas-phase species with the silicon surface has been studied mainly by experimental rather than theoretical methods. This is in contrast with the situation for the gas-phase kinetics, where all the kinetic constants were determined either by quantum chemistry methods and transition-state theory or by fitting to growth rate data. The reason for the different approach is twofold. First, the computational simulation of surface reactions through quantum chemistry is made particularly onerous by the need to include a large number of atoms to reproduce the structure of the surface. For this reason the quantum chemistry investigations of the surface energetics and kinetics were restricted to the study of the reaction of desorption of molecular hydrogen from the surface.

Experimental investigations of the surface reactivity have been made possible by the progress and refinement of surface science techniques such as TPD, LITD, FTIR, mass spectrometry, AES, XPS, and LEED. The kinetics of adsorption and desorption of SiH_2Cl_2, $SiHCl_3$, and $SiCl_4$ on epitaxial silicon has been investigated in works by George and co-workers [15, 19, 35]. Similar experimental apparatuses were adopted in all cases. The experimental setup consisted essentially of an UHV chamber equipped for temperature-programmed desorption (TPD) and laser-induced thermal desorption (LITD). The chemical species desorbed from the surface could be detected by a low-energy electron diffraction (LEED) spectrometer and a quadrupole mass spectrometer. Fourier transform infrared (FTIR) vibrational spectroscopy was adopted to study the structure of surface and gas-phase chemical species, while the surface cleanliness was checked with auger electron spectroscopy (AES). The experiments were usually performed adopting silicon substrates obtained from crystals grown using the Czochralski method. The substrate surfaces were either (111) or (100) oriented. The samples were about 10×10 mm and 250 to 400 μm thick. A thin film of tantalum could be deposited on the back of the sample to ensure uniform heating and therefore obtain better desorption efficiency.

The temperature of the silicon crystal was measured with a thermocouple. The contact between the sample and the chamber was realized by means of two tantalum foils, which were pressed between a copper clamp and a copper support on two sides of the silicon sample. Copper wires were attached to the copper supports for resistive heating. Before being introduced into the chamber the sample was rinsed with methanol and air-dried. Then the samples were cleaned in vacuum by annealing. Si–H–Cl species were adsorbed on the sample by backfilling the chamber.

TPD was adopted to investigate the adsorption of SiH_2Cl_2 on Si(111). The species that desorb from the surface were found to be H_2, HCl, Cl, SiCl, and $SiCl_2$. SiH_2Cl_2 was not detected as a major desorption product, which means that it is completely dissociated after adsorption. HCl was the first species that desorbed from the surface at a temperature of 850 K, while Cl, SiCl, and $SiCl_2$ were observed between 950 and 1000 K. The presence of SiCl in the gas phase was attributed to electron-induced fragmentation of $SiCl_2$ in the mass spectrometer rather than to direct desorption of SiCl from the surface. The kinetic constants for the reactions of desorption of HCl and $SiCl_2$ were determined from the TPD spectra and were calculated from the position and intensity of the temperature peak and from the width of the spectra. It was found that the desorption of HCl proceeds through

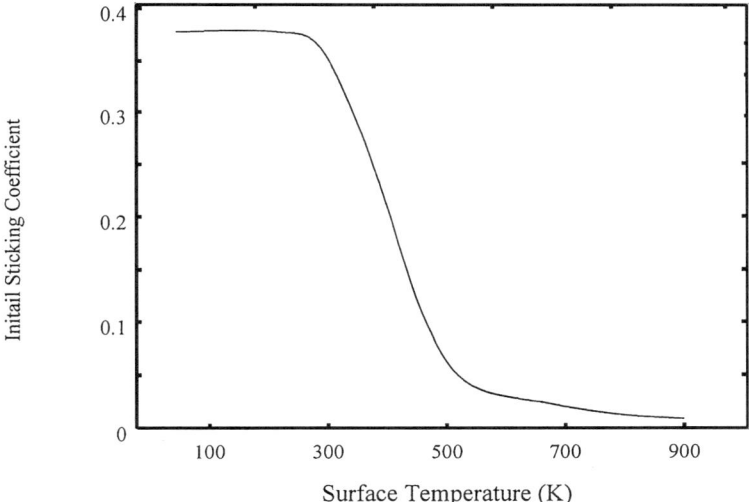

FIG. 4. Measured sticking coefficient on a Si(100) surface measured as a function of surface temperature.

a second-order kinetics with a ν_{des} of $0.25 \text{ m}^2 \cdot \text{s}^{-1}$ and an activation energy of 301.4 kJ/mol. The desorption reaction appears to proceed with an activation energy equal to the enthalpy change for the desorption reaction. The enthalpy change was calculated adopting hydrogen and chloride bond energies of 347.4 [52] and 385.1 kJ/mol [35] and a bond energy for HCl in the gas phase of 431.2 kJ/mol.

The adsorption rate of SiH_2Cl_2 was studied adopting LITD. The sticking coefficient could be calculated by monitoring the surface chloride coverage as a function of time. The amount of chloride present on the surface was determined by measuring the intensity of the SiCl signal during the exposure to SiH_2Cl_2. The decrease in the sticking coefficient with the increase in surface temperature is consistent with a precursor-mediated adsorption process. As shown in Fig. 4 the sticking coefficient has a constant value of about 0.36 from 100 to 200 K, which decreases to 0.1 as the temperature is raised to 350 K and drops to 0.0025 at 850 K. Coon et al. [15] proposed the following series of reactions to explain the adsorption mechanism:

$$SiH_2Cl_2(g) + \text{surf} \rightarrow SiH_2Cl_2(\text{ads}) \tag{26}$$

$$SiH_2Cl_2(\text{ads}) \rightarrow SiH_2Cl_2(g) + \text{surf} \tag{27}$$

$$SiH_2Cl_2(\text{ads}) + \text{sites} \rightarrow SiCl_s + Cl_s + 2H_s. \tag{28}$$

If reactions (26) and (28) proceed without activation energy (as is likely, since they are exothermic processes), the activation energy of the overall process $SiH_2Cl_2 + \text{sites} \rightarrow SiCl_s + Cl_s + 2H_s$ will be equal to the activation energy of reaction (27) changed by sign. This observation, which is not intuitive, can be easily verified by applying the steady-state condition to the surface species concentration. The sticking coefficient obtained by interpolating the data reported in Fig. 4 is $1.7 \times 10^{-3} e^{-1900/T}$.

In a more recent work Mendicino and Seebauer [61] reexamined the surface kinetics of HCl. The experimental apparatus is similar to that adopted by George *et al.* and consisted in a UHV chamber equipped with AES, LEED, a quadrupole mass spectrometer, an ionization gauge, and an ion gun for surface cleaning. The experiments were performed on a Si(100) epitaxial surface. Chlorine and hydrogen were adsorbed on the surface adopting $TiCl_4$ and SiH_4 as gas-phase precursors. It was found that the desorption kinetics of HCl is a function of the surface coverage. When the surface is covered with H and Cl atoms from 5 to 90%, the major desorption products are $SiCl_2$ and H_2. Under these conditions HCl desorbs from the surface following a second-order kinetics with a frequency factor of $1.18 \times 10^{-2 \pm 1.3}$ m^2/s and an activation energy of 298.5 kJ/mol. The activation energy is very similar to that previously calculated by Coon *et al.* (i.e., 301.4 kJ/mol) [15], while the preexponential factor is slightly smaller. It must be noted, however, that Coon *et al.* performed their experiments on a Si(111) crystal, while Mendicino and Seebauer worked on a Si(100) surface, which might explain the small differences found between the two experimental results.

When the surface coverage is very low (i.e., <0.14%) the desorption kinetics changes from double to single site. A possible explanation for this behavior is that the desorption reaction proceeds from a silicon surface site on which H and Cl are present contemporaneously. The preexponential factor and activation energy determined for this reaction were $1.18 \times 10^{-3 \pm 1.5}$ s^{-1} and 202.2 kJ/mol, respectively.

Mendicino and Seebauer also provide a measure of the sticking coefficient of HCl on a low-coverage silicon surface. At 300 K the sticking coefficient for HCl is 0.1, and it decreases toward zero as the surface temperature is raised.

The adsorption of $SiHCl_3$ on a Si(100) $2x1$ surface was studied by Dillon *et al.* [19] using FTIR spectroscopy and TPD. All experiments were performed in the UHV chamber described previously. It was found that, similarly to what was found for SiH_2Cl_2, H_2, HCl, and $SiCl_2$ desorb at 810, 850, and 970 K, respectively. As expected SiCl was not found among the desorbing species, which confirms the hypothesis that in the previous study it was produced by electron fragmentation reactions, which are avoided in the experimental setup adopted by Dillon *et al.* since the gas-phase composition was detected with FTIR. The measured sticking coefficient for $SiHCl_3$ was 0.019 at 500 K.

The surface chemical composition was investigated us FTIR immediately after the adsorption of $SiHCl_3$. It was found that SiH, $SiCl_x$, ClSiH, and Cl_2SiH are present on the surface. The presence of Cl_2SiH, which was detected at low surface temperatures (i.e., 200 K), is indicative of partial dissociation of $SiHCl_3$ upon adsorption. Its presence on the surface was hypothesized because of the presence of an absorption frequency at 775 cm^{-1}, which was attributed to the bending frequency of Cl_2Si-H. As the temperature of the surface was increased, only SiH and SiCl were found to be present on the surface.

The adsorption of $SiCl_4$ on a Si(111) 7x7 surface was studied by Gupta et al. [36] using LITD and TPD. Similarly to what was found for SiH_2Cl_2 the sticking coefficient was found to decrease from 0.18 to 0.03 as the surface temperature increased from 160 K to 600 K. Also in this case this behavior suggests that the adsorption of $SiCl_4$ proceeds through the formation of a surface intermediate species. A mechanism consistent with this finding is the following:

$$SiCl_4(g) + surf \rightarrow SiCl_4(ads) \tag{29}$$

$$SiCl_4(ads) \rightarrow SiCl_4(g) + surf \tag{30}$$

$$SiCl_4(ads) + sites \rightarrow SiCl_s + 3Cl_s. \tag{31}$$

It was also found that the sticking coefficient diminishes if the surface is partially covered with hydrogen, which can be explained by observing that the surface reaction probably requires a certain number of free surface sites to proceed.

From analysis of the TPD spectra the desorption kinetics of $SiCl_2$ could be determined. This was found to be second order in chlorine coverage, with an activation barrier of 280.5 ± 20.9 kJ/mol and a preexponential factor ν_d of $3.2 \times 10^{-4\pm0.1}$ $m^2 \cdot s^{-1}$.

The dependence of the desorption kinetics of $SiCl_2$ on the surface structure was examined by Szabó et al. [82]. The experimental apparatus consisted of a UHV chamber equipped with a molecular beam line, two differentially pumped quadrupole mass spectrometers, an ion gun, an X-ray source, and a hemispherical energy analyzer. The temperature of the crystal could be varied between 130 and 1300 K and was measured through a thermocouple inserted at the edge of the sample. The sample was exposed to beams of molecular and atomic chlorine. TPD spectra were obtained on heating the sample linearly at 10 K/s. The chemical species detected with the quadrupole mass spectrometer were $SiCl_2$ and $SiCl_4$.

It was found that the desorption kinetics of $SiCl_2$ is a function of the surface temperature. Below 800 K the desorption kinetics is first order in chlorine coverage from both the Si(100) and the Si(111) surface. Under these conditions the preexponential factor has a value between 10^{12} and 10^{16} s^{-1} and the activation energy ranges between 126 and 180 kJ/mol. As the temperature passes 900 K the desorption reaction is second order in chlorine coverage if the molar fraction of chlorine adsorbed on the surface is between 0.3 and 0.75. The preexponential factor under these conditions is 0.2 $m^2 \cdot s^{-1}$ and the activation energy is 348 kJ/mol. This kinetic constant is in agreement with that proposed by Gupta et al. [36], 2.0×10^{-15} $m^2 \cdot s^{-1}$ at 1300 K, which is very close to the value of 2.7×10^{-15} $m^2 \cdot s^{-1}$ that can be calculated using the expression of Szabó et al. However, the kinetic constants proposed by the two groups differ considerably for the values of the preexponential factors and the activation energies. This disagreement can be ascribed more to the mathematical interpolation process adopted to fit the kinetic constant expression than to the different experimental setup. This observation suggests that much care must be adopted in using activation energies measured through TPD experiments to determine bond energies of surface species.

TPD spectra realized on the Si(100) surface at temperatures higher than 900 K showed a behavior different from that observed on the Si(111) surface. For a chlorine coverage (θ_{Cl}) between 0.15 and 0.4 the desorption kinetics is first order in chlorine, while for $\theta_{Cl} > 0.5$ the desorption rate becomes independent of the surface concentration and, finally, decreases with increasing surface coverage. The first-order desorption kinetics of $SiCl_2$ from the Si(100) surface observed for a low surface coverage can be interestingly related to the results of Mendicino and Seebauer [61] on the desorption of HCl, which was also first order in surface concentration at very low surface coverage. This may be indicative of the presence of particular surface structures or reconstructions when the concentration of adsorbed species is very low.

V. Gas-Phase Precursors to Deposition and Overall Kinetic Scheme

The identification of the gas-phase chemical species whose adsorption determines the growth of the epitaxial silicon film is a fundamental point for assessing the impact of the gas-phase reaction kinetics on the overall process. We consider the gas-phase precursor to deposition the gas-phase molecule or radical containing a silicon atom that reacts with the growing surface and, upon the release of silicon to the film, determines the growth of the epitaxial layer. For silanes the gas-phase precursor was identified as SiH_4 if the system considered was thermally activated and SiH_3 if the activation of the gas phase occurred through a plasma discharge.

In deposition systems adopting chlorosilanes the identification of the gas-phase precursor to deposition is still a matter of discussion. Two gas-phase species can contribute to the film growth: the nondissociated SiH_xCl_{4-x} molecule and $SiCl_2$. Their relative amounts are strongly dependent on the gas-phase temperature and pressure, the SiH_xCl_{4-x} species being most abundant at low temperatures and high pressures. On the other hand, the sticking coefficient of $SiCl_2$ is greater than that of SiH_xCl_{4-x} because of its higher reactivity. To complicate the problem further, $SiCl_2$ is also the principal product of desorption from the silicon surface, as explained in the previous chapter.

Computer simulations, which might have helped to discriminate between the two gas-phase species, stated that good results in terms of fitting of growth rate and gas phase composition could be obtained considering both $SiCl_2$ and SiH_xCl_{4-x} as deposition precursors [40, 43, 65, 66]. In the following, fluid dynamic and kinetic models adopted to study deposition reactors for the growth of epitaxial silicon are described.

Narusawa investigated the deposition of epitaxial Si from $SiCl_4$ and $SiHCl_3$ diluted in a hydrogen carrier gas and carried out in a horizontal reactor [65, 66].

The experimental conditions investigated covered a surface temperature ranging between 1100 and 1200°C, while the pressure was kept constant at 1 atm. The model equations comprised the energy and mass conservation equations, with appropriate boundary conditions. The fluid dynamic model of the reactor was linked with a simplified kinetic model of the gas phase. The two reactions that were considered representative of the gas phase chemistry were

$$SiCl_4 + H_2 \leftrightarrow SiHCl_3 + HCl \tag{32}$$

$$SiHCl_3 \leftrightarrow SiCl_2 + HCl. \tag{33}$$

The reactions were considered reversible, with backward kinetic constants estimated imposing the thermodynamic consistence condition. At the basis of the model is the assumption that the film growth is determined mainly by the adsorption of $SiCl_2$ and only marginally by that of $SiHCl_3$. Accordingly the boundary condition representative of the concentration of $SiCl_2$ at the deposition surface was assumed to be equal to zero. The adsorption rates of $SiHCl_3$ and $SiCl_2$ were calculated through collision theory adopting a unary sticking coefficient for $SiCl_2$ and 10^{-4} for $SiHCl_3$. The kinetic constants for reactions (32) and (33) were considered unknown parameters and were therefore fitted over experimental growth rate data.

This approach appears to be strongly conditioned by the initial assumptions on the sticking coefficients for $SiCl_2$ and $SiHCl_3$ and by the hypothesis that $SiCl_2$ is the growth species. The kinetic constants so determined should therefore be adopted with care. It is hence noteworthy that the fitted kinetic constants are in good agreement (within factor of an order of magnitude) with those calculated through quantum chemistry, as observed in the preceding section.

A sensitivity analysis was performed to determine the influence of the sticking coefficient of $SiHCl_3$ on the growth rate. It was found that the contribution of $SiHCl_3$ to the total growth rate becomes significant if it has a sticking coefficient higher than 10^{-4}.

Experimental measurements [19] showed that $SiHCl_3$ has a sticking coefficient of 10^{-2} at 500 K, which decreases with increasing temperature. If an activation energy equal to that measured for the SiH_2Cl_2 adsorption reaction is adopted, then the sticking probability of $SiHCl_3$ on a Si surface at 1000°C is 1.7×10^{-3}. It can therefore be concluded that the contribution of $SiHCl_3$ to the Si growth cannot be ruled out a priori.

Habuka et al. [37] have investigated the deposition of Si from $SiHCl_3$ theoretically. Their work consisted of realizing a model of an experimental deposition reactor consisting of a detailed two-dimensional fluid dynamic description with a gas-phase kinetics simplified as much as possible. In particular, because of its thermodynamic stability, $SiHCl_3$ was considered to be the most abundant gas-phase silicon species and therefore it was assumed to be the principal contributor to the film growth. Accordingly the following assumptions were made: no gas-phase reactions are active or can influence the film growth rate; the effect of HCl on the growing film can be accounted for through the following kinetic scheme:

$$SiHCl_3 + \theta \rightarrow SiCl_2^* + HCl \quad (34)$$

$$SiCl_2^* + H_2 \rightarrow Si_{bulk} + 2HCl + \theta \quad (35)$$

$$SiCl_2^* + HCl \rightarrow SiHCl_3 + \theta. \quad (36)$$

Implicit in the formulation of this surface kinetic scheme is that only two species are present on the surface during the growth process: $SiCl_2$ and the free surface sites θ. This assumption is in disagreement with surface science experiments, which showed that silicon monochloride is the most stable surface species.

Reaction (36) was assumed to proceed at a slower rate than reactions (37) and (38) and was therefore ignored. The rates of reactions (34) and (35) were determined through fitting to experimental data measured at temperatures ranging between 1000 and 1400 K and, therefore, within the kinetically controlled regime. The calculated values were $k_{34} = 10^{6.436} e^{-20,688/T}$ and $k_{35} = 10^{3.748} e^{-21,495/T} \text{m} \cdot \text{s}^{-1}$. From what was reported in Section IV.2 it follows that the activation energy of reaction (34) is without physical meaning, since the adsorption of $SiHCl_3$ on the Si surface is known to proceed without barriers. Similarly reaction (38) is an exothermic process and it therefore seems unlikely that it has so high an activation energy. Thus it seems reasonable that the activation energy found by Habuka is due to a surface reaction different from reactions (34), (35), and (36). Analysis of the activation energies of different surface reactions has shown that the value of ~170 kJ/mol calculated by Habuka et al. is close to the activation energy of 197 kJ/mol measured experimentally by Flowers et al. [22] for the desorption of molecular hydrogen.

The possibility that the desorption of hydrogen from the silicon surface is the rate-determining step in film growth in a kinetics-controlled regime was first suggested by Bloem and Giling [6]. Observing experimental growth rate data, they found that the activation energy for the growth of Si from chlorosilane is always near 190 kJ/mol and proposed the desorption of hydrogen from the surface to be the rate-determining step. The effect of hydrogen can be either to occupy the adsorption sites for Si gas-phase species or to react with chloride to yield a desorbing HCl molecule. In this case the role of hydrogen would be to remove the chloride adsorbed on the surface.

Modeling work similar to that of Habuka et al. was performed more recently by Angermaier et al. [1], who compared calculated results with experimental data measured in a rapid thermal CVD (RTCVD) reactor. This work was essentially a fluid dynamic study where the growth of the film was attributed to the diffusion of $SiHCl_3$ from the bulk of the gas toward the deposition surface. Accordingly the gas-phase reactivity was neglected. The main result of the study was the demonstration that a relatively good agreement between experimental and calculated growth rates can be obtained adopting $SiHCl_3$ as the only deposition precursor when the growth is carried out in a diffusion-controlled regime.

Models more complete than those proposed above were realized by Hierlemann et al. [40] and Ho et al. [43]. The aim of these works was to link a kinetic model constituted of elementary reactions representing the essential features of the chemistry of the deposition process with a rigorous model of the reactor fluid dynamics.

In the mechanism proposed by Hierlemann *et al.* the kinetic model adopted was the following:

$$SiH_2Cl_2 + 4Si_{(s)} \rightarrow 2H_{(s)} + 2Cl_{(s)} + 5Si_{(b)} \tag{37}$$

$$2H_{(s)} + 2Si_{(b)} \leftrightarrow H_2 + 2Si_{(s)} \tag{38}$$

$$H_{(s)} + Cl_{(s)} + 2Si_{(b)} \leftrightarrow HCl + 2Si_{(s)} \tag{39}$$

$$2Cl_{(s)} + 3Si_{(b)} \rightarrow SiCl_2 + 2Si_{(s)} \tag{40}$$

$$2Cl_{(s)} + H_2 + 2Si_{(b)} \rightarrow 2Si_{(s)} + 2HCl \tag{41}$$

$$SiH_2Cl_2 + M \leftrightarrow SiCl_2 + H_2 + M \tag{42}$$

$$SiCl_2 + 2Si_{(s)} \rightarrow 2Cl_{(s)} + 3Si_{(b)}. \tag{43}$$

In this scheme the subscripts (s) and (b) refer to surface and bulk species, respectively.

To simplify the model it was assumed that $Si_{(s)}$, $H_{(s)}$, and $Cl_{(s)}$ were the only species present on the surface. $Si_{(s)}$ represents a free surface site available for adsorption, $H_{(s)}$ is an hydrogen atom adsorbed over a Si atom and $Cl_{(s)}$ is a chlorine atom adsorbed over a Si atom.

The kinetic constants for reactions (37)–(43) were found in the literature. To improve the agreement with experimental data, their values were slightly tuned within their uncertainty limits. In particular, the rate of reaction (41) was fitted over growth rate data measured at 10 Torr. This reaction, first proposed by Oshita *et al.* [67, 68] was introduced to account for the dependence of the deposition rate from the hydrogen partial pressure. The kinetic constant so calculated was $k_{41} = 10^{17.477} e^{-36500/T} m^3 \cdot s^{-1} \cdot mol^{-1}$. The collision kinetic constant for hydrogen on a Si(100) surface has a value of 7×10^{16} at 1300 K, which is lower than the Arrhenius factor of reaction (41). Since the collision kinetic constant represents the upper limit for a reaction between a gas-phase molecule and a surface species, reaction (41) appears to be without physical meaning. An alternative pathway that might explain the effect of hydrogen on the silicon surface growth rate can be obtained by considering reaction (41) as the sum of two distinct surface reactions:

$$H_2 + 2Si_{(s)} \rightarrow 2H_{(s)} + 2Si_{(b)} \tag{44}$$

$$2 \times [H_{(s)} + Cl_{(s)} + 2Si_{(b)} \rightarrow HCl + 2Si_{(s)}] \tag{45}$$

$$H_2 + 2Cl_{(s)} + 2Si_{(b)} \rightarrow 2HCl + 2Si_{(s)}. \tag{41}$$

The sum of reactions (44) and (45) equals the global stoichiometry of reaction (41). Reaction (44) is an exothermic process and therefore it is likely to proceed without a significant activation energy, while the experimentally measured activation energy of reaction (45) is 301.5 kJ/mol. This value is very close to that fitted for reaction (41), which is consistent with the hypothesis that the positive effect of hydrogen on the growth rate might be explained by reactions (44) and (45). According to this mechanism the positive effect of hydrogen on the growth rate of silicon is determined by its ability to remove chlorine adsorbed on the surface through the formation of hydrogen chloride.

If we link this piece of information with the possibility, advanced above, that in some cases the desorption of hydrogen from the Si surface is rate determining, we see that gas-phase hydrogen plays an important role in the growth of epitaxial Si from chlorosilanes. Unfortunately the work of Hierlemann *et al.* was not conclusive on which was the gas-phase precursor to deposition since both SiH_2Cl_2 and $SiCl_2$ were found to contribute substantially to the film growth in an amount which depended on the deposition temperature.

A similar approach was adopted by Ho *et al.* [43] to study the deposition of Si from SiH_2Cl_2, $SiHCl_3$, and $SiCl_4$. The kinetic model was very similar to that of Hierlemann *et al.* except for reactions (41) and (42), which were not considered. In addition, the adsorption reactions for $SiHCl_3$ and $SiCl_4$ were introduced as

$$SiHCl_3 + 4Si_{(s)} \rightarrow H_{(s)} + 3Cl_{(s)} + 5Si_{(b)} \qquad (46)$$

$$SiCl_4 + 4Si_{(s)} \rightarrow 4Cl_{(s)} + 5Si_{(b)}. \qquad (47)$$

As in the kinetic scheme proposed by Hierlemann *et al.* also in this case the surface species considered were free silicon sites $[Si_{(s)}]$, adsorbed hydrogen atoms $[H_{(s)}]$, and adsorbed chlorine atoms $[Cl_{(s)}]$.

All the kinetic constants were found in the literature except for that for the adsorption of HCl, which was fitted over experimental data. The predictive capability of the kinetic scheme was tested through the simulations of a horizontal deposition reactor. Unfortunately the data simulated were collected under operating conditions where the growth rate was controlled by diffusion, in which case the gas-phase and surface kinetics play a minor role. Further calculations are therefore needed to test the predictive capability of the proposed model. In contrast with the results of Hierlemann *et al.* Ho *et al.* found that $SiCl_2$ is produced rather than consumed at the surface because of the high rate of the desorption reaction.

VI. Overall Kinetic Scheme and Concluding Remarks

In conclusion, we have shown how the level of detail of the kinetic models adopted to describe the growth of Si films has progressively increased, helped greatly by the results of experimental and first principal investigations.

The gas-phase and surface chemistry of silanes and disilanes can be considered to be well represented by the kinetic schemes reported in Tables II and IV. These schemes were in fact adopted several times with success to model the performances of experimental reactors.

In the case of chlorosilanes much work has been done to understand the growth kinetics of these films, but a definitive kinetic model has not yet been proposed. In

TABLE V

SURFACE REACTIONS FOR THE DEPOSITION OF SILICON FROM CHLOROSILANE'S PRECURSORS[a]

	Reaction	$\log A$	α	E	Ref. No.(s.)[b]
G1	$SiHCl_3 \leftrightarrow SiCl_2 + HCl$	14.69	0	308.5	80
G2	$SiH_2Cl_2 \leftrightarrow SiCl_2 + H_2$	13.92	0	324.0	80
G3	$SiH_2Cl_2 \leftrightarrow HsiCl + HCl$	14.84	0	317.3	80
G4	$SiCl_4 \leftrightarrow Cl + SiCl_3$	15.68	0	465.4	(a)
G5	$SiHCl_3 + H \leftrightarrow SiCl_3 + H_2$	6.39	0	10.6	2
G6	$HCl \leftrightarrow H + Cl$	13.64	0	342.2	3
S1	$SiHCl_3 + 4\sigma \rightarrow SiCl^* + H^* + 2Cl^*$	2.05	0.5	-15.9	19, 43
S2	$SiH_2Cl_2 + 4\sigma \rightarrow SiCl^* + 2H^* + Cl^*$	2.58	0.5	-15.9	15, 43
S3	$SiCl_4 + 4\sigma \rightarrow SiCl^* + 3Cl^*$	3.84	0.5	-2.8	36, 43
S4	$H_2 + 2\sigma \rightarrow 2H^*$	5.36	0.5	72.2	7.b
S5	$HCl + 2\sigma \rightarrow H^* + Cl^*$	4.73	0.5	0	61 (a)
S6	$SiCl_2 + 2\sigma \rightarrow SiCl^* + Cl^*$	4.51	0.5	0	(b)
S7	$SiCl^* + Cl^* \rightarrow SiCl_2 + 2\sigma$	20.20	0	280.5	36
S8	$2Cl^* + Si_b \rightarrow SiCl_2 + 2\sigma$	20.20	0	280.5	36
S9	$2H^* \rightarrow H_2 + 2\sigma$	20.42	0	196.7	22
S9*	$H^* \rightarrow 1/2 H_2 + \sigma$	15.30	0	238.6	22
S10	$H^* + Cl^* \rightarrow HCl + 2\sigma$	21.85	0	298.5	135
S11	$H^* + SiCl^* \rightarrow HCl + 2\sigma + Si_{(b)}$	21.85	0	298.5	61

[a] $k = AT^\alpha e^{-E/RT}$, expressed as s^{-1}, $m^2 \cdot mol^{-1} \cdot s^{-1}$, or $m^3 \cdot mol^{-1} \cdot s^{-1}$; T, in K; and E, in kJ/mol.
[b] (a) Collisional theory with sticking coefficient 0.1, from Ref. 61. (b) Collisional theory with sticking coefficient 0.1, assumed.

particular, it is not known whether $SiCl_2$ or SiH_xCl_{4-x} is the gas-phase precursor to deposition, since a good reproduction of experimental data could be obtained considering either species as the precursor of the growth process. Another point that needs to be fully understood is the effect of hydrogen on the deposition surface.

Based on the information reported in the preceding sections, a kinetic scheme of the gas-phase and surface processes that determine the growth of epitaxial Si films is reported in Table V. With this scheme we do not mean to propose a new kinetic model for the deposition of Si from chlorosilanes, and in fact the predictions of this model were not tested directly through the simulation of experimental reactors. Rather we are trying to collect in a unique picture all the experimental and theoretical results described in this chapter. Consequently, should this model be adopted to simulate experimental data, we suggest reading the previous sections, where the process followed to evaluate each kinetic constant is discussed.

The scheme is divided into two parts, the first dealing only with homogeneous gas-phase reactions and the second considering reactions between gas-phase and surface species and among surface species. Kinetic constants for gas-phase reactions G1–G3 were taken from the quantum chemistry work of Su and Schlegel

[80]. The kinetic constant of reaction G4 was estimated as explained above using Eq. (13). Reaction rates for G5 and G6 were evaluated experimentally [2, 3].

The surface kinetic scheme in Table V comprises 11 nonreversible reactions. The first six represent the adsorption of gas-phase species on the silicon surface. The kinetic constant values we report were calculated using Eq. (4), adopting experimentally measured sticking coefficients. For reactions S1 and S2 a negative activation energy for adsorption of -15.9 kJ/mol is proposed. The physical reason to adopt a negative activation energy is that the adsorption process is probably precursor mediated, which, as explained in Section IV.2b, is directly related to an adsorption rate that decreases with increasing surface temperature. The activation energy and sticking coefficient for reaction S3 were obtained by fitting the experimental data of Gupta et al. [35].

The adsorption processes were considered to have first-order dependence on the free surface sites (i.e., a single-site adsorption mechanism was assumed) and a surface site concentration of 6.8×10^{18} at/m^2 was adopted, which corresponds to a Si(100) surface. Four species were considered to be present on the surface during the growth process: Cl*, H*, SiCl*, and σ. σ represents the free surface sites, Cl* and H* are a chlorine and a hydrogen atom bonded to a Si surface site σ, and SiCl* is a Si atom bonded to a chlorine atom, which faces the gas phase, and to a free surface site σ.

Reactions S7–S11 are desorption reactions whose kinetic constants were determined through experimental TPD studies. Reactions S9 and S9* represent the molecular desorption of H$_2$ from the (100) surface and the proposed kinetic constants are derived from the work of Flowers et al. [22]. Two kinetic constants are proposed, corresponding to the single- and double-site desorption kinetics that were observed.

As a concluding remark we observe that all the surface reactions we propose must be considered only in the forward direction. In fact here we explicitly reported the backward reaction, since its kinetic constant has been often measured directly through experimental studies, independently from the process of evaluation adopted for the corresponding forward reaction.

References

1. Angermaier, D., R. Monna, A. Slaoui, and J. C. Muller, *J. Electrochem. Soc.* **144**, 3256 (1997).
2. Arthur, N. L., P. Potzinger, B. Reimann, and H. P. Steenbergen, *J. Chem. Soc., Faraday Trans.* **85**, 1447 (1989).
3. Baulch, D. L., J. Duxbury, S. J. Grant, and D. C. Montague, *J. Phys. Chem. Ref. Data* **10**, 1 (1981).
4. Bennett, S. L., C. L. Greenwood, and E. M. Williams, *Surf. Sci.* **290**, 267 (1993).
5. Benson, S. W., *Thermochemical Kinetics,* Wiley, New York 1976.
6. Bloem, J., and L. J. Giling, in *Current Topics in Materials Science,* edited by E. Kaldis, Vol. 1, pp. 147–341, North-Holland, Amsterdam, 1978.
7. Boudart, M. G., and G. Diege-Mariadassou, *Kinetics of Heterogeneous Catalytic Reactions,* Princeton University Press, Princeton, NJ 1984.
7b. Bratu, P., K. L. Kompa, and U. Hofer, *Chem. Phys. Lett.* **251**, 1 (1996).

8. Buss, R. J., P. Ho, W. G. Breiland, and M. E. Coltrin, *J. Appl. Phys.* **63**, 2808 (1988).
9. Chase, M. W., C. A. Davies, J. R. Downey, D. J. Frurip, R. A. McDonald, and A. N. Szverud, *J. Phys. Chem. Ref. Data* **14**, (1985).
10. Chatham, H., D. Hills, R. Robertson, and A. Gallagher, *J. Chem. Phys.* **81**, 1770 (1984).
11. Coltrin, M. E., R. J. Kee, and G. H. Evans, *J. Electrochem. Soc.* **136**, 819 (1989).
12. Coltrin, M. E., R. J. Kee, and J. A. Miller, *J. Electrochem. Soc.* **131**, 425 (1984).
13. Coltrin, M. E., R. J. Kee, and J. A. Miller, *J. Electrochem. Soc.* **133**, 1206 (1986).
14. Cowfer, J. A., K. P. Linch, and J. V. Michael, *J. Phys. Chem.* **79**, 1139 (1975).
15. Coon, P. A., P. Gupta, M. L. Wise, and S. M. George, *J. Vac. Sci. Tech.* **A10**, 324 (1992).
16. Curtiss, L. A., J. E. Carpenter, K. Rachavagari, and J. A. Pople, *J. Chem. Phys.* **96**, 9030 (1992).
17. Curtiss, L. A., K. Rachavagari, G. W. Trucks, and J. A. Pople, *J. Chem. Phys.* **94**, 7221 (1991).
18. Czyzewski, J. J., T. E. Madey, and J. T. Yates, Jr., *Phys. Rev. Lett.* **32**, 777 (1974).
19. Dillon, A. C., M. L. Wise, M. B. Robinson, and S. M. George, *J. Vac. Sci. Tech.* **A13**, 1 (1995).
20. Doughty, D. A., and A. Gallagher, *Phys. Rev. A* **42,**, 6166 (1990).
21. Eliason, M. A., and J. O. Hirschfelder, *J. Chem. Phys.* **30**, 1426 (1959).
22. Flowers, M. C., N. B. H. Jonathan, Y. Liu, and A. Morris, *J. Chem. Phys.* **99**, 7038 (1993).
23. Foresman, J. B., and A. Frisch, *Exploring Chemistry with Electronic Structure Methods,* Gaussian, Pittsburgh, (1996).
24. Gao, Q., C. C. Cheng, P. J. Chen, W. J. Choyke, and J. T. Yates, Jr., *J. Chem. Phys.* **98**, 8308 (1993).
25. Gates, S. M., *Chem. Rev.* **96**, 1519 (1996).
26. Gates, S. M., and K. Kulkarni, *Appl. Phys. Lett.* **60**, 53 (1992).
27. Gates, S. M., and K. Kulkarni, *Appl. Phys. Lett.* **58**, 2963 (1991).
28. Gates, S. M., C. M. Greenlief, and D. B. Beach, *J. Chem. Phys.* **93**, 7493 (1990).
29. Gates, S. M., C. M. Greenlief, S. K. Kulkarni, and H. H. Sawin, *J. Vac. Sci. Technol. A* **8**, 2965 (1990).
30. Gilbert, R. G., K. Luther, and J. Troe, *Ber. Bunsen-Ges. Phys. Chem.* **87**, 169 (1983).
31. Gilbert, R. G., and S. C. Smith, *Theory of Unimolecular and Recombination Reactions,* Blackwell, Oxford, 1990.
32. Gilbert, R. G., S. C. Smith, and M. J. T. Jordan, UNIMOL program suite 1993.
33. Golant, V. E., A. P. Zilinskij, and S. E. Sacharov, *Osnovi Fiziki Plasmy* Mir, Moscow 1983.
34. Graves, D. B., and K. F. Jensen, *IEEE Trans. Plasma Sci.* **14**, 78 (1986).
35. Giunta, C. J., R. J. McCurdy, J. D. Chapple-Sokol, and R. G. Gordon, *J. Appl. Phys.* **67**, 1062 (1990).
36. Gupta, P., P. A. Coon, B. G. Koehler, and S. M. George, *J. Chem. Phys.* **93**, 2827 (1990).
37. Habuka, H., T. Nagoya, M. Mayusumi, M. Katayama, M. Shimada, and K. Okuyama, *J. Cryst. Growth* **169**, 61 (1996).
38. Hammond, G. S., *J. Am. Chem. Soc.* **77**, 334 (1955).
39. Head-Gordon, M., *J. Phys. Chem.* **100**, 13213 (1996).
40. Hierlemann, M., A. Kersch, C. Werner, and H. Schäfer, *J. Electrochem. Soc.* **142**, 259 (1995).
41. Hitchmann, M. L., and K. F. Jensen, in *Chemical Vapor Deposition—Principles and Applications,* edited by M. L. Hitchmann and K. F. Jensen, Academic Press, London 1993.
42. Ho, P., M. E. Coltrin, and W. G. Breiland, *J. Phys. Chem.* **98**, 10138 (1994).
43. Ho, P., A. Balakrishna, J. M. Chacin, A. Thilderkvist, B. Haas, and P. B. Comita, *in Proceedings, Electrochemical Society, 98-23, Fundamental Gas-Phase and Surface Chemistry of Vapor-Phase Materials Synthesis,* edited by T. J. Mountziaris, M. D. Allendorf, K. F. Jensen, R. K. Ulrich, M. R. Zachariah, and M. Meyyappan, 1999, pp. 117–122.
44. Ho, P., M. E. Coltrin, J. S. Binkley, and C. F. Melius, *J. Phys. Chem.* **90**, 3399 (1986).
45. Ho, P., M. E. Coltrin, J. S. Binkley, and C. F. Melius, *J. Phys. Chem.* **89**, 4647 (1985).
46. Holleman, J., and J. F. Verweij, *J. Electrochem. Soc.* **140**, 2089 (1993).
47. Hwang J.-T., *Int. J. Chem. Kinet.* **15**, 959 (1983).
48. Jasinski, J. M., in *Frontiers of Organosilicon Chemistry*, edited by A. R. Bassindale and P. P. Gaspar, Royal Society of Chemistry, London 1991.
49. Jasinski, J. M., and S. M. Gates, *Acc. Chem Res.* **24**, 9 (1991).

50. Kee, R. J., F. M. Rupley, and J. A. Miller, Sandia National Laboratories Report SAND87-8215B, 1990.
51. Kleijn, C. R., Th. H. van der Meer, and C. J. Hoogendoor, *J. Electrochem. Soc.* **136**, 3423 (1989).
52. Koheler, B. G., C. H. Mak, D. A. Arthur, P. A. Coon, and S. M. George, *J. Chem. Phys.* **89**, 1709 (1988).
53. Kruppa, G. H., S. K. Shin, and J. L. Beauchamp, *J. Phys. Chem.* **94**, 327 (1990).
54. Kushner, M., *J. Appl. Phys.* **63**, 2532 (1988).
55. Kushner, M., *J. Appl. Phys.* **71**, 4173 (1992).
56. Lavrushenko, B. B., A. V. Baklanov, and V. P. Strunin, *Spectrochim. Acta* **46**, 479 (1990).
57. Madey, T. E., *Science* **234**, 316 (1986).
58. Masi, M., C. Cavallotti, and S. Carrà, *Chem. Eng. Sci.* **53**, 3875 (1998).
59. Masi, M., R. Zonca, and S. Carrà, *J. Electrochem. Soc.* **146**, 103 (1999).
60. Masi, M., G. Besana, L. Canzi, and S. Carrà, *Chem. Eng. Sci.* **49**, 669 (1994).
61. Mendicino, M. A., and E. G. Seebauer, *J. Electrochem. Soc.* **140**, 1786 (1993).
62. Moffat, H. K., K. F. Jensen, and R. W. Carr, *J. Phys. Chem.* **96**, 7683 (1992).
63. Moffat, H. K., K. F. Jensen, and R. W. Carr, *J. Phys. Chem.* **95**, 145 (1991).
64. Moffat, H. K., K. F. Jensen, and R. W. Carr, *J. Phys. Chem.* **96**, 7695 (1992).
65. Narusawa, U., *J. Electrochem. Soc.* **141**, 2072 (1994).
66. Narusawa, U., *J. Electrochem. Soc.* **141**, 2078 (1994).
67. Ohshita, Y., and Hosoi N., *J. Cryst. Growth* **131**, 495 (1993).
68. Ohshita, Y., Ishitani A., and Takada T., *J. Cryst. Growth* **108**, 499 (1991).
69. Perrin, J., Schmidt J. P. M., G. DeRosnay, B. Oevillan, and A. Lioret, *Chem. Phys.* **73**, 393 (1982).
70. Price, G., *Thermodynamics of Chemical Processes*, Oxford University Press, Oxford 1998.
71. Rabitz, H., Kramer M. A., and Dacol D., *Annu. Rev. Phys. Chem.* **34**, 419 (1983).
72. Raghavachari, K., and J. B. Anderson, *J. Phys. Chem.* **100**, 12960 (1996).
73. Sandler, S. I., *Chemical and Engineering Thermodynamics*, Wiley, New York 1989.
74. Sausa, R. C., and A. M. Ronn, *Chem. Phys.* **96**, 183 (1985).
75. Senkan, S. M., *Adv. Chem. Eng.* **18**, 95 (1992).
76. Sinniah, K., M. G. Sherman, L. B. Lewis, W. H. Weinberg, J. T. Yates, Jr., and K. C. Janda, *J. Chem. Phys.* **92**, 5700 (1990).
77. Steinfield, J. J., J. S. Francisco, and W. L. Hase, *Chemical Kinetics and Dynamics*, Prentice–Hall, Englewood Cliffs, NJ 1989.
78. Sterrat, D., C. L. Greenwood, E. M. Williams, C. A. Muryn, P. L. Wincott, G. Thornton, and E. Roman, *Surf. Sci.* **307–309**, 269 (1994).
79. Su, M.-D., and H. B. Schlegel, *J. Phys. Chem.* **97**, 8732 (1993).
80. Su, M.-D., and H. B. Schlegel, *J. Phys. Chem.* **97**, 9981 (1993).
81. Swihart, M. T., and R. W. Carr, *J. Phys. Chem. A* **102**, 1542 (1998).
82. Szabó, A., P. D. Farrall, and T. Engel, *Surf. Sci.* **312**, 284 (1994).
83. Tilden, J. W., V. Costanza, G. J. McRae, and J. H. Seinfield, in *Modeling Chemical Reacting Systems*, edited by K. H. Ebert, P. Deuflhard, and W. Jaeger, Springer-Verlag, New York 1981.
84. Violette, K. E., P. A. O'Neal, M. C. Öztürk, K. Christensen, and D. M. Maher, *J. Electrochem. Soc.* **143**, 3290 (1996).
85. Wittbrodt, J. M., and H. B. Schlegel, *Chem. Phys. Lett.* **265**, 527 (1997).
86. Yates, J. T. Jr., M. D. Alvey, M. J. Dresser, M. A. Henderson, M. K. Kiskinova, R. D. Ramsier, and A. Szabó, *Science* **255**, 1397 (1992).

CHAPTER 3

Epitaxial Growth Facilities, Equipment, and Supplies

V. Pozzetti [1]

CONSULTANT
BRUGHERIO, ITALY

ABBREVIATIONS USED	90
I. INTRODUCTION	90
II. THE EPITAXIAL REACTOR	91
1. *Radiant Heating*	94
2. *Induction Heating*	95
3. *Single-Wafer Reactors*	96
4. *Susceptors*	96
5. *Quartzware*	98
6. *Process Pressure: Atmospheric and Reduced Pressure*	98
III. FACILITIES	100
1. *General Facilities Requirements*	102
2. *Power Supply*	102
3. *Cooling Water*	103
4. *The Clean Room*	103
IV. PROCESS GAS AND DELIVERY	104
1. *Process Plumbing*	104
2. *Process Gases*	104
3. *Gas Purity*	105
4. *Gas Filtration*	107
5. *Delivery Systems*	108
V. EXHAUST TREATMENT	108
1. *Wet Scrubbers*	110
2. *Burn Systems*	110
3. *Thermal Decomposition*	110
VI. EQUIPMENT FOR POWER EPITAXY	111
1. *High-Thickness Deposition Capability*	111
2. *Susceptor Design and the Slip Problem*	113
3. *Run-to-Run Repeatability*	113
4. *Robust Equipment Requirements*	119
5. *High-Resistivity Control*	119
6. *Temperature—Uniformity and Profiling*	121
7. *Reaction Chamber Design*	121
8. *Gas Panel*	122
9. *Gas Injection and Flow Patterns*	123

[1] Present address: Via Volturno 80-P3, 20047 Brugherio (MI), Italy.

VII. ADVANCED APPLICATIONS	. . . .	123
VIII. CONCLUSION	. . . .	124
REFERENCES	. . . .	125

ABBREVIATIONS USED

COO	Cost of ownership
CPU	Central processing unit
CVD	Chemical vapor deposition
DCS	Dichlorosilane (SiH_2Cl_2)
epi	Epitaxy, epitaxial
HF	High frequency
IGBT	Insulated gate bipolar transistor
IR	Infrared
LF	Low frequency
LTO	Low-temperature oxidation
LVC	Liquid vaporizer controller
MBTC	Model-based temperature control
MFC	Mass flow controller
MOS	Metal–oxide–semiconductor
MPU	Microprocessor unit
PCW	Process cooling water
POU	Point of use
RF	Radio frequency
RP	Reduced pressure
RTP	Rapid thermal process
sccm	Standard cubic centimeters per minute
SCR	Silicon controlled rectifier
slm	Standard liters per minute
SMIF	Standard mechanical interface
TCS	Trichlorosilane ($SiHCl_3$)
TET	Silicon tetrachloride ($SiCl_4$)
TTV	Total thickness variation
UPS	Uninterruptible power supply

I. Introduction

The core of an epitaxial (epi) reactor is the reaction chamber, typically made of quartz. Inside the chamber a holder for the silicon substrates, typically made of graphite coated with silicon carbide, is heated up to 900–1250°C while gases flow inside. These gases contain a volatile silicon compound and some dopant compounds carried by a chemical reducing or inert main gas flow. The chemical reactions that occur in the chamber are detailed in other chapters in this book.

The walls of the reaction chamber must be kept cool, and the temperature set point and gradients value are critical. Depending on the chlorosilane species used in the process, either a temperature that is too high may cause silicon deposition or a temperature that is too low may deposit some species of chlorosilane polymers on the chamber walls. These problems may result in a lot of particles on the wafer surface or other process issues.

We explain here some aspects of the machines and, mostly, of what is needed around the reactor to obtain a good epitaxial layer at a reasonable cost. In fact, epitaxial reactors of various suppliers have various sizes, geometrical shapes, heating methods, and gas distributions, all to pursue the same goals. No reactor can run well and economically all the recipes and processes for which device designers and engineers are asking. Also, each model of reactor has its typical advantages and its relevant drawbacks. Individuals approaching epitaxy should bear this in mind, so that they do not waste time and effort trying to run all processes using a sole reactor model.

II. The Epitaxial Reactor

This section is a short overview of the epi reactor describing all types of reactors available on the market. Further aspects are detailed in Section VI.

Reactor geometry is generally classified into two shapes of wafer holders, called susceptors: the disk (pancake or single-wafer type) (Fig. 1) and the pyramid-shaped body (barrel type) (Fig. 2). In the former, the wafer lies horizontally on a disk which has one (single) or more receptacles, while in the latter a truncated pyramid accepts the wafers into cavities located on the faces, close to vertical; depending on the wafer size and bell-jar dimension, the susceptor may accept one or more rows in the pyramid faces. The disk susceptor geometry is a simpler way to robotize the loading/unloading of wafers. This has been tried unsuccessfully in the case of barrel reactors. On the other hand, accidental particles may fall down onto wafers lying horizontally on a pancake. This is less probable in the barrel geometry, where wafers are nearly vertical.

One important parameter is the way in which gases are brought onto the wafer surface: In an orthogonal way for reactors with the pancake configuration, which are therefore called "vertical reactors"; and parallel to the wafer surface in barrel reactors, which are hence called "horizontal reactors." Generally, in the latest models of single- or multiple-wafer reactors (ASM-Epsilon, LPE-PE3061; Applied Materials–Centura), the wafers are located horizontally and the gas flows nearly parallel to the wafer surface (Fig. 3). Because the direction of gas flow can determine the characteristics of epi growth (mainly the profile of epi thickness across the wafer), this must be seriously considered in the choice of a reactor model.

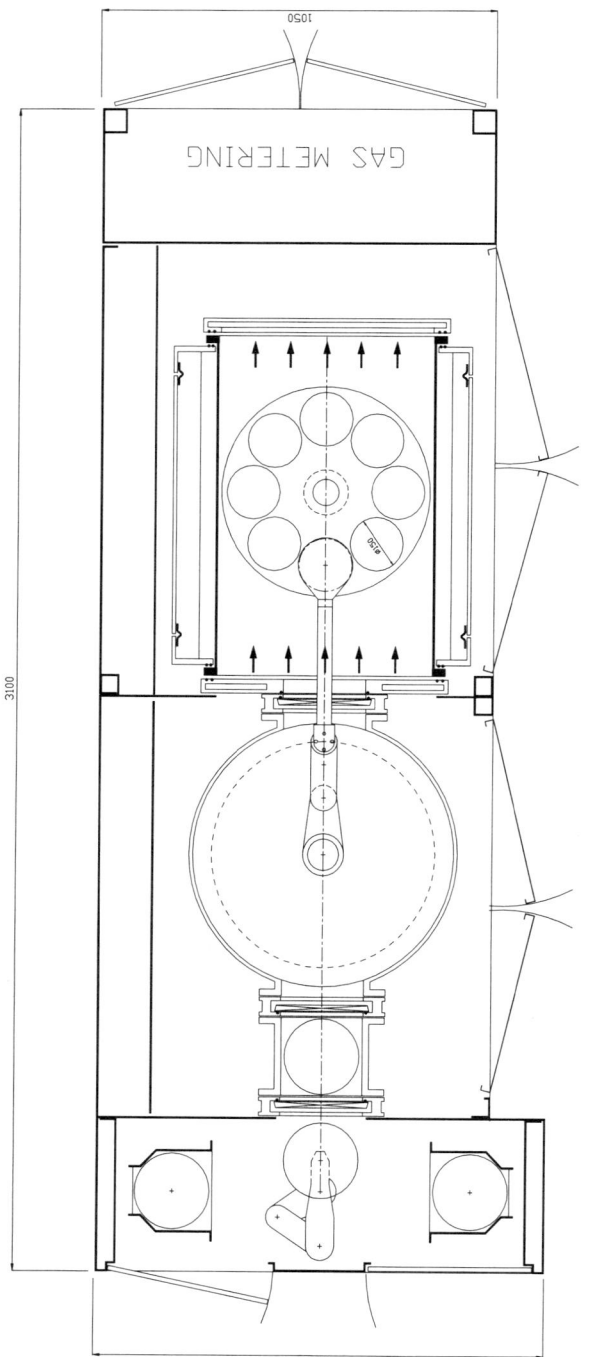

FIG. 1. Layout of the robotized small-batch reactor from LPE.

3 EPITAXIAL GROWTH FACILITIES, EQUIPMENT, AND SUPPLIES 93

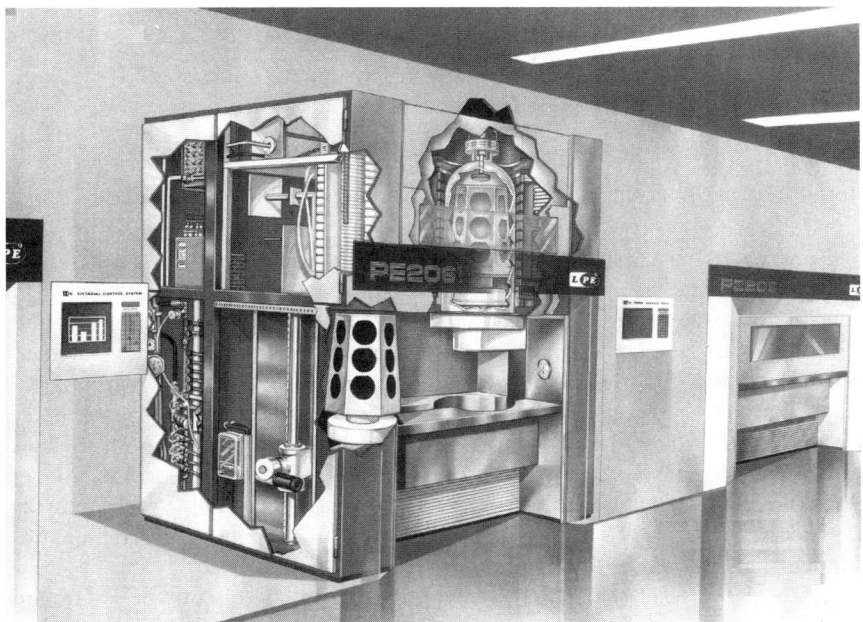

FIG. 2. An installed barrel reactor in a clean room.

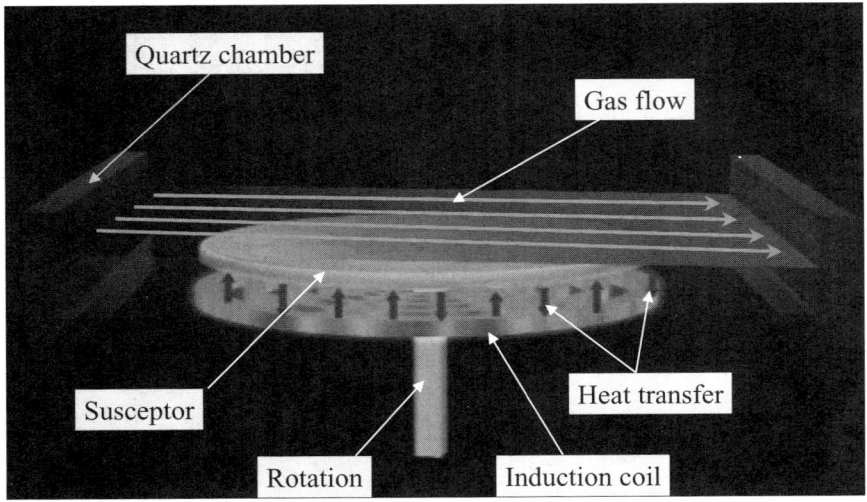

FIG. 3. Reaction chamber of a small-batch reactor.

Another important parameter is the way in which power is transferred to bring the wafers to the proper process temperature, as we will see.

Chapter 1 of this book gives an exhaustive overview of almost all reactor types and processes for CVD deposition including epi.

1. RADIANT HEATING

Some packs of high-energy IR lamps, peaking at a wavelength of some micrometers, which is the peak emission of the blackbody at 1000–1200°C, face closely toward the susceptor and transfer energy to the wafers by radiation. To prevent burning by the lamps, forced air-cooling is needed, and gold-plated reflectors are placed behind the packs to enhance radiation toward the wafers (Fig. 4a). This heating system is widely used in barrel, pancake, and single-wafer reactors, because it is easy to build and provides some advantages, mainly for low-thickness epi. In this system, the energy transfer occurs from the front of the wafer to the back (i.e., the front temperature is higher than the back temperature). This provides some useful effects: The wafer is bowed, its front protruding, and the minor stress on the front peel causes fewer slip-lines.

On the contrary, mass transfer of silicon occurs from the back of the wafer to the susceptor, causing the problem of wafer sticking when epi growth is medium high (i.e., more than 30–50 μm). In addition, *in situ* back-sealing of the wafer via mass transfer is not possible; therefore, expensive protection of the back of the wafer is needed to prevent outdoping (CVD deposition of SiO_2; see Chapter 1).

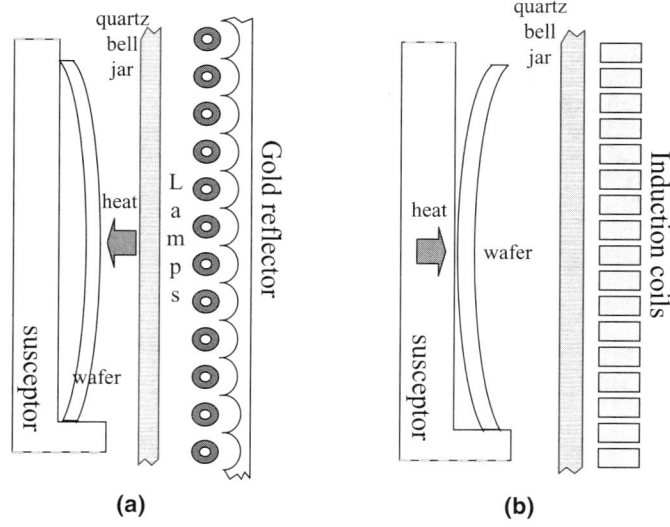

FIG. 4. Heating configurations: (a) lamps; (b) RF.

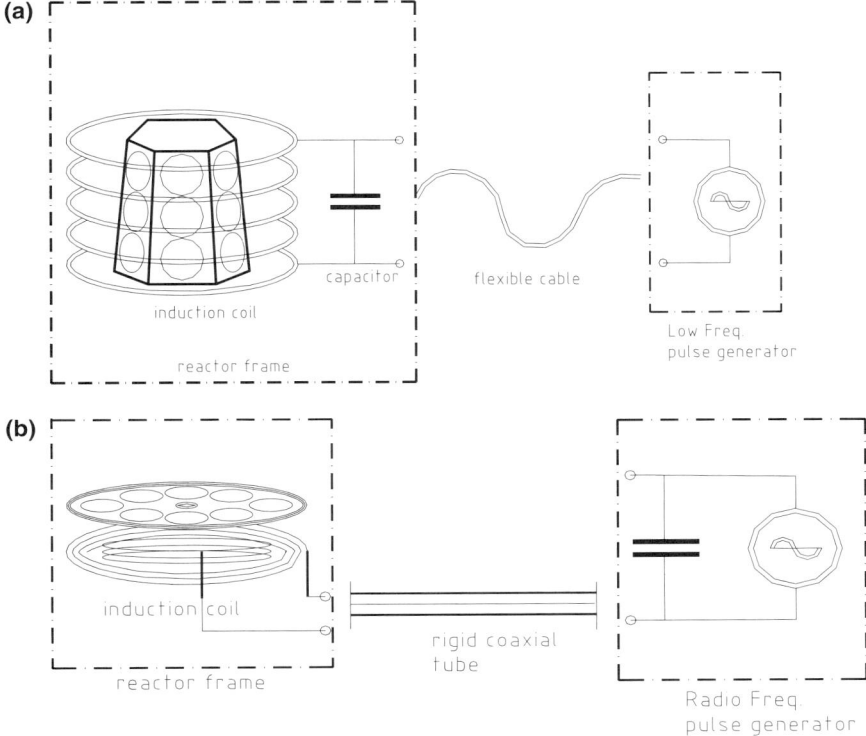

FIG. 5. (a) RF heating system for barrel-type reactors. (b) RF heating system for pancake-type reactors.

2. INDUCTION HEATING

This was the main heating method for cold-wall reactors before IR lamps were developed and it is still the best method for high-power density transfer and for high-temperature processes (Fig. 4b). This approach has advantages and defects opposite those of lamp heating (Figs. 5a and 5b).

- It is suitable for high-thickness epi.
- Energy transfer occurs from the back to the front of wafers (i.e., the front temperature is lower than the back temperature).
- The wafer is bowed with its edges protruding.
- It is more prone to slip-lines.
- *In situ* back-sealing of the wafer via mass transfer is possible, eliminating backside SiO_2 protection to prevent outdoping.

To minimize the tendency toward more frequent slip-line occurrence, induction coils are built and treated to give maximum reflection back to the susceptor,

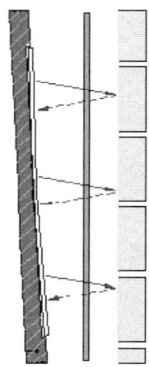 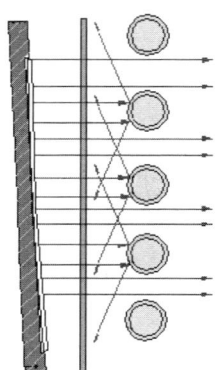

Reflecting coil 96% overlapping:

Power recovery > 85%

Δt<10°C across wafer

Not Reflecting coil small overlapping:

Power recovery < 20%

Δt>30°C across wafer

FIG. 6. Comparison between nonreflective and reflective induction coils.

thus minimizing the temperature gradient throughout the wafer. This also helps reduce power consumption significantly compared to that of a nonreflective system (Fig. 6).

3. SINGLE-WAFER REACTORS

In conventional pancake or barrel reactors, thickness uniformity may barely achieve ±2% as a standard and repeatable value. (See Chapter 7 for details on measurement techniques.) To meet increasing quality requirements, a series of reactors has been developed handling single- or multiple-wafer batches, with greater control of gas flow, which is parallel to the wafer surface in a small reaction chamber: the growth rate ranges between 2 and 5 μm/min, with a good surface quality. In these reactors, the typical uniformity is ±0.5% or better, but in general the throughput is much lower than that of conventional reactors (i.e., the wafer cost is higher, so designers should take these figures into account during the project). Some models use IR lamps, others use induction heating, and some can also work at a reduced pressure (RP).

4. SUSCEPTORS

As stated previously, susceptor geometry varies greatly, depending on the model. However, some aspects are common. The material for susceptors is sintered anisotropic graphite (because of its thermal and electrical conductivity), highly purified, with very close control of grain size to avoid tooling defects due to different grain size. The graphite block, which will become a susceptor, requires a high precision tooling, mainly in recesses for wafer lodging. The surface of receptacles may have a large radius cup tooling (e.g., for 150 mm wafer the reference radius is roughly 25 m) to accompany the wafer bowing due to the thermal gradient across wafer thickness, mainly in induction heated systems, so that the wafer back is in contact with almost the whole recess surface (see Fig. 11). Without this, when only a few points of the pocket surface are in contact with the back of the wafer, they became "hot points" because of the temperature difference, which may increase the probability of bulk defects, such as slip lines.

In the case of induction heating, the electrical conductivity of graphite must be adequate, because the heat is generated by the Joule effect of parasitic currents, and, more important, the uniformity of conductivity should be perfect to avoid hot and cool points on the surface and under the wafer's back.

Graphite cannot be used without protection, for several reasons.

1. Despite purification, it always releases contaminants, metals, and doping elements.
2. Being a porous material, it absorbs dopants from reaction gases and later frees them.
3. Carbon may react with H_2 and HCl gases resulting in the generation of CH_4 and other organic compounds that disturb the epi layer quality.

Thus, the surface of susceptors, after tooling, must be further purified and coated with a layer that is impermeable to reaction gases, typically a layer of silicon carbide (SiC) 50–100 μm thick. Because of the mismatch between the thermal expansion coefficient of SiC and that of graphite, the SiC layer must not be thicker than 100 μm or it will crack. Some special graphite has been developed to lessen this problem. Some major graphite parameters are listed in Table I.

TABLE I

GRAPHITE SPECIFICATIONS[a]

Parameter	Units	Inductive heating
Density	g/cm^3	>1.8
Hardness	shore	>55
Specific resistance	10^{-6} $\Omega \cdot$ m	>12
Porosity	%	<13
Thermal conductivity	WK^{-1} m^{-1}	>80
Thermal expansion coefficient	10^{-6} K^{-1}	>4
Maximum grain size	μm	<40
Ash	ppm	<10

[a]Source: LPE, WQ4034 "graphite specifications" (June 1988).

5. QUARTZWARE

As stated, the reaction chamber is made of pure, clear, fused quartz (SiO_2), as well as the majority of the parts in contact with reaction gases (baffles, shields, etc.). When heat or radiation transmission is not wanted, some sintered, opaque quartz parts may be used and welded to transparent quartz (Fig. 7a).

Parts are manufactured starting from tubes or sheets, softened with flames and shaped or welded in the proper way. To release stresses and avoid breakage, a low-gradient cooling of parts is used.

Oxygen and propane/butane are used for high-temperature flame torches, therefore, this oxidizing reaction produces a lot of water as a reaction product [i.e., oxidrilic groups (OH^-) that are absorbed on the surface of the quartz piece under construction]. Unfortunately, OH^- groups have a peak absorption of radiation in the range of 2.5–3 μm, which corresponds to a peak of emission around the 1000°C operating temperature used for most epi processes (Fig. 7b). Table II shows the transmittance standard for type 214 fused quartz. The optical quality of the quartz wall, mainly for IR lamp heating, may reduce the efficiency of heat transfer, therefore, although more expensive (Fig. 8), treated quartz with low (OH^-) content is used.

Quartz parts must be kept cool, at a temperature below 1250–1350°C; otherwise, the amorphous, vitreous state of quartz starts modifying into crystobalite (a crystal modification of silica), which is an opaque and more fragile state, and the piece can became unusable.

6. PROCESS PRESSURE: ATMOSPHERIC AND REDUCED PRESSURE

Most epi processes are run near atmospheric pressure [more precisely, slightly higher (+50 to +100 mbar) or slightly lower (−50 to −100 mbar) than atmospheric pressure], depending on the reactor and exhaust treatment system (scrubber).

In several other cases (Reduced Pressure: RP), the value of the process pressure is set at about 50–80 Torr (or is supposed to be). In fact, RP reactors require many additional and expensive apparatus to be connected to the reactor's exhaust, typically a vacuum pump (suitable for acid pumping), vacuum valves, and additional process and safety controls. Furthermore, the exhaust gases contain silica, HCl, and unreacted chlorosilanes that polymerize when the exhaust gas temperature decreases, forming a very reactive rubber-like coating in the tubing, valves, and pump. This requires frequent dismounting and acid-cleaning of parts, which is expensive, time-consuming, and sometimes risky.

RP is typically used for some devices with mixed buried layers where tight control of pattern shift and washout is needed (see Chapter 9) and autodoping must be minimized. In fact, silane and DCS, which are suitable for these processes because they are helpful in reducing washout, but they can hardly be used in atmospheric processes because problems in cooling the bell jar.

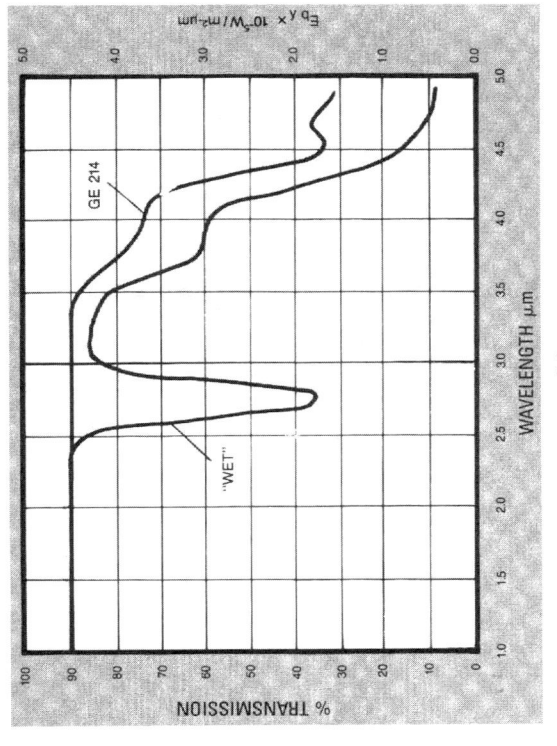

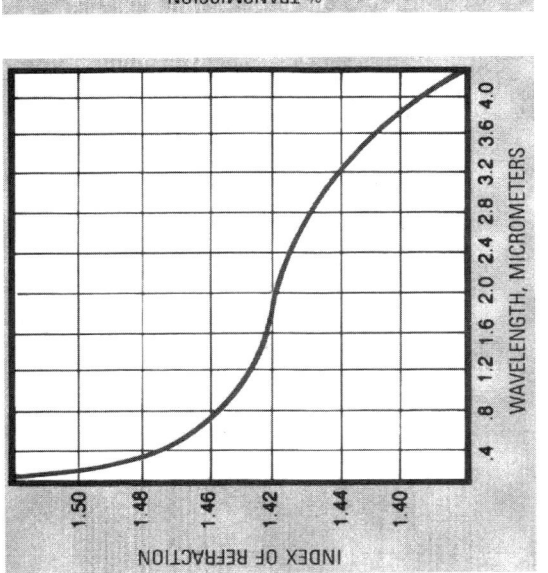

FIG. 7. (a) Index of refraction of fused quartz (Rodney, 1954). (b) Presence of OH^- shown as a dip in the transmission coefficient (General Electric "Fused Quartz Products" brochure).

TABLE II

TYPE 214 FUSED QUARTZ, TRANSMITTANCE STANDARD[a]

Wavelength (μm)	Avg. transmittance (%)	Avg. absorption coefficient (cm^{-1})
0.160	4.6	29.57
0.162	5.8	27.33
0.164	7.4	24.89
0.166	8.4	23.64
0.168	10.9	21.04
0.170	18.5	15.75
0.175	43.6	7.22
0.180	60.4	4.01
0.185	66.1	3.12
0.190	70.4	2.52
0.195	71.3	2.41
0.200	73.4	2.14
0.205	76.1	1.80
0.210	79.4	1.39
0.220	85.3	0.69
0.230	87.3	0.49
0.240	86.5	0.60
0.245	86.6	0.57
0.250	87.7	0.48
0.260	89.5	0.28
0.270	90.2	0.21
0.280	90.7	0.17
0.290	90.9	0.16
0.300	91.1	0.15
0.350	91.7	0.11
0.450	92.2	0.09
0.550	92.5	0.07
0.650	92.7	0.06
0.750	92.9	0.04

[a]Source: General Electric "Fused Quartz Products" brochure.

Thus, a RP process should be used only when strictly necessary, as it implies a higher cost of ownership (COO) due to the higher cost of silicon compounds, RP investment, and system maintenance.

III. Facilities

In semiconductor manufacturing operation, proper facilitation of equipment is critical for process performance. Epitaxial silicon growth is no different. Facilities significantly affect epi wafer cost and downtime. Moreover, contamination resulting from improper facilitation can result in major process issues and system downtime. Manufacturers of epi equipment typically provide both specific

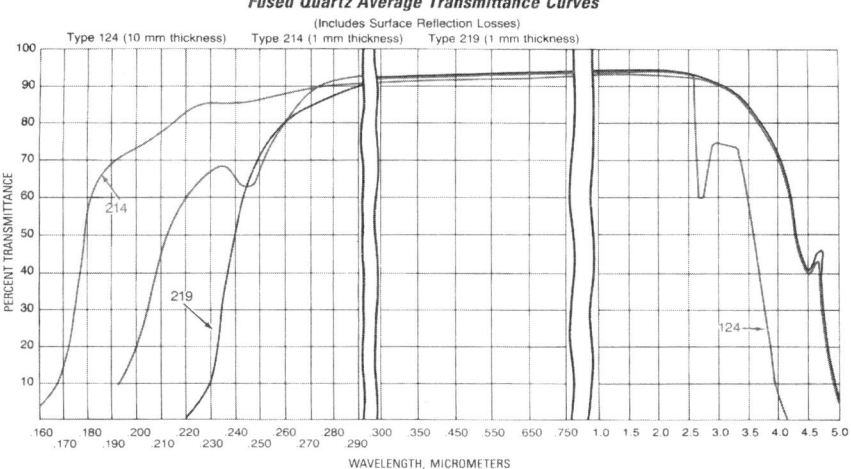

FIG. 8. Fused quartz transmission curves for different commercial products (General Electric "Fused Quartz Products" brochure).

and general information on facilitization of their equipment. In addition to practical concerns such as proper power, gas delivery systems, and effluent treatment, it is important to consider issues such as allowing adequate service space. With continuing focus on improving particle performance and reducing contamination, increasingly stringent requirements are being placed on system facilities. Continuing changes in regulations and industry guidelines mean that more and more

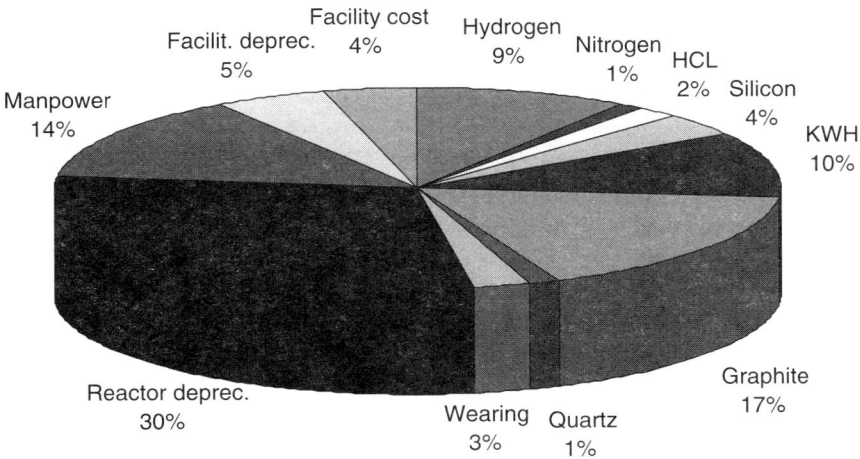

FIG. 9. Cost breakdown for an epitaxial wafer. Facilit. deprec., facilities depreciation.

attention is focused on reducing power, water, and effluents. These changes require the epi facility to be scrutinized constantly for necessary improvements and changes to meet upcoming requirements. For the purposes of this chapter, epi facilities can be described as the space, utilities, and gases used by epi processing equipment.

1. GENERAL FACILITIES REQUIREMENTS

All epi equipment, regardless of the manufacturer, has some overall requirements that must be addressed. They can be summarized as follows.

1. The space used by the equipment itself, with allowable service space.
2. Process gas delivery systems including process plumbing and gas delivery systems.
3. A clean room or space where wafers are handled, measured, and placed in the epi equipment.
4. An effluent treatment system for handling exhaust from the epi process and service space for parts cleaning, storage, and maintenance.

Many variations in the size and quality of these facilities depend on the legal requirements of the location, clean-room philosophy, device generation, and equipment type.

2. POWER SUPPLY

The energy spent in the epi process is one of the most important parameters influencing the cost of the epi process (see Fig. 9). Some electrical power values for epi reactors are listed in Table III. A remarkable amount of power is needed

TABLE III

SOME APPROXIMATE ELECTRICAL POWER CONSUMPTIONS AND YIELDS IN VARIOUS SYSTEMS AT A TEMPERATURE OF 1100°C UNDER STEADY-STATE CONDITIONS[a]

	KVA	
Model	Transmitted	Adsorbed
Induction		
Low frequency—barrel	75	90
High frequency—pancake	100	280
IR heating—barrel	90	105

[a]Data are not homogeneous because of different susceptor dimensions and wafer loads.

during the epi process. Moreover, there is a large difference in electrical yield among different heating systems, which also affects the COO.

3. COOLING WATER

Nearly 80–90% of the power supplied to the reactor must be removed through a water-cooling system, the balance being through reaction gases and air exhaust, resulting in cooling being a remarkable cost too.

Usually, water should be soft treated (softened or D.I. in a closed loop) to prevent scaling inside hot parts. In the case of old-generation RF heating, due to the high voltage applied, high-resistivity water (>5–10 M$\Omega \cdot$ cm) is required to ensure electrical insulation. The water temperature should be not too low, about 18–22°C, to avoid moisture condensation on cooled parts, which can increase corrosion or cause electrical short-circuits. Condensation also depends on the relative humidity of the environment (requirements are discussed in the following section). Many parts of the reactor are exposed to high-temperature radiation from hot pieces, and they are water-cooled. In the case of water supply interruption an emergency water supply system is mandatory to avoid damage or burning of those parts. A well-designed process cooling water PCW delivery system takes care of these aspects of facilities.

4. THE CLEAN ROOM

Particles present on the wafer surface before epi deposition will be transformed into major or minor defects on the epi layer surface or in the bulk. Utmost care should be taken to protect wafers from this detrimental effect. All reactor models have a Class 10 laminar flow environment in the wafer loading area, either manual or robotized. Wafer handling outside the loading area should have a similar level of protection. A clean room of at least Class 100 is mandatory; a Class 10 or a SMIF system inside the reactor loading area is better. It must be remembered that the reactor itself can be a source of particles, especially during the bell-jar opening to the atmosphere, so that laminar flow loading should be carefully designed. A way to avoid contact between the outside air and the inside of the reaction chamber is to isolate the loading area by filling it with nitrogen or by evacuating the air using some dedicated pumps. Gray areas and other places around the reactor should be kept at a comparable cleanliness level (i.e., Class 1,000–10,000).

a. *Temperature and Humidity Control*

Epi process by itself is not very sensitive to the environmental conditions, so that a controlled clean room condition is quite sufficient, but for other kind of problems (see next paragraph), a low RH value is recommended, about 40–45%.

In some cases, when RH rises over 45%, for lack of control or if wafers remain for a long time in the room, some surface problems may also arise ("orange peel" defect).

IV. Process Gas and Delivery

1. Process Plumbing

Stainless steel is often used in epi systems for gas distribution and to provide the support or sealing surfaces for process chambers. Much work has been published on the choice of materials and the importance of purging and desorbing moisture in gas distribution systems (Sugiyama *et al.,* 1989). Because of the corrosive nature of the gases used for epi process, the piping should be at least 316L stainless steel (low carbon). Sometimes Hastelloy C22 is used for HCl lines.

During system maintenance activities, such as chamber cleaning, system components are exposed to air. HCl will condense and corrosion will begin in air containing 1000 ppm HCl at 20°C and 50% relative humidity (Henderson, 1993). For this reason, it is essential to reduce the presence of HCl prior to any maintenance activity in which the chamber will be exposed to air and to control the humidity of an epi system environment.

Concerning the internal surface of stainless-steel piping, it may happen that in a part of the system where gas is flowing (elbows, valves seats, etc.), the velocity may achieve very high values, and it may have some abrasion effects and may transport metal particles to the reaction chamber, contaminating the substrates. The abrasion also depends on the roughness of the inside tube wall. Only electropolished parts are allowed to be in contact with gases. The roughness (Ra) of the inner walls of the tubing ranges from 4 to 10 μin. on average. Also, electropolishing depletes most of the iron from the surface of stainless-steel alloy, leaving exposed mainly Cr and Ni compounds, which are less susceptible to HCl corrosion.

2. Process Gases

Different gases are used in epi reactors. Table IV shows the different sources of gases: in bottles, from liquid, or even in trailers [an example is the unique HCl delivery system from Air Products (www.airproducts.com)], but for all of them the highest purity is mandatory for a good epi layer.

a. Silicon Sources

Silicon sources for epi are chlorosilanes, chemical compounds containing 2 (SiH_2Cl_2 - Dichlorosilane, or DCS) or 3 ($SiHCl_3$ - Trichlorosilane or TCS) or 4 ($SiCl_4$ - Silicon tetrachloride or TET) chlorine atoms versus one silicon atom.

3 EPITAXIAL GROWTH FACILITIES, EQUIPMENT, AND SUPPLIES

TABLE IV
TYPICAL EPI PROCESS GASES

Type	Gas	Source	Comment
Purge	Nitrogen (N_2)	House	From liquid source
Carrier	Hydrogen (H_2)	House	From liquid boil-off or purification
Si source	$SiHCl_3$ (trichlorosilane)	Tank (liquid)	Atmospheric
	SiH_2Cl_2 (dichlorosilane)	Bottle (gas)	Reduced pressure
	SiH_4 (silane)	Bottle (gas)	Reduced pressure
Dopants	AsH_3 (arsine)	Bottle (gas)	Typically 20 to 100 ppm
(balance, H_2)	PH_3 (phosphine)	Bottle	Typically 20 to 100 ppm
	B_2H_6 (diborane)	Bottle	Typically 20 to 100 ppm
Chamber etch	HCl	Bottle/house	Requires high flows (10 to 70 slm)

They are low-boiling liquids at room temperature or/and under low pressure. Their physical properties are summarized in Table VII. Depending from their properties, the method of delivery to the reaction chamber is different. Is worth to notice that SiH_4 is not commonly used in epi reactors.

- $SiHCl_3$ - $SiCl_4$

TCS is preferred to TET, because of lower chlorine contents (less wash-out), but for high temperature process TET is easier to handle and it gives less deposition problem on bell-jar walls. More details on both gasses can be found in Paragraph 21-b.

- SiH_2Cl_2

Dichlorosilane is supplied in cylinders as a liquid, slightly pressurized. It is typically used for RP (reduced pressure) processing. The state of the SiH_2Cl_2 is determined by its temperature and pressure as shown in Fig. 10. In the case of DCS the vapor is directly delivered to the MFC and no bubbler is used.

When pressure in the delivery lines increases over 6 psi, or if a change in temperature occurs, gaseous DCS may return partially into liquid, with problems of flow control. For this, cool zones (air-conditioning ducts, uncontrolled environment rooms) must be avoided along the tube path.

Heating of DCS pipelines is not recommended, it is better to coat with an insulating media and purge the line when it is idle for more than 24 hours, remembering to keep N_2 purging pressure below 6 psi.

3. GAS PURITY

Different processes are more or less sensitive to contaminants present in gases, depending on the process temperature. The industrial trend is to get

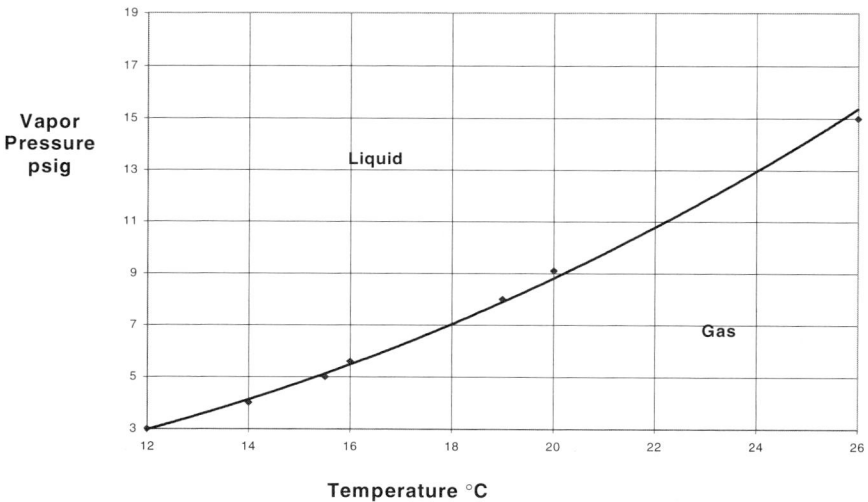

FIG. 10. SiH_2Cl_2 vapor pressure vs temperature.

the best available purity. Following are some values presented in the typical xNy notation and as percentages.

1. H_2: $\geq$6N0, 99.9999%; $H_2O + O_2$ content, <100 ppb.
2. N_2: $\geq$6N0, 99.9999%; O_2 content, <100 ppb.
3. HCl: $\geq$ 4N5, 99.995%.
4. Dopants: 20–100 ppm; balance, H_2, same as above.
5. Chlorosilanes (TET, TCS, DCS): 3N0; intrinsic value, $\geq$3000 $\Omega \cdot$ cm.

Purity is generally the task of gas suppliers, but sometimes further purification at the point of use (POU) may be suitable. Industry can supply a variety of POU purifying cartridges for many gases, but because epi requires a high flow of the main gases, H_2, N_2, and HCl, only bulk purification is economical (www.saesgetters.com). An interesting application is the use of "getters," which are activated at high temperatures to purify gases. The actual capability of a getter-based purifier to clean a gas is to <1 ppt (see Table V).

H_2 and N_2 from liquid sources are intrinsically purer than from cylinders, at least for contaminants that affect the epi process (e.g., H_2O, CH_4, and carbon compounds). But in these cases also, further purification is recommended, for instance, to remove N_2/O_2 solute into liquid H_2. When the H_2 supply is from compressed gas, deoxo and dehydration processes or the use of palladium membrane diffusion is mandatory to achieve the above-stated purity. Other purification media, such as absorption/desorption of gas on hydride compounds, are also suitable, though quite expensive. High-purity HCl as well as chlorosilanes in bulk or in cylinders is

TABLE V

IMPURITY VALUES AFTER PURIFICATION WITH GETTERS[a]

Ar/He		N_2		H_2	
Impurity	Guarantee (ppb)	Impurity	Guarantee (ppb)	Impurity	Guarantee (ppb)
O_2	<1	O_2	<1	O_2	<1
H_2O	<1	H_2O	<1	H_2O	<1
CO	<1	CO	<1	CO	<1
CO_2	<1	CO_2	<1	CO_2	<1
N_2	<1	H_2	<1	N_2	<1
H_2	<1	CH_4	<1		
CH_4	<1				

[a]Courtesy of Saes-Getters (www.saesgetters.com)

very difficult to obtain. First, they are liquified gases and purification is easier only when they have been gasified; second, they are highly corrosive compounds when in contact with moisture.

A comment should be made on the size of high-purity gas containers. Generally, when possible, large bulk containers are preferable, because they have a better purity on delivery, even starting from the same purity at supply. In fact, the ratio of "liquid volume to container wetted surface" is favorable, and the contaminants released by the container surface have been diluted into a larger volume of liquid, thus the contaminant concentration is lower. Moreover, a large container requires fewer connecting operations than many smaller cylinders, and it is obvious that connection is one of the main sources of contamination, even when performed in the best and automatic way.

4. GAS FILTRATION

To prevent particles from reaching the reaction chamber, filtration is mandatory at the starting point of bulk delivery and at the inlet into the reactor. Typically, gas filters give a near-absolute retention value (99.99999%) for a particle size of 0.02 μm or larger. Filtration media can be polymeric membranes in PTFE (Teflon) or similar materials, with the proper porosity and convoluted filtration areas to allow a high gas flow in a relatively small case. In place of membranes, sintered metal grain (Ni, stainless-steel) cartridges are also used.

Both filtering media give good and reliable results, but it seems that membranes have a shorter lifetime and the "memory" of previous gases. It may seem a good solution to place a filter just before the injection of gases into the reaction chamber, but the high gas flow requires very large filters, which means an unbearably large "memory" when the process requires abrupt changes in dopants.

TABLE VI

HCl THERMODYNAMIC VALUES

Density (gas)	1.543 g/sccm
C_p	0.912 kcal/kg
Heat of evaporation	443 kJ/kg
	105.9 kcal/kg

5. DELIVERY SYSTEMS

Delivery of gases is usually made from factory lines or, in the case of dopant gases, by individual cylinders placed in a gas cabinet and handled by automatic change operations. Nevertheless, a note on HCl delivery is worthwhile. HCl, as supplied in cylinders or, recently, in bulk, is a liquefied gas at about 43-bar pressure at room temperature and is used in epi for two main purposes. The first, called "wafer etch," is a process to clean the wafer surface before epi growth, removing the native silicon oxide layer (a few atomic layers) resulting from previous wet processes or simply from exposure to air. A few standard liters per minute of HCl are needed, and no problems arise from the use of gas, except that it must be of the highest purity. On the contrary, in the second use, a high HCl flow is needed to clean out the reaction chamber, susceptor, and quartz components, removing silicon deposited in previous runs.

To achieve a reasonable etching rate ($>6-8$ μm/min) a high concentration of HCl is required, and, depending on the chamber volume, the flow required ranges from 30 to 80 slm. At that flow, the adiabatic evaporation of liquid HCl into the pressure regulators may drop the gas temperature to far below 0°C, generating chemical compounds that are strongly abrasive; in other words, the gas corrodes the metal of which pressure regulators are made and it transports the metal particles into the reaction chamber. To avoid this, a preheating device placed before the pressure regulator can be used along with two or more steps of pressure reduction (from 43 to 4 bars), dividing the heat drops into different regulators.

Some HCl thermodynamic values are listed in Table VI.

V. Exhaust Treatment

Because of differences in country, state, and local codes as well as individual corporate requirements, it is difficult to describe all the requirements and issues that are associated with all epi installations. During the epi process, only a small portion of the entering gas is deposited onto the wafer and the chamber walls (typically between 10 and 30%). The remainder, along with the newly formed by-products, requires some kind of treatment to prevent the release of any hazardous element into the atmosphere. National and local codes will define the treatment requirements. In the United States, abatement equipment must meet the regulations of the

Occupational Safety and Health Act (OSHA), Toxic Gas Ordinance, Uniform Fire Code, and Clean Air Act. State Regional Air Districts often impose additional regulatory requirements (Hayes and Woods, 1996).

The *Book of Semi Standards 1995* details the methods for treating exhaust from epi systems. The standard that categorizes epi exhaust is Semi F5-90. Epi is categorized as producing Category 3 pollutants. The methods listed as acceptable for treatment are

(a) incineration,
(b) incineration plus filtration,
(c) incineration plus scrubbing, and
(d) scrubbing with an alkaline or oxidizing solution.

Section 7.1.2 of the Semi guideline is a good summary of the issues faced when choosing the proper treatment system. It states, "The appropriate control technology for this category is ever changing. The type of gas emitted has total bearing on the type of control which can be used. Some controls treat multiple gases, while others are 'gas specific.' Different control techniques may or may not meet government criteria." Additionally, in Section 7.2.4, it is stated that "design and selection of gas effluent treatment systems must include knowledge of the flows, properties, and concentrations of the gases to be treated. Each gas must be considered individually as well as part of a multitude of gases in the same stream." Finally, an additional complexity is noted in Section 7.6.2: "In most cases, the gases in the exhaust stream are treatable by a single technique and will require the use of multiple treatment methods."

Wet scrubber technology has been the dominant technology for epi exhaust treatment. Today the majority of epi systems (both single wafer and batch systems) utilize a wet-scrubbed exhaust. Wet scrubbers for epi systems are probably the best-understood choice for epi exhaust. Wet scrubbers used in industry may be individual per system, or exhaust from many systems may be combined to enter one large house scrubber. Applied Materials has always recommended that customers dedicate one scrubber per epi chamber to provide the highest level of safety and cause the fewest facilities-related issues. Despite this recommendation, many customers still facilitize all epi reactors to a common scrubber. This can be successful with proper knowledge, safety precautions, and planning, but it is more difficult and may involve multiple equipment vendors.

Some local codes now require treatment in addition to wet scrubbing for epi systems flowing toxic dopants (e.g., arsine) even at very dilute concentrations. Fewer solutions are available when this requirement is added; these solutions and the companies providing them are new to the market. For this reason, additional difficulties in facilitization are faced when the systems are utilized.

Some effluent treatment companies are small and often have difficulties providing service worldwide; making uniform recommendations throughout the globe is difficult. These companies tend to merge, form, and cease to exist at a moment's notice. This means that what was a good company providing good products and services one month may not be alive the next month. An example of this is that

Delatech, the maker of a chemical decomposition/oxidation treatment system, was formed as a combination of Airproteck and Innovative Engineering (according to Delatech's "Exhaust Gas Conditioning" brochure). Guardian, a burning system, recently became a part of EcoSys, a subsidiary of Advanced Technology Materials, Inc., which now includes another wet-scrubber selection, Vector. Vendors and solutions seem to change frequently.

New safety codes often require treatment for dopants, not only dilution, so that some other type of treatment system is used in conjunction with wet scrubbers.

1. WET SCRUBBERS

Wet scrubbers generally have a good capability of treating HCl and chlorosilanes; H_2 passes through and is diluted with air, far below the explosure concentration value. These systems have a very low capability for treating dopants, which are simply diluted into the air flow. Because SiO_2 is a by-product of chlorosilanes decomposition, the scrubber requires a periodic, maybe frequent, maintenance and cleaning. If silane is to be handled by a wet scrubber, care should be taken to provide adequate dilution with N_2, not air. Some key considerations for the choice of a proper scrubber are the total gas flow, including purges, and the large amount of HCl flowed during chamber etch. Another sensitive issue is the velocity of response to abrupt changes in flow and in composition of gases, i.e., the stability of the pressure faced to the reaction chamber, fast pressure fluctuations may trouble process and give a poor surface quality.

2. BURN SYSTEMS

Burn systems perform the burning of H_2, at high temperature, oxidizing dopants to washable compounds. It is advisable to use a flashback arrestor with burn systems. Because burning systems do not treat HCl, a wet scrubber also must be used. An example of a burning system is the Guardian system. There may be issues with atmospheric process and some burning systems because of the large negative draw required. All entry ports should be purged with N_2 to avoid clogging. This includes ports which are used only in fault situations, such as the system/chamber over pressure lines. Weekly preventive maintenance is required to ensure proper operation of a burning system. This type of system is effective for treating dopants.

3. THERMAL DECOMPOSITION

These systems utilize a burn followed by thermal decomposition and must include a water scrubber for HCl treatment. An example of this type of system is from Delatech. The critical issue with this type of system is the amount of total flow that the unit can handle.

It is important to know from the epi equipment vendor the total gas flows which must be treated (not just the typical), otherwise errors in effluent treatment sizing may occur, nullifying the treatment; a common mistake is to forget the excess H_2 diluent's flow. Thermal decomposition systems are effective for treating dopants. The system design including a water scrubber is a nice package for handling epi waste products.

VI. Equipment for Power Epitaxy

When an epi reactor is to be designed, there are some choices to make. There are different kinds of wafer foundries with different needs. We may, however, summarize three main categories.

(1) *Commercial epi services, i.e., epi wafer commercial suppliers.* The wafer manufacturer, without an internal epi facility, addresses himself to external commercial epi suppliers. Usally the microchip is differentiated by circuitry, not by epi. So, epi process is well-defined and standardized, allows high-volume production, and little or no flexibility is required in epi process tuning (CMOS, memories and microprocessor).

(2) *Single-product or single-technology factories.* Here the needs are most often similar to those of case (1), and often the epi processes are to be performed "in line" on semifinished wafers rather than on bare substrates.

(3) *Multiproduct and multitechnology factories with a centralized, internal epi service (in-house epi).* This is the typical case where epi process is highly involved in the device functioning and designing (power, discrete devices, IGBT, etc); often R&D activities may be required too. In the latter case the highest processing flexibility and ease of process recipe conversion is one of the main issues. This extended suitability for many different processes allows the best saturation of the machines, greatly reducing the cost of processing and COO. To extend the reactor capability to high-thickness and high-resistivity epi requires some additional design considerations.

1. High-Thickness Deposition Capability

a. Heating Techniques

A high growth speed value (1.5–3 μm/min) must be taken into account in the case of high epi thickness layers (50–200 μm), for obvious throughput reasons, as well as the duration of the process (over 90 min of deposition). In addition, high process temperatures (1100–1200°C or more) are required to improve the kinetics of chemical reactions and to obtain better stacking and readjustment of the single-crystal lattice. Summarizing, a large amount of power must be transferred into the reaction chamber for a long time to keep the susceptor and wafers at the proper temperature. The most common heating techniques are radiation and induction heating.

Radiant heating is described in Section II. There are geometrical limitations on the maximum number of lamps that can be packed around the reaction chamber, and there are some cooling and lifetime difficulties too.

The induction heating system does not have these limitations and the electrical yield is generally higher, more so when utilizing solid-state pulse generators, preferably at low frequencies. Moreover, the inverted thermal gradient resulting from induction heating avoids the wafer's sticking to the susceptor and allows for mass-transfer *in situ* back-sealing, which can be the only viable protection against autodoping when a very high resistivity on highly doped substrates is required.

The induction heating principle consists of a pulse generator and an oscillating circuit composed of a capacitor and a coil in parallel as seen previously in Fig. 5. The oscillating electromagnetic field couples the susceptor, which is made of a conductive material, SiC-coated graphite, inducing parasitic currents, which, by the Joule effect, heat the body of the susceptor. The pulse generator may operate at a high frequency (radio frequency; RF), about 100–220 kHz, or at a low frequency (LF), about 4–20 kHz. In the former case, a high voltage, in the range of 10–15 kV, is applied to the oscillating circuit and the pulse is provided by transistors or a triode, while in the latter the voltage across the induction coil is in the range of 200–300 V and the pulse is supplied by solid-state power electronic devices such as transistors, SCR, and IGBT.

The RF system is more difficult to handle, because of the high voltages and frequencies. Power transmission between the pulse generator and the coil requires special connecting lines, made of rigid coaxial tubes. Furthermore, the triode has a limited life, then decays and requires replacement or regeneration. In the LF system the coil and capacitors should be very close. Inside the reactor frame, the connection is made with flexible coaxial cables and the relatively low voltage applied to the coil allows the design of very compact induction coils, keeping the distance between wounds in the range of a few millimeters without risk of sparks and discharges.

Besides the component characteristics, the major difference between the two induction heating frequencies consists in the penetration of the electromagnetic field into the susceptor, which is dependent on the frequency (the skin effect). In the RF system the susceptor volume involved in the parasitic current is very shallow; the depth is some tenths of millimeters down the surface facing the induction coil. In contrast, the LF penetration is about 10–20 mm, which means that it involves the wall of the susceptor body completely. The heating is thus softer in LF, and this allows a higher current to be applied, reducing the ramp-up time and reducing the stress for the susceptor and the wafers.

The configuration of induction heating systems has two basic shapes: cylindrical coil (barrel susceptor) and flat spiral (pancake-like susceptor). In some arrangements of the barrel type the induction coil is placed inside the susceptor, but this makes the mechanical design of the reactor and its maintenance more difficult.

As mentioned before, the most remarkable difference between IR and induction heating regards the direction of heat flow.

- In IR systems, the radiant heating is applied first to the wafers and then to the susceptor, generating a thermal gradient through the wafer bulk, higher in the front than in the back, but around the process temperature, the gradient drops to near zero.
- In induction heating, the situation is the opposite, because the heat is transmitted from the susceptor surface to the back of the wafer, generating a thermal gradient higher in the back than in the front; but this gradient can be lowered to a few degrees by means of reflecting media facing the wafers.

2. Susceptor Design and the Slip Problem

The inverted thermal gradient in induction-heated susceptors can be a cause of slip in crystal planes if the slip origin is not well understood and taken into account when the susceptor is designed. The temperature difference of a few degrees between the back and the front of the wafer, because of thermal expansion, causes an elastic bowing of the wafer, i.e., the wafer bows to a shallow cup shape until heated, returning to a flat shape when cooled to room temperature. The bow ranges from some 10 to about 100 μm, according to the temperature gradient applied as mentioned before. This is a quite reversible elastic deformation. Normally the warp value in a wafer does not increase (it almost always decreases a little) after epi processing, so no slip originates from it. But in a flat wafer pocket, the bow causes the wafer edges to be no more in contact with the susceptor itself, becoming slightly colder than the center. This radial thermal gradient causes the wafer edge to shrink slightly, and the resulting radial stress can be released by the formation of slip. To prevent this slipping, some curvature (Fig. 11) should be calculated and tooled out in the susceptor wafer pockets, if thermal gradient value across the wafer is known, but normally it is easier to figure out the best shape for the pockets on an experimental susceptor with many differently shaped pockets. In epi process, temperature may vary some hundred degrees from recipe to recipe and the temperature gradient varies too, so the bowing of receptacle should be a reasonable compromise.

3. Run-to-Run Repeatability

The repeatability depends on how much the parameters that affect the growth rate are kept stable and assume exactly the same values in two consecutive batches

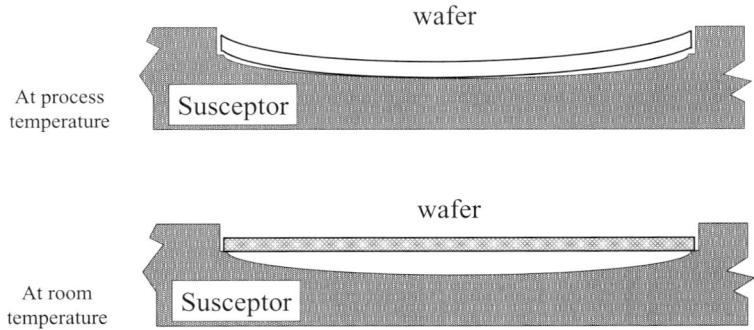

FIG. 11. Susceptor pocket design shaped recess on susceptor for improved heat transmission.

or repeating the same recipe after a period. Usually the target thickness of an epi layer is obtained through a "time-controlled" process, i.e., the final thickness depends on the deposition time stated in the recipe assuming constant parameters. This implies that the growth rate must be constant and known as accurately as possible. The main parameters affecting the growth rate are the process temperature control, the chlorosilane molar ratio, and some other parameters, minor but not negligible.

a. Process Temperature

Measuring the temperature in the susceptor is not as easy and reliable as it might appear; it is usually done with an optical pyrometer (induction-heated reactors) or with photodiodes and thermocouples (lamp reactors). The temperature signal pilots the heating system through a feedback circuitry and this part can be optimized to very high levels.

Problems arise from the reliability of pyrometers, which is affected mostly by the emission coefficient (emissivity e) of the targeted material and by other factors such as the orientation of the sensing head and/or by the quartz wall of the reaction chamber. In the case of batch reactors, the rotating susceptor presents to the pyrometer, alternatively, the SiC-coated surface ($e = 0.9$) and the wafer surface ($e = 0.6-0.7$), thus the signal should be interpolated by circuitry or algorithms to give properly averaged values. The error of a properly averaged value is the square root of $(n - 1)$ less than the error of a single measurement. The reliability of the measured value is affected by geometry (wafer size) and by the calibration of the emissivity. Alternatively the pyrometer may spot susceptor areas not covered with wafers. Dual-band pyrometers improve the reliability, but do not completely solve the problem.

It must be emphasized that the absolute value of temperature is, of course, important, but much more important is the *repeatability* of the reading during the process and from one run to another. When photodiodes/thermocouples are used in IR-heated reactors, they are placed internally to the susceptor because the lamps may affect the measurement (the lamp filament can be read in place of the wafer surface). These systems measure the body of the susceptor, and not the silicon substrates directly, which may affect the accuracy. In addition, they are subject to the "aging" of the sensors, which necessitates frequent calibration. Other factors regarding the heating system and its temperature control that may affect the repeatability of the process are the induction coil dimensional stability in one case and the burnout of IR lamps in the other. Both can modify the temperature profile of the susceptor, and consequently the run-to-run repeatability, in a period of days or weeks, so they need frequent checks and calibrations.

In LF LPE PE2061 induction heating, the coil is a monolithic body which does not change dimensions and overcomes the problems; fine-tuning of the temperature profile is achieved with electric media, a small parasitic coil in parallel with the inductor.

b. Chlorosilanes Molar Ratio

$SiHCl_3$ (TCS) is the most widely used silicon source, both for atmospheric and RP epi processes, but its delivery and measure worth require some more details, which are valid for $SiCl_4$ (TET) also. Some data are listed in Table VII.

TCS vaporization by heating is not recommended, because of the risk of condensation in the lines at room temperature. The most usual way of delivery is bubbling hydrogen through liquid TCS and then driving the resulting mixture of H_2 and TCS to epi reactor, through a measuring system.

The ratio H_2:$SiHCl_3$ is a function of several geometrical and physical factors (temperature, pressure, bubble size, path of gas in the liquid, gas flow amount, etc.), so that it is not easy to predict. Unless special precautions are taken (see Fig. 12 for calculating this amount) it is not simple to keep stable the saturation

TABLE VII

TET, TCS, AND DCS PHYSICAL PROPERTIES

	$SiCL_4$	$SiHCl_3$	SiH_2Cl_2
Molecular weight	170	135	101
Boiling point (°C)	57.6	31.8	8.4
Liquid density (kg/liter)	1.48	1.335	0.878
Gas density at 0°C (kg/m^3)	7.859	6.043	4.397
Heat of evaporation (kcal/kg)	56.55	57.72	73.02
Heat of evaporation (kJ/kg)	193	197.2	249
MFC conversion factor (ref. N_2)	0.28	0.33	—

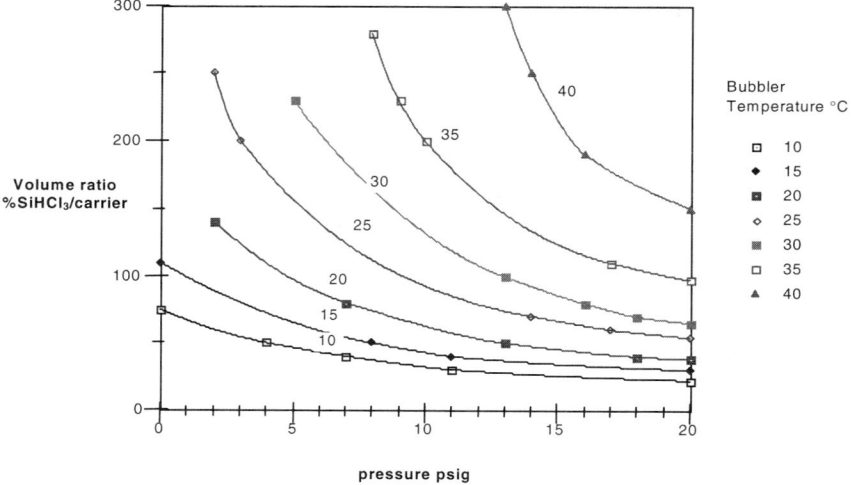

FIG. 12. SiHCl₃ in carrier (H₂).

level and consequently the epi growth rate. Furthermore, epi is a batch process (i.e. discontinuous) thus during the deposition time the chlorosilane delivery is running and then, for a shorter or longer time, it is idle. During this latter period, the saturation values inside the chlorosilanes container may vary greatly, so that at the start of a new deposition cycle, the conditions may be quite different from those in the previous cycle. After an idle time, despite a preventive purge, a high concentration of chlorosilane is driven to the reactor and that results in a very high growth rate, just at the beginning of deposition. It may generate a first epi layer affected by many crystallographic defects (mainly dislocations), which may "print" the quality of the entire epi layer.

To eliminate or lessen this problem, it is common to transfer a small amount of TCS from the big commercial tank into a smaller vessel (bubbler) (Fig. 13), where:

a. the control of temperature is easier
b. the refilling to a stated level is performed (automatically) before any growth cycle
c. the small volume of vapors (1–3 liters) is easily removed and stabilized during usual prepurging steps.

The TCS molar ratio depends mainly on evaporation temperature, so for the best repeatability of the growth rate, a good thermal control of the bubbler is mandatory. This system, if properly applied, allows reaching a standard deviation of about 0.5% in growth rate. Most measuring apparatuses are MFCs placed on carrier gas upstream to the bubbler, or downstream, on the mixture of carrier and chlorosilane vapors (Fig. 14).

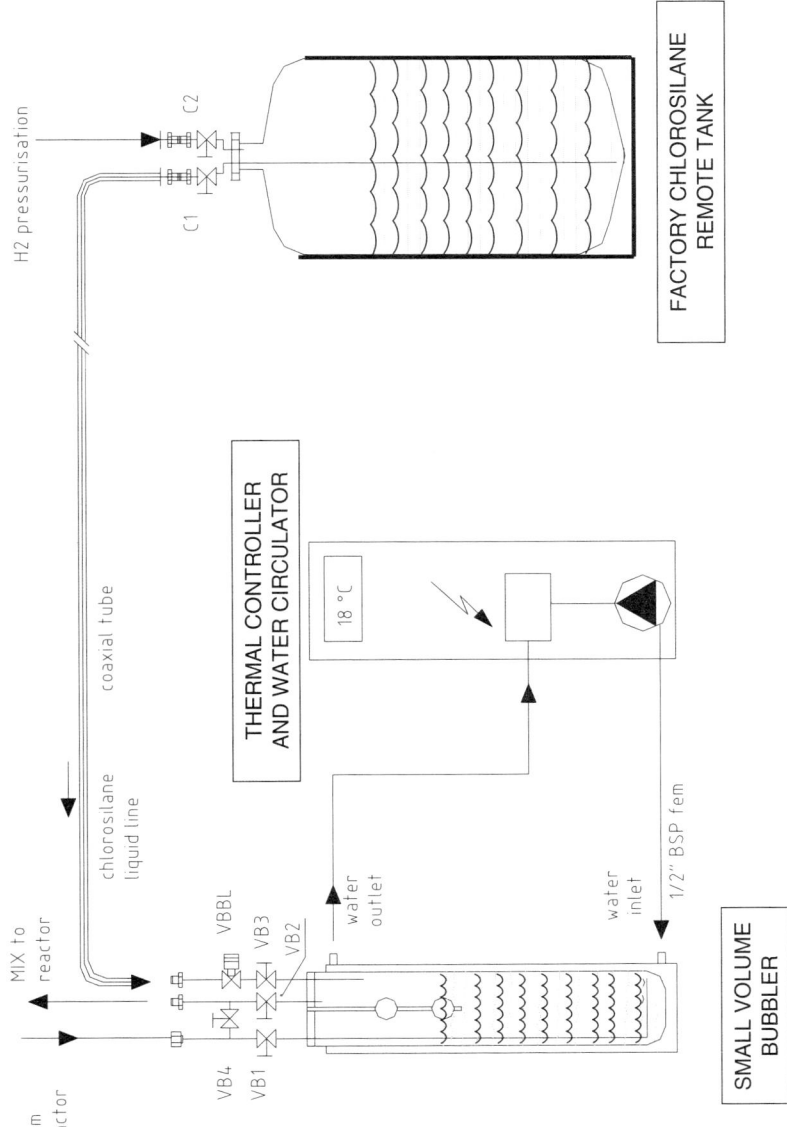

FIG. 13. Optimized chlorosilane delivery system.

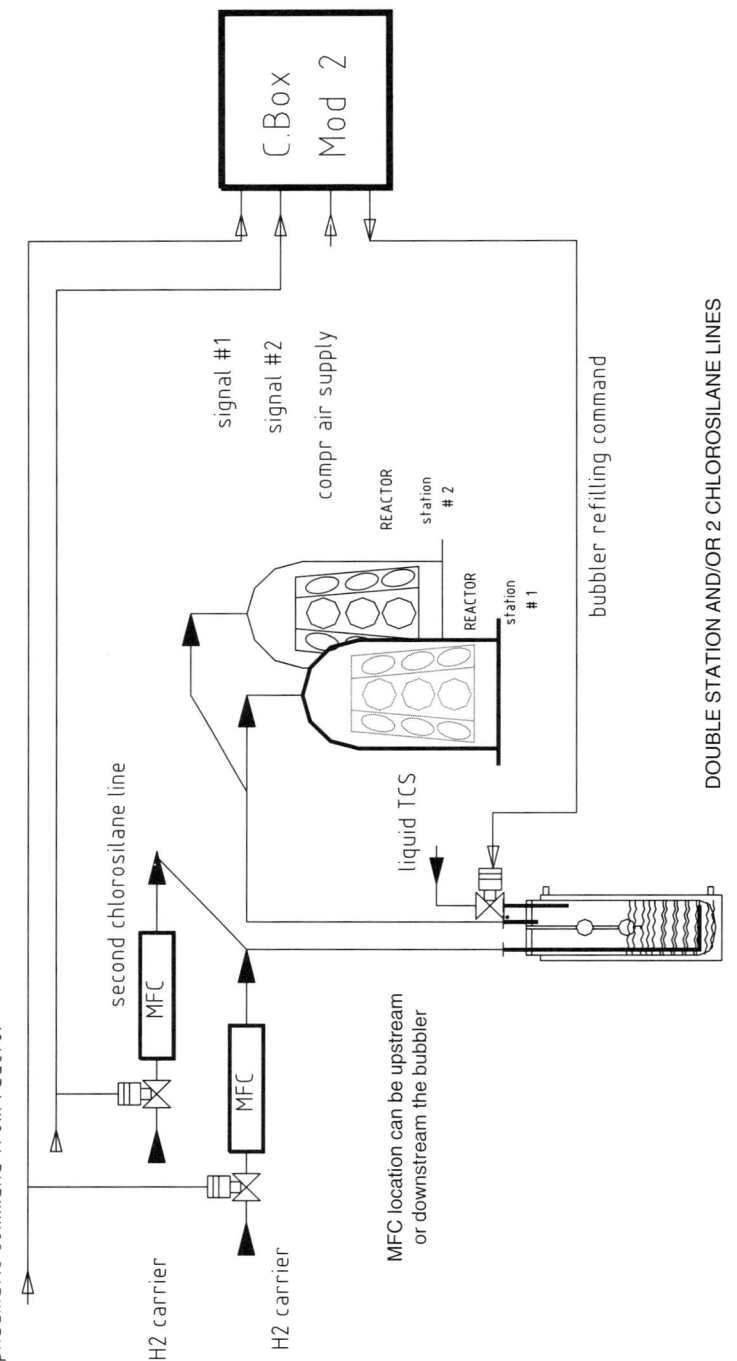

FIG. 14. Schematic of the TCS delivery system.

A source vaporizer can be used: it consists of an MFC coupled with a sensor, which measures a factor related to the value of saturation. This device may take care of small temperature and pressure variations in TCS mixture, but it is not easy to calibrate and has limited ranges of carrier flow (Fig. 15). A similar method, studied for improving the reliability of the saturation, consists of two bubblers connected in series, where carrier gas pass through the first bubbler, then the mixture pass through the second bubbler, increasing the molar ratio.

In a different method, the chlorosilane is delivered as a liquid to the gas panel, measured into a special MFC, and then evaporated into the tubing. In this case, special attention must be paid to the complete degassing of liquid TCS before measurement and to the injection of liquid (droplet size) to prevent unreliability of the system.

c. Other, Minor Factors Affecting Run-to-Run Repeatability

The following items are as important as those discussed previously, but generally they cause no or few problems because they are more easily kept strict and precise control: the flow of gases and their injection into the reaction chamber, the rotation speed of the susceptor, the geometry of the reaction chamber (especially of quartz and graphite parts), and the quartz wall cooling.

4. ROBUST EQUIPMENT REQUIREMENTS

The need for a high epi thickness at long growth times means that the cooling of the quartz walls of the reaction chamber, and in general of the entire reactor, must be highly optimized. The use of LF induction greatly facilitates the design of air-cooling systems because the inductor wounds can be strictly packed together to form a tube that conveys the cooling air, which can be recycled after passing through a controlled heat exchanger. Even for highly stressful recipes preventive maintenance operations can be performed at intervals of several weeks or more than a month.

5. HIGH-RESISTIVITY CONTROL

The incorporation of the doping species into the epi layer and thus its resistivity assume different values depending on the parameters, as described in other parts of this book. In the case where a very high-resistivity epi layer (from 75 up to 300 $\Omega \cdot$ cm), is required, special care in the design of the reaction chamber is necessary.

The volume of the reaction chamber and the gas flow path are the most important parameters. Moreover, because of the very small amount of dopant that must be

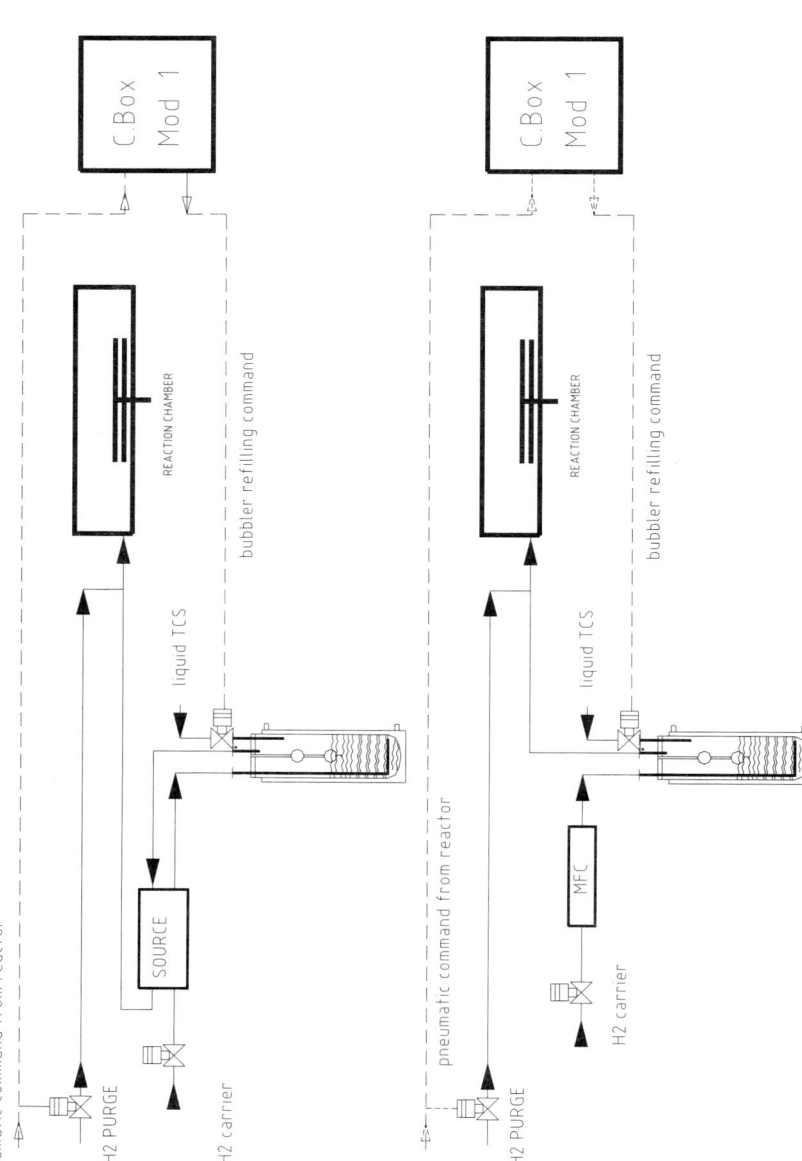

FIG. 15. Comparison between the LVC (liquid vaporizer controller; top) and the MFC (mass flow controller; bottom) configurations.

injected, the gas panel design has great relevance. Furthermore, it is mandatory primarily to achieve a high intrinsic background value, which involves chiefly the reactor (geometry, material in the reaction chamber–gas panel–susceptors) but also other strong factors outside of the reactor itself (gas purity and lines, gas cabinets, chlorosilane supply, careful maintenance, and cleaning procedures). Also, the usual ways of protecting the back of wafers are necessary, such as LTO and high-thickness back-transfer. The back-transfer process is the cheapest and cleanest way to protect the back of the wafer, and it is possible only with induction-heated reactors, but, on the other hand, this method may give poor values of flatness [mainly the total thickness variation (TTV) parameter].

6. Temperature—Uniformity and Profiling

The doping effect is very sensitive to temperature, so that a correct and "flat" temperature profiling of the susceptor is mandatory to keep the resistivity variation, under control, more so when a high resistivity is required. For instance, a $10°C$ variation from point to point can produce about a 3% resistivity variation. On the other hand, chlorosilane must reach the proper decomposition temperature all along the susceptor and, for this, a "flat" temperature profile is not always the most desirable in terms of thickness uniformity. All reactor manufacturers provide the system to reach a satisfactory compromise.

The temperature profiling is done in HF induction by adjusting the turn spacing in the induction coil or, in the case of the rigid coils used in LF induction, through auxiliary adjustable coils applied to selected zones of the main coil, allowing ever more accurate and repetitive profiling. In IR-heated reactor is possible to adjust independently the temperature offset of the lamps rows and reach a suitable temperature profile, but this adjustment must be checked quite frequently due to lamp aging and burnout.

7. Reaction Chamber Design

To assure the lowest level of metal contamination, the first aspect of chamber design should be to minimize the surface of metals parts exposed inside the chamber to reactive process gases, especially if these parts are upstream of the susceptor. Of course, the gases meet a lot of metals during pass-through in gas panels and pipelines, but as long as they are dry and at room temperature and the metal surfaces are passivated or electropolished, a negligible amount of metal ions and/or particles is released. The metal parts in the reaction chamber can, to some extent, be etched (in terms of amount of metal ions released) by gases and by exposure to the atmosphere during wafer loading or maintenance operations.

The amount of metal ions may also affect the lifetime of minority carriers in the device and thus it must be taken into consideration when designing the reaction chamber. All moving parts (e.g., rotation and relevant movable connections or seals) should be avoided or, if impossible, carefully segregated from the gas flow.

Also, the volume of the chamber involved with the gases should be reduced to the minimum necessary for containing parts to achieve the fastest replacement of contained gases. At the same time, dead gas zones and entrapment must be avoided or otherwise separately purged to prevent uncontrollable "memory" of the system, which can reduce the control of the proper resistivity value.

All these factors must be carefully considered when selecting a reactor with the intent to manufacture high-thickness/low-resistivity epi layers for power devices.

8. Gas Panel

In the design of gas panels the selection of components is very important. They should have a very high finishing grade of the internal surface wetted by gas and the lowest internal volumes, avoiding, when possible, entrapped volumes (bellows, dead-end lines, branches that are not well purged). Besides this hardware aspect, some arrangements are usually needed to maintain control of the very low concentration of dopant into the gas flow to the reaction chamber.

Dopants are commonly supplied in cylinders containing mixtures in H_2 or Ar, where the dopant concentration ranges between 10 and 50 ppm (for high-resistivity application). MFCs for dopant lines can cover a large range of resistivity, but in the case of very high resistivity (>100 Ω cm) the MFC should be set at a very low set point, or it may be necessary to replace the dopant cylinder with another at a lower concentration.

An easy way to overcome these problems consists of a MFC dilution system, represented schematically in Fig. 16, which allows the coverage of a wide range of resistivities, from a low value of about 0.3 $\Omega \cdot$ cm to a high value of 300 $\Omega \cdot$ cm, without changing either the cylinder or the MFC. This arrangement allows modification of the concentration of the mixture injected into the reaction chamber from near zero dopant to 100% dopant permitted by the DOPE MFC, by simply setting the three MFCs at proper values. Table VIII lists examples of resistivity obtained with common MFC rates and cylinder concentrations.

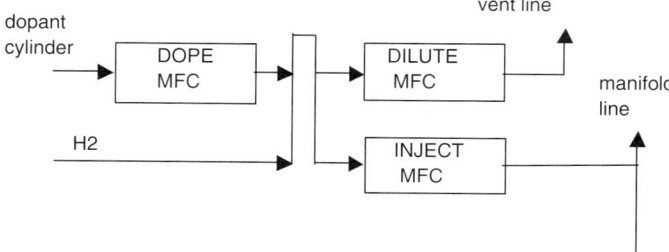

Fig. 16. Schematic of the dopant dilution system.

TABLE VIII
RANGES OBTAINED WITH COMMON MFC RATES AND CYLINDER CONCENTRATIONS[a]

Dopant	ppm	MFC full rate flow (cm³/min)			Max. conc. (cm³/min)		Min. conc. (cm³/min)		Resistivity range (Ω·cm)	
		Dope	Dilute	Inject	Dilute	Inject	Dilute	Inject	Min.	Max.
PH_3	30	100	5000	100	250	100	5000	10	0.3	470
B_2H_6	100	100	5000	100	250	100	5000	10	0.3	1700

[a]Courtesy of LPE, Milan, Italy.

9. Gas Injection and Flow Patterns

Though not in all devices, a high resistivity layer is usually associated with a double or multiple epi layer, and in general the resistivity change is some orders of magnitude and often of opposite type (N ↔ P). In this case, besides the concentration of dopants under steady-state conditions, one must focus on the transition time, i.e., on very fast changes in dopant concentration and/or polarity. For this, the significant points concern the volume of the reaction chamber (related to the flow of carrier gas); a large dome-shaped chamber has a big volume, which is generally a disadvantage.

The velocity of gas flow injection should remain in the laminar range and the direction should not change too much with respect to the wafer surface, to move the previous gases like a piston, pushing them out without turbulence and mixing. In horizontal-flow reactors (where the average gas direction is approximately parallel to the wafer surface, as in barrel-type or single-wafer or new-generation small-batch reactors), that piston action is generally obtained.

The barrel model can act in different ways according to the design of the gas injection system and it may be seen as intermediate between the piston-like and the turbulent gas path. Of course, if injection is vertical, piston-like behavior is prevalent, whereas horizontal injection favors turbulent behavior. Vertical-flow reactors, such as the pancake model, generally have a large chamber volume and the gases are injected at a low velocity toward the top of the chamber, so that the piston effect is poor and in a radial direction. The latter two models may pose some problems regarding high resistivity and fast change of dopant concentration.

VII. Advanced Applications

For the emerging low-temperature epi applications such as SiGe and SEG for an elevated source drain, additional facilities concerns exist. The lower the intended deposition temperature, the more sensitive the process will be to background oxygen. The current generation of RP-CVD epi systems utilizing load-locks and with excellent leak integrity provide the system solutions necessary for these low-temperature applications, but facilities that are adequate for traditional epi films will result in the incorporation of oxygen at low temperatures. If low-temperature processes are intended, the following guidelines should be followed.

1. POU purification may be required for H_2 delivery. While liquid boil-off H_2 is the purest H_2 available, small leaks in fittings and line drops in the facility will often reduce the quality of the H_2 prior to its delivery to the epi system.
2. Because these low-temperature processes often utilize higher concentrations of dopants, with the addition of GeH_4 to the system, thermal abatement may be required in addition to wet scrubbing.
3. POU purification may be required on N_2 and silicon sources. Be careful when ordering POU purification to specify the complete mixture on mixed gases. For example, if 1% B_2H_6 is utilized, be sure to specify that the balance is H_2, since the gas composition is critical for resin-based purification systems.
4. It is desirable to use DISS (diameter index safety system) fittings on gas tanks. These fittings will result in a leak rate of $<1 \times 10^{-9}$ scc (He)/s inboard and 1×10^{-7} scc (He)/scc at 2000 psig helium outboard, when the connection is tightened to 35 ft-lb (Air Products Guidelines and Recommendations).
5. A helium leak-check is required on all lines after maintenance.

VIII. Conclusion

This chapter gives a general overview of the epi reactor and all that is needed to run the process. However, the reader should realize that the knowledge of many technical fields such as epi process, relevant facilities and the reactor itself, is mandatory. Fluids engineering, mechanics, robotics, gas chemistry, thermal physics, software, electronics, high-power electrical (low- and high-voltage), science of material, and environmental science are just the major items involved.

It is hard for anybody to know and manage even a reasonable amount of all of these items, and for that the design of an epi reactor, but also the top utilization in production, must be a task involving a team of skilled people. Today the "super expert" is no longer a suitable approach. Only a team of people can hit the target.

Acknowledgments

Thanks go to Franco Preti of LPE Epitaxial Technology for epi reactor information, to Norma Riley of Applied Materials for her collaboration on exhaust systems and facilities, and to Giorgio Vergani of Saes-Getters for discussion on purifiers and filtration.

REFERENCES

Book of Semi Standards 1995. Facilities and Safety Guidelines, Semiconductor Equipment and Materials International, Mountain View, CA, 1995.
Hayes and Woods, *Solid State Technol.* **Oct**. (1996).
Henderson, P., HCl Manufacture, Purification and Analysis, Semi-Forum on Process Gases, Epitaxial Deposition, 1993.
Rodney, S., *J. Opt. Soc. Am.* **Sept**. (1954).
Sugiyama, Ohmi Okumura, and Nakahara, *Microcontamination* 1–89 (1889).
Westbrook, J. H., *Physics and Chemistry of Glasses, Vol. 1,* 1960.

CHAPTER 4

Epitaxial Growth Techniques: Low-Temperature Epitaxy

J. Murota

RESEARCH INSTITUTE OF ELECTRICAL COMMUNICATION
LABORATORY FOR ELECTRONIC INTELLIGENT SYSTEMS
TOHOKU UNIVERSITY
SENDAI, JAPAN

ABBREVIATIONS USED	127
I. INTRODUCTION	128
II. CVD MACHINES	128
III. SURFACE TREATMENT	133
IV. EPITAXIAL GROWTH MECHANISMS	135
1. Adsorption and Reaction in Undoped $Si_{1-x}Ge_x$ Epitaxial Growth on a (100) Surface	135
2. In Situ Doping of B and P in $Si_{1-x}Ge_x$ Epitaxial Growth	138
V. HIGH-QUALITY Si/$Si_{1-x}Ge_x$/Si HETEROSTRUCTURE GROWTH AT HIGH Ge FRACTIONS	143
VI. CONCLUSIONS	147
REFERENCES	148

ABBREVIATIONS USED

AP-CVD	Atmospheric-pressure CVD
CVD	Chemical vapor deposition
epi	Epitaxial
IR	Infrared
LP-CVD	Low-pressure CVD
LRP	Limited reaction processing
MBE	Molecular beam epitaxy
MOSFET	Metal–oxide–semiconductor field effect transistor
RF	Radiofrequency
RTA	Rapid thermal annealing
RT-CVD	Rapid thermal CVD
STM	Scanning tunneling microscope

TEM   Transmission electron microscope
UHV-CVD   Ultrahigh vacuum CVD

I. Introduction

High-temperature epitaxial (epi) growth of Si on Si, above 1000°C, by CVD has been employed on the industrial level for a long time [1] because lowering the growth temperature affords polycrystalline or amorphous growth, while MBE offers the advantage of low-temperature epi growth (see Chapter 5). Recently, low-temperature epi growth of Si and $Si_{1-x}Ge_x$ has become increasingly important for the fabrication of novel Si devices, i.e., minimized Si MOS devices and Si-based heterodevices, because high-performance devices require abrupt steps in the doping profile [2–5]. So far, MBE has mainly been used for producing such devices, and remarkable progress has been achieved (see Chapter 5). However, CVD offers many advantages, such as a high throughput, *in situ* doping, and selective deposition. Improvements in CVD machines, the quality of the gases used, and CVD processing have enabled low-temperature epi growth [6–9].

In this chapter we focus on the low-temperature epi growth of Si and $Si_{1-x}Ge_x$ on Si, CVD machines, growth mechanisms, surface treatment, and *in situ* doping growth of $Si_{1-x}Ge_x/Si$ heterostructures.

II. CVD Machines

Since lowering the growth temperature generates more unexpected contaminations on the Si surface, it is very important to avoid any contamination on the Si surface during transport of the wafer into the reactor and during the wafer's stay in the reactor prior to deposition. Meyerson *et al.* [6, 10] developed a hot-wall ultrahigh vacuum CVD (UHV-CVD) to avoid oxidation of the Si surface and achieved low-temperature epi growth of Si using a SiH_4–H_2 gas mixture. The concept is based on the basic chemical equilibrium data for the $Si/O_2/H_2O/SiO_2$ system reported by Smith and Ghidini [11, 12]. Meyerson [6] designed a machine with a base pressure of the order of 10^{-7} Pa and a deposition pressure of about 0.1 Pa to achieve an oxide-free surface at very low temperatures, below 800°C, extrapolating the high-temperature data [12] for the formation of oxide on the Si(100) surface as shown in Fig. 1 and assuming the existence of 1 ppm contaminating species. On the other hand, Donahue and Reif [7] found that lowering the epi growth temperature is achieved by Ar plasma cleaning of the Si surface in the reactor. Murota *et al.* [8] developed an ultraclean hot-wall low-pressure CVD (LP-CVD) in the deposition pressure region around 10 Pa by improving the quality of the gases used at the reactor inlet and avoiding air contamination into the reactor during loading and unloading of wafers; achieved low-temperature Si selective deposition and epi

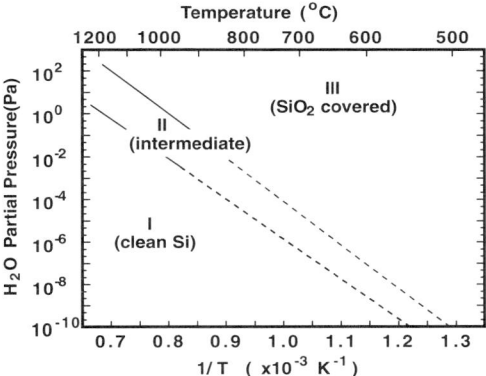

FIG. 1. Chemical equilibrium data for the $Si/H_2O/SiO_2$ system [12]. The dashed lines show the extrapolation from the experimental data.

growth at 650–850°C using a SiH_4–H_2 gas mixture in an ultraclean environment, and found that the SiH_4 reduction of native oxide occurs even at 650°C. Sedgwick et al. [9] developed an ultraclean atmospheric-pressure CVD (AP-CVD) with a H_2 point-of-use purifier; achieved low-temperature selective epi growth at 600–850°C using a SiH_2Cl_2–H_2 gas mixture, despite the partial pressures of O_2 and H_2O ($\sim 10^{-3}$ Pa) being higher than that expected above; and also showed that the epi quality is degraded by Si oxidation due to the oxidant in the carrier gas prior to growth [13]. All recent data [6–9, 13–16] show that Si epi growth proceeds on Si single crystals at low temperatures above 500°C, at least in an oxidation-free environment, as shown in Fig. 2.

Si substrate heating methods include the hot-wall and cold-wall types. In the hot-wall type, the reactor is a cylindrical quartz tube inserted in a furnace. Since temperature control is easy for many wafers, a high throughput and good uniformity of the growth can be obtained. The disadvantages are the limitation of deposition pressure and deposition temperature for the polymerization of Si source gases such as SiH_4 and the slow temperature response compared with that in the cold wall-type. In the cold-wall type, the wafers are generally set on a SiC-coated graphite susceptor heated by RF power or IR lamp irradiation outside of a reactor cooled by air or water. Gibbons et al. [17] developed limited reaction processing (LRP) using rapid thermal annealing (RTA), in which the wafer is heated up directly through quartz window by IR lamps. In this method, the deposition process is controlled by ramping the temperature up and down and flowing the reactant gases continuously, while this machine was employed in the conventional deposition process as a rapid thermal CVD (RT-CVD) [16, 18, 19]. The merits of RT-CVD are the wide range of deposition pressures, from UHV to AP, and the very fast switching of temperature. The disadvantages of the cold-wall type are the poor temperature control and low throughput for low growth rate conditions of the film because many wafers cannot be set in the reactor compared with the hot-wall type.

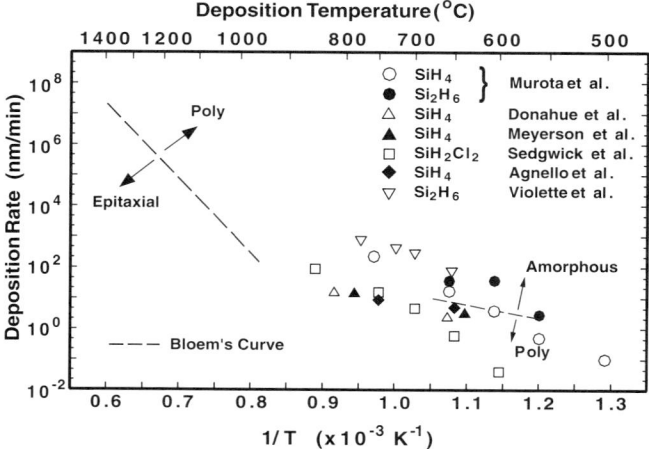

FIG. 2. Relationship between deposition temperature and deposition rate. The dashed curves show the boundaries among epitaxial, polymorphous, and amorphous Si deposition by conventional CVD technology. All data points refer to Si epitaxial growth.

As an example of a low-temperature epi growth machine, an ultraclean hot-wall LP-CVD machine [8, 14, 15] is shown schematically in Fig. 3. The reactor structure of a cylindrical quartz tube is the same as that of the conventional hot-wall LP-CVD machine [20]. This machine was made UHV compatible with gate valves and a turbomolecular pump system. The front side of the turbomolecular pump is connected to the reactor with several sizes of vacuum exhaust tubes, and the reactor can be directly evacuated by the turbomolecular pump from AP. The back pump of the turbomolecular pump was changed from a rotary pump to a dry pump [21]. These are keys to preventing any contamination from the exhaust line. To minimize air contamination in the reactor during loading and unloading of the wafer, a N_2-purged transfer chamber was combined with the reactor inlet.

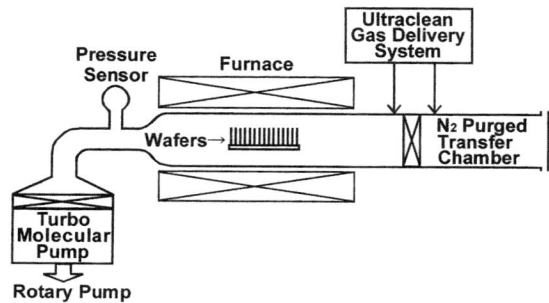

FIG. 3. Schematic diagram of an ultraclean hot-wall LP-CVD system.

The wafers, placed on a quartz boat, were transported into the reactor in an ultraclean N_2 atmosphere through the transfer chamber. After closing the gate valve, the N_2 flow was stopped and the reactor tube was purged with high-purity H_2. To prevent any contamination from the exhaust line, the wafers were transported into the reactor at the reactor temperature of about 100°C, and then heated to the deposition temperature while purging with H_2 gas under a pressure of about 200 Pa. Subsequently, high-purity reactive gas was added and the Si and/or $Si_{1-x}Ge_x$ deposition began. Here, the moisture level of the N_2, H_2, and SiH_4 gases used was 10 ppb or lower and that of GeH_4 was 23 ppb or lower at the reactor inlet [21]. Before loading the wafers into the transfer chamber, they were cleaned in several cycles in a 4:1 solution of $H_2SO_4:H_2O_2$ or in hot $NH_4OH-H_2O_2-H_2O$ and hot $HCl-H_2O_2-H_2O$ solutions, high-purity de-ionized water, and 2% HF.

The influence of air contamination on the Si deposition was simulated by changing the degree of air contamination inside the reactor during wafer transport using the N_2 purge transfer chamber. The results [8] are shown in Figs. 4 and 5. In the case of ultraclean deposition with the N_2 purge transfer chamber, there is an incubation period for Si deposition on SiO_2, the deposition rate is dependent on the substrate material (Fig. 4a), and Si epitaxial growth on Si(100) proceeds as shown in Fig. 5a, even with a deposition rate of 20 nm/min at 650°C, at which polycrystalline Si would be formed in conventional processing [23, 24]. In the case of air-contaminated deposition without the N_2 purge transfer chamber, similar to that in a conventional hot-wall CVD machine, the incubation period and deposition rate become independent of the substrate material (Fig. 4b), and the structure of the air-contaminated film on Si(100) is polycrystalline as shown in Fig. 5b. These results for air-contaminated deposition mean that the Si substrate surface is oxidized and the number of reaction sites at the SiO_2 surface increases by the adsorption of volatile SiO formed by the reaction of residual Si-containing species and residual air in the reactor.

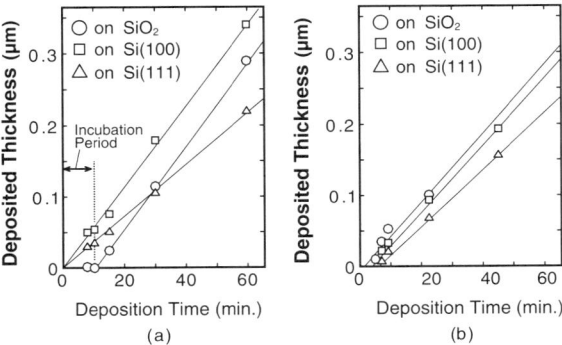

FIG. 4. Deposition time dependence of the deposited film thickness at a deposition temperature of 700°C and a SiH_4 partial pressure of 0.15 Pa. (a) Ultraclean deposition; (b) air-contaminated deposition.

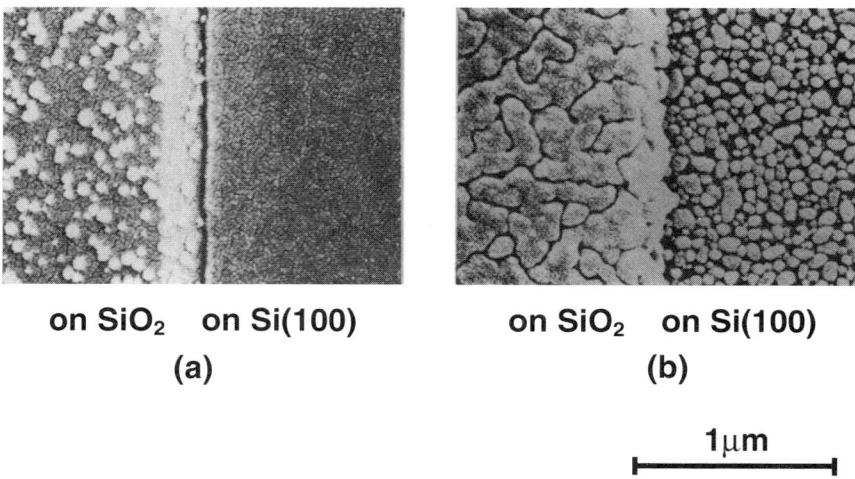

on SiO₂   on Si(100)
(a)

on SiO₂   on Si(100)
(b)

|— 1μm —|

FIG. 5. SEM micrographs of sample surfaces grown at a temperature of 700°C and a SiH$_4$ partial pressure of 0.15 Pa. (a) Ultraclean deposition; (b) air-contaminated deposition.

Next, the contaminations of O and C from exhaust were evaluated by the impurity pileup among the substrate, the first layer, and the second layer. The results are shown in Fig. 6 [25]. With 40-min evacuation during the first and second layer depositions, impurity pileups are found (Fig. 6a), while, with H$_2$ gas purging under a pressure of about 3Pa between the first and the second layer depositions, the pileups are not observed. Since the pileups are caused by contaminations from the exhaust line, the molecular flow region should not be employed as a

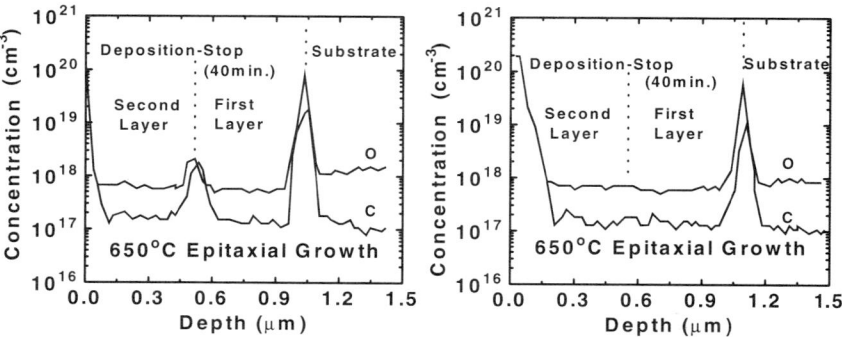

FIG. 6. The contaminations of O and C by (a) vacuum evacuation and (b) H$_2$ gas purging of about 3 Pa during the first and second layer deposition. The first and second epilayers were deposited at a temperature of 650°C, a SiH$_4$ partial pressure of 6 Pa, a H$_2$ partial pressure of 3 Pa, and a total deposition pressure of 9 Pa.

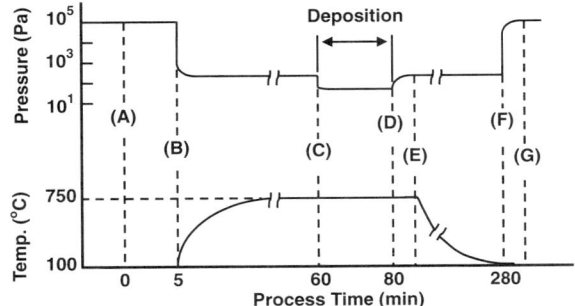

FIG. 7. Typical epitaxial layer deposition process sequence. (A) Sample load in N_2 atmosphere; (B) H_2 input, vacuum evacuation start, and heat-up; (C) SiH_4 input; (D) SiH_4 stop; (E) cooldown; (F) vacuum-evacuation stop; (G) sample unload in a N_2 atmosphere.

mass-productive deposition condition when long maintenance times should not be expended. Impurity pileups at the interface between the epi layer and the Si substrate can be removed by the process sequence shown in Fig. 7, where the wafers were transported into the reactor at a reactor temperature of about 100°C, and then heat-up to a deposition temperature of 750°C was carried out with H_2 gas purging under a pressure of about 200 Pa. By this method, no pileups of O and C at the interface between the epi layer and the Si substrate were found (see Section III, Fig. 9) at a background level. It is considered that, at temperatures between 100°C and the deposition temperature, the impurities adsorbed on the substrate surface are desorbed and/or the reaction between the substrate surface and impurities diffused from the exhaust line during wafer loading is suppressed by the decreasing reactor temperature. The total process time can be reduced by using a high-speed heat-up and cool-down heater [26].

In conclusion, low-temperature epi growth requires CVD machines to avoid air contamination to the reactor from outside and any contamination entering the reactor from the exhaust line, to use high-purity gases, and to ramp the temperature up and down.

III. Surface Treatment

To achieve high-quality epi growth of Si, it is essential to remove any contamination on the Si substrate by surface treatment before epi growth. In the MBE process, organic and metallic impurities on the Si surface were removed by wet cleaning, a thin protective oxide layer was formed to passivate the surface against recontamination during wafer transport into the reactor, and then the oxide layer was removed by thermal heating in a UHV environment, Fig. 8 (see Chapter 5). However, in the CVD process, the formation of a UHV environment is not practical

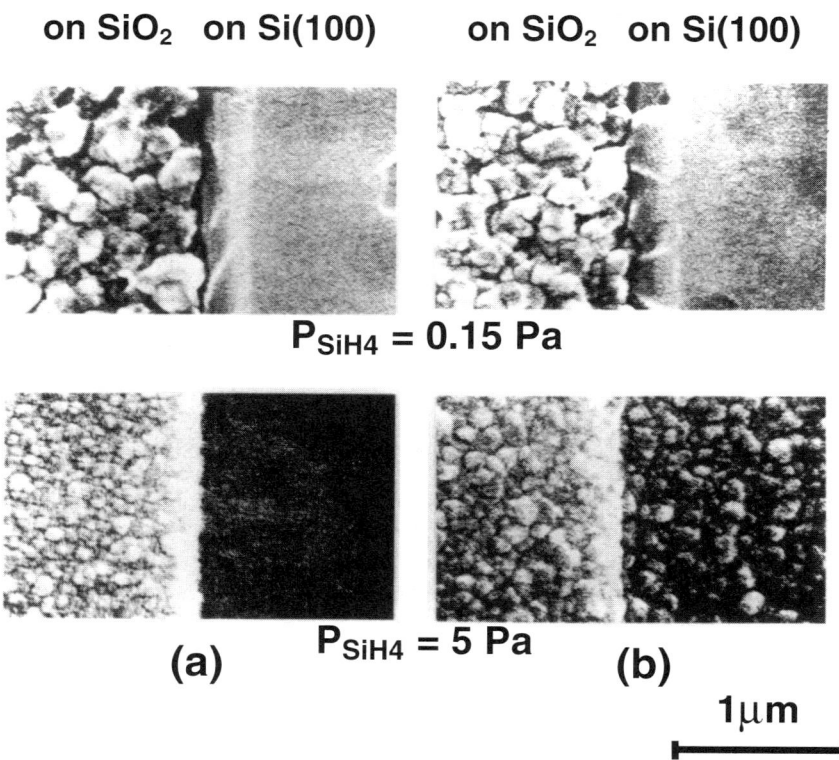

FIG. 8. SEM micrographs of sample surfaces grown at 650°C on Si(100) substrates treated in (a) a diluted HF solution and (b) a NH$_4$OH–H$_2$O$_2$-H$_2$O solution at 90°C.

because of unexpected contamination in the reactor. Reduction of the protective oxide layer by SiH$_4$ was reported [26]. Flat surface epi growth occurs even at 650°C for lower SiH$_4$ partial pressure deposition in an ultraclean LP-CVD (Fig. 8) [27]. However, there are some oxygen and carbon contaminations at the interface between the epi layer and the Si substrate. Oxygen contamination, especially, results in defect formation in the epi layer [13]. It is clear that high-temperature H$_2$ baking and high-temperature SiH$_4$ exposure should not be employed to fabricate the recent ultrasmall Si devices, and plasma cleaning induces defects in the epi layer.

To avoid recontamination during wafer transport into the reactor, hydrogen surface passivation was proposed [28]. Hydrogen termination is obtained using conventional diluted HF etching of SiO$_2$ [29, 30], but it is not perfect due to the final rinse with DI water. For epi growth at 600 or 650°C on the HF-treated Si surface, oxygen and carbon pileups of the order of 10^{13} cm^{-2} were found [15, 31] as shown in Fig. 9a [15]. To remove such contamination, Murota and Ono [15] proposed thermal desorption at 750°C in a H$_2$ environment of 200 Pa (Fig. 9b) and

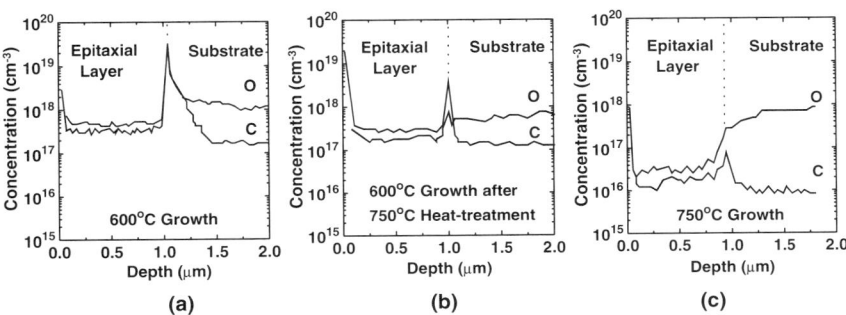

FIG. 9. Depth profiles of oxygen and carbon in the Si epitaxial growth sample, evaluated by secondary ion mass spectrometry. (a, c) Results for epitaxial growth at 600 and 750°C, respectively, after heating up from about 100°C. (b) Result for epitaxial growth at 600°C after preheating at 750°C. Oxygen and carbon pileups at the interface between the epitaxial layer and the Si substrate are observed, although the background levels are different between the samples.

subsequent buffer Si epi growth (Fig. 9c) using SiH_4 by ultraclean LP-CVD. Sanganeria et al. [31] proposed thermal desorption in a 1-atm H_2 environment at 750°C and subsequent buffer Si epi growth using Si_2H_6 by AP/RT-CVD. Their results show that oxygen and carbon pileups at the interface between the epi layer and the Si substrate are drastically reduced by both the thermal desorption and the reduction due to the Si source gas at 750°C.

IV. Epitaxial Growth Mechanisms

1. Adsorption and Reaction in Undoped $Si_{1-x}Ge_x$ Epitaxial Growth on a (100) Surface

As the products of the deposition rates for the Ge fraction and the Si fraction in undoped $Si_{1-x}Ge_x$, the reaction rates of GeH_4 and SiH_4 are obtained, respectively. These results are shown in Fig. 10 [15, 32]. With increasing GeH_4 partial pressure, the GeH_4 reaction rate increases monotonically, while the SiH_4 reaction rate increases up to the maximum value and then decreases (Fig. 10a). With increasing SiH_4 partial pressure, the GeH_4 and SiH_4 reaction rates increase up to the maximum value and then decrease (Fig. 10b). The dependence of these reaction rates on the gas partial pressure can be explained based on a Langmuir-type adsorption and reaction scheme [15, 32]. Since the thickness and the Ge fraction had a uniformity of better than 5% within a wafer and from wafer to wafer for 1.25-in.-diameter wafers set as shown in Fig. 1, and scarcely depended on the wafer spacing in the range 6–20 mm and the gas flow rate, the reaction rate was confirmed to be controlled by the surface reaction.

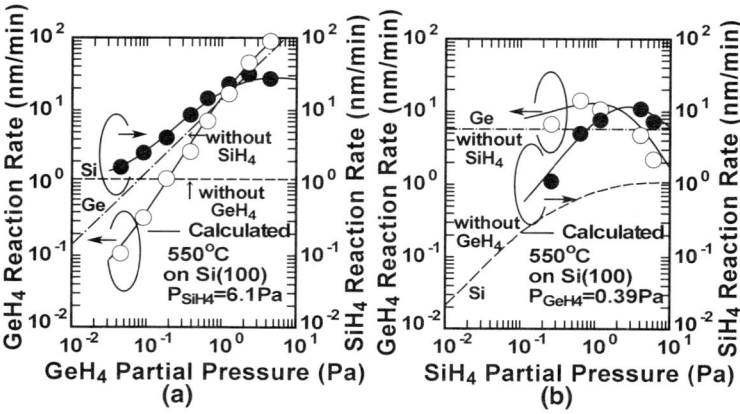

FIG. 10. Dependences of the GeH$_4$ and SiH$_4$ reaction rates on the (a) GeH$_4$ and (b) SiH$_4$ partial pressures in the undoped Si$_{1-x}$Ge$_x$ deposition. The H$_2$ partial pressure is 25 Pa. The solid lines are calculated from Eqs. (3)–(6) using the parameters in Table I.

If it is assumed that one SiH$_4$ or GeH$_4$ molecule is adsorbed at a single adsorption site, according to Langmuir's adsorption isotherm, and decomposes there [33, 34], the SiH$_4$ and GeH$_4$ reaction rates R_{Si} and R_{Ge} are given by

$$R_{Si} = \frac{k_{Si}[k_1/(k_{Si}+k_{-1})]P_{SiH_4}n_0}{1+[k_1/(k_{Si}+k_{-1})]P_{SiH_4}+[k_2/(k_{Ge}+k_{-2})]P_{GeH_4}}, \quad (1)$$

$$R_{Ge} = \frac{k_{Ge}[k_2/(k_{Ge}+k_{-2})]P_{GeH_4}n_0}{1+[k_1/(k_{Si}+k_{-1})]P_{SiH_4}+[k_2/(k_{Ge}+k_{-2})]P_{GeH_4}}, \quad (2)$$

where k_{Si} and k_{Ge} are the SiH$_4$ and GeH$_4$ reaction rate constants, k_1 and k_2 the SiH$_4$ and GeH$_4$ adsorption rate constants, and k_{-1} and k_{-2} the SiH$_4$ and GeH$_4$ desorption rate constants, respectively, n_0 is the total adsorption site density, and P_{SiH_4} and P_{GeH_4} are the SiH$_4$ and GeH$_4$ partial pressures, respectively. The expressions $K_{Si} = k_1/(k_{Si}+k_{-1})$ and $K_{Ge} = k_2/(k_{Ge}+k_{-2})$ denote the SiH$_4$ and GeH$_4$ effective adsorption equilibrium constants, respectively. Under some reasonable assumptions such as $[k_1/(k_{Si}+k_{-1})]P_{SiH_4} \gg [k_2/(k_{Ge}+k_{-2})]P_{GeH_4}$, $k_{Si} \gg k_{-1}$, and $k_{Ge} \gg k_{-2}$ [15, 32], Eqs. (1) and (2) become

$$R_{Si} = \frac{k_1 P_{SiH_4} n_0}{1+(k_1/k_{Si})P_{SiH_4}}, \quad (3)$$

$$R_{Ge} = \frac{k_2 P_{GeH_4} n_0}{1+(k_1/k_{Si})P_{SiH_4}}. \quad (4)$$

The Ge fraction x is given by

$$x = \frac{R_{Ge}}{R_{Si}+R_{Ge}} = \frac{k_2}{k_1} \cdot \frac{P_{GeH_4}}{P_{SiH_4}+(k_2/k_1)P_{GeH_4}}. \quad (5)$$

4   EPITAXIAL GROWTH TECHNIQUES: LOW-TEMPERATURE EPITAXY   137

TABLE I

THE FITTING PARAMETERS $k_{1i}n_0$, $k_{2i}n_0$, AND $k_{Si_i}n_0$ ON THE (100) SURFACE,
CALCULATED FROM EQS. (3)–(6) USING DATA IN FIGS. 8 AND 9

i^a	$k_{1i}n_0$ (cm min^{-1} Pa^{-1})	$k_{2i}n_0$ (cm min^{-1} Pa^{-1})	$k_{Si_i}n_0$ (cm min^{-1})
1	24×10^{-8}	400×10^{-8}	12×10^{-8}
2	140×10^{-8}	600×10^{-8}	12×10^{-8}
3	40×10^{-8}	150×10^{-8}	1200×10^{-8}

$^a i = 1$, 2, and 3 correspond to the Si–Si, Si–Ge, and Ge–Ge pair sites, respectively.

To explain the results shown in Fig. 8, we introduced the adsorption and reaction site dependence of the rate constants [15, 32]. Namely, in the case of the reaction on a (100) surface, there are Si–Si, Si–Ge, and Ge–Ge pair sites, whose densities are proportional to $n_0(1-x)^2$, $2n_0 x(1-x)$, and $n_0 x^2$, respectively. Thus $k_a n_0$ ($a = 1$, 2, Si) can be expressed as

$$k_a n_0 = \sum_{i=1}^{3} k_{ai} n_0 c_i, \quad c_1 = (1-x)^2, \quad c_2 = 2x(1-x), \quad c_3 = x^2, \quad (6)$$

where k_{ai} is the rate constant at each pair site, respectively, on the (100) surface. Details on how the values of $k_{ai}n_0$ ($a = 1$, 2, Si; $i = 1$, 2, 3) were determined, how accurate they are, and what they mean are given elsewhere [15]. Very good agreement is obtained between the experimental data and the reaction rate (solid lines in Fig. 10) calculated from Eqs. (3)–(6) and the values listed in Table I. It should be noted that some values in Table I are revised a little from those in the previous report [15]. Although both sets of parameters result in similar agreements with the experimental results in Fig. 10, the present set is better for describing the doping characteristics discussed in the next Section (IV.2). From Table I, with this accuracy, it is found that the SiH$_4$ and GeH$_4$ adsorption rate constants become larger at the bond site of the Si–Ge pair than those at the bond sites of the other pairs on the (100) surface, while the SiH$_4$ surface reaction rate constant becomes the largest at the bond site of the Ge–Ge pair. The enhancement of the SiH$_4$ reaction rate constant at the bond site of the Ge–Ge pair may result from Ge acting as a desorption center for hydrogen atoms of SiH$_4$ on the surface [35].

In the case of the present pure Si deposition without GeH$_4$, the activation energy of the SiH$_4$ reaction rate constant listed in Table I is about 41 kcal/mol, which is in good agreement with that of the deposition rate under the investigated conditions shown in Fig. 4 and, also, with that in other investigations [33, 36]. It has been reported that Si deposition is limited by SiH decomposition, in other words, hydrogen desorption from SiH [37, 38]. It is found that the SiH$_4$ reaction rate constant, $k_{Si_1}n_0$ listed in Table I agrees with the SiH constant estimated from Ref. 39 within 50%. Also, the equilibrium surface SiH$_4$ coverage,

$(k_{11}/k_{Si1})P_{SiH_4}/[1+(k_{11}/k_{Si1})P_{SiH_4}]$, during pure Si deposition calculated from Table I is in good agreement with the data on equilibrium surface hydrogen coverage in Ref. 39. Therefore, it is suggested that $k_{Si1}n_0$ is the SiH decomposition rate constant and $k_{11}n_0$ is the total SiH formation rate constant at the surface from SiH_4 in the gas phase. For the case of GeH_4 decomposition, it has been reported that Si atoms play the role of decomposition centers of GeH_4 with SiH formation, even though the decomposition rate of GeH_4 is higher at the Ge bond sites than at the Si bond sites, and SiH decomposition is enhanced by the existence of Ge atoms on Si(100) [40]. Based on this reported result, it is considered for the values in Table I that the SiH_4 reaction rate constant becomes large at the bond site of Ge, and the SiH_4 adsorption rate constant $k_{12}n_0$; in other words, the SiH formation rate constant at the bond site of the Si–Ge pair is increased by SiH generation from GeH_4 decomposition, and the GeH_4 adsorption rate constants $k_{21}n_0$, and $k_{22}n_0$; in other words, the GeH formation rate constants include the transition to GeH due to SiH generation from GeH_4 in the gas phase. Further investigations on such decomposition steps are necessary.

2. *In Situ* DOPING OF B AND P IN $S_{1-x}Ge_x$ EPITAXIAL GROWTH

In the case of B doping [32], the reduction of the deposition rate occurs only at higher GeH_4 partial pressures (Fig. 11a). The Ge fraction scarcely changes with B_2H_6 addition (Fig. 12a), although the lattice constant of the film decreases with increasing B_2H_6 partial pressure. The B concentration increases nearly proportionally with the B_2H_6 partial pressure up to the 10^{22} cm^{-3} range (Fig. 13a).

In the case of P doping, the reduction of the deposition rate shifts to higher PH_3 partial pressures with increasing GeH_4 partial pressure (Fig. 11b). The decrease in the deposition rate at the higher PH_3 partial pressure is considered to be caused by the decrease in the density of free surface sites due to surface adsorption of

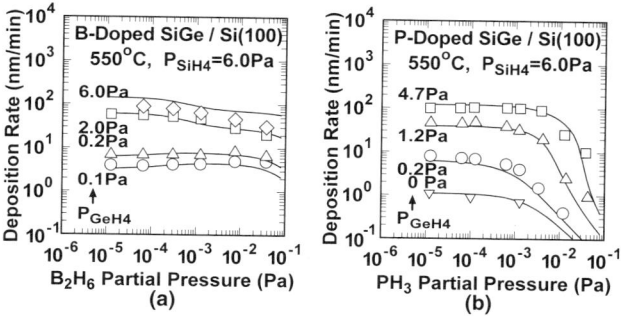

FIG. 11. Dependences of the deposition rate on the (a) B_2H_6 and (b) PH_3 partial pressures. The solid lines are calculated from Eqs. (3)–(5) and (7)–(9) using the parameters in Tables I–III.

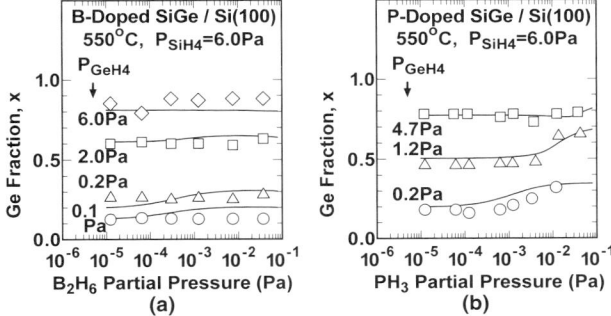

FIG. 12. Dependences of the Ge fraction on the (a) B_2H_6 and (b) PH_3 partial pressures. The solid lines are calculated from Eqs. (3)–(5) and (7)–(9) using the parameters in Tables I–III.

PH_3, and also it is assumed that neither the SiH_4 nor the GeH_4 molecules can react effectively on the P or PH_x adsorbed site. The increase in the Ge fraction is delayed at the higher GeH_4 partial pressure. It is considered that high-concentration P doping induces more suppression of the SiH_4 reaction compared with the GeH_4 reaction on the surface. The Ge fraction x increases in the higher PH_3 partial pressure region (Fig. 12b). The P concentration increases up to a maximum value and then decreases with increasing PH_3 partial pressure (Fig. 13b). The PH_3 partial pressure at the maximum also increases at the higher GeH_4 partial pressure, like the deposition rate shown in Fig. 10b.

If it is assumed that one dopant molecule (B-hydride or P-hydride) occupies one free surface site according to Langmuir's adsorption isotherm, and the site where

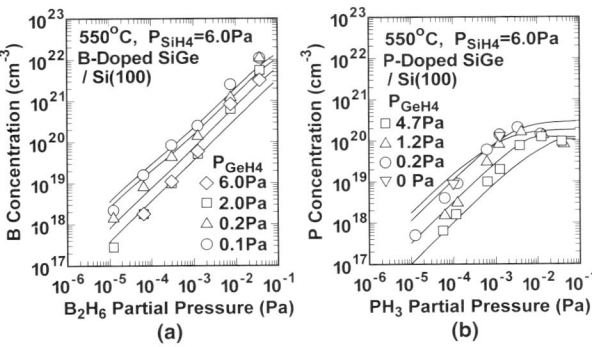

FIG. 13. Dependences of the dopant concentrations in $Si_{1-x}Ge_x$ films on the (a) B_2H_6 and (b) PH_3 partial pressures. The solid lines are calculated from Eqs. (3)–(5) and (7)–(9) using the parameters in Tables I–III.

the dopant molecule is adsorbed becomes inactive for $Si_{1-x}Ge_x$ deposition, Eq. (6) can be modified as

$$k_a n_0 = \sum_{i=1}^{3} k_{ai} n_0 c_i \frac{1}{1 + K_{Di} P_D} \quad (a = 1, 2, Si; D = B, P), \quad (7)$$

where K_{Di} is the effective adsorption equilibrium constant of dopants at each pair site (Si–Si, Si–Ge, and Ge–Ge, respectively) and P_D is the dopant partial pressure. The K_{Di} is given by

$$K_{Di} = \frac{k_{Di}}{K_{SDi} R c_i + k_{-Di}} \quad (D = B, P), \quad (8)$$

where k_{Di} is the dopant adsorption rate constant, k_{-Di} is the dopant desorption rate constant, K_{SDi} is the effective segregation coefficient between the surface coverage of dopant molecules and the concentration of dopant incorporated in the depositing film at each pair site, respectively, and R is the $Si_{1-x}Ge_x$ deposition rate, $R = R_{Si} + R_{Ge}$. Here the dopant incorporation is assumed to be conducted by Henry's law and the reaction rate constant in Eq. (1) or (2) is replaced by the dopant incorporation rate at each site. The dopant concentration C_D in the deposited film is given by

$$C_D = \sum_{i=1}^{3} K_{SDi} c_i \frac{K_{Di} P_D}{1 + K_{Di} P_D}. \quad (9)$$

It is also assumed that B_2H_6 molecules are decomposed completely into BH_3 in the gas phase [41]. The values of k_{Di}, k_{-Di}, and K_{SDi} ($i = 1, 2, 3$; $D = B, P$) are determined by fitting the data shown in Figs. 9–11 to Eqs. (3)–(5) and (7)–(9) with the values listed in Table I. The best values determined by the method described below are listed in Tables II and III. Fairly good agreement is obtained between all the experimental results and the calculated values (solid lines in Figs. 11–13). This supports the feasibility of the present formalism.

In the case of B doping, because a B concentration higher than 10^{22} cm^{-3} is available for Si and Ge deposition and the experimental results (Fig. 13b) are almost-linear (not saturated yet) to the dopant partial pressure P_D, a certain value

TABLE II

THE FITTING PARAMETERS k_{Bi}, k_{-Bi}, AND K_{SBi} ON THE (100) SURFACE AT 550°C, CALCULATED FROM EQS. (7)–(9) AND TABLE I

i^a	k_{Bi} (cm^{-2} min^{-1} Pa^{-1})	k_{-Bi} (cm^{-2} min^{-1})	k_{SBi} (cm^{-3})
1	1×10^{17}	1×10^{15}	300×10^{20}
2	3×10^{17}	0.02×10^{15}	0.3×10^{20}
3	2×10^{17}	20×10^{15}	300×10^{20}

$^a i = 1, 2$, and 3 correspond to the Si–Si, Si–Ge, and Ge–Ge pair sites, respectively.

TABLE III

THE FITTING PARAMETERS k_{Pi}, k_{-Pi}, AND K_{SPi} ON THE (100) SURFACE AT 550°C, CALCULATED FROM EQS. (7)–(9) AND TABLE I

i^a	k_{Pi} (cm^{-2} min^{-1} Pa^{-1})	k_{-Pi} (cm^{-2} min^{-1})	k_{SPi} (cm^{-3})
1	5×10^{16}	100×10^{12}	3×10^{20}
2	8×10^{16}	1×10^{12}	1×10^{20}
3	1×10^{16}	1×10^{12}	1×10^{20}

$^a i = 1$, 2, and 3 correspond to the Si–Si, Si–Ge, and Ge–Ge pair sites, respectively.

larger than 10^{22} cm^{-3} was set for K_{SB1} and K_{SB3}. Then, from the deposition rate and the doping characteristics, the k_{B1}, k_{-B1}, k_{B3} and k_{-B3} values were determined. Next, the k_{B2}, k_{-B2}, and K_{SB2} values were determined so that the calculations with Eqs. (3)–(5) and (7)–(9) could fit the experimental results shown in Figs. 11a, 12a, and 13a. Although the best-fitting values listed in Table II are not very accurate, it is certain that the segregation coefficient K_{SB} becomes the smallest at the Si–Ge pair site. It is known from evaluation by Eq. (9) that K_{B2} at the Si–Ge pair site is the largest, which means that the B-hydride adsorbs most easily at the Si–Ge pair site, and as a result, with increasing B_2H_6 partial pressure the weighted sums of k_1 and k_2 evaluated by Eq. (7) and Table I decrease more than that of k_{Si}. This, with the aid of Eqs. (3) and (4), explains the decrease in the deposition rate at high B_2H_6 partial pressures.

In the case of P doping, considering the P coverage to be ~ 1 at the maximum values of the P concentration in Fig. 13b, the K_{SPi} values were determined from Eq. (9). Next, the K_{Pi} values were determined from the data at lower PH_3 partial pressures the k_{Pi} and k_{-Pi} values were related by Eq. (8). The k_{P1} and k_{-P1} values were then determined from the deposition rate and the doping characteristics in P-doped Si deposition. The other parameters were determined so that the calculations with Eqs. (3)–(5) and (7)–(9) could fit the experimental results shown in Figs. 11b, 12b, and 13b.

From the best-fitting values listed in Table III and evaluation of the parameters using Eqs. (5), (7), and (8) with the aid of the parameters in Table I, it is known that with PH_3 addition, adsorption of SiH_4 and GeH_4 becomes greatest at the Si–Si pair site and the k_2/k_1 value becomes larger, which results in an increase in x [see Eq. (5)]. Namely, the increase in the Ge fraction in the higher PH_3 partial pressure region observed in Fig. 12b is caused by more suppression of SiH_4 and GeH_4 adsorption/reaction at the Si–Ge and Ge–Ge pair sites than at the Si–Si pair site. Existence of the maximum P concentration in Fig. 13b is explained by the saturation of PH_3 adsorption and the lower solubility of P at the Si–Ge and Ge–Ge pair sites than at the Si–Si pair site (see the K_{SPi} values).

Next, the relationship between dopant and carrier concentrations and resistivity in the films is discussed. In the case of B-doped films, the carrier concentration

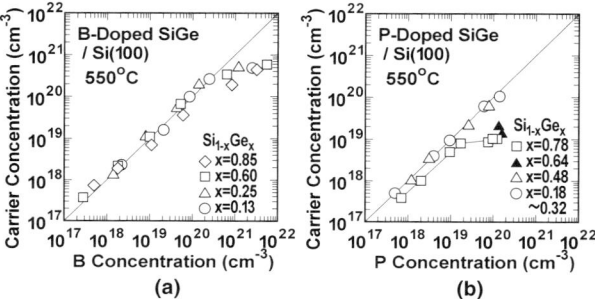

FIG. 14. Dependences of the carrier concentrations in $Si_{1-x}Ge_x$ films on the (a) B and (b) P concentrations.

is nearly equal to the B concentration up to about 2×10^{20} cm^{-3}, and it tends to saturate to about 5×10^{20} cm^{-3} at a B concentration below 10^{22} cm^{-3} (Fig. 14a). Discrepancy of the lattice constants from Vegard's law was observed at a higher B concentration, of the order of 10^{20} cm^{-3} and above, which corresponds with saturation of the carrier concentration. The resistivity decreases with increasing carrier concentration down to about 5×10^{-4} Ω-cm (Fig. 15a), where the data are close to Irvin's curve of Si [42]. To state more details of the Ge fraction dependence, the resistivity exhibits a maximum around $x = 0.25$ at carrier concentrations below 10^{20} cm^{-3} and then decreases with increasing Ge fraction, but these characteristics are not apparent at carrier concentrations above 10^{20} cm^{-3}. It is considered that resistivity is influenced by alloy scattering. In the case of P-doped films with a Ge fraction higher than 0.5, the electrically inactive P atoms are observed independently of the P concentration and the carrier concentration tends to saturate to about 10^{19} cm^{-3} at a higher P concentration, up to about 2×10^{20} cm^{-3} (Fig. 14b). This means that the P-doped $Si_{1-x}Ge_x$ film with the higher Ge fraction has the lower solid solubility of electrically active P. Thus, a Ge fraction x lower than 0.5

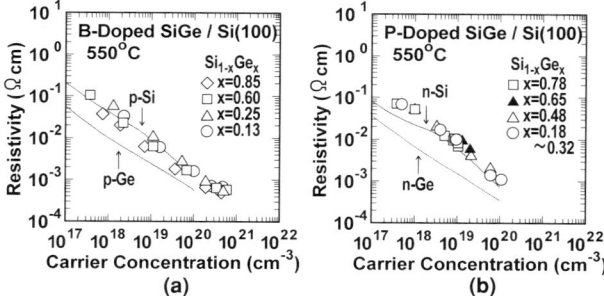

FIG. 15. Dependences of the resisitivity of $Si_{1-x}Ge_x$ films on the (a) B and (b) P concentrations.

should be used for the lower electrical contact resistance between $Si_{1-x}Ge_x$ and metal. The resistivity in the film with a Ge fraction above 0.18 is independent of the Ge fraction and is higher than that in Si (Fig. 15b). This result is in marked contrast with that of the B-doped film.

V. High-Quality $Si/Si_{1-x}Ge_x/Si$ Heterostructure Growth at High Ge Fractions

The low-temperature epi growth of $Si/Si_{1-x}Ge_x/Si$ heterostructures has become increasingly important for the fabrication of novel Si devices [2–5]. For excellent heterodevices, high Ge fractions in the strained $Si_{1-x}Ge_x$ layer as well as atomically flat surfaces and interfaces are necessary. At a higher Ge fraction, the heterojunction appears to be degraded by island growth as well as by generation of misfit dislocations due to the larger lattice mismatch between the Si and the $Si_{1-x}Ge_x$ layers [9]. Here the epitaxial growth conditions of the $Si/Si_{1-x}Ge_x/Si$ heterostructure with atomically flat surfaces and interfaces at high Ge fractions are optimized.

The high-quality epi growth of $Si_{1-x}Ge_x$ films on Si(100) substrates has been performed in a SiH_4–GeH_4–H_2 gas mixture using the ultraclean hot-wall LP-CVD machine mentioned in Section II. The typical process sequence for $Si/Si_{1-x}Ge_x/Si$ heterostructure growth is shown in Fig. 16 [15] (see Figs. 7 and 9). A lower temperature during $Si_{1-x}Ge_x$ and Si capping layer depositions results in a smoother surface. At a Ge fraction of about 0.2, atomically flat surfaces and interfaces can be obtained by depositing the $Si_{1-x}Ge_x$ and Si capping layers at 550°C. For higher Ge fractions, however, much lower deposition temperatures are suitable, namely, 500°C for a $Si_{0.5}Ge_{0.5}$ layer and 450°C for a $Si_{0.3}Ge_{0.7}$ layer, although no change of heterostructure surfaces with a Ge fraction of about 0.7 was detected after capping layer deposition at 550°C or lower [15, 44]. Among the surface morphologies of the

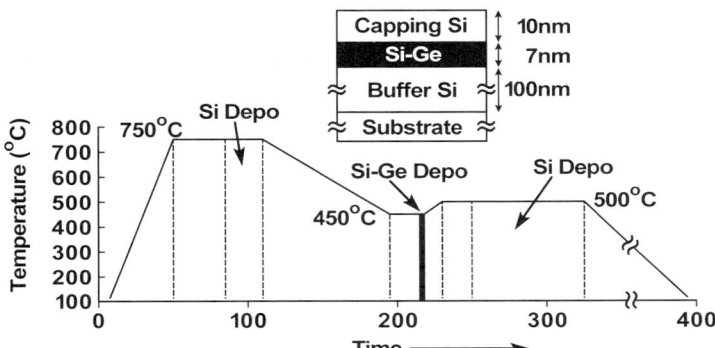

FIG. 16. Typical process sequence for $Si/Si_{1-x}Ge_x/Si$ heterostructure growth in the ultraclean hot-wall LP-CVD system.

heterostructures containing $Si_{1-x}Ge_x$ layers, with $x = 0.2$, 0.5, and 0.7, deposited at 550, 500, and 450°C, respectively, as well as the $Si_{0.3}Ge_{0.7}$ layer and Si substrate just after deposition of the Si buffer layer, there was no visible difference within the detection limit of STM (average surface microroughness, <0.4 nm). These results clearly show that lowering the deposition temperature of the $Si_{1-x}Ge_x$ layers is necessary with increasing Ge fraction to prevent island growth of the heterostructure.

Cross-sectional TEM images of $Si/Si_{1-x}Ge_x/Si$ heterostructures are shown in Fig. 17 [15]. In the case of $Si/Si_{0.8}Ge_{0.2}/Si$ and $Si/Si_{0.5}Ge_{0.5}/Si$ heterostructures deposited at 500°C with a flat surface, atomically flat interfaces are observed and the lattice image has no misfit dislocations, as shown in Figs. 17a and b. In the case of the $Si/Si_{0.3}Ge_{0.7}/Si$ heterostructure deposited at 500°C, island growth of the $Si_{0.3}Ge_{0.7}$ layer and misfit dislocations in the island growth region are observed, as shown in Fig. 17c. The island growth may induce the generation of defects. In the case of the $Si/Si_{0.3}Ge_{0.7}/Si$ heterostructure containing a $Si_{0.3}Ge_{0.7}$ layer deposited at 450°C with a flat surface, regular atomic alignment at the interface is observed, but the interface position is not clearly distinguishable, as shown in Fig. 17d. This feature of the TEM image contrast may be induced by the strain between the $Si_{0.3}Ge_{0.7}$ layers and the Si layer and/or interdiffusions of Si and Ge during $Si/Si_{0.3}Ge_{0.7}/Si$ heterostructure growth. With wet oxidation for 2 h at 700°C to form gate oxide, in the case of the $Si/Si_{0.5}Ge_{0.5}/Si$ heterostructure, the interface position tends to become vaguer than that without oxidation, as compared in Figs. 17e and b. This result indicates the possibility of interdiffusion of Si and Ge during oxidation processes. In the case of the $Si/Si_{0.3}Ge_{0.7}/Si$ heterostructure, the interface position after oxidation tends to be clearer than that without oxidation, as compared in Figs. 17f and d. It is considered that the strain is dispersed by the interdiffusion of Si and Ge.

For the above samples, the Ge fraction x in the very thin $Si_{1-x}Ge_x$ layers was estimated from the lattice constant of a thicker relaxed $Si_{1-x}Ge_x$ layer, deposited under the same conditions. The Raman intensity ratio of the Ge–Ge peak to the Si–Ge peak for the heterostructure is nearly the same as that for the thicker layer; in other words, the Ge fraction in the heterostructure is nearly equal to that in the thicker layer [45]. Upon wet oxidation for 2 h at 700°C, the Raman spectrum of the heterostructure sample shown in Fig. 17b scarcely changed, but a larger decrease in Raman intensity was observed for thinner $Si_{1-x}Ge_x$ layers with a higher Ge fraction [45]. Further investigations of the interdiffusions are necessary.

The long-channel pMOSFETs (Fig. 18) were fabricated on the $Si/Si_{1-x}Ge_x/Si$ heterostructures deposited under the same conditions as those for the sample shown in Fig. 17 using the self-aligned Si gate process [46]. A 10-nm-thick gate oxide was thermally grown by wet oxidation at 700°C. All annealing processes were performed at 700°C or lower. The threshold voltage and the peak field-effect mobility of $Si_{1-x}Ge_x$-channel MOSFETs are shown in Fig. 19 [46]. The $Si_{0.5}Ge_{0.5}$-channel MOSFET having a flat surface has the highest peak field-effect mobility, resulting in a large mobility enhancement, about 70% at 300 K and over 150% at 77 K, compared with those of the MOSFET without a $Si_{1-x}Ge_x$ channel. This mobility enhancement is excellent compared with those reported by other investigators [4, 5].

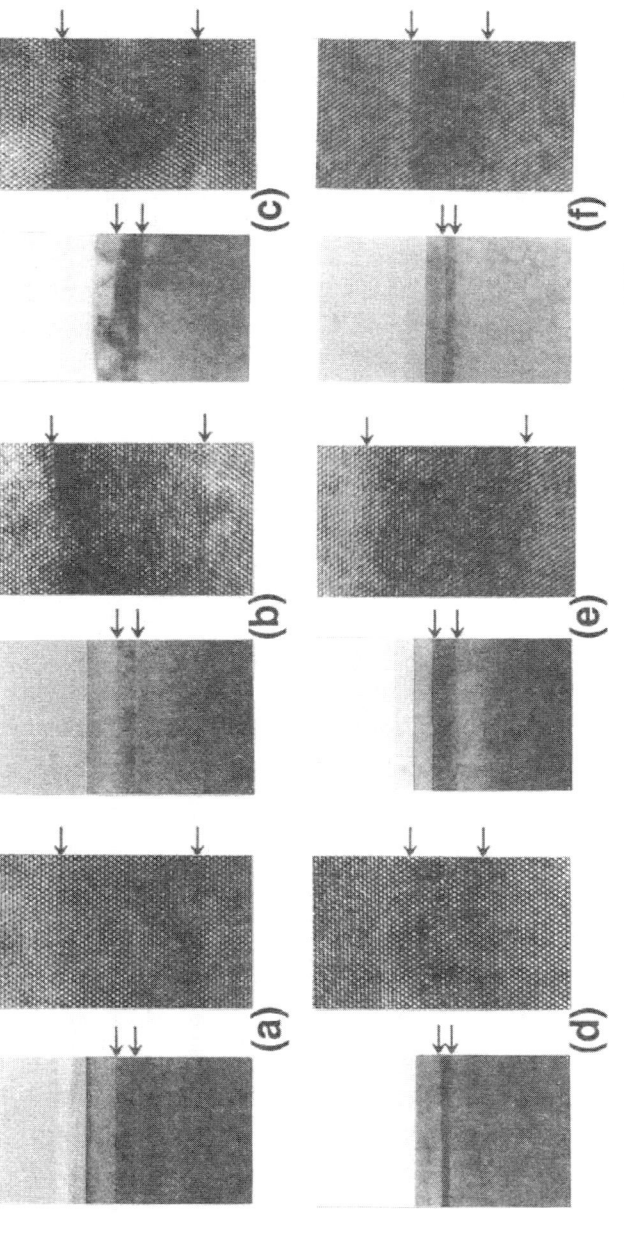

FIG. 17. Cross-sectional TEM images observed from the [011] azimuth of Si/Si$_{1-x}$Ge$_x$/Si heterostructures grown on Si(100), where Si$_{1-x}$Ge$_x$ layers with a Ge fraction x of (a) 0.2, (b) 0.5, and (c) 0.7 were deposited at 500°C, at a deposition rate of 0.46, 4.6, and 22 nm/min, respectively, and one (d) with $x = 0.5$ was deposited at 450°C at a deposition rate of 3.7 nm/min. The Si capping layer was deposited at 500°C at a deposition rate of 0.11 nm/min. For deposition of the Si$_{1-x}$Ge$_x$ and Si capping layers, the total gas pressure was about 25 Pa, and the partial pressures of SiH$_4$ and GeH$_4$ were in the ranges 1–1.4 and 0.03–0.75 Pa, respectively, with H$_2$ or Ar as the carrier gas. (e, f) The samples shown in b and d, respectively, after wet oxidation for 2 h at 700°C. Arrows indicate the interfaces in the heterostructure.

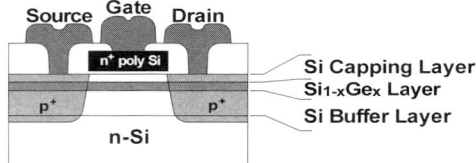

FIG. 18. Schematic cross section of the $Si_{1-x}Ge_x$-channel pMOSFET.

The subthreshold slopes (about 80 mV/decade at 300 K and about 30 mV/ decade at 77 K) of the $Si_{0.5}Ge_{0.5}$-channel MOSFET were comparable to those of the MOSFET without a $Si_{1-x}Ge_x$ channel. This indicates that defect density in the $Si_{0.5}Ge_{0.5}$ layer is low. On the other hand, the $Si_{0.3}Ge_{0.7}$-channel MOSFET has a higher threshold voltage and a lower peak field-effect mobility than the $Si_{0.5}Ge_{0.5}$-channel MOSFET. Since the band gap of an unstrained $Si_{1-x}Ge_x$ layer increases with decreasing Ge fraction and is larger than that of a strained layer [47], as is well known, it is considered that the Ge fraction decreases and/or the heterostructure becomes unstrained by generating defects, although such defects were not observed in Fig. 17f, during device fabrication. It should be noted that the $Si_{0.3}Ge_{0.7}$ layer has the possibility for a more drastic decrease in the Ge fraction during interdiffusion since it is thinner than the $Si_{0.5}Ge_{0.5}$ layer, as shown in Fig. 17.

The leakage current between source and drain at zero gate voltage also depends on the thickness of the $Si_{1-x}Ge_x$ layer even with a flat surface (Fig. 20). Although the leakage current scarcely depends on the thickness for $x = 0.2$, it becomes high at a thickness above 7 nm for $x = 0.5$, and above 4 nm for $x = 0.7$ [48]. Since the STM measurements showed that the surface roughness of these degraded samples

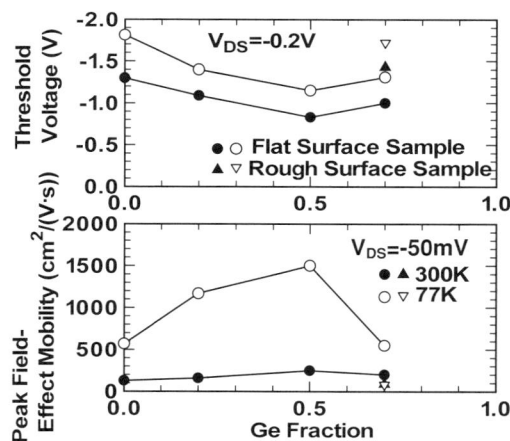

FIG. 19. Threshold voltage and peak field-effect mobility of the $Si_{1-x}Ge_x$-channel MOSFET fabricated on n-type Si(100) of 1–3 Ω-cm.

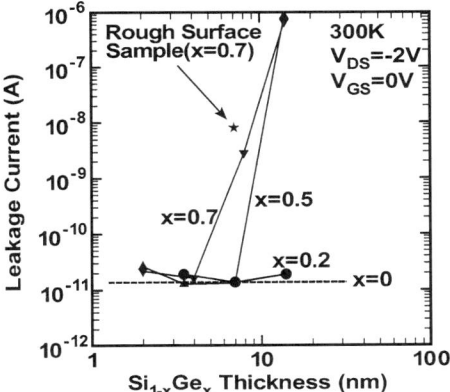

FIG. 20. $Si_{1-x}Ge_x$ thickness dependence of the leakage current of the $Si_{1-x}Ge_x$-channel MOSFET.

was almost the same as that of the nondegraded samples [44], it is suggested that the leakage current is caused mainly by misfit-dislocation generation in $Si_{1-x}Ge_x$ layers thicker than the critical thickness. On the other hand, it is found that the $Si_{0.3}Ge_{0.7}$-channel MOSFET with a rough surface, where the $Si_{0.3}Ge_{0.7}$ layer was deposited at 500°C, has a remarkably low mobility and a high leakage current as shown in Figs. 19 and 20. Such a sample has many misfit dislocations, which were observed by TEM as mentioned above. These results indicate that the nanometer-order surface roughness of the heterostructure and the generation of the misfit dislocation by exceeding the critical thickness of the $Si_{1-x}Ge_x$ layer degrade the device performance. The results mentioned above clearly show that lowering the deposition temperatures of the $Si_{1-x}Ge_x$ layers and optimizing the $Si_{1-x}Ge_x$ layer thickness are essential to prevent island growth and generation of the misfit dislocation in the heterostructures at a higher Ge fraction.

VI. Conclusions

Low-temperature epitaxy of Si and $Si_{1-x}Ge_x$ on Si(100) is achieved using UHV-CVD, LP-CVD and AP-CVD machines. The key points are the wafer surface treatment in the reactor and the ultraclean environment. In the case of LP-CVD using SiH_4, GeH_4, and dopant gas (B_2H_6 and PH_3) in the surface reaction-limited regime, the deposition rate, the Ge fraction and the *in situ* doping characteristics are expressed based on a modified Langmuir-type adsorption and reaction scheme, assuming that the reactant gas adsorption/reaction depends on the surface materials. In high-quality $Si/Si_{1-x}Ge_x/Si(100)$ heterostructure growth for a high Ge fraction with atomically flat surfaces and interfaces, relatively low deposition temperatures such as 450–500°C and optimization of the layer thickness are suitable. $Si_{1-x}Ge_x$ heterodevices with high Ge fractions are attractive for high performance.

But to apply a $Si_{1-x}Ge_x$ layer containing a Ge fraction above 0.5 to heterodevices, further investigations are necessary, especially of interdiffusion of Si and Ge during epitaxial growth, the low solid solubility of electrically active P, and thermal degradation in the device fabrication process. To suppress interdiffusion and break through the solid solubility of the electrically active P, atomic layer-by-layer processing is promising, where Langmuir-type monolayer adsorption and reaction kinetics conduct complete separation of surface adsorption and reaction of the reactant gas [15, 49, 50]. To suppress the thermal degradation, carbon doping is one of the candidates [50, 52].

Acknowledgments

The author thanks Drs. M. Sakuraba and T. Matsuura for helpful discussions. This study was partially supported by the Public Participation Program for the Promotion of Information Communications Technology R&D of the Telecommunications Advancement Organization of Japan and a Grant-in-Aid for Scientific Research on Priority Area "New Group Semiconductor: Control of Proporties and Application to Ultrahigh Speed Opto-Electronic Devices" (Area No. 739) from the Ministry of Education, Science, Sports and Culture of Japan.

References

1. Bloem J., and L. G. Giling, *Current Topics in Materials Science*, Vol. 1, edited by E. Kaldis, Chap. 4, North-Holland, Amsterdam, 1978.
2. Iyer S. S., G. L. Patton, J. M. C. Stork, B. S. Meyerson, and D. L. Harame, *IEEE Trans. Electron Devices* **ED-36**, 2043 (1989).
3. Kasper E., H. Kibbel, H. J. Herzog, and A. Gruhle, *Ext. Abstr., 1993 Int. Conf. Solid State Devices Mater.*, Makuhari, Tokyo, p. 419, 1993.
4. Verdonckt-Vandebroek S., E. F. Crabbe, B. S. Meyerson, D. L. Harame, P. J. Restle, J. M. C. Stork, A. C. Megdanis, C. L. Stanis, A. A. Bright, G. M. W. Kroesen, and A. C. Warren, *IEEE Electron Device Lett.* **EDL-12**, 447 (1991).
5. Garone P. M., V. Venkataraman, and J. C. Sturm, *IEEE Electron Device Lett.* **EDL-13**, 56 (1992).
6. Meyerson B. S., *Appl. Phys. Lett.* **48**, 797 (1986).
7. Donahue T. J., and R. Reif, *J. Appl. Phys.* **57**, 2757 (1985).
8. Murota J., N. Nakamura, M. Kato, N. Mikoshiba, and T. Ohmi, *Appl. Phys. Lett.* **54**, 1007 (1989).
9. Sedgwick T. O., M. Berkenblitz, and T. J. Kuan, *Appl. Phys. Lett.* **54**, 2689 (1989).
10. Meyerson B. S., E. Ganin, D. A. Smith, and T. N. Nguyen, *J. Electrochem. Soc.* **133**, 1232 (1986).
11. Smith F. W., and G. Ghidini, *J. Electrochem. Soc.* **129**, 1301 (1982).
12. Ghidini G., and F. W. Smith, *J. Electrochem. Soc.* **131**, 2934 (1984).
13. Angello P. D., and T. O. Sedgwick, *J. Electrochem. Soc.* **139**, 1140 (1992).
14. Murota J., M. Kato, R. Kircher, and S. Ono, *J. Phys. IV (France)* **1**, C2–795 (1991).
15. Murota J., and S. Ono, *Jpn. J. Appl. Phys.* **33**, 2290 (1994).
16. Violette K. E., M. K. Sangareria, M. C. Öztürk, G. Harris, and D. M. Maher, *J. Electrochem. Soc.* **141**, 3269 (1994).
17. Gibbons J. F., C. M. Gronet, and K. E. Williams, *Appl. Phys. Lett.* **47**, 721 (1985).

18. Green M. L., D. Brasen, H. Luftman, and V. C. Kannan, *J. Appl. Phys.* **65**, 2558 (1989).
19. Sturm J. C., P. V. Schwartz, E. J. Prinz, and H. Manoharan, *J. Vac. Sci. Technol.* **B9**, 2011 (1991).
20. Rosler R. S., *Solid State Technol.* **20**, 63 (1977).
21. Noda T., D. Lee, H. Shim, M. Sakuraba, T. Matsuura, and J. Murota, *Thin Solid Films* **380**, 57 (2000).
22. Ohmi T., J. Murota, Y. Kanno, Y. Mitsui, K. Sugiyama, T. Kawasaki, and H. Kawano, *Proc. 1st Int. Symp. ULSI Sci. Technol.*, Philadelphia, p. 807, Electrochem. Soc., Pennington, NJ, 1987.
23. Wong C. Y., A. E. Michel, R. D. Isaac, R. H. Kasl, and S. R. Mader, *J. Appl. Phys.* **55**, 1131 (1987).
24. Probst V., H. Böhm, H. Schaber, H. Oppolzer, and I. Weitzel, *J. Electrochem. Soc.* **135**, 671 (1998).
25. Murota J., M. Furuno, M. Kato, N. Mikoshiba, S. Ono, H. Kurokawa, T. Sato, N. Nakamura, and F. Ikeda, *The Electrochemical Society Extended Abstracts,* Fall Meeting, Seattle, WA, Oct. 14–19, 1990, Abstr. No. 399, p. 576.
26. Kunii Y., and Y. Sakakibara, *Jpn. J. Appl. Phys.* **26**, 1816 (1987).
27. Murota J., N. Nakamura, M. Kato, N. Mikoshiba, and T. Ohmi, *Proc. 1st Int. Symp. Adv. Mater. ULSI,* Atlanta, p. 103, Electrochem. Soc., Pennington, NJ, 1988.
28. Meyerson B. S., F. J. Himpsel, and K. Uram, *J. Appl. Phys. Lett.* **57**, 1034 (1990).
29. Takagi T., I. Nagai, A. Ishitani, H. Kuroda, and Y. Nagasawa, *J. Appl. Phys.* **64**, 3516 (1988).
30. Burrows V. A., Y. J. Chabal, G. S. Higashi, K. Raghavachari, and S. B. Christman, *Appl. Phys. Lett.* **53**, 998 (1988).
31. Sanganeria M. K., M. C. Öztürk, G. Harris, K. E. Violetle, I. Ban, C. A. Lee, and D. M. Maher, *J. Electrochem. Soc.* **142**, 3961 (1995).
32. Murota J., M. Sakuraba, and T. Matsuura, in *Defects in Silicon/1999,* edited by T. Abe, W. M. Bullis, S. Kobayashi, W. Lin, and P. Wagner, PV99-1, p. 189, Electrochem. Soc., Pennington, NJ, 1999.
33. Comfort J. H., and R. Reif, *J. Electrochem. Soc.* **136**, 2386 (1989).
34. Claassen W. A. P., and J. Bloem, *J. Electrochem. Soc.* **127**, 194 (1980).
35. Meyerson B. S., K. C. Uram, and F. K. LeGoues, *Appl. Phys. Lett.* **53**, 2555 (1988).
36. Claassen W. A. P., J. Bloem, W. G. J. N. Valkenburg, and C. H. J. van den Brekel, *J. Cryst. Growth* **57**, 259 (1982).
37. Liehr M., C. M. Greenlief, S. R. Kasi, and M. Offenberg, *Appl. Phys. Lett.* **56**, 629 (1990).
38. Gates S. M., and S. K. Kurkarni, *Appl. Phys. Lett.* **58**, 2963 (1991).
39. Liehr M., C. M. Greenlief, M. Offenberg, and S. R. Kasi, *J. Vac. Sci. Technol.* **A8**, 2960 (1990).
40. Greenlief G. M., P. C. Wankum, D.-A. Klug, and L. A. Keeling, *J. Vac. Sci. Technol.* **A10**, 2465 (1992).
41. Bauer S. H., *J. Am. Chem. Soc.* **78**, 5775 (1996).
42. Irvin J. C., *Bell Syst. Technol.* **41**, 387 (1962).
43. Bean J. C., L. C. Feldman, A. T. Fiory, S. Nakahara, and I. K. Robinson, *J. Vac. Sci. Technol.* **A2**, 436 (1984).
44. Schütz R., J. Murota, T. Maeda, R. Kircher, K. Yokoo, and S. Ono, *Appl. Phys. Lett.* **61**, 2674 (1992).
45. Sakamoto K., J. Murota, T. Maeda, K. Goto, S. Ushida, and S. Ono, *Proc. Int. Conf. Adv. Microelectron. Devices Proc.* Research Institute of Electrical Communication, Tohoku University, Sendai, p. 449, 1994.
46. Goto K., J. Murota, T. Maeda, R. Schütz, K. Aizawa, R. Kircher, K. Yokoo, and S. Ono, *Jpn. J. Appl. Phys.* **32**, 438 (1993).
47. People R., *Phys. Rev. B* **32**, 1405 (1985).
48. Tsuchiya T., K. Goto, M. Sakuraba, T. Matsuura, and J. Murota, *Thin Solid Films* **369**, 379 (2000).
49. Murota J., T. Matsuura, and M. Sakuraba, *Future Trends in Microelectronics,* edited by S. Luryi, J. Xu, and A. Zaslavsky, p.79, John Wiley & Sons, New York, 1999.
50. Tillack B., B. Heinemann, and D. Knoll, *Thin Solid Films* **369**, 189 (2000).
51. Shimamune Y., M. Sakuraba, T. Matsuura, and J. Murota, *Thin Solid Films* **380**, 134 (2001).
52. Knoll D., B. Heinemann, K.-E. Ehwald, B. Tillack, P. Schley, and H. J. Osten, *Thin Solid Films* **369**, 342 (2000).

CHAPTER 5

Epitaxial Growth Techniques: Molecular Beam Epitaxy

Y. Shiraki[1]

RESEARCH CENTER FOR ADVANCED SCIENCE AND TECHNOLOGY (RCAST)
UNIVERSITY OF TOKYO
TOKYO, JAPAN

ABBREVIATIONS USED	151
I. INTRODUCTION	152
II. MBE MACHINES	153
1. *Solid-Source MBE Machines*	153
2. *Gas-Source MBE Machines*	154
III. SURFACE TREATMENT	155
IV. GROWTH MECHANISMS	157
1. *Solid-Source MBE*	157
2. *Gas-Source MBE*	158
V. DOPING	162
1. *Solid-Source MBE*	162
2. *Gas-Source MBE*	163
VI. GROWTH OF SiGe(C) ALLOYS	166
1. *Solid-Source MBE*	166
2. *Formation and Control of SiGe Alloys by Gas-Source MBE*	166
VII. FORMATION OF Si/Ge HETEROSTRUCTURES	169
1. *Band Structures of Si/Ge Heterostructures*	169
2. *The Critical Thickness*	170
3. *Surface Segregation*	172
4. *Surfactant-Mediated Growth*	173
VIII. GROWTH OF SiGe BUFFER LAYERS	174
IX. SELECTIVE GROWTH	175
X. FORMATION OF SUPERLATTICES, QUANTUM WIRES, AND QUANTUM DOTS	177
XI. CONCLUSION	181
REFERENCES	182

ABBREVIATIONS USED

AES	Auger electron spectroscopy
AFM	Atomic force microscopy
CVD	Chemical vapor deposition

[1]Current address: Department of Applied Physics, University of Tokyo, 7-3-1 Homgo, Bunkyo–ku, Tokyo 113-8656, Japan.

DP	Diffusion pump
EB	Electron beam
FET	Field effect transistor
GS-MBE	Gas-source molecular beam epitaxy
HBT	Heterostructure bipolar transistor
LEED	Low-energy electron diffraction
LPE	Liquid-phase epitaxy
LT	Low temperature
MBE	Molecular beam epitaxy
MOSFET	Metal–oxide–semiconductor FET
PL	Photoluminescence
QW	Quantum well
QWR	Quantum wire
RBS	Rutherford backscattering
RHEED	Reflection high-energy electron diffraction
SEG	Selective epitaxial growth
SIMS	Secondary-ion mass spectroscopy
SK	Stranski–Krastanov
SMG	Surfactant-mediated growth
SPE	Solid-phase epitaxy
SS-MBE	Solid-source molecular beam epitaxy
TDS	Temperature-programmed desorption spectrum
III–V	Compounds such as gallium arsenide (GaAs)
TMP	Turbo mulecular pump
UHV	Ultrahigh vacuum
UPS	Ultraviolet photoemission spectroscopy
VPE	Vapor-phase epitaxy
XPS	X-Ray photoemission spectroscopy

I. Introduction

Silicon epitaxial growth can be done by vapor-phase (VPE), liquid-phase (LPE), or solid-phase (SPE) epitaxy techniques. LPE is done by crystallization of epitaxial layers on Si substrates from In melt in which Si and other necessary materials are solved. Crystallization by thermal annealing from amorphous silicon formed by heavy ion implantation or vacuum deposition is a typical example of Si SPE. However, LPE and SPE are employed mainly for scientific studies or purposes and are not commonly used to obtain silicon epitaxial layers for device applications. On the other hand, VPE is the key technology used to form epitaxial Si layers nowadays. Si VPE is classified into two groups: chemical vapor deposition (CVD) and vacuum deposition or evaporation, which is called molecular beam epitaxy (MBE). Both techniques can provide not only simple epitaxial growth but also precise control of thickness with an atomic scale of grown layers including heterostructures. However, the growth mechanisms are very different between the techniques. Chemical reaction of source materials both in the gas phase and on substrate surfaces governs the growth in CVD, while reaction on the substrate surface is important only in the case of MBE. In this chapter, the details of MBE, i.e., machines, processes, and growth mechanisms, are described.

II. MBE Machines

An outline of MBE machines is given in this section. There are two methods in MBE, depending on which source materials are used, solid or gas. They are called solid-source (SS) and gas-source (GS) MBE, respectively.

1. SOLID-SOURCE MBE MACHINES

Since MBE is a method to deposit materials on crystalline substrates under UHV, the system consists of a UHV growth chamber and a first entry-lock, by which samples can be changed without disturbing UHV conditions as shown in Fig. 1. Except for an evaporation source of Si and a sample holder, almost all equipment is similar to that for III–V MBE systems. The temperature at which silicon melts and gives rise to a significant evaporation rate is so high ($\sim 1600°C$)

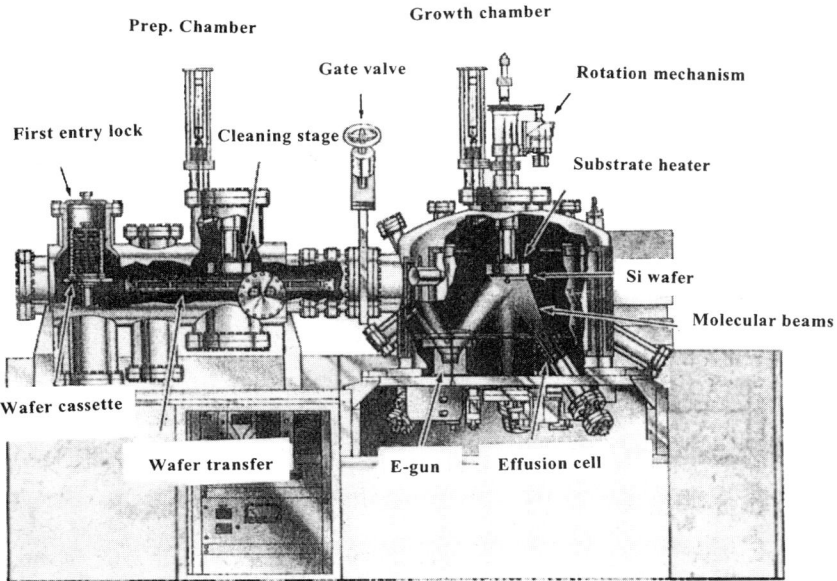

FIG. 1. Schematic illustration of a solid-source MBE machine for Si and Ge growth.

that a Knudsen cell, which is usually used in III–V MBE machines, cannot easily be used to heat the silicon source. Therefore, electron beam evaporation sources are normally employed. Very pure Si charges are used and the center portion of the charge should be melted so that the colder outer Si shell acts as a crucible. Since stray electron beams tend to hit the crucible and surrounding walls to cause outgassing and contamination, some cautions should be paid for protection.

An e-gun Si evaporator may be considered as a point source. The flux is roughly proportional to $\cos^2 \theta L^{-2}$, where L is the distance between the source and the substrate and θ is the angular separation of the substrate from a line perpendicular to the center of the source. A typical deposition rate is 1–20 Å and a sample is usually rotated to obtain a high uniformity, better than $\pm 5\%$.

A quartz thickness monitor or Centinel III is used as the Si thickness monitor. The power supply of an e-gun can be controlled by the feedback loop of the thickness monitor, but the stability of the deposition rate is poorer than in conventional effusion cells. Moreover, discharge due to dust in the chamber sometimes occurs, resulting in damage to normal crystal growth. To overcome these drawbacks of e-gun evaporators, a specially designed high-temperature effusion cell has recently been developed [1]. Although high-quality epitaxial layers, both Si and SiGe alloy, are obtained, the deposition rate of the cell is lower than that of e-guns and large-volume cells are now under development.

Silicon wafers are held by a specially designed sample holder in which any direct contact between Si and metal components is avoided and Si or graphite rings are used. This is because silicon is very reactive with metals at high temperatures.

2. Gas-Source MBE Machines

A simple apparatus used for studies on GS-MBE growth is basically a SS-MBE machine equipped with gas inlets. Chambers are evacuated by large pumps such as chemical-type turbomolecular pumps (TMPs), cryopumps, or UHV diffusion pumps (DPs), and special precaution should be taken for safety since these gases are toxic and/or combustible.

Figure 2 shows a GS dedicated MBE system. The center plate, on the center part of which an Si wafer is loaded, realizes differential pumping, more than two orders of magnitude, with the two series of UHV DPs. The gases are supplied through the manifold, which has many small holes in the top plate to achieve a high uniformity of the growth without sample rotation. Cryoshroud is completely eliminated from this system and running water cools the chamber wall. Source gases are controlled by the mass flow controllers and gas switching valves. The base pressure is of the order of 10^{-10} Torr and the working pressure is in the range of 10^{-3}–10^{-4} Torr. The typical growth rate is in the range of 1–10 Å/s.

This machine is sometimes called a UHV-CVD or cold-wall UHV-CVD. However, the growth takes place in the regime of molecular flow, and therefore it should be noted that it is MBE and is different from the UHV-CVD method developed by the IBM group [2–4].

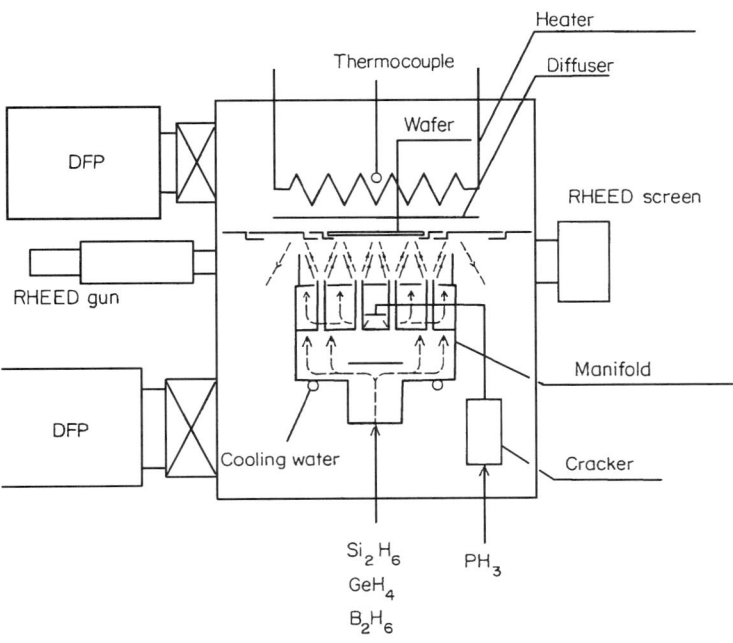

FIG. 2. Schematic illustration of a gas-source dedicated MBE machine (Daido–Hoxan VCE S2020).

For Si, SiGe, and SiC growth, 100% Si_2H_6, 100% GeH_4, and C_2H_2 or SiH_3CH_3 are used as source gases. For doping, B_2H_6 for p-type and PH_3 or AsH_3 for n-type are usually used.

III. Surface Treatment

Atomically clean surfaces are essential for achievement of high-quality epitaxial growth. Contaminants on substrates prevent surface migration of Si atoms and act as impurities and/or nucleation centers of lattice defects. Figure 3 shows the dislocation density in MBE films as a function of the thermal cleaning time on Si (100) substrates. It is obvious that the defects depend strongly on the surface cleanliness and shows the importance of clean surfaces for epitaxial growth. Although it is not always easy to know whether or not substrate surfaces are sufficiently clean, at least the following requirements should be satisfied.

1. Superstructures must be observed in reflection high-energy electron diffraction (RHEED) or low-energy electron diffraction (LEED) measurements.
2. No foreign elements should be detectable by surface analytical measurements, such as Auger electron spectroscopy (AES), X-ray photoemission

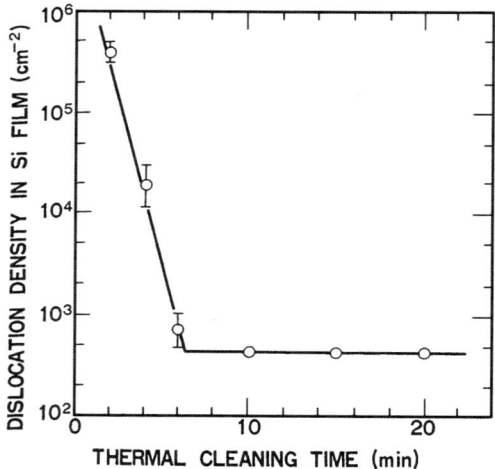

FIG. 3. Dislocation density in Si (100) MBE films as a function of thermal cleaning time.

spectroscopy (XPS), ultraviolet photoemission spectroscopy (UPS), and Rutherford backscattering (RBS).
3. No slip-lines or dislocation networks should be observable by optical microscopy, atomic force microscopy (AFM), or X-ray topography.

Chemical etching procedures used in conventional Si technology are necessary as the first step in sample preparation. A typical cleaning procedure known as "peroxide" cleaning consists of $NH_4OH-H_2O_2$ cleaning and oxidation, HF oxide removal, and $HCl-H_2O_2-H_2O$ (4:1:1) cleaning and oxidation. This method provides a final oxide layer with a thickness of 50 Å.

This oxide can be removed by thermal heating in a MBE machine. Above 800°C, the oxide layer decomposes rapidly and oxygen is desorbed. However, heating above 1200°C is still necessary because carbon contaminants cannot be eliminated at low temperatures. To overcome this drawback, a modified thermal cleaning method is now widely employed [5]. This method is essentially similar to peroxide cleaning and careful repetition of oxidation and oxide removal is performed. This procedure is known effectively to eliminate C and heavy metal contaminants in the final oxide layer, which is formed in the solvent $HCl-H_2O_2-H_2O$ (3:1:1) at 90–100°C. Since the thickness of this oxide is less than 10 Å, clean surfaces without any contamination are obtained below 800°C as shown in Fig. 4.

The sputter and anneal method is also used to remove the protection oxide. However, ion bombardment causes entrapment of incident Ar atoms and contamination due to the simultaneous sputtering-out of impurities from the sample holder and surroundings. To reduce these effects lighter elements such as nitrogen and hydrogen should be used.

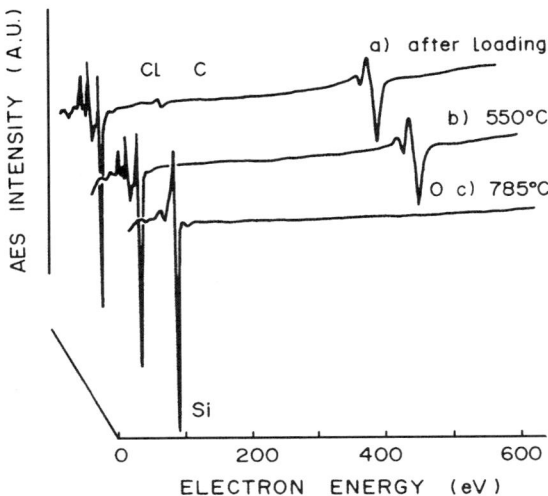

FIG. 4. Auger spectra of Si surfaces: (a) room-temperature spectrum of a freshly oxidized surface; (b) spectrum after heating at 550°C; (c) spectrum after heating at 785°C.

Si or Ga beams are sometimes used to perform low-temperature surface cleaning. This is a kind of dry chemical etching, and impinging Si or Ga atoms react with silicon oxide to form volatile SiO or Ga_2O species which are desorbed at temperatures below 1000°C.

Very recently hydrogenated surfaces have begun to be prepared instead of oxide layers by using HF treatment as the final cleaning step. Careful dipping in a HF solution can provide clean Si surfaces only with hydrogen atoms, which is good enough for MBE growth, especially for GS-MBE.

Another serious contamination problem is piling-up of boron contamination at interfaces between Si substrates and epitaxial films. This B impurity is believed to be incorporated during sample transfer in clean rooms, though the source has not been well identified yet. Although the contamination cannot be completely eliminated, the impurity can be reduced to a level low enough for device applications by paying attention to the materials used in clean rooms.

IV. Growth Mechanisms

1. SOLID-SOURCE MBE

Solid-source MBE of silicon is rather straightforward compared with other methods such as CVD, and the growth rate is simply dependent on the arrival rate of source materials. The sticking coefficient of silicon is almost 100% and does not depend on the substrate temperature. A typical growth rate is 1–10 Å/s and this rate can easily be controlled using conventional e-gun facilities.

A particular advantage of MBE is that it permits crystal growth at temperatures lower than those used in conventional techniques. The minimum temperature for crystal growth depends strongly on the vacuum conditions and the UHV condition is very effective for reducing the growth temperature. The epitaxial temperature also depends on the deposition rate and thickness of grown layers. Jona reported that a few monolayers deposited at room temperature exhibit a bulk structure in LEED observations [6]. This temperature is strikingly low compared with that for conventional methods. To date, however, Si device-quality layers have been grown only at temperatures higher than 300°C, which is still sufficiently low for thermal diffusion to be ignored.

2. GAS-SOURCE MBE

Gas-source MBE growth is governed by dissociative adsorption of the gas molecules and the following desorption of dissociated products which leave Si atoms on Si substrates. The thermal decomposition of source molecules does not occur in the gas phase under the molecular flow conditions of MBE and therefore the reaction of molecules with substrate surfaces is essential.

At high temperatures where MBE growth takes place, impinging silane molecules immediately dissociate into SiH_x fragments and hydrogen desorbs, leaving Si atoms on the growing surface. In other words, hydrogen desorption is the process governing Si GS-MBE growth. Figure 5 shows a temperature-programmed desorption spectrum (TDS) of hydrogen from Si (111) surfaces [7]. There are two desorption peaks, which correspond to desorption from the Si dihydride phase and the monohydride phase for the lower and higher temperature peaks, respectively. This suggests that the MBE growth temperature should be above the higher TDS peak temperature and above 800 K.

There are two distinctive growth modes in GS-MBE, resulting from competition between the dissociative adsorption of hydrogenated molecules and the hydrogen desorption from relevant segments. If the hydrogen desorption rate is high enough compared with the gas adsorption rate, the growth is determined by the source supply and the growth rate becomes proportional to the supply. This is called supply-limited growth. If the growth temperature is relatively low and the hydrogen desorption rate is not high enough, on the other hand, the growth is limited by the hydrogen desorption; this is called reaction-limited growth.

Figure 6 shows a typical example of the temperature dependence of the SiGS-MBE growth rate with Si_2H_6 [8]. At temperatures higher than 700°C, the growth rate is not significantly dependent on the temperature, while the growth rate decreases exponentially with decreasing temperature below 700°C. The latter

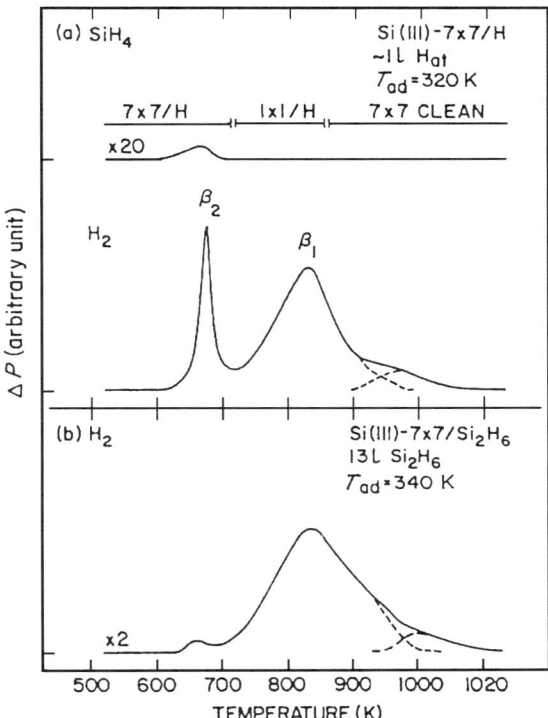

FIG. 5. Temperature-programmed desorption spectra (TDS) of hydrogen from H (a) and Si_2H_6 (b) adsorbed Si (111) surfaces. A TDS spectrum of SiH_4 is also shown in the upper part of a.

temperature dependence gives an activation energy of 47 kcal/mol, which agrees with the hydrogen desorption energy from Si (100) surfaces, and the growth is in the reaction-limited regime. The growth rate increases linearly with increasing Si_2H_6 flow rate at a certain temperature and gives rise to a saturation tendency at high flow rates as shown in Fig. 7 [8]. Here the former corresponds to the supply-limited region again, while the saturation behavior corresponds to the reaction-limited region. Therefore, supply-limited growth proceeds at a lower Si_2H_6 flow rate and/or higher temperatures, while reaction-limited growth occurs under the opposite conditions.

It is also shown in Fig. 7 that the growth rate depends strongly on the Si crystallographic orientation. This probably reflects the fact that the hydrogen desorption temperature on (100) is higher than that on other surfaces.

Surface migration of adsorbates is very important for MBE growth, and hydrogen on surfaces should be noted generally to impede the migration of adsorbed Si atoms and deteriorate the crystal quality. Figure 8 shows the growth temperature

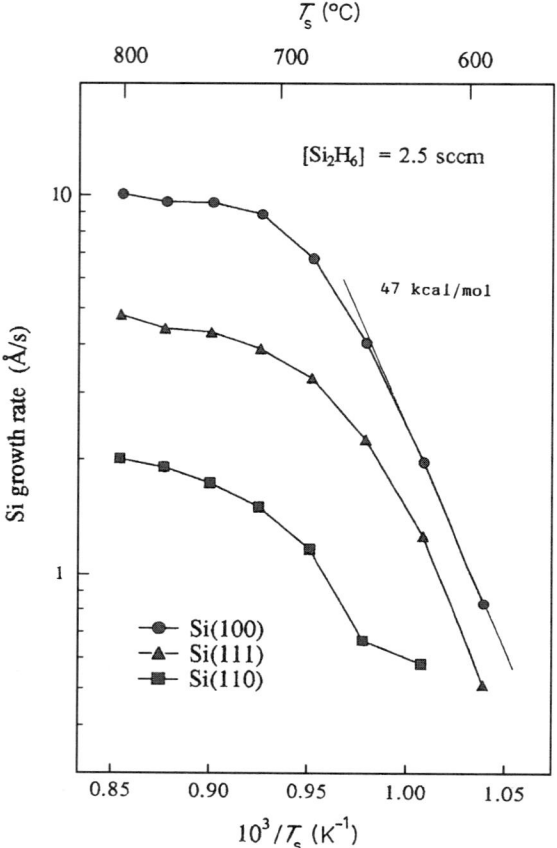

FIG. 6. Si growth rate with Si_2H_6 gases as a function of substrate temperature T_s on various Si surfaces. The gas flow rate is constant at 2.5 sccm, and the activation energy of 47 kcal/mol, corresponding to hydrogen desorption from Si (100), is deduced from the temperature dependence in the lower-temperature region.

dependence of PL spectra of SiGe layers [9]. It is seen that GS-MBE provides highly luminescent layers and the efficiency increases with increasing growth temperature until 700°C, above which it almost saturates. This tendency corresponds well with the temperature dependence of the growth rate and the luminescence efficiency is seen to be strongly temperature dependent in the reaction-limited regime. This is because hydrogen residing on the surface impedes the surface migration of atoms, which is essential for the formation of high-quality epitaxial layers. Therefore, hydrogen desorption must occur so that the surface migration of Si atoms is not prevented. That is, the growth temperature should be as high as possible to eliminate hydrogen completely from the growing surface.

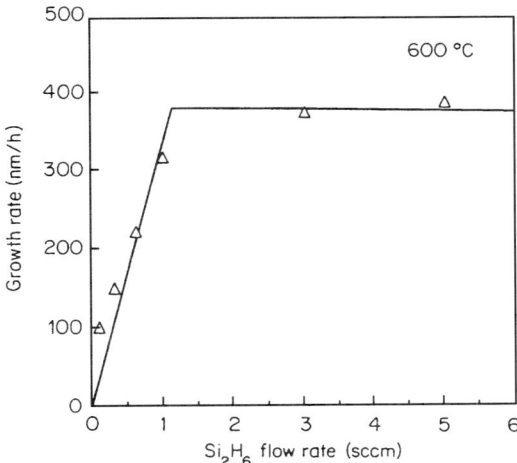

FIG. 7. Flow rate dependence of Si epitaxial growth rate with Si_2H_6 on Si (100) substrates at 600°C. Supply-limited and reaction-limited regimes are seen, respectively, in the low- and high-flow rate regions.

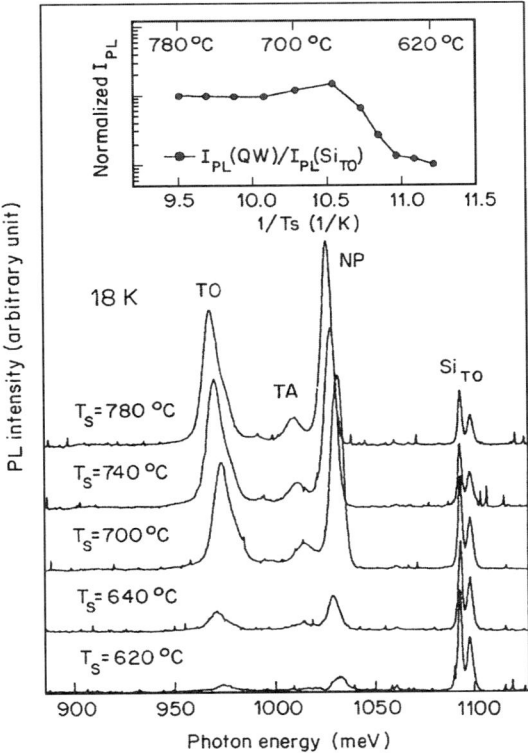

FIG. 8. Growth temperature dependence of PL spectra of 34-Å thick $Si_{0.82}Ge_{0.18}$. Inset: The normalized PL intensity as a function of the growth temperature.

V. Doping

1. SOLID-SOURCE MBE

Doping of Si SS-MBE can be achieved by means of coevaporation of dopant atoms just as in III–V compound MBE growth. Almost all dopants for Si can be supplied by thermal evaporation, but there are practical limitations that make certain elements suitable and others unsuitable for doping. For practical use, dopant arrival rates of 10^{10} to 10^{16} atoms/cm^2 s are necessary. Since a temperature range between 200 and 1300°C is acceptable for cell handling, Sb for n-type and Ga and Al for p-type are usually used in Si SS-MBE. Although P and As are commonly used n-type dopants, their partial pressures are too high to be controlled. When compounds such as GaAs are used as source materials, however, the partial pressure lies in the preferential operating region and becomes acceptable.

Figure 9 shows the sticking probabilities for Ga, Al, and Sb. As shown in this figure, the sticking probability depends strongly on the substrate temperature, with an almost-exponential relationship. This is characteristic of evaporation doping, which complicates MBE growth.

The doping level is commonly controlled by varying the dopant oven temperature. Figure 10 shows C–V profiles for Ga-doped films where profile control was

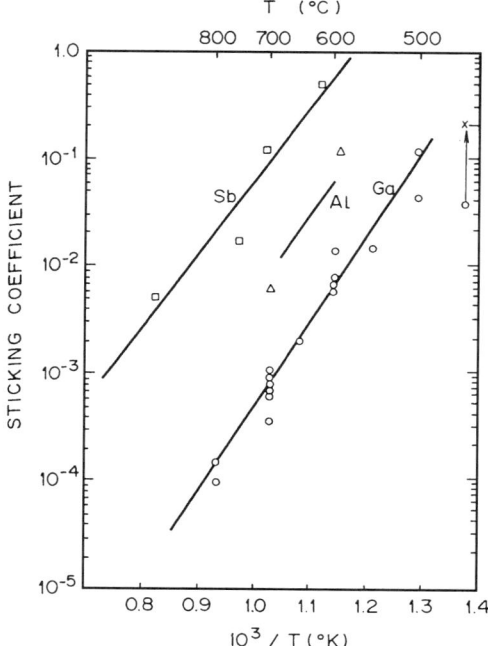

FIG. 9. Sticking probabilities for Sb, Al, and Ga as a function of Si substrate temperature.

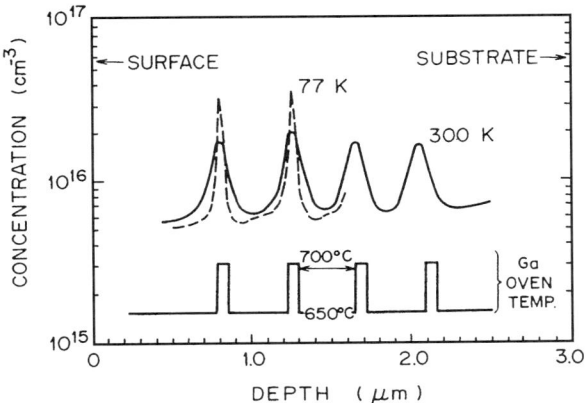

FIG. 10. Carrier concentration profiles determined by a C–V profiler and Ga oven temperature program.

tested by periodically changing the dopant temperature. Changes in the Ga oven temperature are seen to be faithfully reproduced as changes in the doping level.

The mobility of Si MBE layers at room temperature is comparable to that of bulk crystal. The upper limit of MBE doping levels is in the range of $1-2 \times 10^{18}$ cm^{-3} for both n(Sb) and p(Ga), which is lower than the solubility of dopants in Si. This is because the sticking coefficient of dopants drops at a high dopant flux. More importantly, surface segregation of dopants brings about a doping limit and causes smearing of doping profiles, the details of which are discussed later.

A possible way to overcome doping-associated problems is ionization doping. Instead of effusion cells, ion sources with or without mass separators are equipped in MBE machines. The doping level can easily be monitored and controlled through measurement of the ion current, and the sticking probabilities increase to near unity and the doping level reaches $\sim 10^{20}$ cm^{-3} for Sb. Figure 11 shows the relationship between carrier concentration and Sb$^+$/Si flux ratio, and it is seen that accurate control of the Sb doping level over the range of $10^{16}-10^{20}$ cm^{-3} is achieved. Extremely sharp doping profiles can be obtained by this method. It is also interesting that partially ionized beams are effective and show effects almost equal to those of pure ion sources. The mobility of these films is known to be almost equal to the bulk mobility.

2. GAS-SOURCE MBE

In the case of GS-MBE, doping of n- and p-type impurities is performed using PH$_3$ or AsH$_3$ gas and B$_2$H$_6$ gas, respectively.

Figure 12 shows the relationship between B concentration in Si (100) epitaxial layers and B$_2$H$_6$ flow rate with a fixed Si$_2$H$_6$ flow rate at 687°C [10, 11].

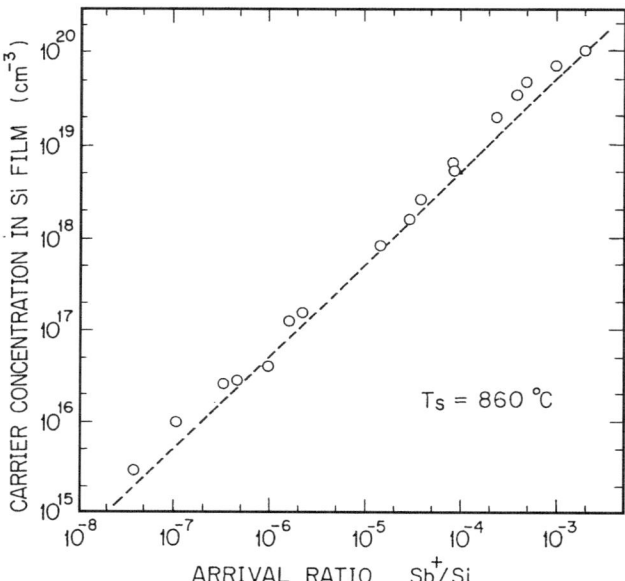

FIG. 11. Carrier concentration in Sb ion-doped films as a function of Sb^+/Si flux ratio. The dashed line indicates 100% Sb doping efficiency.

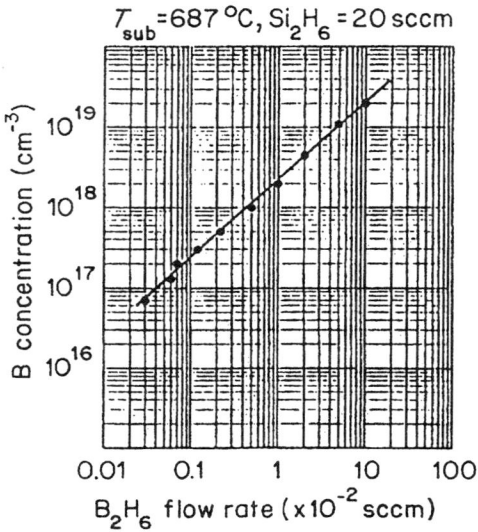

FIG. 12. B concentration as a function of B_2H_6 flow rate.

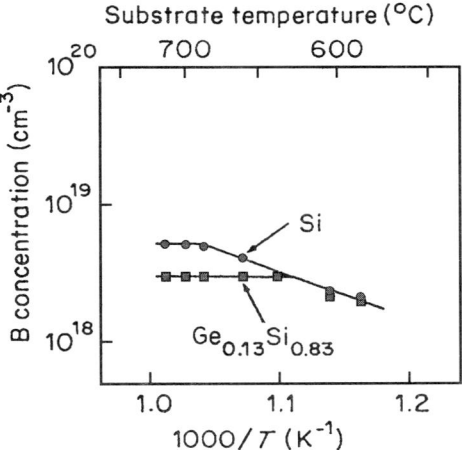

FIG. 13. B concentration in Si and SiGe layers as a function of substrate temperature.

B concentration is seen to increase linearly with increasing flow rate. When the Si_2H_6 flow rate is changed, with a fixed gas ratio for B_2H_6 gas, it is known that the B concentration is constant, although the growth rate changes depending on the flow rate [10]. B doping is determined solely by the B_2H_6/Si_2H_6 flow ratio, provided that the growth temperature is constant, indicating that the sticking coefficient of B_2H_6 has the same tendency as that of Si_2H_6 in both the supply-limited and the reaction-limited growth mode regimes.

Although abrupt profiles are realized by switching the gas valve, it should be noted that the memory effect of doping occurs in the range of 10^{16} cm^{-3} once B_2H_6 gas is introduced into the growth chamber. This memory effect is a disadvantage of GS-MBE, and baking or introducing Cl_2 gas into the chamber is necessary to eliminate the effect [10].

Figure 13 indicates the temperature dependence of B doping in both Si and $Si_{0.87}Ge_{0.13}$ layers [10] and shows that two temperature regimes exist, where the B concentration is temperature dependent and independent, respectively. The transition temperature is seen to decrease when GeH_4 is added. However, the doping is seen to be precisely controlled by using these gas species.

For n-type doping, the carrier concentration is also controlled by the PH3 or AsH_3 flow rate. However, it is noteworthy that the growth rate is decreased by PH_3 gas and that the crystal quality deteriorates, especially in the high doping range [12, 13]. This is because lone pairs of P surface state are doubly occupied and no dangling bonds exist on the P-adsorbed Si surfaces, causing a reduction of the sticking probabilities of gas species and impeding the surface migration of Si atoms. Moreover, PH_3 is known to have a more serious memory effect than B doping.

VI. Growth of SiGe(C) Alloys

1. SOLID-SOURCE MBE

SiGe alloys can be grown on silicon substrates in a straightforward manner by SS-MBE and the alloy composition can be controlled by changing the arrival rates of source materials. Ge vapor is generated by both conventional effusion cells and e-guns. However, since the Si and Ge lattice mismatch is about 4.2%, the growth temperature is very crucial. That is, there exists a critical thickness below which coherent or pseudomorphic growth of SiGe alloys takes place on Si substrates and the thickness is strongly dependent on the temperature. The details of the critical thickness are discussed later. The growth temperature of SiGe alloys is set to be much lower than that for Si homoepitaxy to avoid the generation of misfit dislocations. The alloy composition does not depend on the growth temperature strongly, unlike GS-MBE, as described in the following section.

SiGeC alloys are formed by adding C with specially designed high-temperature effusion cells. However, since the solubility of C in Si(Ge) is extremely low, only alloys with a C concentration of less than several percent can be grown by conventional SS-MBE.

2. FORMATION AND CONTROL OF SiGe ALLOYS BY GAS-SOURCE MBE

In the case of GS-MBE, SiGe alloys can be formed by simply supplying both Si_2H_6 and GeH_4 gases. However, the alloy composition is not linearly proportional to the gas mixture ratio but has the relationship shown in Fig. 14 [9]. It is well established that for vapor-phase growth the following relationship holds between the solid Ge content (nGe) and the vapor-phase Ge-source fraction:

$$nGe/(1 - nGe) = c[GeH_4]/([Si_2H_6] + [GeH_4]) \qquad (1)$$

where c is the incorporation coefficient and is generally dependent on the growth temperature [9, 14]. Ge concentration in the bulk phase is seen to be represented by Eq. (1), with $c = 0.15$ for 600°C. Since this relationship is reproducible, all kinds of SiGe alloys are basically grown by GS-MBE.

Figure 15 shows the growth temperature dependence of the Ge concentration of grown layers [15]. The SiGe alloy composition is seen to be independent of the growth temperature unless the gas flow ratio is not changed. Figure 16a is an Arrhenius plot of the growth rate of Si and SiGe alloys [15], and it shows that both are linearly proportional to the inverse of the growth temperature, giving activation energies of 47 and 24 kcal/mol for Si and $Si_{1-x}Ge_x$, respectively. It is noted that the addition of GeH_4 gas increases the growth rate, especially at low temperatures, which is consistent with the result of UHV-CVD growth. Figure 17

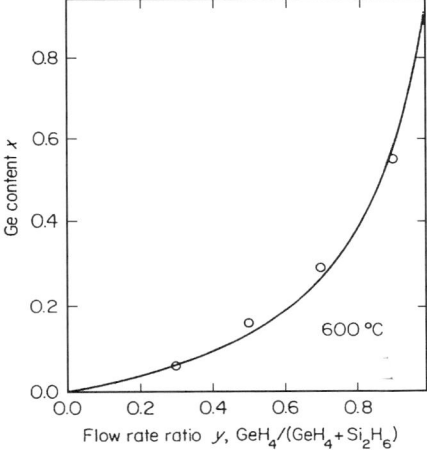

FIG. 14. Ge content of grown SiGe films as a function of supply gas mixture ratio at 600°C.

shows the Si_2H_6 flow rate dependence of the growth rate as a function of the mixing ratio of source gases at 587°C [16]. It is clear that the growth mode changes from the supply-limited region to the reaction-limited region with increasing gas flow rate. When GeH_4 is added, the transition region is widened and the growth rate increases with increasing GeH_4 content in the reaction-limited region. The increase in growth rate and the decrease in activation energy due to GeH_4 gas

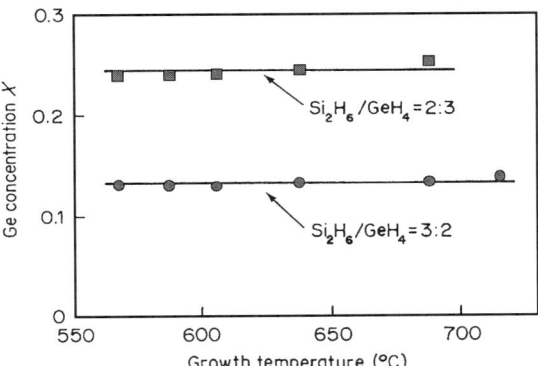

FIG. 15. Growth temperature dependence of the Ge concentration of grown SiGe layers at fixed gas flow ratios.

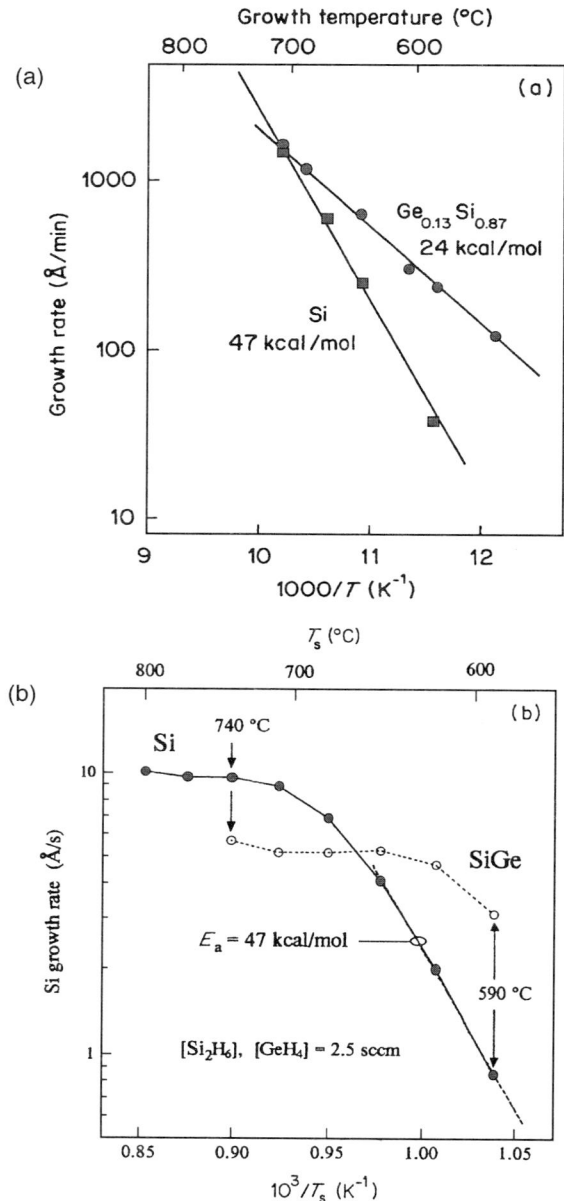

FIG. 16. Growth temperature dependence of growth rate of Si and SiGe layers in the low (a) and high (b) temperature regions.

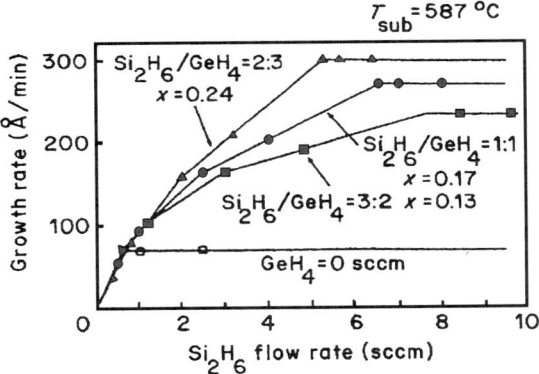

FIG. 17. Si₂H₆ flow rate dependence of the growth rate of Si and SiGe layers as a function of the mixing ratio of source gases at 587°C.

are thought to be caused by the enhancement of hydrogen desorption from the SiGe surface compared with the Si surface.

VII. Formation of Si/Ge Heterostructures

1. BAND STRUCTURES OF Si/Ge HETEROSTRUCTURES

Band structures of SiGe layers are modified owing to the strain induced by the lattice mismatch. This provides us another possibility for band engineering in Si/Ge heterostructures. The band gap of SiGe layers grown on Si substrates is decreased from the bulk. The lowest conduction band of the strained SiGe is the delta valley, the same as Si crystals, over almost the whole composition range. The degenerated heavy and light hole bands of the valence band are separated and the energy difference of the spin-orbit splitting band is changed. This change effectively modifies the transport properties of electrons and holes in the layers, giving chances to improve transistor performances, especially those of p-MOSFETs and p-HEMTs.

The band alignment at Si/Ge heterointerfaces is also significantly modified and type I and type II alignments are realized by changing the strain distribution. When SiGe ($x < 0.5$) layers are grown on Si substrates and are laterally compressed, the band alignment becomes type I and electrons and holes are confined in the same SiGe region. On the other hand, when Si layers are grown on the unstrained SiGe layers and are under lateral tensile strain, type II alignment is realized and electrons and holes are separately confined. Both band lineups are very much useful from the point of view of device applications. Especially, since the band discontinuity at

2. THE CRITICAL THICKNESS

Since SiGe/Si is a lattice-mismatched system, the layer thickness where heteroepitaxial growth occurs coherently is limited and it is called the critical thickness. The coherent growth is deteriorated above the thickness due to the accumulating strain energy, and the strain is released mainly by generating misfit dislocations. Calculation of the critical thickness in thin heteroepitaxial layers on thick substrates was carried out by Matthews and Blakeslee (MB) based on the force balance consideration [17, 18]. There are two forces which act on the dislocation and they balance each other at the critical thickness. That is, the force associated with the strain is opposed by the line tension of the misfit dislocation, which gives the critical thickness. The Ge composition or misfit dependence of the critical thickness calculated by this method is shown in Fig. 18 together with experimental data for SS-MBE. As shown in the figure, the SiGe/Si system exhibits a large deviation from the thermal equilibrium calculation when the growth temperature is low [19]. This means that a large metastable regime exists, especially in the low-Ge composition region. It is reported that the critical thickness is

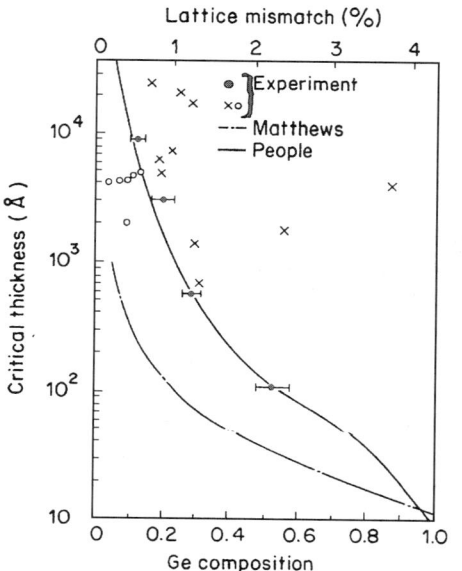

FIG. 18. Critical thickness of SiGe layers on Si (100) substrates as a function of Ge composition. (X) Misfit dislocations; (O, ●) coherent growth.

strongly temperature dependent and that it exceeds more than 50 times that theoretically predicted at 330°C for $Si_{0.7}Ge_{0.3}$ [20].

The solid line in Fig. 18 was obtained by People and Bean to describe the critical thickness obtained experimentally and it seems to represent the experimental data well [21]. However, it should be kept in mind that the physics underlying their model is not clear and a conceptual mistake exists, though their expression is frequently cited to explain the critical thickness in various strained systems.

A more sophisticated approach to this problem is given by Dodson and Tsao, who introduce a concept of excess stress to describe the temperature dependence of the critical thickness [22, 23]. Although the excess stress, which is defined as the difference between the forces acting on the dislocation presented in the MB formulation, is zero at the critical thickness given by the equilibrium theory, dislocations are assumed to glide via thermal activation over the Peierls barrier when strained layers are in the metastable regime. They move with a velocity which is linear to the excess stress, giving rise to the temperature dependence of the critical thickness. This formalism is confirmed to express the Ge composition dependence as well as the temperature dependence adequately [24, 25].

In the high-Ge content region, island formation is another pathway of strain release above the critical thickness. Figure 19 shows the growth process at 500°C in the initial stage of SiGe films on (100)Si-(2 × 1) as a function of Ge content [14]. It is evident in this figure that (811)-faceted islands having a doubly ordered (111) stacking structure composed of Si and Ge atoms start to be formed after the formation of (2 × 1) superstructures on the top of two-dimensional SiGe layers. Formation of more prominent three-dimensional islands with (311) facets is seen to follow when the film thickness is increased. The thickness at which the

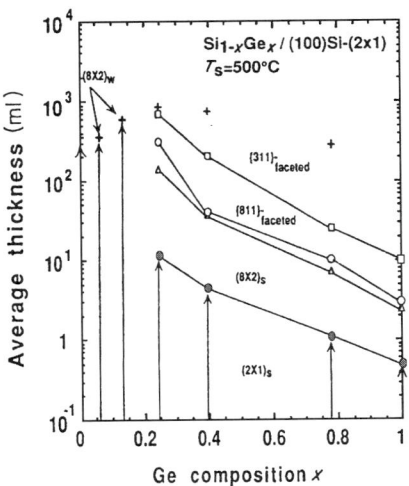

FIG. 19. Growth mode change of SiGe layers as a function of Ge composition.

(311)-faceted islands start to grow coincides with the critical thickness obtained experimentally for SS-MBE, shown in Fig. 18.

3. Surface Segregation

In the case of SS-MBE, interface smearing is recognized to be a problem in SiGe/Si heterostructures. Surface segregation, which is an exchange reaction between impinging atoms and surface atoms of substrates, is known to be the main cause of interface smearing. This phenomenon was first recognized to be important when MBE layers were doped with various kinds of impurities [26, 27]. More or less, however, surface segregation generally takes place in cases not only of doping but also of heterointerface formation. Ge and In are now well known to segregate when forming Si-on-(Si)Ge and GaAs-on-In(Ga)As heterostructures, respectively.

Figure 20 shows how surface segregation occurs when a Si overlayer is grown on the Ge layers [28]. The dashed line in the figure shows the X-ray photoemission spectroscopy (XPS) intensity of Ge atoms as a function of the Si overlayer thickness without surface segregation. It is clearly shown that the XPS intensity (filled circles) does not follow an exponential decay but is much stronger than the expectation. This phenomenon is well described in terms of the two-state exchange model, where only exchange of atoms in surface and subsurface states is taken into account [26, 29]. The solid line in Fig. 20 is the Si overlayer thickness dependence of Ge XPS intensity calculated by the two-state exchange model, which indicates that the model describes the surface segregation well.

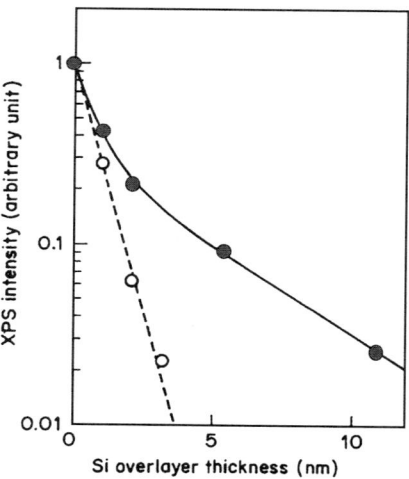

FIG. 20. Ge X-ray photoemission (XPS) signals as a function of the thickness of the Si overlayer on Ge atoms. Filled and open circles are data on Si growth without and with Sb atoms on Ge layers, respectively. Solid and dashed lines represent the thickness dependence of Ge XPS intensity with and without surface segregation of Ge atoms, respectively.

4. SURFACTANT-MEDIATED GROWTH

Ge on Si is a typical example showing the Stranski–Krastanov (SK) growth mode, and it is difficult to obtain Ge films with flat surfaces on Si substrates. A small amount of atoms such as As deposited on Si substrates prior to Ge deposition is known to reduce the surface migration of Ge atoms and prevent island formation [30], which is called surfactant-mediated growth (SMG). Surfactant is known to prevent surface segregation as well. In Fig. 20, the open circles show the XPS intensity of Ge atoms when 0.75-ML Sb atoms are introduced as suppressors of segregation at the heterointerface before Si layers are overgrown [28]. It is shown that the XPS intensity follows an exponential decay and that surface segregation is effectively suppressed. This is understood as follows: when Si atoms are deposited on the Sb-covered Ge surface, the position of Sb becomes the subsurface state and Sb exchanges position with impinging Si atoms sitting in the surface state. However, site exchange between Si and Ge atoms does not follow, since Ge atoms do not occupy the surface state or the subsurface state, that is, they are in the bulk state once Si is deposited, and therefore they do not participate in the surface segregation.

It is known that As, Ga, and Bi [31] also effectively suppress Ge segregation and that abrupt heterointerfaces can be formed by the same SMG method at temperatures higher than 500°C.

GS-MBE is another important example of SMG, or pseudo-SMG, where atomic hydrogen decomposed from hydrogenated gases acts as a suppressor of Ge segregation. Figure 21 shows the well width dependence of the PL peak energy of samples

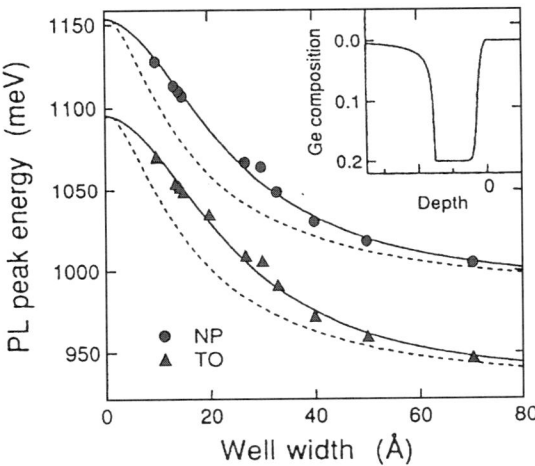

FIG. 21. Well width dependence of quantum well PL peaks. Solid and dashed lines represent calculated results based on a modified well shape due to surface segregation and a square potential well, respectively. Inset: Schematic of the well shape. NP, no phonon; TO, (transverse optic) phonon replica.

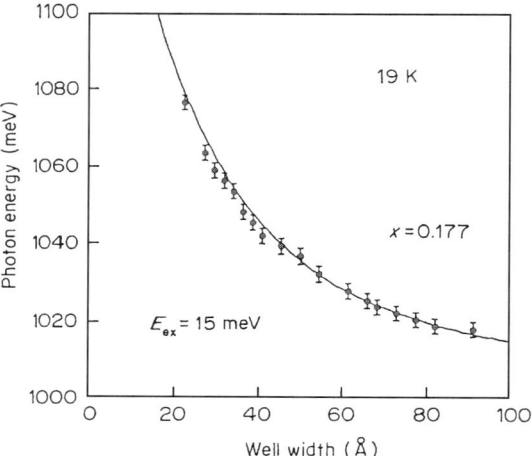

FIG. 22. Well width dependence of the no-phonon PL peak energy of GS-MBE-grown quantum wells. The exciton binding energy E_{ex} is assumed to be 15 meV.

grown by SS-MBE [32]. There is significant deviation from the calculation, which comes from deformation of the well shape due to Ge surface segregation. The well width dependence of the PL peaks for the GS-MBE sample, on the other hand, is seen to coincide well with the square-well potential calculation without surface segregation of Ge as shown in Fig. 22 [9]. The profile of SiGe layers was also examined by means of SIMS and it was found that the SS-MBE sample has a long tail toward the surface due to segregation, while the GS-MBE sample gives rise to a very sharp profile. These results clearly indicate that GS-MBE provides heterostructures without significant interfacial smearing. It is also reported that Ge segregation is almost absent in UHV-CVD growth, and hydrogen and/or hydrogenated compounds are speculated to act as suppressors [33]. Introduction of atomic hydrogen into the solid SS-MBE chamber confirmed the role of atomic hydrogen; it was found that surface segregation is suppressed even at temperatures above 500°C, where hydrogen is considered to desorb from Si surfaces [34, 35]. This suggests that hydrogen has a relatively long resident time on Si surfaces and then can stay to act as a surfactant in a period of monolayer growth.

VIII. Growth of SiGe Buffer Layers

To realize various band configurations, tensile Si(Ge) and compressive (Si)Ge layers are formed on completely relaxed SiGe buffer layers, or virtual substrates, with a high quality. Formation of fully relaxed SiGe layers with flat surfaces is, however, not straightforward; strain tends to make the surface rough and crosshatched patterns are frequently observed on the grown surfaces. To obtain flat surfaces, graded buffers are usually employed. Fully relaxed SiGe layers grown on a buffer layer with nine steps from 0 to 18% provide sharp and intense band edge emissions, suggesting a high quality of the top region of the buffer, although it is relaxed and many misfit dislocations are accommodated at the interface between the buffer layer and the Si substrate [36].

Recently, a more promising technique for obtaining good SiGe buffer layers was developed [37] and its usefulness demonstrated [38]. In this method, 50-nm-thick Si layers are deposited on Si substrates at 400°C before the formation of SiGe buffer layers. This low-temperature (LT) Si layer may act as an absorber of dislocations, and therefore high-quality 500-nm SiGe layers can be grown on this LT Si. Not only is the dislocation density of the buffer obtained by this method much lower, but also the surface roughness is much better, than those of conventional graded buffers. Modulation doping structures with p-type Ge channels were fabricated on the pseudo-substrates grown by this LT method, and their mobility was reported to be about 1300 cm^2/V s at room temperature, which is about 10 times higher than that of p-channel Si MOSFETs [38].

IX. Selective Growth

GS-MBE has a big advantage for selective epitaxial growth (SEG). Si epitaxial growth takes place only on Si surfaces when epitaxy is carried out on patterned Si substrates with films such as SiO$_2$. This selectivity comes from the difference in sticking probability of silane molecules between Si and SiO$_2$, and dissociation of molecules hardly occurs on SiO$_2$ surfaces.

However, it should be noted that this selective growth can be achieved only under limited conditions [39]. Selective growth hardly occurs under conditions of high temperatures and/or high growth rates. Moreover, long growth times reduce the selectivity even under selective growth conditions, and polycrystalline Si grows on SiO$_2$ substrates. This is because, although the sticking probability of silane is low on SiO$_2$ layers, it is not zero, and small nuclei are formed even on SiO$_2$ surfaces. If the nucleus size is very small, atoms leave the nucleus due to the energy balance and crystal growth does not take place. When a nucleus exceeds the critical size, the nucleus grows rapidly and the selectivity diminishes. A high temperature and high flow rate enhance the probability of large nuclei being formed. The formation of nuclei depends on surface properties such as the surface energy and density of dangling bonds of substrates, and it is known that the selectivity on Si$_3$N$_4$ patterned substrates is much poorer as shown in Fig. 23 [40]. There, the vertical axis is the critical gas level at which selectivity disappears, and it is seen to be much lower on Si$_3$N$_4$ than on SiO$_2$. To enhance the selectivity, therefore, nuclei should be removed before their size exceeds the critical value. For this purpose, the addition of halogen

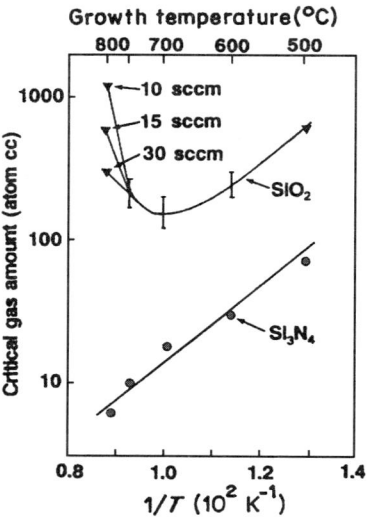

FIG. 23. Critical gas amount for SiO_2 and Si_3N_4 masking as a function of the growth temperature.

gas to the source gas is very effective [40]. Figure 24 shows the critical flow of selectivity when chlorine gas is added [40]. It is seen that the effect is significant on both SiO_2 and Si_3N_4 and that the critical flow is increased about 6 and 40 times on SiO_2 and Si_3N_4, respectively, by adding 0.003 sccm Cl_2, although the growth rate is decreased.

A problem concerning selective growth is faceting. That is, Si (111), (311), and (511) facet surfaces tend to appear at the edge of Si epitaxial films. The faceting

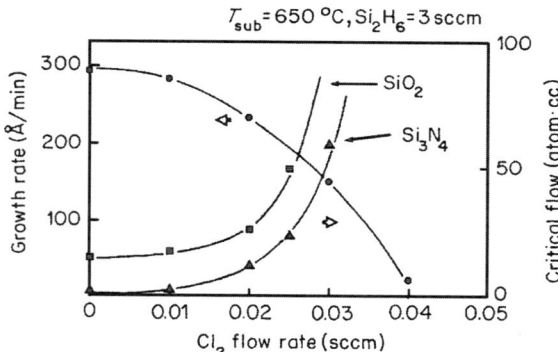

FIG. 24. Critical gas amount for SiO_2 and Si_3N_4 masking and growth rate dependences on Cl_2 flow rate at 650°C.

occurs to minimize the surface energy and the (111) surface is known to be the most stable for diamond-type crystals, resulting in the (111) faceting. Although they are metastable, (311) and (511) surfaces are also energetically favorable and faceting may occur on these surfaces under a nonequilibrium condition such as MBE growth [41].

To suppress faceting and obtain a selective growth layer with vertical sidewalls, reaction-limited growth conditions should be employed. As described in the previous section, the growth rate is proportional to the gas flow rate in the low-flow region (supply-limited growth), while it is independent at high flow rates (reaction-limited growth). This holds not only on (100) surfaces, but also on (511), (311), and (111) surfaces. Although the growth depends strongly on the surface orientation due to the different reactivities among the surfaces in the supply-limited regime, the growth rate is limited by the hydrogen desorption in the reaction-limited regime and the difference in the growth rate becomes small. This results in weakening the facet formation.

X. Formation of Superlattices, Quantum Wires, and Quantum Dots

The coupling of quantum wells (QWs), which is essential for the formation of superlattices, is well understood based on effective mass approximation by taking the band alignment precisely into account. Figure 25 shows PL spectra of symmetrically coupled QWs with various barrier widths grown on Si substrates [42].

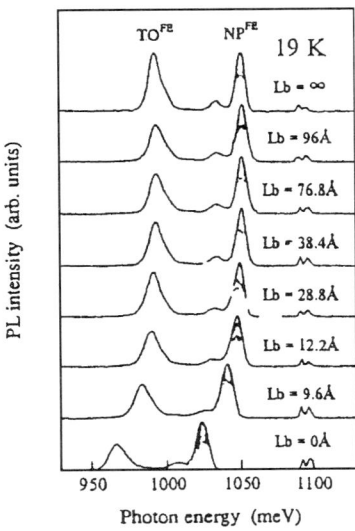

FIG. 25. PL spectra of symmetrically coupled quantum wells with various barrier widths.

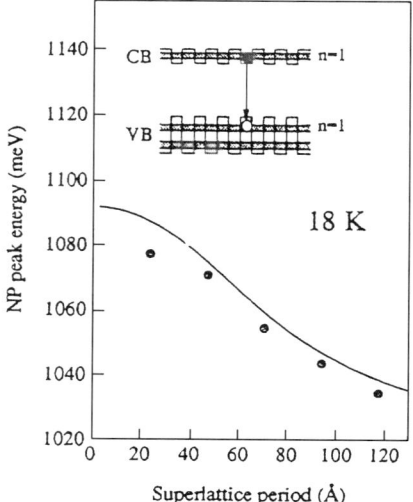

FIG. 26. PL peak energy as a function of superlattice period in 99-period SiGe/Si superlattices.

As shown in the figure, the peaks shift to higher energies with increasing barrier width. The peak energies of coupled QWs agree well with the calculated results over the whole range, in accordance with the theoretical prediction [43] that the band alignment is type I when the Ge content is lower than 0.5 and indicating that ideally square-shaped wells are formed by GS-MBE.

The evolution of superlattices is seen when the number of coupled wells is increased [42]. Since an increase in the number of coupled wells lowers the ground-state energy and finally forms a miniband, the PL peak energy, corresponding to the miniband edge, decreases with increasing well number. This situation was confirmed and an effective mass calculation was found to give reasonable values for the energy shift. Figure 26 shows the PL peak energy shift as a function of the superlattice period in 99-period strained SiGe/Si superlattices [44]. It is seen that the peak energy follows a simple Kronig–Penny-type calculation.

One of the advantages of GS-MBE is that selective growth can be realized between Si and SiO_2 substrates and epitaxy occurs only on Si as described before. By exploiting this advantage, a wire geometry of SiGe/Si QWs was formed on V-groove patterned Si (100) substrates with (111) facets [45, 46]. At the bottom of the V-groove, crescent-shaped SiGe features, the lateral and vertical wire sizes of which were approximately 300 and 80 Å, respectively, were found to be grown. Figure 27 shows PL spectra of quantum structures grown (a) on a planar Si (100) substrate and (b) on a V-groove patterned Si substrate [46]. It is significant that a large blue shift is seen in the spectrum of the latter sample. Polarization experiments show that the wire structure in the V-grooved sample is likely to be a quantum

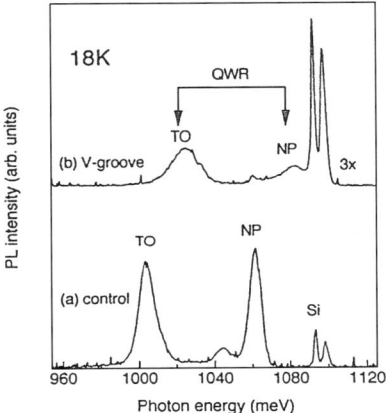

FIG. 27. PL spectra of quantum structures grown on (a) planar Si (100) and (b) V-groove patterned Si substrates.

wire (QWR). However, the energy shift in QWRs is too big and does not agree with a simple estimation from the size. The main cause of the energy shift may be the change in the strain distribution. A wire surrounded by Si crystal is considered to be under hydrostatic-like pressure and the pressure causes bandgap broadening which is comparable to the observed energy shift. Because spatial variation of the Ge composition in the wire is also likely to occur, more detailed study is required to clarify the nature of QWRs in the SiGe/Si system.

Figure 28 shows the thickness dependence of PL spectra of Si/pure Ge/Si QW structures [47]. Up to 3.7 ML, the PL shows a conventional quantum confinement effect of QWs, and the peaks shift to lower energies with increasing Ge layer thickness. Above 3.7 ML, however, the peaks originating from QWs stop the energy shift and a new broad peak appears. The appearance of this broad peak corresponds well to the formation of Ge islands. That is, above the critical thickness, Ge begins to form islands on Ge wetting layers to release the strain energy. It is remarkable that these Ge dots give rise to significant luminescence, and therefore, their applications as quantum dots are very much expected. However, the size and position of these dots are random. So, how to control their size and position is a crucial issue of regarding quantum dot formation.

Combination of selective epitaxial growth and electron beam (EB) lithography is a promising approach to controlling the dots. As described before, GS-MBE offers the advantage of providing selectivity between Si and SiO$_2$ surfaces and epitaxial growth takes place only on Si surfaces. If the window size is small enough, only one Ge dot grows in a window and the dot size decreases with decreasing window size as shown in Fig. 29. A Ge dot is formed on the Ge wetting layer, that is, the SK growth mode also occurs even in SEG. These controlled Ge dots give rise to

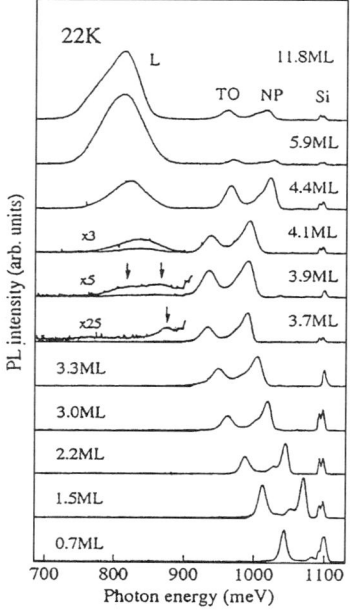

FIG. 28. Thickness dependence of PL spectra of Si/pure Ge/Si quantum wells.

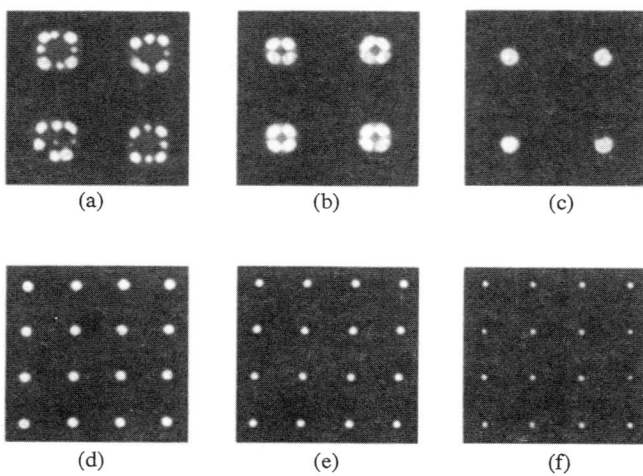

FIG. 29. Plane-view AFM images of Ge dots grown in windows of different sizes. The images are $2 \times 2\ \mu m^2$ in area. The window sizes are (a) 580 nm, (b) 440 nm, (c) 300 nm, (d) 180 nm, (e) 130 nm, and (f) 90 nm.

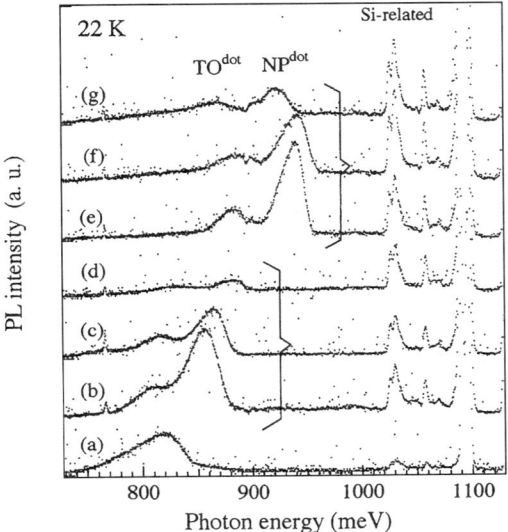

FIG. 30. PL spectra of Ge dots in windows of various sizes. (a) Without SiO_2 masks. The window sizes are (b) 580 nm, (c) 440 nm, (d) 300 nm, (e) 180 nm, (f) 130 nm, and (g) 90 nm.

luminescence as shown in Fig. 30 and the peak position shifts to higher energies with decreasing dot size. However, the size of Ge dots is still too big to provide a significant quantum confinement effect, and therefore, the energy shift observed is considered to come from the change in the strain distribution around the dots. To enhance the quantum effects, several methods of reducing Ge dot size are now under investigation, and predeposition of elements such as C [48] and B [49] has been shown to be very effective.

XI. Conclusion

Silicon epitaxial growth by the MBE technique has been reviewed, and its high potential to provide high-quality films and heterostructures with atomic-scale accuracy demonstrated. It is characteristic of this method that the epitaxy is done at relatively low temperatures without deterioration of crystal qualities. It is also a big advantage that MBE easily realizes new structures, especially quantum structures, which are essential for new devices based on silicon materials. Not only heterostructure bipolar transistors HBTs and new field effect transistors FETs using SiGe(C)/Si heterostructures but also optical devices including light-emitting devices will be fabricated on silicon substrates by this technique in the near-future.

References

1. Yaguchi, H., T. Yamamoto, and Y. Shiraki, *Thin Solid Films* **321**, 241 (1998).
2. Meyerson, B. S., *Appl. Phys. Lett.* **48**, 797 (1986).
3. Meyerson, B. S., F. K. LeGoues, T. N. Nguyen, and D. L. Harame, *Appl. Phys. Lett.* **50**, 113 (1987).
4. Meyerson, B. S., *Proc. IEEE* **80**, 1592 (1992).
5. Ishizaka, A., and Y. J. Shiraki, *Electrochem. Soc.* **133**, 666 (1986).
6. Jona, F., in *Surfaces and Interfaces/Chemical and Physical Characterisitcs*, edited J. J. Burke et al., Chap. 18, Syracuse University Press, Syracuse, NY 1967.
7. Imbihl, R., J. E. Demuth, S. M. Gates, and B. A. Scott, *Phys. Rev.* **B39**, 5222 (1989).
8. Yoshida, H., S. Fukatsu, and Y. Shiraki, unpublished.
9. Fukatsu, S., N. Usami, Y. Kato, H. Sunamura, Y. Shiraki, H. Oku, T. Ohnishi, Y. Ohmori, and K. Okumura, *J. Cryst. Growth* **136**, 315 (1994).
10. Tatsumi, T., *Cho LSI Gijutu 17, Debaisu to purosesu No.7; Handoutai Kenkyu 38*, edited by J. Nishizawa, Kogyochousakai (Japanese).
11. Hirayama, H., M. Hiroi, and K. Koyama, *Appl. Phys. Lett.* **58**, 1991 (1991).
12. Hirayama, H., and T. Tatsumi, *Appl. Phys. Lett.* **55**, 131 (1989).
13. Hirayama, H., and T. Tatsumi, *Thin Solid Films* **184**, 125 (1990).
14. Yasuda, Y., Y. Koide, A. Fujukawa, N. Ohshima, and S. Zaima, *J. Appl. Phys.* **73**, 2288 (1993).
15. Aketagawa, K., T. Tatsumi, M. Hiroi, T. Niino, and J. Sakai, *Jpn. J. Appl. Phys.* **31**, 1432 (1992).
16. Tatsumi, T., K. Aketagawa, K. Miyanaga, and M. Hiroi, in *Ext. Abstr., 1993 Int. Conf. Solid State Devices Mater.*, Makuhari, p. 225, 1993.
17. Matthews, J. W., and A. E. Blakeslee, *J. Cryst. Growth* **27**, 118 (1974); *J. Cryst. Growth* **29**, 273 (1975); *J. Cryst. Growth* **32**, 265 (1976).
18. Mattews, J. W., *J. Vac. Sci. Technol.* **12**, 126 (1975).
19. Bean, J. C., L. C. Feldmann, A. T. Fiory, S. Nakahara, and I. K. Robinson, *J. Vac. Sci. Technol.* **A2**, 436 (1984).
20. Chern, C. H., K. L. Wang, G. Bai, and M.-A. Nicolet, *Mater. Res. Soc. Proc.* (*Silicon Mol. Beam Epi.*) **220**, 135 (1991).
21. People, R., and J. C. Bean, *Appl. Phys. Lett.* **47**, 322 (1985); *Appl. Phys. Lett.* **49**, 229 (1986).
22. Dodson, B. W., and J. Y. Tsao, *Appl. Phys. Lett.* **51**, 1325 (1987).
23. Tsao, J. Y., B. W. Dodson, S. T. Picraux, and D. M. Cornelison, *Phys. Rev. Lett.* **59**, 2455 (1987).
24. Hull, R., J. C. Bean, D. J. Eaglesham, J. M. Bonar, and C. Buescher, *Thin Solid Films* **183**, 117 (1989).
25. Yaguchi, H., K. Fujita, S. Fukatsu, Y. Shiraki, and R. Ito, *Jpn. J. Appl. Phys.* **30**, L1450 (1991).
26. Harris, J. J., D. E. Ashenford, C. T. Foxon, P. J. Dobson, and B. A. Joyce, *Appl. Phys. A* **33**, 87 (1984).
27. Metzger, R. A., and F. G. Allen, *J. Appl. Phys.* **55**, 931 (1984).
28. Fujita, K., S. Fukatsu, H. Yaguchi, T. Igarashi, Y. Shiraki, and R. Ito, *Jpn. J. Appl. Phys.* **29**, L1981 (1990).
29. Fukatsu, S., K. Fujita, H. Yaguchi, Y. Shiraki, and R. Ito, *Appl. Phys. Lett.* **59**, 2103 (1991).
30. Copel, M., C. Reuter, E. Kaxiras, and R. M. Tromp, *Phys. Rev. Lett.* **63**, 632 (1989).
31. Sakamoto, K., K. Kyoya, K. Miki, H. Matsuhara, and T. Sakamoto, *Jpn. J. Appl. Phys.* **32**, L204 (1993).
32. Fukatsu, S., Doctoral thesis.
33. Copel, M., and R. M. Tromp, *Appl. Phys. Lett.* **58**, 2648 (1991).
34. Ota, G., S. Fukatsu, Y. Ebuchi, T. Hattori, N. Usami, and Y. Shiraki, *Appl. Phys. Lett.* **65**, 2975 (1994).
35. Nakagawa, K., A. Nishida, Y. Kimura, and T. Shimada, *Jpn. J. Appl. Phys.* **33**, L1331 (1994).
36. Nayak, D. K., N. Usami, S. Fukatsu, and Y. Shiraki, *Appl. Phys. Lett.* **63**, 3509 (1993).
37. Li, J. H., C. S. Peng, Y. Wu, D. Y. Dai, J. M. Zhou, and Z. H. Mai, *Appl. Phys. Lett.* **71**, 3132 (1997).
38. Ueno, T., T. Irisawa, and Y. Shiraki, *Physica E* **7**, 790 (2000).

39. Hirayama, H., T. Tatsumi, and N. Aizaki, *Appl. Phys. Lett.* **52**, 2242 (1988).
40. Tatsumi, T., K. Aketagawa, M. Hiroi, and J. Sakai, *J. Cryst. Growth* **129**, 275 (1992).
41. Hirayama, H., M. Hiroi, and T. Ide, *Phys. Rev.* **B48**, 17331 (1993).
42. Fukatsu, S., and Y. Shiraki, *Ext. Abstr., 1993 Int. Conf. Solid State Devices Mater.,* Makuhari, p. 895, 1993.
43. Van de Walle, C. G., and R. M. Martin, *Phys. Rev.* **B34**, 5621 (1986).
44. Fukatsu, S., *Solid-State Electron.* **37**, 817 (1994).
45. Usami, N., T. Mine, S. Fukatsu, and Y. Shiraki, *Appl. Phys. Lett.* **63**, 2789 (1993).
46. Usami, N., T. Mine, S. Fukatsu, and Y. Shiraki, *Appl. Phys. Lett.* **64**, 1126 (1994).
47. Sunamura, H., N. Usami, Y. Shiraki, and S. Fukatsu, *Appl. Phys. Lett.* **66**, 3024 (1995).
48. Schmidt, O. G., C. Lange, K. Eberl, O. Kienzle, and F. Ernst, *Appl. Phys. Lett.* **71**, 2340 (1997).
49. Takamiya, H., M. Miura, N. Usami, and Y. Shiraki, *Thin Solid Films* **369**, 84 (2000).

CHAPTER 6

Epitaxial Growth Modeling

M. Masi

DIPARTIMENTO DI CHIMICA FISICA APPLICATA
POLITECNICO DI MILANO
MILANO, ITALY

S. Kommu

MEMC ELECTRONIC MATERIALS INC.
EPI TECHNOLOGY GROUP
ST. PETERS, MISSOURI

NOMENCLATURE	185
I. INTRODUCTION	187
1. *Physical and Chemical Aspects of Silicon Epitaxy*	189
2. *The Multiscale Modeling Concept*	190
3. *Conditions for Epitaxial Growth: The Terrace Step Kink Mechanism*	192
II. DETAILED MODELING OF EPITAXIAL REACTORS	193
1. *Conservation Equations*	193
2. *Computational Issues*	196
3. *Estimation of Involved Parameters*	197
4. *Examples of Simulations*	198
III. REDUCED-ORDER MODELS FOR EPITAXIAL SILICON DEPOSITION	209
1. *Model Equations*	209
2. *Boundary Layer Relationships for Mass and Heat Transport Parameters*	210
3. *Examples of Simulations*	214
IV. ATOMISTIC ASPECTS: CONTROL OF CRYSTAL MORPHOLOGY	215
1. *Modeling of Epitaxial Silicon Growth on the Terrace Scale*	215
2. *Linking the Atomic Scale with the Reactor Scale*	217
3. *Stability of Epitaxial Growth*	218
V. SUMMARY	220
REFERENCES	220

NOMENCLATURE

a_0 Lattice constant (m)
a_i Number of film atoms in the *i*th species
C_i Gas-phase mole concentration (mol/m^3)

C_p	Fluid heat capacity (J/kg k)
d_{eq}	Equivalent diameter; see Eq. (17) (m)
D_i	Diffusion coefficient of the ith species with respect to the carrier (m²/s)
D_{ij}	Binary diffusivity of the ith species with respect to the jth one (m²/s)
D_s	Adatom surface diffusivity (m²/s)
E_D	Activation energy for surface adatom diffusion (j/mol)
E_k	Activation energy for adatom insertion into terrace kinks (J/mol)
E_I	Activation energy for adatom incorporation processes (J/mol)
F	Overall mass flow rate (kg/s)
g	Acceleration of gravity (m/s²)
G	Film growth rate (m/s)
h_s	Heat transport coefficient (W/m²/K)
H	Reactor section height (m)
I	Identity matrix
k_B	Boltzman's constant (J/atom/K)
$k_{c,i}$	Mass transport coefficient (m/s)
k_G	Rate constant of the adatom nucleation reaction (m⁵/mol/atom/s)
k_k	Rate constant of adatom insertion into terrace kinks (m/s)
k_s	Rate constant of adatom coalescence (m²/atom/s)
k_T	Fluid thermal conductivity (W/m/K)
L	Reactor or terrace length (m)
M_i	Molecular weight of the ith species (kg/mol)
M_s	Molecular weight of deposited solid (kg/mol)
M_T	Mixture average molecular weight (kg/mol)
n	Versor orthogonal to the deposition surface
$\mathbf{N}_i$	Mass flux of the ith species (kg/m²/s)
N_I	Adatom surface density (atoms/m²)
N_s	s-cluster surface density (clusters/m²)
N_σ	Density of terrace free dangling bonds (atoms/m²)
P	Pressure (Pa)
q	Wave number (1/m)
r	Radial coordinate (m)
$\tilde{r}_{S,k}$	Rate of the kth gas-phase reaction (mol/m³/s)
$\tilde{r}_{G,k}$	Rate of the kth surface reaction (mol/m²/s)
R	Gas constant (J/mol/K)
$\tilde{R}_I$	Adatom nucleation rate (atoms/m²/s)
$\tilde{R}_{G,i}$	Production rate of the ith species in the gas phase (mol/m³/s)
$\tilde{R}_{S,i}$	Production rate of the ith species on the surface (mol/m³/s)
s	Cluster numeral dimension
S_e	External surface per unit reactor volume (1/m)
S_v	Deposition surface per unit reactor volume (1/m)
T	Temperature (K)
t	Time (s)
u	Gas velocity vector (m/s)

u_0	Unperturbed advancement velocity of the step; see Eq. (25) (m/s)
U	Overall heat transfer coefficient (W/m²/K)
V	Reactor volume (m³)
x_s	Diffusion length (m)
y_i	Mole fraction of the ith species
z	Axial coordinate (m)

GREEK LETTERS

$\alpha_{T,i}$	Thermal diffusion factor
β	Step stiffness (J/m)
β_T	Volume expansion coefficient (1/K)
$\Delta\tilde{H}_k$	Enthalpy variation associated with the kth reaction (J/mol)
φ	Atomic flux to the deposition surface (atoms/m²/s)
ρ	Fluid mass density (kg/m³)
ρ_s	Deposited solid mass density (kg/m³)
λ	Mean free path (m)
λ^*	Critical perturbation size (m)
μ	Fluid viscosity (kg/m/s)
ν	Silicon vibration frequency (1/s)
$\nu_{G,ik}$	Stoichiometric coefficient of the ith species in the kth gas-phase reaction
$\nu_{S,ik}$	Stoichiometric coefficient of the ith species in the kth surface reaction
ξ	Terrace step fluctuation (m)
τ	Characteristic time (s)
ω	Susceptor rotation speed (rad/s)
ω_q	Frequency of the terrace step fluctuation (1/s)
ω_i	Weight fraction of the ith species

The superscripts and T $\sim$ indicates quantities expressed on a molar, instead of a mass, basis. The subscripts c and T indicate all the quantities referred to the carrier gas and the transposed vectors, respectively.

I. Introduction

The aim of this chapter is mainly to describe the various aspects related to the modeling of chemical vapor deposition (CVD) reactors used for the industrial growth of epitaxial silicon. The process silicon epitaxy is essentially a CVD process used for depositing thin films of single-crystal silicon on single crystal silicon substrate and it is used extensively in the microelectronics and semiconductor industries. The requirements of the industry from this process are highly demanding, i.e., epitaxial silicon films must have an excellent thickness uniformity and excellent quality (minimum defects in the epitaxial layer). Therefore, for the production of epitaxial films, sophisticated control as well as fundamental understanding of the deposition process is required.

The growth of epitaxial silicon is usually performed by means of a thermally activated process, where the substrates are held over a heated susceptor placed in a quartz or a stainless-steel chamber under atmospheric- or reduced-pressure conditions. Quartz-wall reactors (popularly known as cold-wall reactors), whose external walls are continuously cooled by recirculating air or water, are, however, more popularly used. These reactors are characterized by a very high temperature difference between the susceptor and the reactor external walls. A number of cold-wall reactors are commercially available: horizontal, vertical (rotating disk or stagnation point configurations), and barrel reactors. The reactant feed to these reactors usually contains a deposition precursor (SiH_xCl_{4-x}, $x = 0, 1, 2, 4$) highly diluted in a carrier gas (H_2, N_2, etc.) and a controlled amount of a dopant compound [42, 45, 79]. The temperature at which the deposition process is performed and the sensitivity of the deposition process to oxidizing impurities are dependent on the hydrogen content in the precursor, i.e., the lower the hydrogen content, the higher the deposition temperature and the lower the process sensitivity to oxidizing impurties (see Table I) [48, 95]. The reducing sensitivity of higher chlorine-containing chlorosilanes (i.e., lower hydrogen-containing chlorosilanes) to oxidizing impurities is due to the etching action by the chlorine-containing species, particularly HCl, formed during the deposition process. In addition, these chlorine-containing species enhance the film quality by etching

TABLE I

COMPARISON OF CHLOROSILANE PRECURSORS FOR EPITAXIAL SILICON DEPOSITION WITH RESPECT TO PROCESS FEATURES[a]

Source	Pressure (atm)	Temperature (°C)	Growth rate (μm/min)	Sensitivity to oxidizing impurity	Reactor-wall deposition	Selective growth	Pattern shift
$SiCl_4$	1.0	1150–1300	0.2–2.0	E	E	—	P
	0.1	1100–1250	0.1–0.6			E	VG
$SiHCl_3$	1.0	1100–1250	0.2–2.0	VG	VG	—	F
	0.1	1050–1200	0.1–0.8			VG	VG
SiH_2Cl_2	1.0	1050–1150	0.1–1.0	F	G	—	G
	0.1	1000–1100	0.1–0.6			G	VG
SiH_4	1.0 0.1	850–1100	0.1–0.5	P	P	F	E

[a]E, excellent; VG, very good; G, good; F, fair; P, poor. From Refs. 48 and 95.

high-energy surface states [6]. Moreover, the etching action is also responsible for the reduced wall deposition and good selectivity obtained with chlorosilanes.

The design and optimization of CVD processes and reactors to fulfill the increasing demand in the semiconductor industry are very difficult and time-consuming tasks, and the lack of detailed fundamental models has forced industrial CVD practitioners to rely on methods of trial and error as well as on statistical methods to create purely empirical models of reactor behavior [115]. The empirical relations that are produced following these procedures are difficult to use if the reactants or the reactor geometry is changed. On the other hand, mathematical modeling and simulation provide an excellent economic alternative to trial and error-based experimental techniques. The immediate benefits to be realized are fewer experiments, reduction in waste during experimentation, and the ability to deal with different reactive species and reactor geometries. Accordingly, hereafter a mechanistic description of all the physical and chemical phenomena involved in the epitaxial process and the fundamental equations necessary for their description are reported. Different models having varying computational complexities are considered. First, detailed two-dimensional (2D) and 3D models suitable for process and reactor design are discussed. However, due to the huge amount of computational resources that they require, their use in the day-to-day operations of deposition systems is not feasible. Thus, the derivation of simple reduced models is then examined. These reduced-order models have the advantage of being readily used for on-line process control. Finally, the link between the reactor scale and the atomic scale models is examined. This is due to the fact that the goal in the modeling deposition process is not only to control the uniformity of the epitaxial film thickness and dopant concentration in the film but also to predict and control the epitaxial film morphology.

1. PHYSICAL AND CHEMICAL ASPECTS OF SILICON EPITAXY

A schematic representation of the silicon epitaxy process in shown in Fig. 1 [8, 47, 66]. In general, the several steps that occur in a CVD process are as follows: transport of gaseous species toward the growing surface in a nonisothermal flow field; surface processes, such as the adsorption of precursors, the surface diffusion of adatoms over the terrace, and their incorporation into step kinks or island clusters; desorption of reactants and by-products from the deposition surface; and, finally, transport of these species back into the bulk gas phase.

The temperature dependence of film growth rate has a general behavior (see Fig. 2) and this behavior is established depending on which of the above steps are the rate-determining step for epitaxial silicon deposition [6, 8, 24]. At low temperatures the reaction kinetics is usually the rate-limiting step, leading to growth rates that are strongly dependent on temperature, but at higher temperatures transport is the rate-limiting step, leading to growth rates that are not very strongly dependent on temperature. The transition temperature between one regime and

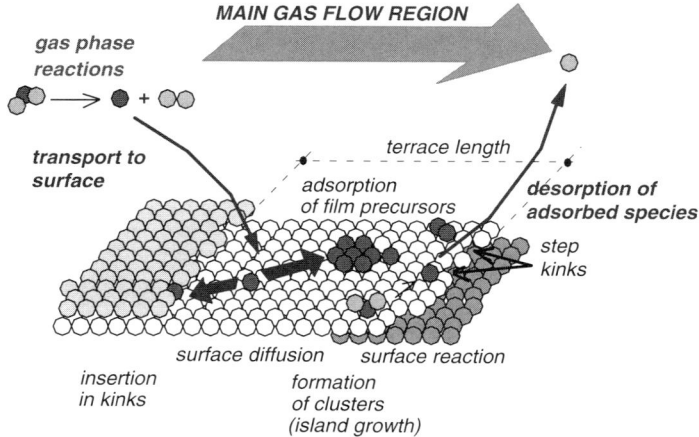

FIG. 1. Schematic of fundamental mass transport and reaction processes in CVD.

the next is strongly dependent on the type of precursor used and, also, on the geometrical configuration of the considered reactor. For example, in novel types of deposition reactors where a high rotation speed of the susceptor significantly increases the mass transfer rate, the transition from the kinetic- to the transport-controlled regime occurs at a higher temperature than in traditional reactors with a static or slowly rotating susceptor. In some systems (see Fig. 2b) a further increase in the deposition temperature leads to a growth regime again controlled by kinetic aspects, i.e., at these high temperatures desorption and etching phenomena become the rate-determining steps [2, 47, 71].

2. THE MULTISCALE MODELING CONCEPT

From the above discussion it is evident that silicon epitaxy is a complex dynamical process, and to characterize the epitaxy process fully, the phenomena occurring on different length scales must be understood. As an illustration Fig. 3 demonstrates the interconnection between the substrate scale (whose characteristic length is of the order of centimeters) and the microscopic scale (whose characteristic length is of the order of nanometers) [8, 50]. Therefore, to describe the phenomena occurring on the microscopic scale an analysis of the atomic details of the surface is required. On reactor length scales (where the phenomena belong to the domain of fluid dynamics and reaction kinetics), the analysis can be performed using a traditional chemical engineering approach i.e., by using continuum conservation equations. The atomistic scale, however, involves an analysis of the elementary surface processes and requires a physical–chemical approach. The above-mentioned interconnection arises from the fact that the growth rate is determined due to the

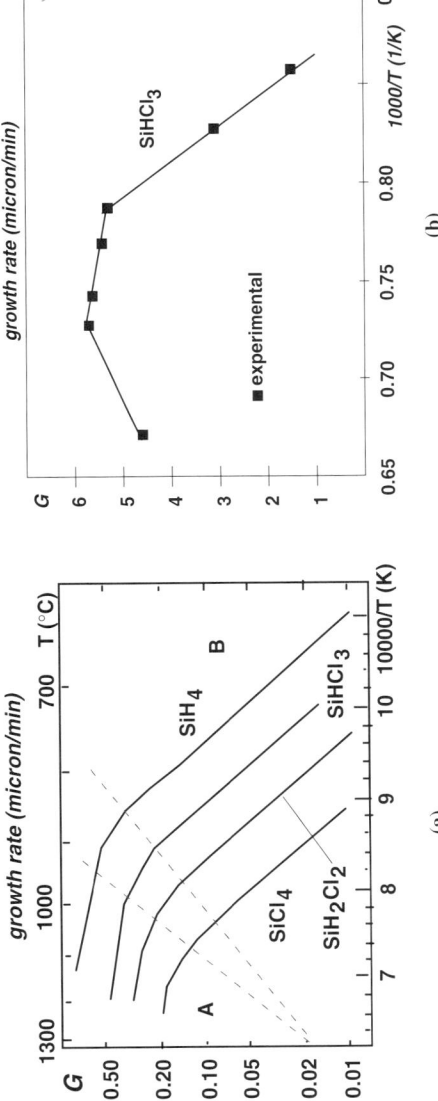

FIG. 2. (a) Silicon growth rate as a function of temperature for different Si precursors. $P = 1$ atm; precursor concentration, 1% (vol). A, transport-limited region; B, kinetically controlled region. Data from Refs. 6 and 24. (b) Silicon growth rate as a function of temperature. $SiHCl_3/H_2$, 15% (vol). Horizontal reactor with a rotating susceptor; data taken from Ref. 2.

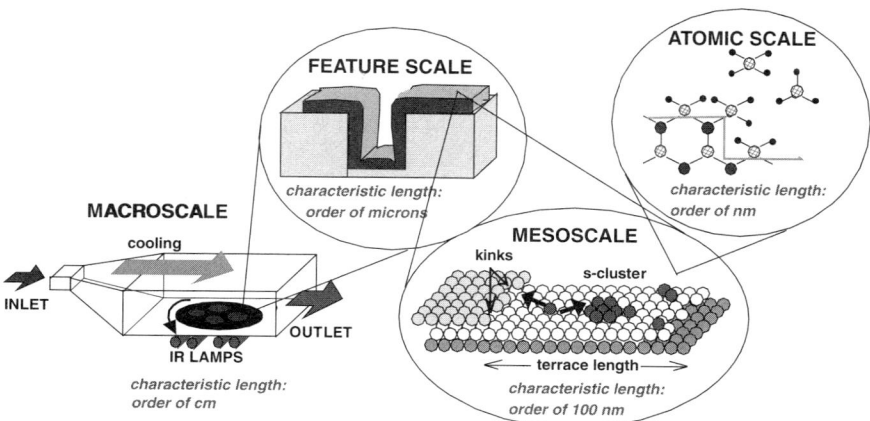

FIG. 3. Schematic of the different scales and characteristic lengths in modeling CVD processes: macroscale, feature scale, mesoscale, and atomic scale (nanoscale).

combined effect of the rates of surface processes and the rates at which the reactive species diffuse to and from the surface. Moreover, the surface kinetics influences the deposition selectivity as well as the surface morphology.

Solution of the complete set of equations describing the phenomena during silicon epitaxy on both macroscopic and microscopic length scales in complicated reactor geometries is a very complex and time-consuming task. Therefore, to understand the microscopic phenomena underlying the formation of thin films it is necessary to use models that link a simplified description of the fluid dynamics of the reactor to detailed reaction kinetics involved in the deposition. The information so obtained on the microscopic scale can then be coupled with the macroscopic set of equations.

3. Conditions for Epitaxial Growth: The Terrace Step Kink Mechanism

Epitaxial film growth is very commonly explained by the terrace step kink (TSK) growth mechanism, originally proposed by Burton *et al.* [7, 88]. According to this mechanism adatoms nucleated over the deposition surface, viewed as an evolving terrace, diffuse onto the terrace till their insertion into kinks present at the terrace edge (step). Essentially, the time scales of transport of molecules to the substrate surface and their diffusion over the surface determine the surface morphology. For example, if the time scale of transport of molecules to the substrate surface is faster than the time scale of adatom diffusion, then aggregates or clusters of atoms form. If the cluster size exceeds a critical dimension, the growth proceeds through the formation of islands leading to polycrystalline silicon [6]. Control of the epitaxial silicon film morphology is traditionally performed using empirical relationships, but as discussed later it is possible to describe quantitatively all the processes involved in the TSK growth mechanism.

II. Detailed Modeling of Epitaxial Reactors

In this section simulation studies of macroscale reacting flows for predicting film growth rate and film composition are discussed. Under typical CVD operating conditions the epitaxial silicon deposition process is a steady-state process and hence the governing set of equations is derived at steady state. However, for determining the phenomena during operations such as a change in the initial reactant composition or a temperature ramp, transient simulations must be performed. Transient simulations also have to be performed when atomic layer epitaxy or rapid thermal processes are of interest [47].

1. Conservation Equations

The CVD reactor model involves equations describing the conservation of mass, momentum, and energy [5, 96]. Models that are 2D as well as 3D in space are discussed in this section, but simplified 1D models are discussed in Section III. In general the precursors used for CVD are diluted in a carrier gas such as hydrogen. Thus, the simpler Fickian law for diffusion can be adopted instead of the detailed Stefan–Maxwell equations [8, 9, 15, 60]. In addition, mass flux due to temperature gradients (thermal diffusion) must also be considered. Furthermore, due to the low conversions and high dilutions, the heat of reaction and the heat of generation due to viscous dissipation can generally be neglected [15, 100].

The model equations for a generic CVD process are reported below [5, 96].
Mass balance equations:

$$\frac{\partial \rho}{\partial t} = \nabla \cdot (\rho \mathbf{u}) \tag{1}$$

$$\rho \left(\frac{\partial \omega_i}{\partial t} + \mathbf{u} \cdot \nabla \omega_i \right) = \nabla \cdot [\rho D_i (\nabla \omega_i - \alpha_{T,i} \omega_i \omega_c \nabla \ln T)]$$

$$+ M_i \sum_{k=1}^{NRG} \nu_{G,ik} \tilde{r}_{G,k} \tag{2}$$

Momentum balance equation:

$$\rho \left(\frac{\partial \mathbf{u}}{\partial t} + \mathbf{u} \cdot \nabla \mathbf{u} \right) = \rho \mathbf{g} - \nabla P + \nabla \cdot \mu \left(\nabla \mathbf{u} + \nabla \mathbf{u}^T - \frac{2}{3} \mathbf{I} \nabla \cdot \mathbf{u} \right) \tag{3}$$

Energy balance equation:

$$\rho C_{\mathrm{p}}\left(\frac{\partial T}{\partial t}+\mathbf{u}\cdot\nabla T\right)=\nabla\cdot(k_T\nabla T)+\sum_{k=1}^{\mathrm{NRG}}\tilde{r}_{\mathrm{G},k}(-\Delta\tilde{H}_k) \qquad (4)$$

All the symbols and notations in the above equations are explained in the Nomenclature section at the beginning of the chapter. The gas density, ρ, has to be calculated using the equation of state for an ideal gas ($\rho = PM_{\mathrm{T}}/RT$) because the use of the Boussinesq approximation is not adequate in CVD systems (due to the presence of steep temperature gradients). Furthermore, the dependence on temperature and composition of all the physical and chemical properties must be included [29, 48, 49, 58].

The mass balance of the ith chemical species [Eq. (2)] considers contributions from diffusion, convection, and gas-phase production (or consumption) due to chemical reactions in the gas phase $\tilde{r}_{\mathrm{G},k}$. As mentioned, from a rigorous point of view the diffusion contribution should be calculated using the Stefan–Maxwell relations [5, 60]. For epitaxial CVD reactors that operate under dilute conditions the rigorous approach above can be substituted with simpler expressions:

$$\mathbf{N}_i = -\rho D_i[\nabla\omega_i - \alpha_{\mathrm{T},i}\omega_i\omega_{\mathrm{c}}\nabla\ln T]. \qquad (5)$$

The use of simplified flux expressions in place of the general Stefan–Maxwell formulation in reacting flows computations was investigated extensively for combustion systems [47], and in CVD systems the use of approximate relationships was discussed by Coltrin *et al.* [15] and by Kleijn *et al.* [60]. Under conditions where the precursor is dilute in the carrier gas, the effective diffusion coefficient D_i can be set equal to the binary diffusion coefficient or it can be estimated by the relation given by Blanc's formula [5]:

$$D_i = \sum_{j\neq i}\frac{1-y_i}{y_j/D_{ij}} \qquad (6)$$

where y_i represents the mole fraction of the ith species in the gas phase ($y_i = \omega_i M_t/M_i$). The thermal diffusion factor $\alpha_{\mathrm{T},i}$ can be estimated through relationships based on the kinetic theory of gases [41].

In the energy balance equation, the viscous dissipation [5] and Dufour flux [44] are neglected because their contribution is negligible under typical CVD conditions. Furthermore, for the low gas flow rates (with respect to the speed of sound) typically used in CVD reactors, the pressure effects on energy conservation are ignored.

The boundary conditions for Eqs. (1)–(4) are specific to the reactor configuration, but some generally adopted conditions are no slip for gas velocity and no-penetration conditions at solid walls, i.e., susceptor and reactor external reactor walls. The flow velocity at the inlet is usually specified and fully developed flow conditions are imposed at the outlet section. The thermal boundary conditions significantly influence the flow, and therefore to avoid additional numerical complication of the model represented by Eqs. (1)–(4), simple assumptions of fixed wall temperatures can be assigned provided that such temperatures were determined a priori by solving simplified models of the reactor [70, 72]. One other boundary condition usually used is the energy flux at the wall, but if radiation issues become important, more detailed boundary conditions should be adopted. A discussion of this problem is reported by Fotiadis *et al.* [29]. For the boundary condition on species mass conservation, no net flux at nonreacting surfaces is imposed. On the reacting surfaces (i.e., the susceptor and walls with deposits) the mass flux is equated to the net rate of incorporation due to surface chemical reactions $r_{S,k}$,

$$-\rho D_i \mathbf{n} \cdot [\nabla \omega_i - \alpha_{T,i} \omega_i \omega_c \nabla \ln T] = M_i \sum_{k=1}^{NRS} \nu_{S,ik} \tilde{r}_{S,k} = M_i \tilde{R}_{S,i}, \qquad (7)$$

where $\nu_{S,ik}$ is the stoichiometric coefficient for the ith species in the kth surface reaction. For mass transfer-controlled reactor systems, the analysis of the reaction chemistry can be significantly reduced. In particular, it is sufficient to consider only the mass balance of the main precursors, and for the boundary condition on reacting surfaces a fast reaction (relative to transport of the precursors to the depositing surface) leading to $\omega_i = 0$ along the deposition surface is imposed.

The film growth rate can then be estimated by the net incorporation rate of the film species,

$$G = \frac{M_s}{\rho_s} \sum_{i=1}^{NS} \sum_{k=1}^{NRS} a_i \nu_{S,ik} \tilde{r}_{S,k}, \qquad (8)$$

where a_i is the number of silicon atoms in the ith precursor species and M_S and ρ_S are the molecular weight and the mass density of the film, respectively.

To enhance the numerical solution of the model and to identify the key parameters of the system, it is convenient to nondimensionalize the governing set of equations. Generally, all the variables and model parameters are scaled using a characteristic scale under certain reference conditions. For instance, the scaling values for gas velocity, gas temperature, and mass fraction of the ith species are the inlet gas velocity, the wafer temperature, and the inlet mass fraction, respectively. The nondimensionalization leads to different important dimensionless groups such as the Reynolds, Grashof, Prandtl, Schmidt, Gay Lussac, and thermal diffusion numbers (see Table II), and depending on the magnitude of these dimensionless groups, the importance of different phenomena can be established. For example, from the magnitude of the Damkohler number, the importance of transport of species relative to the reaction kinetics leading to epitaxial silicon deposition can be determined.

TABLE II
Relevant Dimensionless Groups in Silicon Epitaxy CVD Modeling

Number	Definition	Physical meaning	Typical values
Damkoler Gas	$Da = \dfrac{\tilde{R}_{G,i} L}{C_i u_{ref}}$	$\dfrac{\text{Residence time}}{\text{Characteristic reaction time}}$	10^{-3}–10^{+3}
Damkoler Surface	$Da = \dfrac{\tilde{R}_{G,i} L}{C_i D_s}$	$\dfrac{\text{Characteristic diffusion time}}{\text{Characteristic reaction time}}$	10^{-3}–10^{+3}
Grashof	$Gr = \dfrac{g \beta_T \rho_{ref}^2 L^2 T_{wafer}}{\mu_{ref}^2}$	$\dfrac{\text{Buoyancy forces}}{\text{Viscous forces}}$	1–10^{+5}
Knudsen	$Kn = \dfrac{\lambda}{L}$	$\dfrac{\text{Mean free path}}{\text{Characteristic length}}$	$<10^{-2}$
Peclet	$Pe = ReSc$	$\dfrac{\text{Mass transfer by convection}}{\text{Mass transfer by diffusion}}$	10^{-1}–10^{+3}
Prandtl	$Pr = \dfrac{\mu_{ref} C_{P,ref}}{k_{T,ref}}$	$\dfrac{\text{Momentum diffusivity}}{\text{Thermal diffusivity}}$	0.7
Rayleigh	$Ra = GrPr$	$\dfrac{\text{Buoyancy force}}{\text{Viscous force}}$	1–10^{+5}
Reynolds	$Re = \dfrac{\rho_{ref} u_{ref} L}{\mu_{ref}}$	$\dfrac{\text{Inertia forces}}{\text{Viscous forces}}$	0.1–100
Schmidt	$Sc = \dfrac{\mu_{ref}}{\rho_{ref} D_{i,ref}}$	$\dfrac{\text{Momentum diffusivity}}{\text{Mass diffusivity}}$	1–10

2. Computational Issues

In the solution of CVD reactor models for dilute systems, the chemical species equations (2) are weakly coupled with the main flow. The system fluid dynamics is thus driven by the gas flow rate and by the temperature difference existing between the hot susceptor and the outer cold walls. The schematic sequence of computations needed to solve the simulation of a CVD epitaxial reactor is illustrated in Fig. 4.

Due to the highly nonlinear nature of CVD model equations, numerical techniques such as finite differences or finite elements have to be used to solve the equations simultaneously. Nowadays, commercial general-purpose numerical codes for solving the mass, momentum, and energy balance equations simultaneously are available (e.g., FIDAP [26], FLUENT [27], CFX [11], PHOENICS-CVD [90], and MP-SALSA [81]). For simpler geometries with complex reaction kinetics and for complex reactor geometries under transport limiting conditions and under conditions where a reduced-order kinetic model is used, very robust simulation models have been developed [14, 16, 22, 33, 35, 58, 62, 64, 65, 76, 83, 100]. The memory and speed of the current-generation computers, however, are still bottlenecks when dealing with 3D complex reactor geometries with arbitrary kinetics involving arbitrary numbers of chemical species. Considering the dramatic growth of computer capabilities in recent times, such a limit hopefully will not last long.

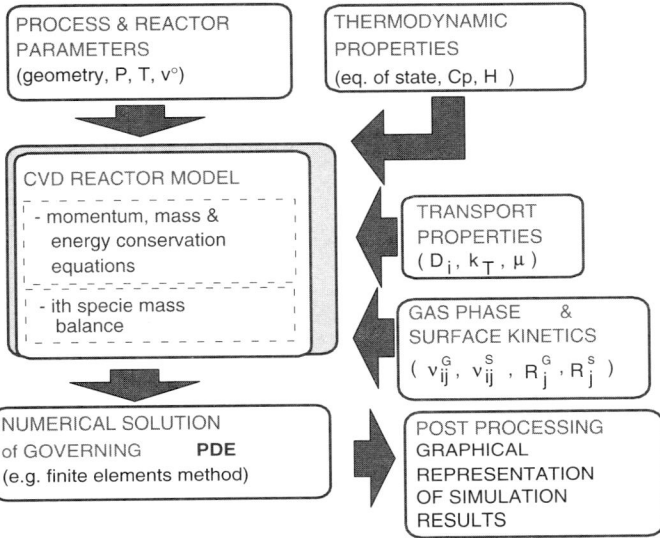

FIG. 4. Schematic illustration of the elements involved in solving a CVD reactor model (dilute reactants in inert carrier gas). (Reprinted with permission from S. Carrà and M. Masi, *Progress in Crystal Growth Characteristics* 1998; 37: 1.)

Whenever a numerical solution is predicted, it is always important to verify the solution against a proper benchmark case available in the literature [33, 58]. This is especially important when different reaction chemistry numerical codes and multidimensional computational fluid dynamics (CFD) codes are coupled to solve problems. This is because, in a SEMATECH benchmarking effort for CVD tailored codes [33], differences of about one order of magnitude were observed between species concentrations as predicted by different codes. It was concluded that the robustness of the numerical codes when handling complex chemistries was not very satisfactory, but on the contrary, the reliability of only fluid dynamics numerical codes was found to be satisfactory. Therefore, for situations when complex chemistry cannot be avoided, improvements in the performance of the numerical codes is required.

3. ESTIMATION OF INVOLVED PARAMETERS

Besides the computational difficulties, the main problem concerning the simulation of epitaxial reactors is related to the estimation of the model parameters. The estimation of thermodynamic (e.g., enthalpy and heat capacity) and transport (e.g., viscosity, thermal conductivity, and diffusivity) parameters is relatively simple compared to the estimation of the kinetic parameters (e.g., reaction rate parameters). For this reason, Chapter 2 of this book is specifically addressed to the latter issue. For determining a detailed chemical mechanism, a combined experimental and theoretical analysis is required. The use of transition state theory coupled with suitable *ab initio* quantum-chemical programs such as AMPAC [1], GAUSSIAN 98 [32], and CERIUS2 [10] is becoming popular for theoretically determining detailed reaction mechanisms. Due to their industrial importance, most of the silicon precursors have been studied in detail, and detailed reaction mechanism precursors such as silane are now available and included in the data file of standard software [17, 53].

A large collection of thermodynamic parameters for all the molecular and radical species involved in silicon deposition is now available and the values of these parameters were determined either experimentally or by means of quantum chemical calculations [4, 12, 14, 15, 28, 54, 114]. Regarding the transport parameters, they can be estimated with consolidated relationships derived from the molecular theory of gases as a function of Lennard–Jones parameters [5, 41, 92]. In fact, transport data for most common silicon precursors as well as software for estimating the transport properties of different species are available nowadays [21, 55].

4. EXAMPLES OF SIMULATIONS

This section illustrates some examples of simulations of reactors for epitaxial silicon depositions using detailed 2D or 3D models. Different silicon precursors (i.e., silane and chlorosilanes) and different reactor geometries (e.g., vertical, barrel, and horizontal) were examined. The examined reactors can process a single or a few wafers. Due to the increasing trend in wafer diameter (e.g., from the 6-in. diameter of the late 1980s, to the 12-in. diameter of the late 1990s, up to the 18-in. diameter forecast for 2005), it is expected that the reactors used for silicon epitaxy will operate in a single-wafer mode. On the other hand, reactor configurations able to process a large number of wafers are always considered for study to increase the overall process productivity. All the examined systems operate under cold-wall reactor conditions. The vertical reactor is generally simulated through 2D models taking advantage of symmetry. The geometry of the barrel reactor is intrinsically 3D, but because of its complexity it is often reduced to 2D using fluid dynamics similarities. 3D simulations for barrel reactors have also been performed on simplified geometries. Despite its apparent simplicity, the flow features in horizontal reactors are intrinsically 3D and thus detailed models need to be developed. A summary of the most relevant simulations for epitaxial silicon deposition is presented in Table III together with the main feature of each model.

a. The Horizontal Reactor

The horizontal reactor can be considered a landmark in the field of deposition reactors: The gases flow in a horizontal duct whose heated lower side constitutes the susceptor. Horizontal reactors are also commonly operated under cold-wall

TABLE III
Summary of Relevant Models for Silicon Epitaxial Deposition[a]

Reactor	Silicon precursor	Pressure	Kinetics	Model	Ref. No.(s.)
Horizontal, RD	TCS/H_2	AP	LSK	3D	2
Horizontal, RD	TCS/H_2	AP	SGK, LSK	3D	36, 37
Horizontal, RD	TCS/H_2	AP	LSK	3D	63, 64
Horizontal	SIL/H_2	AP	GPT	3D	82
Horizontal, RD	TCS/H_2	AP	LSK	3D	84, 85
Horizontal	—	AP	GPT	3D	87
Horizontal	—	AP	GPT	3D	94
Horizontal	SIL/H_2	AP	DGK	2D/3D	59
Horizontal	TET	AP	GPT	2D	13
Horizontal	SIL/H_2	AP	DGK, DSK	2D	14, 15
Horizontal	TET/H_2	AP	GPT	2D	97
Horizontal	TET,SIL/H_2	AP	GPT	2D	108
Horizontal	DCS,TET/H_2	AP	GPT	1D	23
Horizontal	SIL/H_2	AP	GPT	1D	24
Horizontal	TET/H_2	AP	GPT	1D	70
Barrel	TET/H_2	AP	GPT	3D	117
Barrel	TCS/H_2	AP	LSK	2D	19
Barrel	TET/H_2	AP	GPT	2D	31
Barrel	TET/H_2	AP	GPT, LSK	2D	51, 52
Barrel	TET/H_2	AP	GPT	2D	69
Barrel	TET/H_2	AP	GPT	2D	72–74
Barrel	TCS/H_2	AP	SGK, LSK	2D	75, 76
Barrel	TCS/H_2	AP	GPT	2D	86
Barrel	—	AP	GPT	2D	103
Barrel	—	AP	GPT	2D	118
Barrel	SIL/H_2	AP	GPT	1D	20
Barrel	TET/H_2	AP	GPT	1D	70
Barrel	DCS/H_2	LP	LSK	1D	109
Vertical, RD	SIL/H_2	AP	DGK, DSK	3D	62, 100
Vertical, RD	TET/H_2	AP	LSK	3D	80
Vertical	SIL/H_2	AP	DGK, DSK	2D	16
Vertical	DCS/H_2	AP	DGK, LSK	2D	39
Vertical	SIL/H_2	AP	GPT	2D	43
Vertical	SIL/H_2	AP	DGK	2D	56
Vertical, RD	SIL/H_2	LP	DGK	2D	60
Vertical	SIL/H_2	LP	LSK	2D	105
Vertical	TET/H_2	AP	GPT, LSK	1D	44
Vertical	SIL/H_2	AP	SGK	1D	61
Vertical, RD	TET/H_2	AP	GPT, LSK	1D	91
Vertical, RD	TET/H_2	AP	GPT	1D	107
Vertical	—	AP	GPT	1D	113

[a]RD, rotating susceptor; LSK, lumped surface kinetics; SGK, simplified gas-phase kinetics; GPT, gas-phase precursor mass transport; DGK, detailed gas-phase kinetics; DSK, detailed surface kinetics; LP, low pressure; AP, atmospheric pressure; TCS, $SiHCl_3$; DCS, SiH_2Cl_2; TET, $SiCl_4$; SIL, SiH_4.

conditions at atmospheric or reduced pressure. From the early stages this configuration has been adopted for growing epitaxial silicon films. The increase in the wafer diameter and the stringent requirements for producing high-quality wafers have led to increased use of horizontal reactors. New-generation reactors are equipped with a rotating susceptor to increase the radial uniformity of the deposited film and with robotized wafer loading–unloading equipment [2, 36–38, 63, 64]. Some examples of horizontal reactors are illustrated in Fig. 5.

An important parameter of horizontal reactor geometry is the height/width aspect ratio, which needs to be kept as low as possible to avoid complex flow recirculation inside the reactor (e.g., the Gr number scales with the cube of the reactor height). An example of the complex flows occurring in these reactors as a function of Gr/Re^2 is illustrated by Jensen [46], where it can be seen that for reactors not equipped with a rotating susceptor the flow features are directly reflected in the nonuniformity of the film thickness.

More interesting from an industrial point of view are the simulations of the new generation of horizontal reactors. All these systems operate with rotating susceptors and with a reduced reactor height to limit the above-mentioned flow problems. The gas inlet apparatus is properly designed to direct the precursors toward the wafers to minimize reactor-wall deposits. In addition, the heating system is carefully designed to ensure a constant wafer temperature. This is because the wafer temperature has to be controlled within a few degrees around the set-point value to minimize slip-line generation and to maximize dopant incorporation uniformity. The latter parameters are obviously more sensitive to the wafer temperature than the silicon growth rate because they are generally controlled by the surface kinetics, while the overall deposition rate is generally mass transport controlled.

Some examples of the simulation of novel horizontal reactors for silicon epitaxial deposition are reported in Refs. 2, 36, 37, and 64. In all of the above-cited systems, the adopted silicon precursor was trichlorosilane diluted in hydrogen carrier. Simplified gas-phase kinetics and lumped Langmuir–Hinshelwood surface kinetics were considered in these studies. The introduction of more insights on surface kinetics was necessary because the growth rate was not completely under mass transfer control. Figure 6 shows a comparison between experimental and theoretical silicon growth rates. The figure clearly demonstrates that an excellent comparison between experimentally measured growth rates and model predictions is obtained when Langmuir–Hinshelwood surface kinetics is considered but not when simple linear kinetics is considered. To analyze the effect of wafer rotation on the precursor distribution within the reactor, the reader can refer to Figs. 7 and 8. The $SiHCl_3$ mass fraction contours along the longitudinal midplane of the ASM reactor for stationary and rotating susceptor are illustrated in Fig. 7. In the absence of wafer rotation, the deposition rate is nonuniform both in the flow direction and in the transverse direction. Hence, wafer rotation is required if a good thickness uniformity is to be achieved since rotation translates any location on the wafer through both high and low deposition rate regimes. It can readily be seen that the wafer rotation distorts the symmetry of the precursor distribution in the X–Y plane.

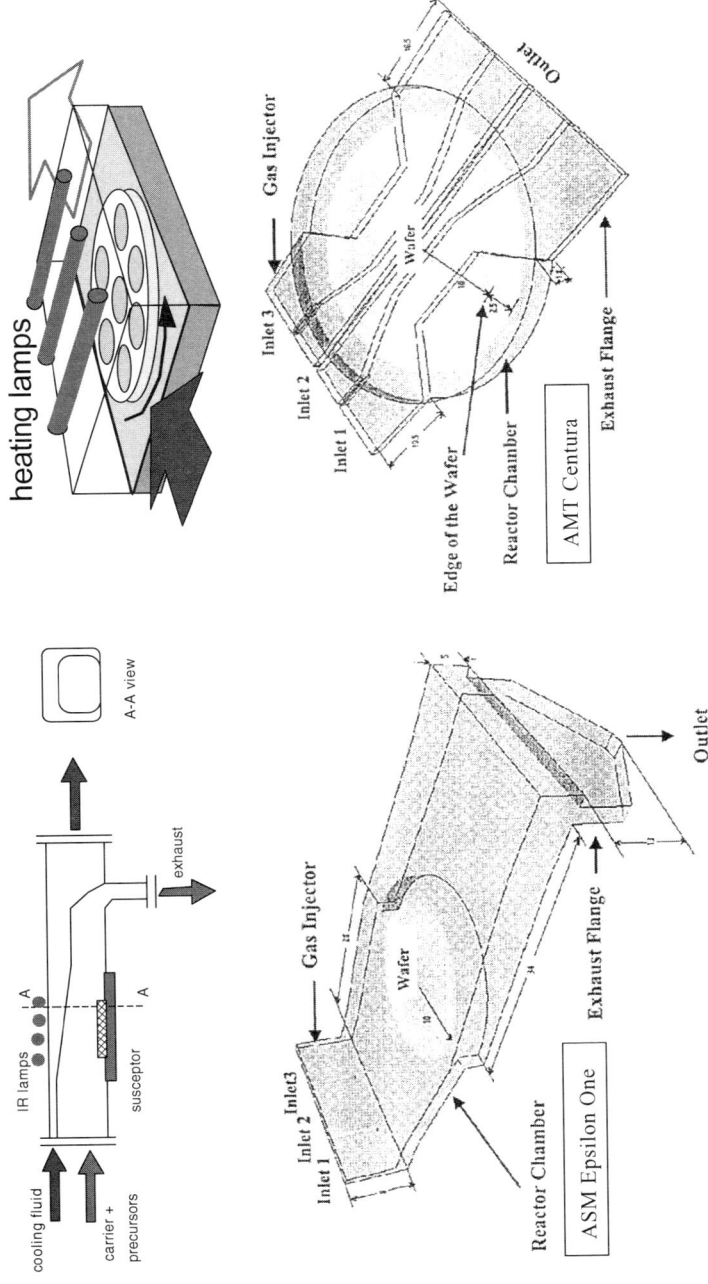

FIG. 5. Schematic illustration of traditional and new-generation horizontal reactors.

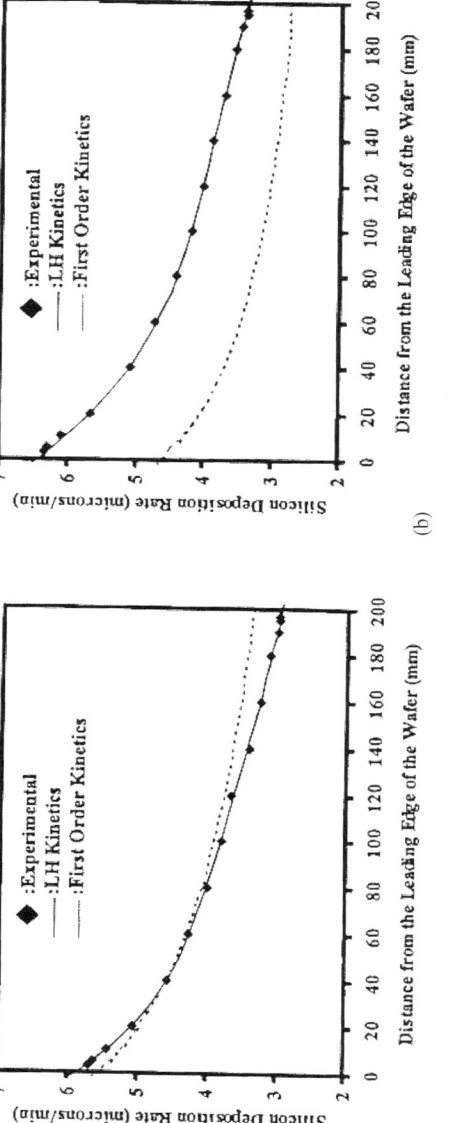

FIG. 6. Theoretically predicted and experimentally measured Si deposition rates in the flow direction along the reactor centerline as a function of the inlet flow rate. (a) $Q = 1.7$ liters/s, (b) $Q = 2.4$ liters/s. $SiHCl_3$, 71% (wt); wafer temperature, 1398 K. Reprinted with permission from Ref. 64.

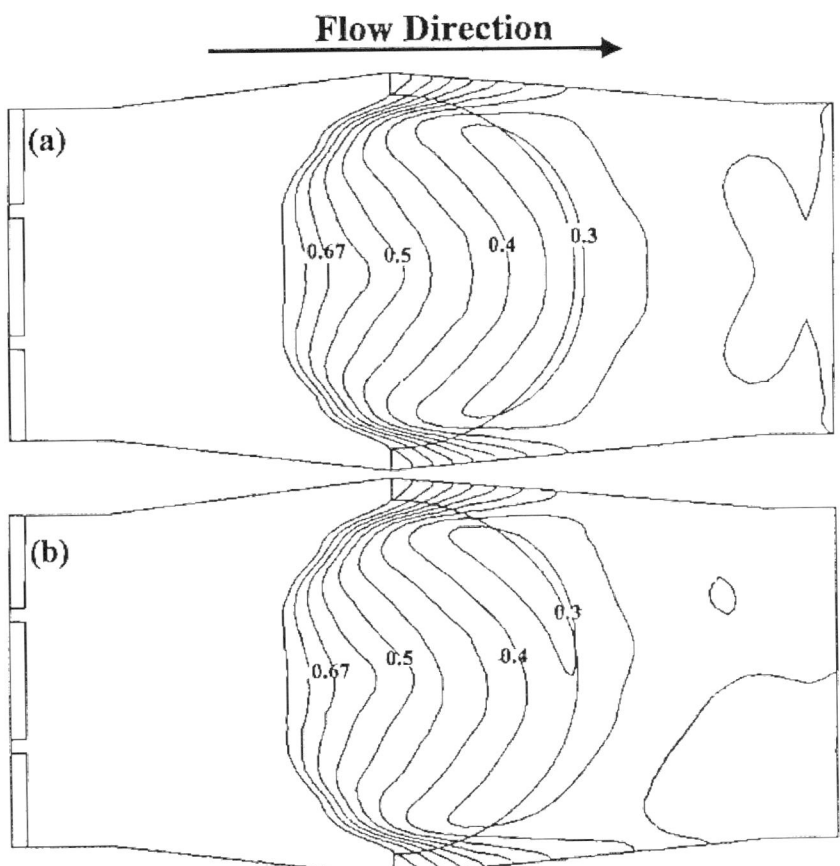

FIG. 7. SiHCl$_3$ mass fraction profile along the X–Y midplane in an ASM horizontal reactor. $Q =$ 1.7 liters/s; wafer temperature, 1398 K; SiHCl$_3$, 61% (wt). (a) Stationary wafer; (b) wafer rotating at 35 rpm. (Reprinted with permission from S. Kommu, G. M. Wilson, and B. Khomami, *Journal of the Electrochemical Society* 2000; 147: 1538.)

In any case, the averaging effect of the wafer rotation in the azimuthal direction coupled with the flow distribution in the reactor aids in the control of the radial uniformity of the deposited film as illustrated in Fig. 8.

As a general conclusion, all the above-cited references show a satisfactory agreement between the calculated and the experimentally measured growth rates. Thus, the success of detailed simulation models for predicting the overall growth rates can be considered very good. However, the standard precautions pointed out by Kleijn [58] should be taken into consideration when more detailed chemical kinetics are introduced into the models to simulate phenomena such as dopant incorporation and film defect formation.

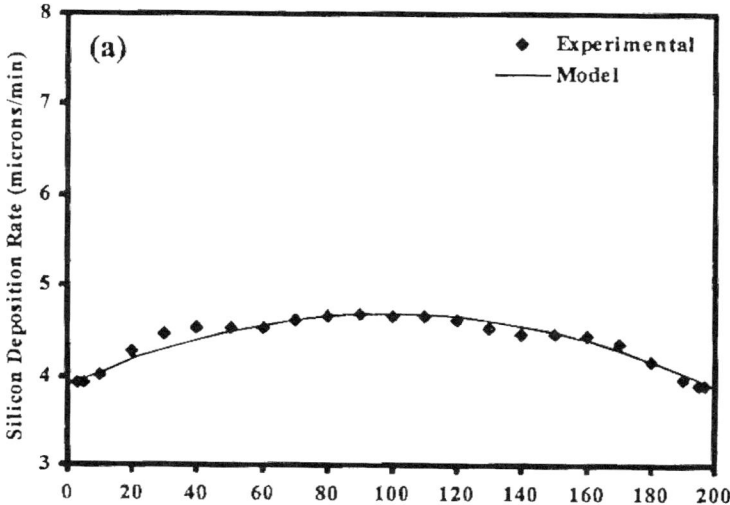

FIG. 8. Comparison between experimental and computed Si deposition rates for a wafer rotating at 35 rpm. $Q = 1.7$ liters/s; wafer temperature, 1398 K; inlet $SiHCl_3$, 71% (wt). (Reprinted with permission from S. Kommu, G. M. Wilson, and B. Khomami, *Journal of the Electrochemical Society* 2000; 147: 1538.)

b. *The Vertical Reactor*

Vertical reactors, whose schematic representation is shown in Fig. 9, hold the wafers horizontally over a heated susceptor that can be kept stationary or rotated at different speeds. The gases are injected vertically down onto the deposition surface. The design effort is usually focused on the gas distribution apparatus, which is mainly of the showerhead kind. Particularly important is the coverage of the center of the susceptor by the impinging gases, which has a strong influence on the radial uniformity of the deposition profile [35, 110, 112, 113, 116]. The control of temperature uniformity of the susceptor through radiative heating is a major

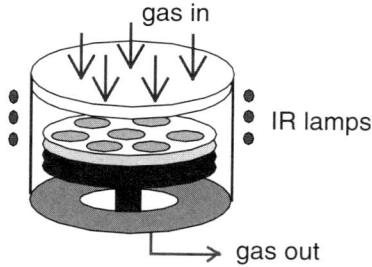

FIG. 9. Schematic illustration of a vertical reactor.

issue for these reactors. Due to symmetry this reactor configuration is relatively easier to model. These reactors are usually simulated using 2D models, but when complex kinetics are under consideration, simpler 1D models are also adopted. Transient 3D simulations have also been performed to analyze flow instabilities and transition to turbulence [62, 100].

Many simulations have been reported for the deposition of compound semiconductors (e.g., III–V and II–VI) [29, 93, 110, 112, 116, 121] but there are also examples devoted to the deposition of silicon, mainly through silane or chlorosilanes diluted in hydrogen [14, 39, 56–58]. These reactors have also been used to determine intrinsic kinetics [14, 56].

Some examples of simulations, taken from Kleijn *et al.* [59], are reported in Figs. 10 and 11. This example considers a vertical reactor, operating with

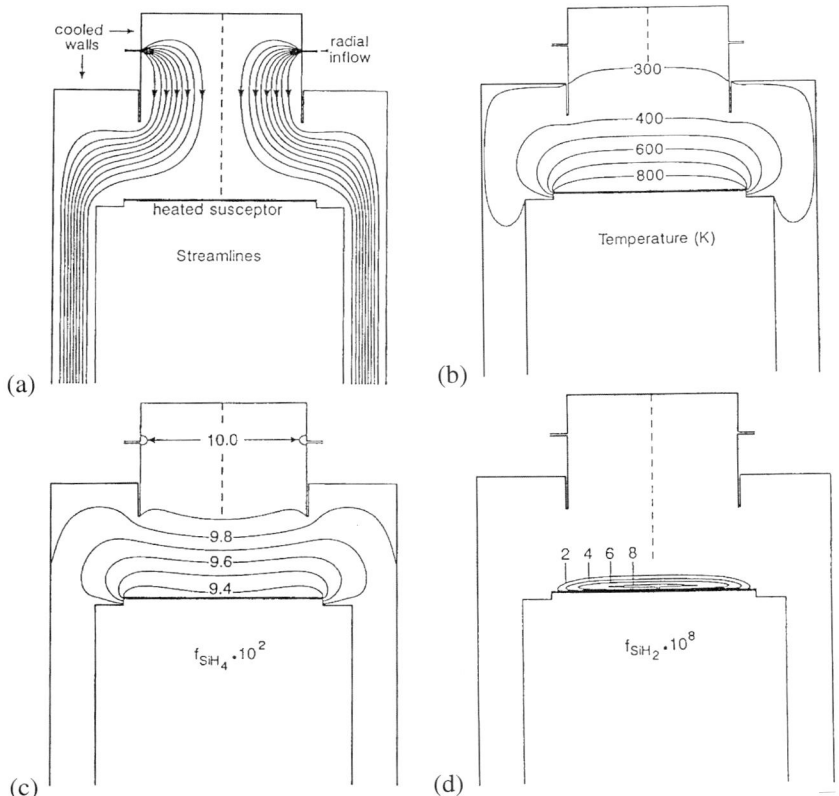

FIG. 10. Calculated (a) streamlines, (b) temperatures, and gas-phase compositions (expressed as mole fraction) of (c) SiH_4 and (d) SiH_2 for a vertical reactor. Reactor pressure, 133 Pa; susceptor temperature, 900 K; inlet temperature, 293 K; gas flow rate, 1000 sccm; silane inlet mole fraction, 0.1; carrier, N_2. 2D simulation with detailed chemistry. (Reprinted with permission from C. R. Kleijn, *Journal of the Electrochemical Society* 1991; 138: 2190.)

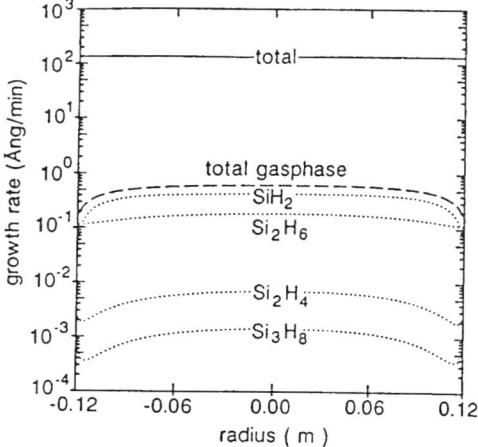

FIG. 11. Contribution of the main silane fragments to the film growth rate in a vertical reactor. Conditions as in Fig. 10. 2D simulation with detailed chemistry. (Reprinted with permission from C. R. Kleijn, *Journal of the Electrochemical Society* 1991; 138: 2190.)

silane diluted in hydrogen operating under low-pressure conditions. The calculated streamlines, isotherms, and mole fractions of main gas species are illustrated in Fig. 10. It can be seen that the spatial distribution of the chemical species is coherent with the progressive silane decomposition and with the gaseous chemistry considered. Figure 11 illustrates the contribution to the film growth rate of each of the main precursors. It can readily be seen that SiH_2 is the main deposition precursor under the examined conditions. Obviously, an alteration in process parameters can easily modify this conclusion. Figure 10 also shows the greater radial uniformity that can be obtained by means of vertical reactors.

From the above discussion it can be concluded that improvements in vertical reactor performance and new process designs can rely on modeling support.

c. *The Barrel Reactor*

Silicon epitaxy performed in barrel reactors has the advantage of processing a large number of wafers per batch. These reactors are, however, difficult to optimize and operate for processing large-diameter wafers, thus justifying the current practice to replace these reactors with horizontal or vertical single-wafer reactors [18, 67]. In a barrel reactor, the wafers are held by a heated prismatic susceptor contained in a quartz bell that is externally cooled by air. Different commercial reactor configurations exist, with the main differences being in the bell-jar shape and in the gas inlet apparatus (e.g., the Applied Material, the LPE, and the Spire types). A schematic representation of a typical barrel reactor is shown in Fig. 12. Barrel

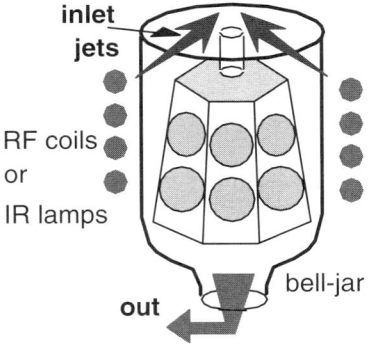

FIG. 12. Sketch of a barrel reactor (AM type).

reactors operate at nearly atmospheric or at reduced pressure and also under high temperature gradients that greatly affect gas flow patterns. Complex buoyancy-driven secondary flows are usually observed, thus generating detrimental effects on film uniformity [117].

A few illustrations of the simulation studies performed in barrel reactors are given below. The reactors considered are the LPE 2061S and the AMT 7700, with the former using $SiCl_4$–H_2 precursors and the latter using $SiHCl_3$–H_2 precursors. The simulation results given in this section were obtained using detailed models where a 3D flow dynamics is solved in a 2D geometry under cylindrical coordinates.

In the AMT reactor the gases are injected by a seal plate and by two lateral jets as illustrated in Fig. 12. The main gas inject nozzles are located approximately 60° to each side of the back center of the reactor (120° apart). They face toward each other, and their flows are opposing. The resulting flow down the susceptor is not symmetrical, and in fact significant channeling occurs. The susceptor rotation reduces the channeling effects, leading to a more uniform deposition profile over the different susceptor faces. Accordingly, some caution should be adopted in reducing the fully 3D system into a simplified 2D geometry, and, also, in the case where all three velocity components are considered [75, 76].

The LPE reactor differs from the AMT reactor in its inlet apparatus; jet nozzles are replaced by a distributor injecting the gases radially and symmetrically over the prismatic susceptor.

Examples of the simulations performed on these reactors are reported in Figs. 13 and 14, which show that there is very good agreement between the experimentally measured growth rates along the longitudinal section of the susceptor and the model predictions. It was determined that the growth rate profile is sensitive to the external wall temperature, external quartz bell shape, and inlet gas flow rate [72–76].

As a general conclusion, the success of modeling barrel reactors in predicting growth rates is lower than that of vertical and horizontal reactors, mainly because due to the geometrical complexity of these reactors, detailed 3D models of these

208   M. Masi and S. Kommu

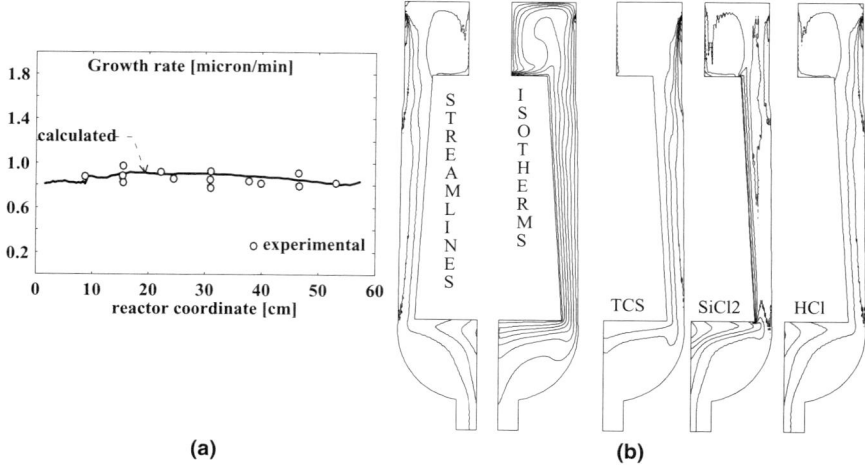

FIG. 13. Simulation of an AMT 7700 barrel reactor. (a) Comparison between calculated and experimental growth rates along the susceptor midline. (b) Calculated streamlines, isotherms, and main species weight fractions. Susceptor temperature, 1423 K; wall temperature, 700 K; H_2 main flow rate, 76 slm; H_2 rotation flow rate, 40 slm; $SiHCl_3$ inlet mole fraction, 0.05; susceptor tilting, 2°; susceptor length, 0.597 m; wafer diameter, 0.153 m. Data taken from Ref. 75.

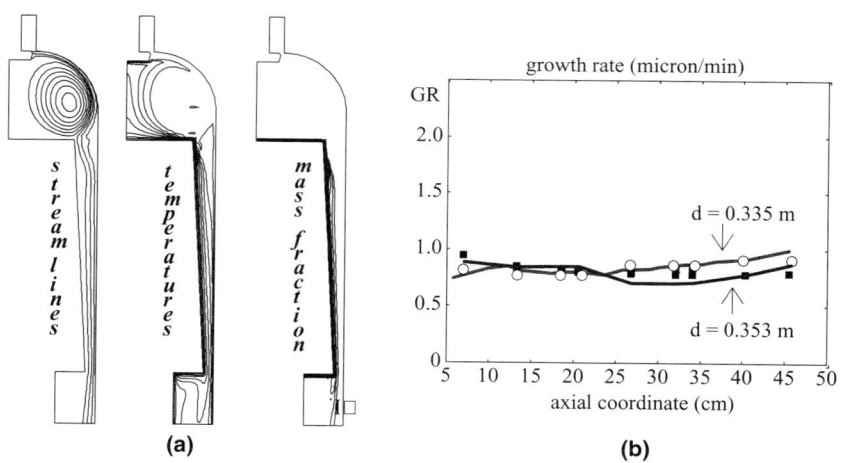

FIG. 14. Simulation of an LPE 861 barrel reactor for silicon epitaxy. Atmospheric pressure. Susceptor temperature, 1523 K; wall temperature, 460 K; H_2 flow rate, 0.41 g/s; $SiCl_4$ flow rate, 0.28 g/s; susceptor tilting, 3°; susceptor length, 0.45 m; wafer diameter, 0.127 m. (a) Example of calculated streamlines, isotherms, and isocomposition lines. (b) Comparison between calculated (lines) and experimental growth rate values along the susceptor longitudinal midline for two external bell diameters. Data taken from Ref. 74.

reactors are yet to be developed. However, the existing simpler models (i.e., 1D and 2D) can be used to provide some insights into the reactor behavior and for guiding experiments.

III. Reduced-Order Models for Epitaxial Silicon Deposition

Considering the computational complexities discussed above, when the system geometry is not very complicated, such as in the case of horizontal reactors or in vertical reactors, a first-approximation analysis can be performed using 1D models. These models can describe the system only along the reactor main coordinate satisfactorily if proper relationships to describe the precursors mass transport toward the susceptor are adopted. Here, such an approach is described with reference to a horizontal reactor but it can be easily extended to vertical and barrel reactors. These models allow a quick analysis of reactor features and optimizations, especially when systems involving complex kinetics are considered. Examples are available for horizontal [24, 70, 102], vertical [91, 113], and barrel [20, 69, 70, 76] reactors.

Other simplified model approaches use the analytical solution of the reactor fluid dynamics available for a simpler framework such as the impinging jet, the rotating disk, or the flow in rectangular ducts, and then the species mass conservation equations is solved by using the prior determined gas velocity field [16, 58, 83, 86, 87]. The latter approach has been widely adopted to simulate MOCVD reactors where complex chemical reactions are involved.

In any case, the main advantage of the simplified model is the possibility quickly to investigate complex kinetics, without the necessity of massive computer facilities. These models can be used at early reactor design stage to derive algorithms for the on-line control of the deposition process and for managing the day-to-day operations.

1. Model Equations

The model equations can be derived directly from the complete set of equations discussed above. Due to the decoupling of the velocity field, which can thus be determined a priori, the reactor model reduces to mass and energy conservation equations.

The mass balances for the ith gaseous species and the overall one can then be written

$$\frac{d(F\omega_i)}{dV} = M_i \sum_{k=1}^{NRG} v_{G,ik} \tilde{r}_{G,k} - S_v N_i, \qquad i = 1, NCG \qquad (9)$$

$$\frac{dF}{dV} = -S_v \sum_{i=1}^{NCG} N_i, \qquad (10)$$

where $F = u\rho\Omega$, V, and S_v are the overall mass flow rate, the reactor volume, and the deposition surface per unit reactor volume, respectively. For the case of a horizontal reactor, the differential volume is proportional to the reactor coordinate z by the reactor free cross section Ω (e.g., $dV = \Omega dz$). The subscripts G and S indicate a quantity referred to the gas phase and to the deposition surface, respectively. The mass flux includes the contribution of ordinary and thermal diffusion and, in a 1D schematization, can be written [8, 70]

$$N_i = k_{c,i}\rho\left[(\omega_i - \omega_{S,i}) - \alpha_{T,i}\omega_i(1 - \omega_i)\ln(T_S/T_G)\right], \quad (11)$$

where $k_{c,i}$, T_S and T_G are the mass transfer coefficient, the susceptor temperature, and the gas temperatures, respectively. The second term in Eq. (11) was derived assuming a linear variation of the temperature within the boundary layer. The mass transfer coefficient $k_{c,i}$ can be evaluated with reference to the gas motion within the considered framework. Then the surface mass fractions ω_i^S can be obtained equating the mass flux toward the deposition surface with the corresponding surface reaction rate:

$$-N_i = M_i \sum_{j=1}^{NRS} \nu_{S,ij}\tilde{r}_{S,j}, \quad i = 1, \text{NCG}. \quad (12)$$

The adsorbed species concentrations can be calculated imposing their no net production:

$$\sum_{j=1}^{NRS} \nu_{S,ij}\tilde{r}_{S,j} = 0, \quad i = 1, \text{NA}. \quad (13)$$

For the energy balance, the following equation can be directly derived from Eq. (4),

$$\frac{d(FC_pT_G)}{dV} = S_v h_s(T_S - T_G) - S_e U(T_G - T_C) + \sum_{k=1}^{NRG} \tilde{r}_{G,k}\left(-\Delta\tilde{H}_k\right), \quad (14)$$

where T_C, S_e, h_s, and U are the cooling gas temperature, the external tube surface per unit reactor volume, the heat transfer coefficient at the susceptor surface, and the overall coefficient with respect to the heat transfer outside the reactor, respectively. The heat of reaction contribution in Eq. (14) can be safely neglected because of the low precursor conversions and the high dilution usually prevailing in the considered systems. Solution of the algebraic and ordinary differential equations system can be pursued by means of any integration subroutine, adopting $\omega_i = \omega_i^e$, $F = F^\circ$, and $T_G = T_G^\circ$ as initial conditions associated with Eqs. (9), (10), and (14), respectively.

2. BOUNDARY LAYER RELATIONSHIPS FOR MASS AND HEAT TRANSPORT PARAMETERS

The proper boundary layer relationship to be used for estimating the mass transport coefficient depends on the geometry considered. Accordingly, relationships are available for rectangular duct, annular duct, impinging jet, and rotating disk configurations. A collection of the available relationships can be found in Refs. 5, 34, 68, 101, and 104. A few examples are presented below to complete the model presentation. Here only equations to estimate the mass transport coefficient under completely mass transport-controlled conditions are considered. Under conditions where kinetics and transport are the rate-determining steps, these relationships are not valid, necessitating detailed numerical solution, followed, if possible, by reduction to simpler physical models. It is also possible to estimate the order of magnitude of different quantities using relationships such as

$$\text{Nu} = \text{Sh}(\text{Pr}/\text{Sc})^{1/3}. \tag{15}$$

In a rectangular duct the mass transport coefficient can be estimated by the following expression of the Sherwood number in terms of a quickly convergent series [104]:

$$\frac{k_{c,i} H}{D_i} = \text{Sh} = \left[\frac{1}{\text{Sh}^\infty} + 0.25 \sum_{j=1}^{\infty} \gamma_j \exp\left(-\frac{8\beta_j^2 z}{3H\,\text{ReSc}}\right)\right]^{-1}, \tag{16}$$

where Sh^∞ is the limiting Sherwood value for fully developed flow [68, 104]; H, z, Re, and Sc are the reactor spacing between the susceptor and the external wall, the longitudinal coordinate, and the Reynolds (evaluated with respect to H) and the Schmidt dimensionless numbers, respectively; and γ_j and β_j are the numerical parameters of the series [104].

For fully developed flow in a vertical annular duct, which is a situation resembling the barrel reactor, the following relationship can be adopted [25]:

$$\text{Sh} = 1.96 \left[\text{Pe}\frac{d_{eq}}{L} + 0.04 \left(\text{Ra}\frac{d_{eq}}{L}\right)^{0.75}\right]^{0.319}, \tag{17}$$

where Pe, Ra, and d_{eq} indicate the Peclet number, the Rayleigh number, and the equivalent diameter of the annular duct in which the reactants are flowing (e.g., $4 \times$ cross section/wetted perimeter), respectively. The latter should be used as a characteristic length in each of the dimensionless numbers in Eq. (17).

For the rotating disk configuration, the mass transport coefficient can be calculated through the relationship [101]

$$\text{Sh} = 0.39\sqrt{\text{Re}}, \tag{18}$$

this value being constant along the radius of the rotating disk. Here the Reynolds number is calculated as a function of the disk rotation velocity ω (i.e., Re = $\rho \omega R^2 / \mu$), where R indicates the susceptor radius.

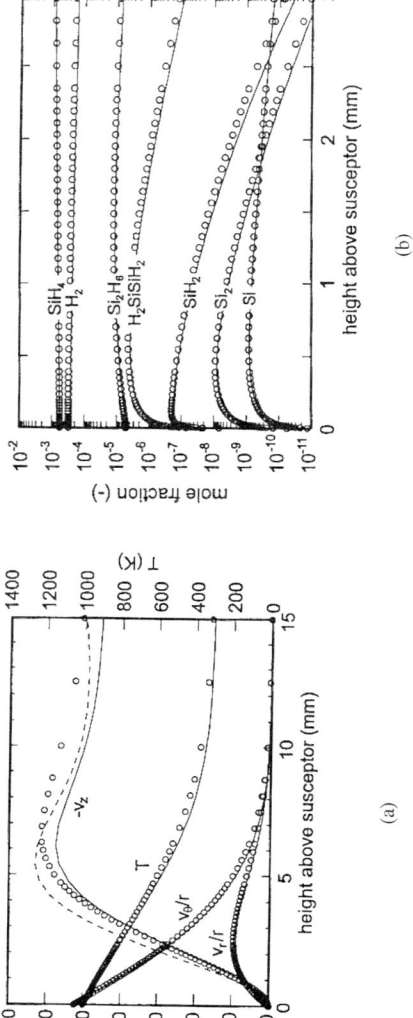

FIG. 15. Simulation of a vertical reactor by means of simplified and detailed codes: rotating disk configuration. (a) Comparison of calculated temperature and fluid velocities as a function of the distance from the susceptor. (b) Comparison of calculated gas-phase composition as a function of the distance from the susceptor. Reprinted with permission from Ref. 58.

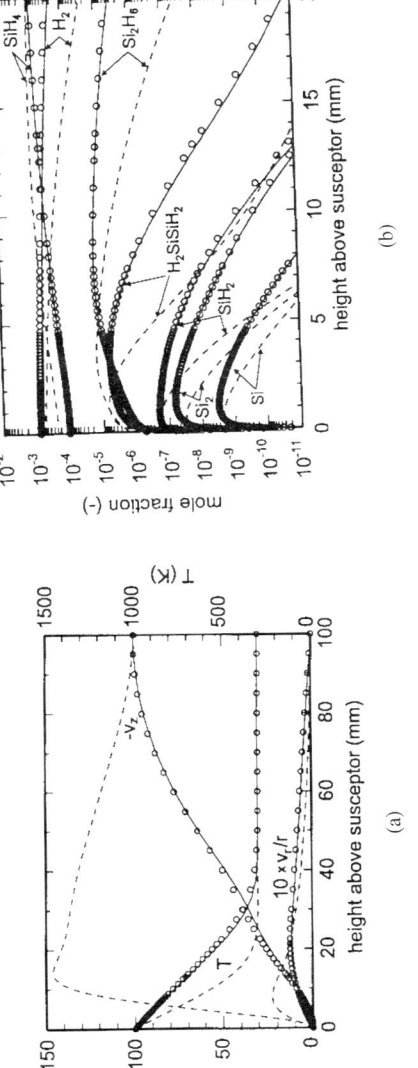

FIG. 16. Simulation of a vertical reactor by means of simplified and detailed codes: impinging jet configuration. (a) Comparison of calculated temperature and fluid velocities as a function of the distance from the susceptor. (b) Comparison of calculated gas-phase composition as a function of the distance from the susceptor. Reprinted with permission from Ref. 58.

3. Examples of Simulations

Only some key features of the reactor behavior can be predicted by means of 1D models, such as the growth rate profile along the main coordinate of the reactor, as well as the composition profiles. Of particular interest are examples where the results of simulations performed through simplified models are compared with those of detailed models.

A quite interesting example is reported in Ref. 58, where the solutions obtained through the simplified SPIN code are summarized with detailed simulations in a vertical reactor geometry in the case of both rotating and nonrotating susceptors. The case examined refers to silicon deposition from silane, a case in which all the kinetic aspects are well understood. From the data plotted in Figs. 15 and 16 it can be deducted that the agreement between the values calculated with simplified models and those calculated with detailed models is closer when the analytical solution for the fluid flow matches the detailed solution well. This is obviously the case for the rotating disk situation, where the analytical solution matches the detailed one as illustrated in Fig. 15a. Consequently, the composition profiles reported in Fig. 15b exhibit a reasonable agreement. For the impinging jet case, where the susceptor is kept stationary, the deviation in the flow description by the analytical solution is greater as illustrated in Fig. 16.

For a barrel reactor, such a comparison is illustrated in Fig. 17, where a 2D solution is compared with experimental data and with two 1D solutions differing only by the relationship adopted to estimate the mass transport coefficient. It can be seen that in this case the discrepancies are greater than those found for the previous case. Here, the geometrical and the flow complexities can be simulated with less accuracy by simplified mass transport relationships combining asymptotically two or more flow modes through the boundary layer theory.

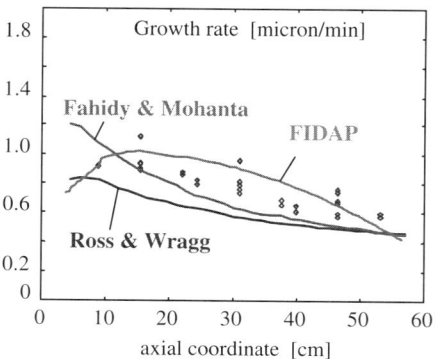

FIG. 17. Simulation of an AMT barrel reactor by means of simplified and detailed codes: 2D simulation (FIDAP) and 1D simulations using different relationships for the estimation of the mass transport coefficient. Data taken from Ref. 76.

IV. Atomistic Aspects: Control of Crystal Morphology

As pointed out, with the improvement in the performances of microelectronic devices, the requirements for the deposited material quality are becoming more and more stringent. However, as described in the previous sections, while the thickness nonuniformity of growth rate profiles is kept almost under control, the surface morphology is usually controlled empirically through extensive experimental studies aimed at characterizing the process variables that affect the surface features [47, 111].

Theoretical approaches to film morphology description reported in the literature generally refer to molecular beam epitaxy deposition processes, whose operating conditions differ significantly from those of the CVD processes adopted for depositing silicon films. A review of the theories available to describe the atomistic aspects of film growth is reported in Refs. 88 and 98. In this section these general theories are extended to the growth of epitaxial silicon in CVD processes. The main feature of this approach is to describe the nucleation of adsorbed silicon atoms on the growing surface (i.e., the so-called adatoms) through chemical reactions. As discussed in Section I.2 the final aim of the modeling work is to describe the correspondence between process parameters and film properties, determined by the mechanism of surface formation. The growth of silicon films occurs mainly through two mechanisms: layer-by-layer growth and island growth [7, 88]. As the latter can be viewed as a degeneration of the former growth mechanism, the analysis presented here focuses on the layer-by-layer one. This mechanism is also called the Frank–van der Merwe [30] or TSK mechanism [7]. The description of this growth process occurs on a scale significantly different from that examined previously for reactor modeling [8, 77, 78]. Here the characteristic length of the system is of the same order of magnitude as the terrace length, i.e., 10–100 nm. Under these conditions the transition from layer-by-layer to island growth can be identified by means of a critical cluster coverage value, which is about 50%.

1. Modeling of Epitaxial Silicon Growth on the Terrace Scale

The description of the atomic mechanisms involved in silicon growth developed in this section is based on the TSK model of the surface [7]. In this framework, the conservation of the adsorbed silicon species (adatoms) is determined by the competition between their production through surface reactions and their diffusion perpendicular to the step edge. The migrating adatoms can also coalesce to form clusters during this time period. As the cluster mobility over the terrace is almost negligible, only the surface diffusion of adatoms is considered, i.e., the clusters are assumed to be nonmigrating. Accordingly, the terrace growth model considers a terrace of length L, and the adatoms are nucleated by surface chemical reactions. Then, the so-formed adatoms migrate up the terrace surface to be incorporated into a terrace kink or into a cluster. An *s-cluster* is formed by the incorporation of a migrating adatom by an *(s−1)-cluster* and it disappears on the incorporation of another adatom. All these processes are illustrated in Fig. 1. Here the analysis of the terrace growth process is developed in only a simplified monodimensional approach, because our aim is only to introduce the possibility of describing the process physics. Accordingly, the steady-state material balances for adatoms and *s-mers* on the terrace are [77, 78]

$$\tilde{R}_1 - N_1 \sum_{s=1}^{S_{\max}} k_s N_s + D_s \frac{\partial^2 N_1}{\partial z^2} = 0, \quad i = 1, \tag{19}$$

$$k_1 \sqrt{s-1} \cdot N_{s-1} N_1 = k_1 \sqrt{s} \cdot N_s N_1, \quad s = 2, S_{\max}, \tag{20}$$

where the first term in Eq. (19) indicates adatom nucleation by the chemical reaction, the second term indicates the consumption of adatoms by cluster formation, and the last term indicates the diffusion of adatoms over the terrace. Equation (20) indicates the balance for the cluster containing s atoms, and it contains only the generation terms because they are assumed to be nonmigrating over the surface. Moreover, $S_{\max}$, D_s, k_1, and k_s are the maximum cluster size over the terrace, the surface diffusion coefficient of adatoms, and the rate constant of the adatom coalescence with adatoms and s-clusters, respectively. The balance equations for the s-mers (20) have been written here assuming a fractal dimension of the cluster $p = 0.5$ (i.e., disk-shaped cluster). Accordingly, the rate constant for the incorporation into an s-cluster becomes $k_s = k_1 \sqrt{s}$ [8]. The other parameters such as the surface diffusivity and the incorporation constant can be estimated by the kinetic theory as [77, 98]

$$D_s = \nu \cdot a_0^2 \cdot \exp(-E_D/RT), \tag{21}$$

$$k_1 = 0.25 \cdot \nu \cdot a_0^2 \cdot \exp(-E_1/RT), \tag{22}$$

where $\nu = 7.52 \cdot 10^{13} \mathrm{s}^{-1}$ is the silicon vibration frequency, $a_0 = 0.408$ nm is the lattice parameter, and E_D and E_1 are the activation energies for the diffusion and the incorporation processes, respectively. For silicon, $E_D \approx 22$ kcal/mol, and E_1 is 2–3 kcal/mol lower than E_D. The boundary conditions for Eq. (19) are the symmetry and the flux conditions at step kinks:

$$N_1(0) = N_1(L) \tag{23a}$$

$$-D_s \frac{\partial N_1}{\partial z}\bigg|_0 = k_k \cdot N_1(0). \tag{23b}$$

The incorporation constant of adatoms into step kinks can be estimated by the kinetic theory as $k_k = 0.25 \cdot \nu \cdot a_0 \cdot \exp(-E_k/RT)$. A first approximation for the activation energy of this process is $E_k \approx E_1/2$. A more detailed analysis of the boundary conditions removes the symmetry condition by introducing a different insertion rate at upper and lower step kinks (Schoebel effect). This is discussed in Refs. 89 and 98. Due to the simplified nature of our analysis, the latter aspect is not analyzed further.

The adatom nucleation rate is determined by the surface reactions that generate single silicon atoms over the surface. As a first approximation, this rate is proportional to the precursor gas-phase concentration and to the density of the free dangling bond over the terrace N_σ, that is, $R_1 = k_G C_i N_\sigma$ or $R_1 = k_G C_i N_\sigma^2$, depending whether the precursor's adsorption follows a single-site or a dual-site mechanism. The accurate expressions, however, require knowledge of the surface kinetics for the considered precursors. The reader can refer to Chapter 2 of this book for a detailed analysis of the chemical aspects involved in epitaxial silicon growth. Equations (19) and (20) can be combined to obtain a nonlinear ordinary differential equation of second order that can be solved numerically.

2. Linking the Atomic Scale with the Reactor Scale

The only external parameter of the terrace model is the maximum cluster dimension, which needs to be determined by the linking condition, which equates the macroscopic value of the growth rate with that calculated at the terrace-scale level (i.e., $R_1 = G \cdot N_A$). This condition can be easily obtained by imposing the average value of the adatom density calculated by the terrace model, that is, the left-hand side of Eq. (24), to be equal to the corresponding value calculated by the reactor model [e.g., through Eq. (13)]:

$$\bar{N}_1 = \frac{1}{L} \int_0^L N_1(z) dz. \tag{24}$$

The outcome of Eq. (24) is the maximum cluster size $S_{\max}$. Equations (19) to (24) constitute a self-compatible model of film growth on the terrace scale. Model outputs are the adatom and cluster coverage of the terrace surface, the distribution function of the cluster dimensions, and an independent estimation of the film growth rate with respect to the estimated growth rate by means of reactor-scale models. As pointed out, the coupling of the growth rates calculated by the two models (reactor and terrace) represents the link between the two modeling scales. By inspection of these numbers it is then possible to check whether or not the film is growing by the layer-by-layer mechanism or by the formation of merging islands. As the maximum island size increases, the probability of abandoning the TSK growth mechanism also increases. This aspect is confirmed by atomic force microscope images of epitaxial silicon growth. These images show that with increasing silane partial pressure, the size of surface silicon clusters decreases. Islands of about 15 nm were observed at 690 K and a silane partial pressure of 2×10^{-7} mbar, while at a silane partial pressure of 10^{-5} mbar only single adatoms

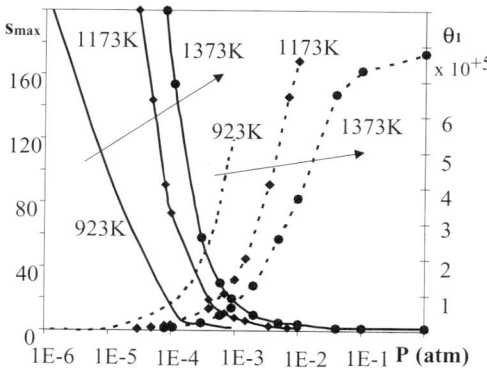

FIG. 18. Terrace-scale modeling. Influence of pressure on growth parameters at different temperatures. (———) $T = 923$ K, $L = 18$ nm; (♦) $T = 1173$ K, $L = 27$ nm; (●) $T = 1373$ K, $L = 35$ nm. $L =$ terrace length. Bold lines, maximum cluster size; dashed lines, mean surface coverage. Silicon deposition by SiH_4/H_2 mixture: reactor volume, 8753 cm^3; deposition surface, 79 cm^2; feed flow rate, 40 sccm; inlet silane mole fraction, 0.8. Reprinted with permission from Ref. 77.

or dimers were present [106, 119, 120]. Simulations performed using the terrace model above describe this trend qualitatively. An example of silicon growth from silane is illustrated in Fig. 18, which shows that the surface coverage, never higher than 1%, increases, while the cluster size decreases, with an increase in the silane partial pressure; the reverse occurs with an increase in the surface temperature. The calculated island density decreases with an increase in the temperature as clusters become larger, and this fact is also in agreement with experimental findings [106]. At higher pressures, the overall surface coverage, as expected, increases, while the growing surface appears to be covered only by the low single adatom concentration necessary for epitaxial silicon growth by the TSK mechanism.

3. STABILITY OF EPITAXIAL GROWTH

As illustrated in Fig. 18, an increase in the silane partial pressure leads to films of good quality, preventing the transition to island growth. At higher pressures mainly monomers and a few dimers cover the terrace, representing the ideal conditions for TSK growth. Unfortunately, an increase in the pressure and then in the deposition rate gives rise to the emergence of step wariness, which can generate instability toward disordered crystal growth. This mechanism of disruption of the epitaxial growth type has been thoroughly investigated for molecular beam epitaxy systems [3, 40, 89, 98, 99]. This aspect is still under discussion for CVD systems, and for this reason it is discussed only briefly in this section. Pimpinelli and Villain [88] have recently published a review of these aspects.

In terms of the linear stability of the step flow growth process that characterizes the layer-by-layer growth mode, the evolution of the step fluctuation around the

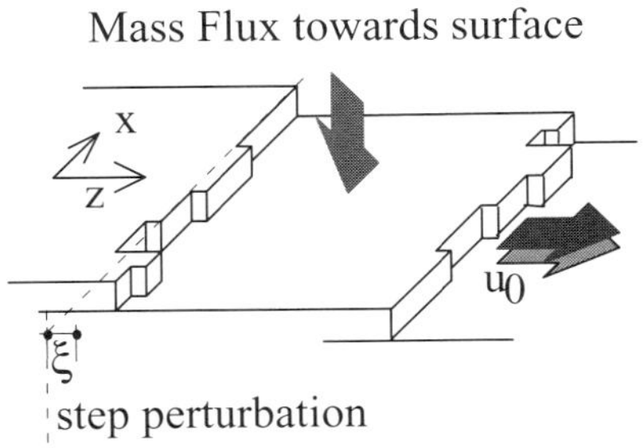

6 EPITAXIAL GROWTH MODELING

FIG. 19. Sketch of a terrace perturbation and growth instability generation.

straight unperturbed shape can be analyzed by checking the sign of the wave number of the perturbation [98, 99]. In this case, the step motion is stable only if the wave number is negative. Only in this case the step motion is able to recover from the occurrence of any perturbation. On the other hand, by imposing a wave number equal to zero, the maximum allowable perturbation can be obtained. Accordingly, in the linear stability analysis framework the evolution of the step fluctuations ξ, sketched in Fig. 19, can be investigated by introducing a small perturbation to the velocity of advancement of the straight step u_0, as a cosinusoidal term with wave number q,

$$\xi(x,t) = u_0 t + a_q e^{\omega_q t} \cos(qx), \qquad (25)$$

where u_0 is the unperturbed advancement velocity of the step, which can be estimated from the rate of any of the surface processes. From the linear stability conditions, the system is stable only if $\omega_q < 0$. As the surface diffusion length in epitaxial systems occurs at high temperatures ($x_s = \sqrt{D_s \tau}$), which results in lengths significantly greater than the terrace length, the critical ω_q value can be calculated as [98, 99]

$$\omega_q \cong u_0 q - \frac{\beta D_s a_0^4 C_{eq}}{k_B T} q^3 \leq 0, \qquad (26)$$

where $C_{eq} = \varphi \cdot \tau$ is the equilibrium concentration of silicon adatoms for a straight step and β is the step stiffness, which accounts for the influence of the surface curvature on the local adatom concentration due to the Gibbs–Thomson effect. In the above equations, τ is the time necessary to saturate the surface with a flux φ (i.e., equal to the surface reaction rate under steady-state conditions). Equation (26) can easily be resolved to obtain the critical perturbation length λ that indicates the maximum allowable deviation from step linearity that can be adsorbed by the growing system:

$$\lambda^* = \frac{1}{q} < \sqrt{\frac{\beta D_s a_0^4 C_{eq}}{L k_B T \varphi}}. \qquad (27)$$

The value of the critical perturbation size λ^* is comparable to a spontaneous perturbation of the order of a few silicon atoms. From Eq. (27), it is determined that the edge instability occurs at an approximate value of λ^* of 2 nm, in agreement with both the value of the pressure at which instability starts and the size of clusters observed with a scanning tunnel microscope. Finally, from this value it is possible to estimate the maximum partial pressure of the deposition precursor below which the epitaxial growth is stable.

V. Summary

This chapter has described the different aspects involved in modeling CVD reactors used for the industrial growth of epitaxial silicon. The chapter has reviewed and systematically summarized different simulation studies conducted on different reactors at different length scales. It is concluded that the models on the reactor scale can be considered highly reliable for the purpose of reactor optimization and design for most reactor configurations. However, detailed intrinsic reaction kinetic models leading to epitaxial silicon deposition for many commonly used precursors, detailed reactor models coupled with complex reaction kinetics, and atomistic models for determining film morphology are yet to be successfully developed. The improved efficiency of computing resources coupled with the emergence of the multiscale modeling concept should lead in the near-future to the emergence of detailed models which can address many fundamental issues on both macroscopic and microscopic scales.

REFERENCES

1. AMPAC, Semichem, Shawnee Mission, KS (www.semichem.com).
2. Angermeier, D., R. Monna, A. Slaoui, and J. C. Muller, *J. Electrochem. Soc.* **144**, 3256 (1997).
3. Bales, G. S., and A. Zangwill, *Phys. Rev.* **B41**, 5500 (1990).
4. Barin, I., O. Knacke, and O. Kubaschewski, *Thermochemical Properties of Inorganic Substances* Springer, Berlin, 1977.
5. Bird, R. B., W. E. Stewart, and E. N. Lightfoot, *Transport Phenomena,* Wiley, New York, 1960.
6. Bloem, J., and L. J. Giling, *Curr. Topics Mater. Sci.* **1**, 147 (1978).
7. Burton, W. K., N. Cabrera, and F. C. Frank, *Phil. Trans. Royal Soc. (London)* **A243**, 299 (1951).
8. Carrà, S., and M. Masi, *Prog. Crystal Growth Charact.* **37**, 1 (1998).
9. Carrà, S., C. Cavallotti, and M. Masi, *Mater. Sci. Forum* **276–277**, 135 (1998).
10. CERIUS2, Molecular Simulation Inc., San Diego, CA (www.msi.com).
11. CFX, AEA Technology, Harwell, UK (www.aeat.com).
12. Chase, M. W., C. A. Davies, J. R. Downey, R. A. McDonald, and A. N. Syverud, *J. Phys. Chem. Ref. Data* **14**, (1985).
13. Chiu, K., and F. Rosemberger, in *Chemical vapor deposition X*, edited by G. W. Cullen, pp. 175–180, Electrochemical Society, Pennington, NJ, 1987.
14. Coltrin, M. E., R. J. Kee, and J. A. Miller, *J. Electrochem. Soc.* **131**, 425 (1984).
15. Coltrin, M. E., R. J. Kee, and J. A. Miller, *J. Electrochem. Soc.* **133**, 1206 (1986).

16. Coltrin, M. E., R. J. Kee, and G. H. Evans, *J. Electrochem. Soc.* **136**, 819 (1989).
17. Coltrin, M. E., R. J. Kee, and F. M. Rupley, *Surface Chemkin: A Fortran Package for Analyzing Heterogeneous Chemical Kinetics at Solid Surface-Gas Phase Interface,* Sandia National Laboratories, Livermore, CA, 1990.
18. Corboy, J. F., and R. Pagliaro, Jr., *RCA Rev.* **44**, 231 (1983).
19. De Paola E., P. Duvernueuil, A. Langlais, and M. Nguyen, *J. Phys. IV* **9**(Pt8), 221 (1999).
20. Dittman, F. W., *Adv. Chem. Ser.* **133**, 463 (1979).
21. Ern, A., and V. Giovangigli, EGLIB, a Multicomponent Transport Software for Fast and Accurate Evaluation Algorithms, www.cmap.polytechnique.fr/www.eglib/, 1997.
22. Ern, A., V. Giovangigli, and M. D. Smooke, *J. Comput. Phys.* **126**, 21 (1996).
23. Evans, E. D., and B. Subramanian, *J. Electrochem. Soc.* **138**, 589 (1991).
24. Everstein, F. C., P. J. W. Severin, C. H. J. van der Brekel, and H. L. Peek, *J. Electrochem. Soc.* **117**, 925 (1970).
25. Fahidy, T. Z., and S. Mohanta, in *Advances in Transport Processes,* edited by A. S. Majumdar, Vol. 1, pp. 83–138, 1980, Wiley Eastern Ltd., New Delhi.
26. FIDAP, Fluent Inc., Evanston, IL (www.fluent.com).
27. FLUENT, Fluent Inc., Evanston, IL (www.fluent.com).
28. Foresman, J. B., and A. Frisch, *Exploring Chemistry with Electronic Structure Methods,* Gaussian Inc., Carnegie, PA, 1998.
29. Fotiadis, D. I., S. Kieda, and K. F. Jensen, *J. Crystal Growth* **102**, 441 (1990).
30. Frank, F. C., and J. H. van der Merwe, *Proc. Roy. Soc. London* **A198**, 205 (1949).
31. Fujii, E., H. Nakamura, K. Haruna, and Y. Koga, *J. Electrochem. Soc.* **119**, 1106 (1972).
32. GAUSSIAN 98, Gaussian Inc. Carnegie, PA (www.gaussian.com).
33. Geyling, F. T., Technical Report 94012188A-ENG, SEMATECH, 1994.
34. Gosse, J., *Technical Guide to Thermal Processes,* Cambridge University Press, Cambridge, 1986.
35. Gupta, V., C. Theodoropoulos, J. D. Peck, and T. J. Mountziaris, in *Semiconductor Processes and Device Performance Modeling,* edited by S. T. Dunham and J. S. Nelson, pp. 161–166, Material Research Society, Warrendale, PA, 1997.
36. Habuka, H., T. Nagoya, M. Katayama, M. Shimada, and K. Okuyama, *J. Electrochem. Soc.* **142**, 4272 (1995).
37. Habuka, H., T. Nagoya, M. Mayusumi, M. Katayama, M. Shimada, and K. Okuyama, *J. Crystal Growth* **169**, 61 (1996).
38. Habuka, H., Y. Ayoama, S. Akiyama, and Q. Toru, *J. Crystal Growth* **207**, 77 (1999).
39. Hierlemann, M., A. Kersch, C. Werner, and H. Schafer, *J. Electrochem. Soc.* **142**, 259 (1995).
40. Hirose, F., M. Suemitsu, and N. Miyamoto, *J. Appl. Phys.* **70**, 5380 (1991).
41. Hirschfelder, J. O., C. F. Curtiss, and R. B. Bird, *Molecular Theory of Gases and Liquids,* Wiley, New York, (1967).
42. Hitchmann, M. L., and K. F. Jensen, in *Chemical Vapor Deposition—Principles and Applications,* edited by M. L. Hitchmann and K. F. Jensen, pp. 1–30, Academic Press, London, 1993.
43. Houtman, C., D. B. Graves, and K. F. Jensen, *J. Electrochem. Soc.* **133**, 961 (1986).
44. Jenkinson, J. P., and R. Pollard, *J. Electrochem. Soc.* **131**, 2911 (1984).
45. Jensen, K. F., *Chem. Eng. Sci.* **42**, 923 (1987).
46. Jensen, K. F., *J. Crystal Growth* **98**, 148 (1989).
47. Jensen, K. F., in *Chemical Vapor Deposition—Principles and Applications,* edited by M. L. Hitchmann and K. F. Jensen, pp. 31–90, Academic Press, London, 1993.
48. Jensen, K. F., and W. Kern, in *Thin Film Process II,* edited by J. L. Vossen and W. Kern, pp. 283–368, Academic Press, San Diego, CA, 1991.
49. Jensen, K. F., E. O. Einset, and D. I. Fotiadis, *Annu. Rev. Fluid. Mech.* **23**, 197 (1991).
50. Jensen, K. F., T. G. Mihopoulos, S. Rodger, and H. Simka, in *Chemical Vapor Deposition XIII,* edited by T. M. Besmann, M. D. Allendorf, McD. Robinson, and R. K. Ulrich, pp. 67–75, Electrochemical Society, Pennington, NJ, 1996.
51. Juza, J., and J. Cermak, *Chem. Eng. Sci.* **35**, 429 (1980).
52. Juza, J., and J. Cermak, *J. Electrochem. Soc.* **129**, 1627 (1982).

53. Kee, R. J., F. M. Rupley, and J. A. Miller, *Chemkin II: A Fortran Chemical Kinetics Package for Analyzing Gas-Phase Chemical Kinetics,* Sandia National Laboratories, Livermore, CA, 1989.
54. Kee, R. J., F. M. Rupley, and J. A. Miller, *The Chemkin Thermodynamic Data Base,* Sandia National Laboratories, Livermore, CA, 1990.
55. Kee, R. J., G. Dixon-Lewis, J. Warnatz, M. E. Coltrin, and J. A. Miller, *A Fortran Computer Code Package for the Evaluation of Gas Phase Multicomponent Transport Properties,* Sandia National Laboratories, Livermore, CA.
56. Kleijn, C. R., in *Chemical Vapor Deposition XIII,* edited by T. M. Besmann, M. D. Allendorf, McD. Robinson, and R. K. Ulrich, pp. 83–88, Electrochemical Society, Pennington, NJ, 1986.
57. Kleijn, C. R., *J. Electrochem. Soc.* **138**, 2190 (1991).
58. Kleijn, C. R., *Thin Solid Films* **365**, 294 (2000).
59. Klein, C. R., and C. J. Hoogendorn, *Chem. Eng. Sci.* **46**, 321 (1991).
60. Kleijn, C. R., Th. H. van der Meer, and C. J. Hoogendoorn, *J. Electrochem. Soc.* **136**, 3423 (1989).
61. Kleijn, C. R., K. J. Kuijlars, and H. E. A. van den Akker, *Chem. Eng. Sci.* **51**, 2119 (1996).
62. Kleijn, C. R., K. J. Kuijlars, M. Okkerse, H. van Santeen, and H. E. A. van den Akker, *J. Phys. IV* **9**(Pt8), 123 (1999).
63. Kommu, S., and G. M. Wilson, in *Chemical Vapor Deposition XIV,* edited by M. D. Allendorf and C. Bernard, pp. 222–229, Electrochemical Society, Pennington, NJ, 1997.
64. Kommu, S., G. M. Wilson, and B. Khomami, *J. Electrochem. Soc.* **147**, 1538 (2000).
65. Kuiper, A. E. T., C. H. J. van der Brekel, J. de Groot, and F. W. Weltkamp, *J. Electrochem. Soc.* **129**, 2288 (1982).
66. Kuech, T., and K. F. Jensen, in *Thin Film Process II,* edited by J. L. Vossen and W. Kern, pp. 369–442. Academic Press, San Diego, CA, 1991.
67. Lord, A. H., *J. Electrochem. Soc.* **134**, 1227 (1987).
68. Luikov, A. V., *Heat and Mass Transfer,* Mir, Moskow R, 1980.
69. Manke, C. W., and L. F. Donaghey, *J. Electrochem. Soc.* **124**, 561 (1977).
70. Masi, M., S. Carrà, M. Morbidelli, V. Scaravaggi, and F. Preti, *Chem. Eng. Sci.* **45**, 3551 (1990).
71. Masi, M., H. Simka, K. F. Jensen, T. F. Kuech, and R. Potemski, *J. Crystal Growth* **124**, 483 (1992).
72. Masi, M., S. Fogliani, and S. Carrà, *J. Phys. IV* **C5**, 261 (1995).
73. Masi, M., S. Fogliani, and S. Carrà, *Mater. Sci. Forum* **203**, 203 (1996).
74. Masi, M., S. Fogliani, and S. Carrà, in *Chemical Vapor Deposition XIII,* edited by T. M. Besmann, M. D. Allendorf, McD. Robinson, and R. K. Ulrich, pp. 125–130, Electrochemical Society, Pennington, NJ, (1996).
75. Masi, M., G. Radaelli, N. Roda, P. Raimondi, S. Carrà, G. Vaccari, and D. Crippa, in *Semiconductor Processes and Device Performance Modeling,* edited by S. T. Dunham and J. S. Nelson, pp. 187–192, Material Research Society, Warrendale, PA, 1998.
76. Masi, M., C. Cavallotti, F. Di Muzio, S. Carrà, D. Crippa, and G. Vaccari, *J. Phys. IV* **9**(Pt8), 273 (1999).
77. Masi, M., V. Bertani, C. Cavallotti, and S. Carrà, *Chem. Vap. Depos.* **6**, 1 (2000).
78. Masi, M., V. Bertani, C. Cavallotti, and S. Carrà, *Mater. Chem. Phys.,* in press (2000).
79. Meyerson, B. S., in *Chemical Vapor Deposition—Principles and Applications,* edited by M. L. Hitchmann and K.F. Jensen, pp. 219–244, Academic Press, London, 1993.
80. Miyauchi, A., Y. Inoue, and S. Takaya, in *Fundamentals of Gas Phase and Surface Chemistry of Vapor Phase Material Synthesis,* edited by M. D. Allendorf, M. R. Zachariah, L. Mountziaris, and A. H. McDaniel, pp. 370–375, Electrochemical Society, Pennington, NJ, 1998.
81. MP-SALSA, Sandia National Laboratories, Albuquerque, NM (www.sandia.com).
82. Moffat, H., and K. F. Jensen, *J. Electrochem. Soc.* **135**, 459 (1988).
83. Mountziaris, T., and K. F. Jensen, *J. Electrochem. Soc.* **138**, 2426 (1991).
84. Narusawa, U., *J. Electrochem. Soc.* **141**, 2072 (1994).

85. Narusawa, U., *J. Electrochem. Soc.* **141**, 2078 (1994).
86. Nyce, T., and F. Rosemberger, in *Chemical Vapor Deposition X,* edited by G. W. Cullen, pp. 53–60, Electrochemical Society, Pennington, NJ, 1987.
87. Nyce, T., J. Ouazzani, A. Duran-Dauubin, and F. Rosemberger, *Int. J. Heat Mass Transfer* **35**, 1481 (1992).
88. Pimpinelli, A., and I. Villain, *Physics of Crystal Growth,* Cambridge University Press, Cambridge, 1998.
89. Pimpinelli, A., and A. Videcoq, *Surf. Sci.* **445**, L23 (2000).
90. PHOENICS-CVD, CHAM Ltd., Wimbledon, UK.
91. Pollard, R., and J. Newman, *J. Electrochem. Soc.* **127**, 744 (1980).
92. Reid, R. C., J. M. Prausnitz, and T. K. Sherwood, *The Properties of Gases and Liquids*, McGraw–Hill, New York, 1977.
93. Ren Sun, Y., D. W. Weyburne, and Q. S. Paduano, in *Semiconductor Processes and Device Performance Modeling*, edited by S. T. Dunham and J. S. Nelson, pp. 193–198, Material Research Society, Warrendale, PA, 1998.
94. Rhee, S., J. Szekely, and O. Ilegbusi, *J. Electrochem. Soc.* **134**, 2552 (1986).
95. Robinson, McD., in *Microelectronic Materials and Processes*, edited by A. Lavy, NATO ASI Series E, Vol. 164, Chap. 2, Kluwer, Boston, 1989.
96. Rosner, D. E., *Transport Processes in Chemically Reacting Systems,* Buttherworths, Boston, 1986.
97. Rundle, P. C., *J. Crystal Growth* **11**, 6 (1971).
98. Saito, Y., *Statistical Physics of Crystal Growth*, World Scientific, Singapore, 1996.
99. Saito, Y., and M. Uwaka, *Phys. Rev.* **B49**, 10677 (1994).
100. Santen, H. V., C. R. Kleijn, and H. E. A. Van Den Akker, in *Chemical Vapor Deposition XIV*, edited by M. D. Allendorf and C. Bernard, pp. 214–221, Electrochemical Society, Pennington, NJ, 1997.
101. Schlichting, H., *Boundary Layer Theory,* McGraw–Hill, New York, 1979.
102. Shepperd, W. H., *J. Electrochem. Soc.* **112**, 988 (1965).
103. Shih, P. H., K. Chen, and Y. Liu, *AIChE Symp. Ser.* **84**, 96 (1988).
104. Skelland, A. H. P., *Diffusional Mass Transfer*, J. Wiley, New York, 1974.
105. Soukane, S., and P. Duverneuil, in *Chemical Vapor Deposition XIV*, edited by M. D. Allendorf and C Bernard, pp. 238–245, Electrochemical Society, Pennington, NJ, 1997.
106. Spitzmüller, J., M. Fehrenbacher, M. Pitter, H. Rauscher, and R. J. Behm, *Phys. Rev. B.* **55**, 4659 (1997).
107. Sugawara, K., *J. Electrochem. Soc.* **119**, 1479 (1972).
108. Takahashi, R., Y. Koga, and K. Sugawara, *J. Electrochem. Soc.* **119**, 1406 (1972).
109. Takoudis, C., and M. M. Kastelic, *Chem. Eng. Sci.* **44**, 2049 (1989).
110. Theodoropoulos, C., H. K. Moffat, and T. J. Mountziaris, in *Semiconductor Processes and Device Performance Modeling*, edited by S. T. Dunham and J. S. Nelson, pp. 155–160, Material Research Society, Warrendale, PA, 1998.
111. Thiart, J. J., V. Hlavacek, and H. J. Viljoen, *Thin Solid Films* **365**, 275 (2000).
112. Thompson, A. G., P. Zawandski, E. Armor, R. A. Stall, and G. H. Evans, in *Chemical Vapor Deposition XIII*, edited by T. M. Besmann, M. D. Allendorf, McD. Robinson, and R. K. Ulrich, pp. 101–106, Electrochemical Society, Pennington, NJ, 1996.
113. Vosen, S. R., in *Chemical Vapor Deposition XIII*, edited by T. M. Besmann, M. D. Allendorf, McD. Robinson, and R. K. Ulrich, pp. 95–100, Electrochemical Society, Pennington, NJ, 1996.
114. Wagman, D. D., W. H. Evans, V. B. Parker, R. H. Schumm, I. Halow, S. M. Bailey, K. L. Churney, and R. L. Nuttal, *J. Phys. Chem. Ref. Data* **11**, (1982).
115. Wang, X. A., and R. L. Mahajan, *J. Electrochem. Soc.* **142**, 3123 (1995).
116. Winters, W. S., and G. H. Evans, in *Chemical Vapor Deposition XIII*, edited by T. M. Besmann, M. D. Allendorf, McD. Robinson, and R. K. Ulrich, pp. 89–94, Electrochemical Society, Pennington, NJ, 1996.

117. Yang, L., B. Farouk, and R. L. Mahajan, *J. Electrochem. Soc.* **139**, 2666 (1992).
118. Young, G. W., S. Hariharan, and R. Carnahan, *SIAM J. Appl. Math* **52**, 1509 (1992).
119. Zhang, Z., F. Wu, H. J. W. Zandvliet, B. Polsema, H. Metiu, and M. G. Lagally, *Phys. Rev. Lett.* **74**, 3644 (1995).
120. Zhang, Z., and M. G. Lagally, *Science* **276**, 377 (1995).
121. Zhou, J., and C. A. Wolden, in *Fundamentals of Gas phase and Surface Chemistry of Vapor Phase Material Synthesis*, edited by M. D. Allendorf, M. R. Zachariah, L. Mountziaris, and A. H. McDaniel, pp. 352–357, Electrochemical Society, Pennington, NJ, 1998.

CHAPTER 7

Epitaxial Layer Characterization and Metrology

V.-M. Airaksinen

R&D DEPARTMENT
OKMETIC OYJ
VANTAA, FINLAND

ABBREVIATIONS USED	226
I. INTRODUCTION	226
II. ON SAMPLING AND ACCURACY	229
1. Sampling and Cost	229
2. Precision and Accuracy	229
3. Measurement Pattern, Mapping, and Variation	230
III. DOPING CONTROL	235
1. General	235
2. The Four-Point Probe	239
3. Mercury Probe CV	241
4. Spreading Resistance Profilometry	243
5. Noncontact Resistivity Measurements: Air-Gap CV, Air-Gap SPV CV, and Air-Gap SPV	245
6. Secondary-Ion Mass Spectrometry	247
IV. THICKNESS MEASUREMENTS	248
1. Standard FTIR	248
2. Model-Based FTIR	251
3. Thickness Measurements on n/N and p/P Structures	252
4. Resistivity and Thickness Measurements on Multilayer Structures	253
V. CONTAMINATION AND SURFACE QUALITY	255
1. General	255
2. Surface Inspection: Particles, Defects, and Microroughness	256
3. Metal and Molecular Contamination	261
VI. FUTURE DEVELOPMENTS AND CONCLUSIONS	270
1. Measurements on and Characterization Requirements for 300-mm Wafers	270
2. Should We Monitor the Wafers or the Process?	271
REFERENCES	272

Abbreviations Used

AAS	Atomic absorption spectroscopy
AES	Auger electron spectroscopy
AFM	Atomic force microscope
ASTM	American Society for Testing and Materials
DLTS	Deep-level capacitance transient spectroscopy
GC-MS	Gas chromatography–mass spectrometry
ELYMAT	Electrolytic metal analyzer
4-PP	Four-point probe
FTIR	Fourier transform infrared spectroscopy
IC	Integrated circuit
ICP-MS	Inductively coupled plasma–mass spectometry
IMS	Ion mobility spectrometry
MIS	Metal–insulator—semiconductor
μ-PCD	Microwave photoconductive decay
NAA	Neutron activation analysis
PLS	Polystyrene latex spheres
RBS	Rutherford backscattering
SCA	Surface charge analyzer
SCP	Surface charge profiler
SIMS	Secondary-ion mass spectrometry
SPV	Surface photovoltage
SRP	Spreading resistance profilometry
SSIS	Scanning surface inspection system
TD GC-MS	Thermal desorption gas chromatography–mass spectrometry
TOF-SIMS	Time-of-flight SIMS
TXRF	Total-reflection X-ray fluorescence
UHV	Ultrahigh vacuum
VPD	Vapor-phase decomposition
XPS	X-Ray photoelectron spectroscopy

I. Introduction

The purpose of epitaxy is to grow a silicon layer of uniform thickness and accurately controlled electrical properties and so provide a perfect substrate for the subsequent device processing. Therefore, it is essential that the epitaxial wafer is free from defects and conforms to the structural and electrical specifications. It is the task of characterization to ensure that this goal is reached. In fact characterization and metrology have four distinct main functions in the production of epitaxial wafers:

- Provide feedback for *process control*.
- Act as a *troubleshooting* aid to reveal and locate problems.
- Perform *filtering* to remove defective and out-of-specification wafers.
- Supply data for *process development*.

These four main tasks of process control, troubleshooting, filtering, and process development set very different requirements for the characterization methods in terms of speed, sampling, precision, invasiveness, and cost. As a result, a whole array of techniques has been developed, ranging from the very simple, imprecise, and cheap to the highly specialized, accurate, and expensive. A typical control scheme for the production of epitaxial wafers is shown in Fig. 1 and some of the main parameters to be measured and characterized in epitaxial wafers are listed in Table I.

The primary characteristics of an epitaxial layer are resistivity and thickness. These two parameters provide the necessary data for process control, therefore requiring measurements that are fast, precise, and cost-effective. The existing electrical and optical methods to provide this information are covered in Sections III–IV, with the main emphasis on the strengths and weaknesses of each method for the purpose of controlling the epitaxy process.

Troubleshooting and filtering usually employ various techniques to check for defects and contamination to remove defective wafers and to identify possible process problems before they start affecting yield. For this purpose, fast, convenient,

TABLE I

IMPORTANT EPITAXIAL WAFER PARAMETERS

	Parameter	Typical sampling frequency (%)
Epi thickness	Average thickness (or centerpoint thickness)	1–4
Thickness variation	Within wafer	1–4
	Wafer to wafer	1–4
Epi resistivity	Average resistivity (or centerpoint resistivity)	1–4
Resistivity variation	Within wafer	1–4
	Wafer to wafer	1–4
	Vertical gradient, transition width	0.1–1
Wafer geometry	Warp, bow	100
	Flatness	100
Surface defects	Stacking faults, mounds, hillocks, pyramids, and spikes	100
	Dislocations and slip	4–100
	Haze, microroughness	100
	Scratches	100
Contamination	Metal contamination	0.1–1
	Organic contamination	0.01–0.1
	Adherent particles	100

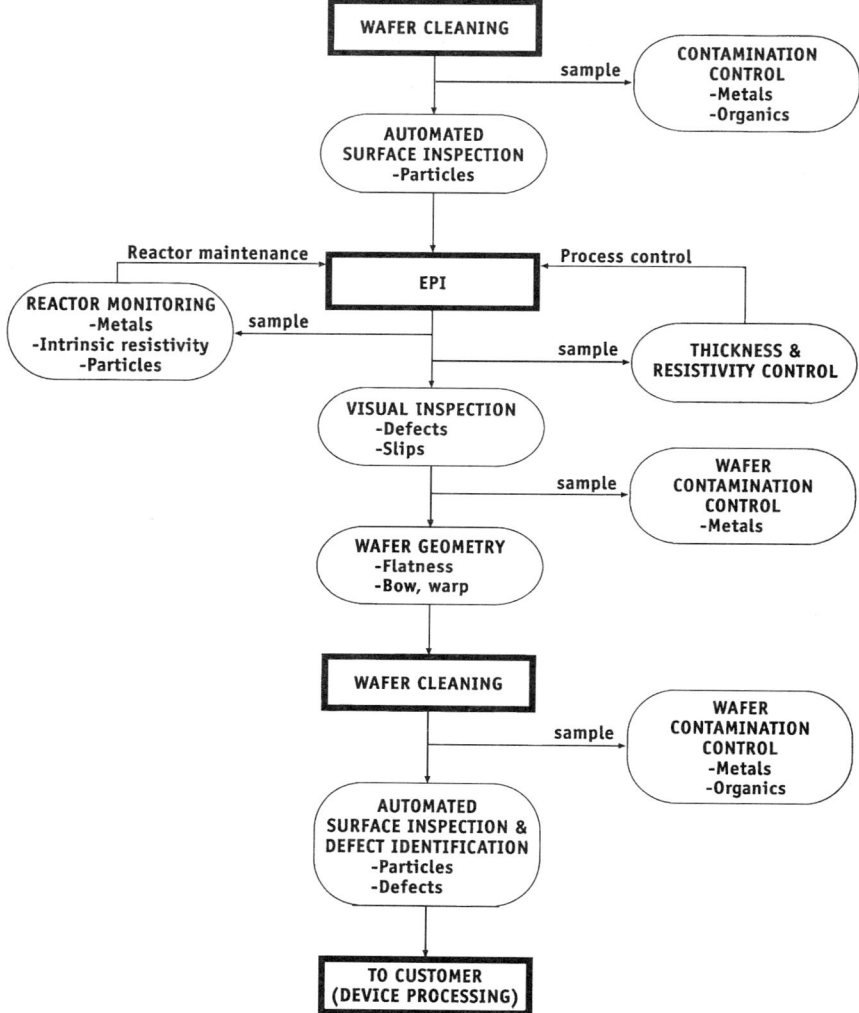

Fig. 1. A flowchart showing some of the possible characterization, inspection, and metrology steps utilized in the production of epitaxial silicon wafers.

and, if possible, noninvasive methods are needed that allow sufficiently frequent sampling and, to some extent, 100% inspection. Surface inspection for defects and surface roughness measurements are included in Section V, as are the highly sensitive methods for characterizing metal and molecular contamination in the bulk and on the wafer surface.

The chapter ends with a discussion of the probable future developments required by the expected uniformity and purity improvements of the epitaxial structures due to the more demanding IC manufacturing processes.

II. On Sampling and Accuracy

1. Sampling and Cost

As measuring instruments are getting more sophisticated due to automation, better precision, and better sensitivity, they are also becoming more expensive, making the fixed cost of many types of measurement quite high. As a result, the throughput and the measurement cost per wafer are inversely proportional. From this it follows that 100% inspection of wafers is normally economical only when the throughput of the measuring instrument is much higher than that of a typical epitaxial reactor: the measurement must take less than 60 s per wafer. Apart from automated surface scanners and instruments for measuring wafer shape, this requirement is usually not met and some kind of sampling is necessary.

In principle, the optimal sampling frequency for a measurement or inspection is easy to determine: the sampling rate should be increased until the marginal benefit of an extra sample equals the marginal cost. In practice it is often difficult to quantify both the marginal benefit and the cost, and the sampling scheme is determined by other means, for instance, by reference to a suitable standard [2], or the well-established seat-of-the-pants method. Typical sampling rates for various wafer parameters are given in Table I.

There is a general tendency to underestimate characterization and metrology costs. It is usually easy to calculate the cost of wafers lost due to invasive measurements since they affect the yield and reactor throughput, but other factors tend to get less attention. These include capital cost, labor, equipment maintenance, consumables, additional processing required (i.e., wafer cleaning), clean room space, and possibly reduced yield due to additional handling. In practice the marginal cost is determined mostly by two factors: wafers and labor, including opportunity costs that are incurred if operators are not available for other urgent tasks when their utilization approaches 100%.

The total capital and labor costs can easily amount to $0.50–$5 per wafer for the most common types of characterization. The main cost item, however, is the cost of samples: invasive methods usually cause a good wafer to be scrapped, at a cost of $20–$200. The economic incentives for minimized sampling frequencies and the use of noninvasive techniques are substantial for the most expensive large-diameter wafers.

2. Precision and Accuracy

For a measurement or inspection to be useful it is important that the results are somehow comparable with measurements done by operators at several locations using different equipment. *Repeatability* is the ability of the instrument to produce the same result when the measurement of one sample is repeated many times within a short period of time. *Reproducibility* is the variability of the average results when the repeatability test is done by several operators, possibly utilizing different instruments. *Precision* describes the combination of repeatability and reproducibility.

Accuracy (sometimes called *bias* or *error*) gives the difference between the average of many measurement results and a known reference value. Long-term changes in bias define *stability:* it is the drift in the average result when the reproducibility test is performed several times over a long period of time.

The capability of a measurement system can be sufficiently described by its range, precision, and accuracy. The procedure for the determination of these parameters is called a gauge repeatability and reproducibility (R&R) study or measurement system analysis (MSA) [1]. Precision and accuracy tend to be intimately connected: if a measurement is reproducible and repeatable, two instruments can be used to obtain essentially identical results and therefore the accuracy will be good if suitable standard samples are available. It is important to note that the determination of accuracy requires the use of standard samples that can be traced back to primary standards. Such standards do not always exist, making it impossible to define the accuracy of an instrument or measurement. An important example is Fourier transform infrared spectroscopy (FTIR), which is almost universally used for the thickness measurement of epitaxial layers but is known to suffer from sample and instrument-dependent bias that is difficult to quantify because suitable standards are unavailable. A measurement can also be accurate but suffer from poor precision. Spreading resistance profiling (SRP) is a good example of such a situation. SRP can be made quite accurate by repeated comparison of standard samples but has a notoriously poor precision due to operator dependency and stability problems.

Accuracy is generally not a requirement for measurements. To be useful for process control the precision-to-tolerance ratio (P/T) of the measurement should be sufficiently low. In other words, the variability of the measurement (σ_m) must be much smaller than the variation caused by the process (σ_p) itself:

$$\frac{P}{T} = \frac{\sigma_m}{\sigma_p} \ll 1. \tag{1}$$

The comparability of measurements at different facilities (i.e., epitaxial production and IC manufacturing) is sometimes achieved by the exchange of internal standards ("golden wafers"). Most of the time even golden wafers are omitted and the different values obtained at each facility are tolerated if the measurements remain stable and precise. For troubleshooting purposes, only good repeatability is needed, whereas for the filtering function a reasonable reproducibility is sufficient.

3. Measurement Pattern, Mapping, and Variation

Variations in the thickness and resistivity of the epitaxial layer include lateral ("within-wafer") variations and variations between wafers ("wafer-to-wafer"

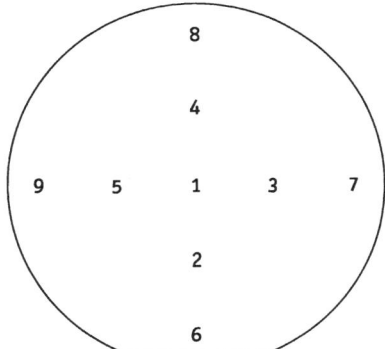

FIG. 2. The nine-point cross pattern for measurement sites. Semi M2: points 1, 2, 3, 4, and 5. Semi M11: points 1, 6, 7, 8, and 9.

variations). In addition, the resistivity profile can also have depth-dependent variations. To quantify within-wafer variations thickness and resistivity have to be measured at several sites on the wafer. The typical 9-point cross pattern is shown in Fig. 2. The use of this pattern is popular for slow measurements such as CV. The lateral within-wafer variations (R) have two definitions according to SEMI standards [69, 70].

SEMI M2-1296:

$$R = (\max - \min)/\text{average}$$

with measurement at the center and four half-radius points (1–5 in Fig. 2).

SEMI M11-0697:

$$R = (\max - \min)/(\max + \min)$$

with measurement at the center and four edge points (1 and 6–9 in Fig. 2).

Wafer-to-wafer variation is defined by SEMI as the variation of the *center value*. The main advantage of separating within-wafer and wafer-to-wafer variations is that this definition facilitates reactor setup and process control. For demanding applications it is also common to use a definition that specifies an upper limit for the standard deviation (σ) of *all measured points* on all wafers.

The five-point calculations used in SEMI definitions will not, in general, produce correct values for either the average or R. Especially, within-wafer variations may be grossly underestimated. Naturally, the measurement will describe the layer better if the number of measurement points in the cross pattern is increased and if all values are used in the calculations. Radial thickness distributions for two wafers grown in different single-wafer reactors are shown in Fig. 3. These distributions have a high degree of rotational symmetry and the diagonals used for the thickness measurement pattern can be chosen freely.

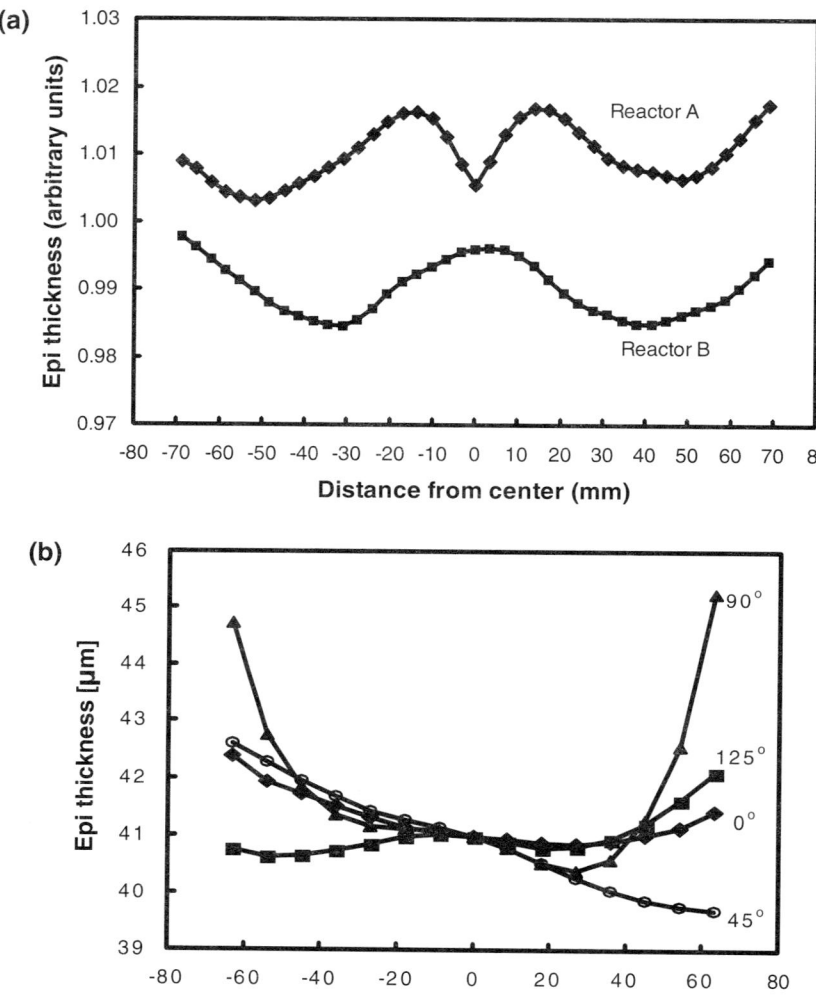

FIG. 3. Typical thickness profiles. (a) Diagonal profiles of two 150-mm wafers grown in different single-wafer reactors. (b) Thickness of a wafer grown in a vertical batch reactor measured along four diagonals at 45° angles. Within-wafer variation is affected by the directions chosen for the calculation: Semi M2, $R = 2.2–4.0\%$; Semi M11, $R = 3.5–4.9\%$. Variation measured from all points, $R = 6.5\%$.

Figure 4 shows how the average thickness and thickness variation values measured from these wafers are affected by the number of measurement points in the cross pattern. It is clear that the nine-point pattern can give erroneous within-wafer variations even for these very symmetrical distributions. In fact, up to 17 points are required to produce a fairly correct value of R. For wafers

7 EPITAXIAL LAYER CHARACTERIZATION AND METROLOGY   233

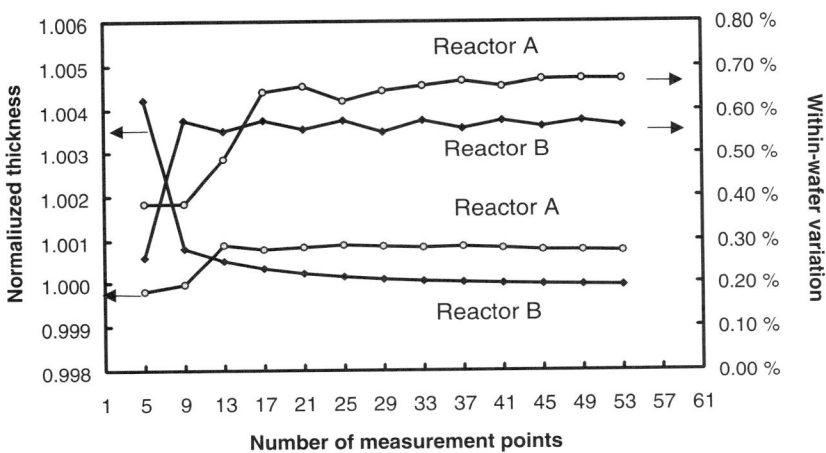

FIG. 4. Effect of measurement pattern on the average thickness and thickness variation. The values are calculated for the two single-wafer profiles in Fig. 3a. Up to 17 points are required in a cross pattern to obtain a valid approximation of the within-wafer variation.

processed in batch reactors the situation is worse because of skewed distributions and a cross-pattern cannot be guaranteed to give proper results unless the measurement axes are correctly aligned with the wafer. For example, for the wafer shown in Fig. 5, the resistivity measurement will not produce good results using the nine-point pattern due to the lack of rotational symmetry. In principle, batch-processed

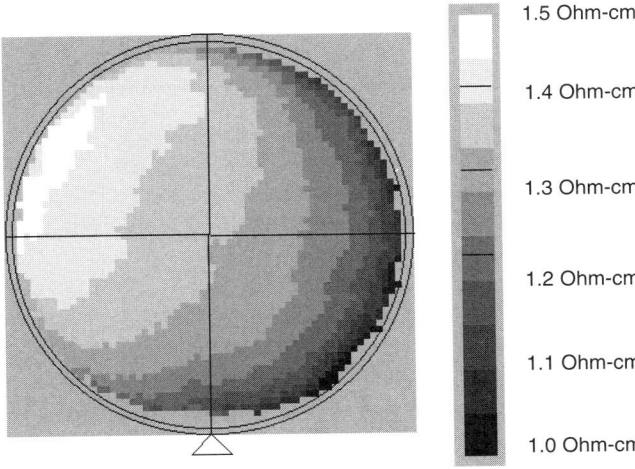

FIG. 5. A resistivity map of a 125-mm epitaxial wafer grown in a barrel-type batch reactor measured with SCP. Note that the normal nine-point cross pattern will miss the region of highest resistivity when measured along the diagonals marked on the map (Picture courtesy of QC Solutions, Inc.)

wafers should be mapped using a grid pattern, but this is rarely done. The minimum requirement for a moderately accurate determination of within-wafer variation is to use a star pattern with 17 points (center, 8 points at half-radius, and 8 points at the edge).

The mapping capability is generally dependent on the measurement speed. If the instrument is slow, true mapping is impossible in practice due to time constraints, except in exceptional circumstances. Mapping can be a powerful tool for discovering process problems. Epitaxial layers processed in single-wafer reactors usually have a very high degree of rotational symmetry, therefore mapping can easily reveal out-of-pocket conditions, asymmetric gas flows, and backstreaming. Vertical batch reactors tend to produce highly asymmetric layers and mapping can provide valuable information on cross-contamination, abnormal flow patterns, and autodoping [60].

In addition to the lateral resistivity variation, variations in the vertical direction must also be measured. Ideally, the resistivity within the epitaxial layer should be constant and the transition at the substrate interface should be abrupt. In reality the transition from the substrate to the epitaxial layer has a finite width due to dopant

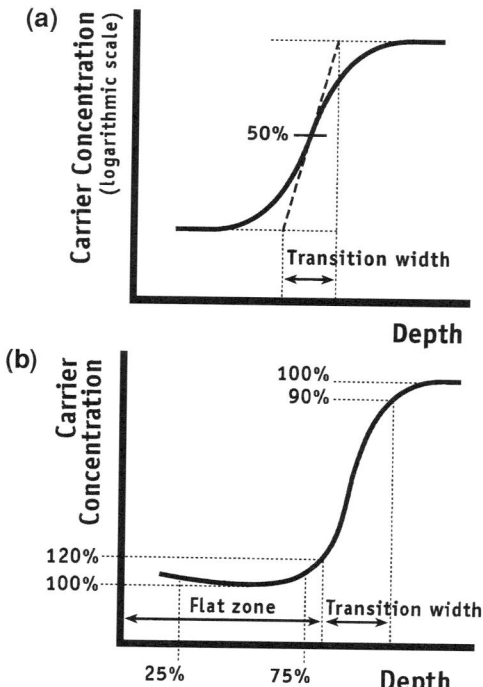

FIG. 6. Two definitions of flat zone and transition width. (a) Tangent method. (b) Transition width equals the difference between the epitaxial layer thickness and the flat-zone thickness.

diffusion and autodoping through the gas phase. Some definitions of the transition width are shown in Fig. 6.

III. Doping Control

1. GENERAL

Epitaxial silicon layers are doped with As or P to achieve n-type conductivity or with B for p-type material. Doping may be characterized by resistivity, carrier concentration, or dopant atom concentration. The ideal technique for doping control would be noninvasive, able to measure the doping profile as a function of depth, and capable of high-speed mapping, would work for any epitaxial layer/substrate combination, would not require surface preparation prior to measurement, would have zero edge exclusion, and would naturally be cheap. From the comparison in Table III it can be seen that no single method can meet all of these requirements. For this reason it is customary to have several instruments available for doping monitoring: for instance, a four-point probe (4-PP), a mercury capacitance–voltage probe (Hg-CV), and a spreading resistance profiler. These three techniques are still the dominant methods for resistivity control. The use of noncontact techniques is becoming more common to achieve savings in wafer costs. Dopant atom concentrations can be measured directly by RBS or SIMS, neither of which is suitable for production control.

The resistivity (ρ) and the concentration of free carriers (n) are roughly inversely proportional, i.e.,

$$\rho = \frac{1}{q \cdot \mu \cdot n}, \tag{2}$$

where q is the unit charge and μ the carrier mobility. μ is nearly constant and resistivity and carrier concentration can be converted into each other with an empirically based mobility model and are often treated interchangeably. Usually the conversion is based on the data of Thurber et al. [77, 78] as given and extended in ASTM F 723-97 [4]. Occasionally, Irwin's data [45] is used for the conversion, which can lead to major errors, especially in p-type material between 0.01 and 1 Ω-cm, compared with the ASTM F 723 values. The free carrier concentration is not identical to the concentration of dopant atoms. There can be substantial local deviations between the two distributions because either dopant atoms are not fully ionized, charge carriers diffuse away from dopant atoms due to concentration gradients, or carriers are swept away by internal electric fields formed at p/n junctions or at the wafer surface. It is possible to calculate the free carrier concentration from the distribution of dopant atoms and the inverse problem can also be solved.

All sample preparation requirements naturally increase the total measurement time and decrease the effective speed, making the measurement more costly. SIMS

TABLE II

COMPARISON OF TECHNIQUES FOR MEASURING CARRIER CONCENTRATION AND RESISTIVITY

	Classic production control techniques			Noncontact techniques			Dopant profiling
	4-Point probe (resistivity)	Hg-CV (carrier concentration)	SRP (resistivity)	Air-gap SPV (carrier concentration)	Air-gap SPV-CV (carrier concentration)	Air-gap CV (carrier concentration)	SIMS (dopant atom concentration)
Reproducibility	++	+	--	++	+	+	+
Accuracy	+	++	--	+	++	++	++
Cost	++	+	-	+	+	+	--
Sample damage	--	-	--	++	+	+	--
Sample preparation	++	-	--	+	++	-	++
Throughput	+	-	--	++	-	-	--
Mapping	+	++	--	++	+	+	-
Depth profiling	--	+	++	--	++	++	++
Sample limitations	-	+	++	+	+	+	++
Advantages	Simple to use		Layer thickness Depth profiling Resistivity range	Fast measurement	No sample preparation Depth profiling	Depth profiling	Measures most elements
Limitations	Thin layers >10 Ω-cm p/n junction required No profiling	Resistivity >1 Ω-cm	Reproducibility Frequent calibration User dependent	Requires inverted surface No profiling	Possibility of dielectric breakdown in air gap Profiling limited	Accurate control of air gap Limited profiling depth	Calibration samples required Requires expert user Detection limits, 10^{14}–10^{15} cm^{-3}

TABLE III

SELECTED CHARACTERIZATION TECHNIQUES FOR DEFECT AND
CONTAMINATION CONTROL

Parameter	Technique
Surface defects and particles	Visual inspection
	SSIS
	Microscopy
Surface microroughness	AFM
	Interference microscope
	Laser scatterometry
Crystallographic defects	Chemical etching and microscopy
Bulk metal contamination	Minority carrier recombination
	DLTS
	SIMS
	NAA
Surface metal contamination	TXRF
	ICP-MS
	SIMS
Surface molecular contamination	IMS
	GC-MS
	TOF-SIMS

uses ultrahigh-vacuum equipment, has a low throughput, and requires an expert user. All these factors combine to make it a costly technique. The ability of SIMS to measure concentration profiles is, however, becoming more important with silicon–germanium technology. Spreading resistance profilometry (SRP) equipment is basically simple and affordable instruments are available, but an experienced technician is needed to get reliable results. Frequent calibration and probe conditioning and slow sample preparation make the measurement labor intensive and quite expensive. Hg-CV is better in this respect even though its throughput is reduced by the slow wafer preparation and measurement procedure. The 4-pp is easy to maintain and fast. Air-gap SPV has the highest throughput of the noncontact techniques but air-gap CV and air-gap SPV-CV are relatively slow measurements.

Noncontact techniques naturally offer the possibility of measurement without sample damage. SRP, SIMS, and 4-pp cause sufficient damage to prevent the wafer from being used for subsequent processing. Properly adjusted, Hg-CV is relatively gentle and does not tend to damage the wafer. The main issue with Hg-CV is Hg contamination. Even though mercury can easily be removed from the wafer surface and also tends to evaporate from the wafer during high-temperature processing due to the high vapor pressure, the reuse of wafers measured with Hg-CV is quite rare due to the possible contamination of wafer cleaning baths.

Apart from the 4-pp, all doping monitoring requires more or less extensive sample preparation. Both Hg- and air-gap CV require a Schottky surface barrier. Therefore, the wafer surface needs to be initially in depletion. The wafer surface condition requirement is even more stringent for SCP, which calculates the carrier density from the maximum depletion width when the surface is in strong inversion. Usually a chemical treatment is not necessary prior to a 4-pp measurement. The exception is high-resistivity samples where the surface must be in depletion (or the flat band condition). SRP of course generally involves angle lapping to allow depth profiling. For SIMS a small sample piece usually has to be cut from the wafer.

SIMS has excellent depth-profiling capabilities and is capable of measuring the concentrations of most elements. The limitation of SIMS is the sensitivity, which limits its use to concentrations above 10^{14}–10^{15} cm^{-3}. SRP can measure carrier concentrations from 10^{12} up to 10^{20} cm^{-3} and is extensively used for resistivity profiling. Hg-CV also has profiling capability but the thickness range is limited by the breakdown voltage. For most common epitaxial structures the thickness range is sufficient for profiling down to the epitaxial layer–substrate interface (Fig. 7). The resistivities probed with CV can range between 1 and 100 Ω-cm (corresponding to carrier concentrations of 10^{14}–10^{16} cm^{-3}), but the technique has been used for monitoring even 1000-Ω-cm n-type layers. Noncontact CV techniques are capable of depth profiling, but the depth range is more limited than for Hg-CV. SCP and 4-pp have no profiling capability.

SCP is very fast and can measure several thousand points in an acceptable time, with essentially no edge exclusion. Hg-CV, air-gap CV, and air-gap SPV-CV are usually used to measure either a five- or a nine-point pattern due to their low speed. Air-gap CV methods also have large edge exclusions. The 4-pp can be used in a

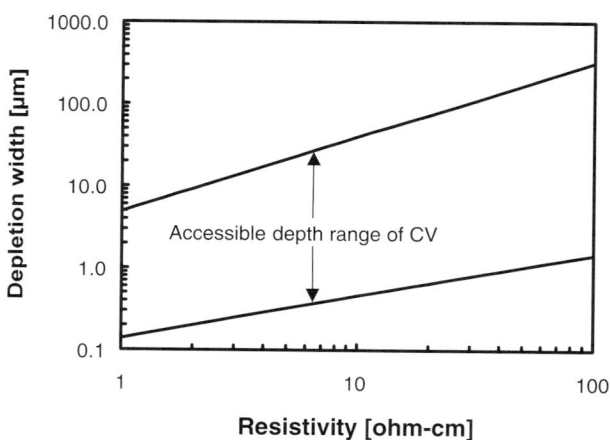

FIG. 7. Accessible depth range in Hg-CV measurement. The minimum depth is limited to a value of about $3 * L_D$ (Debye length), and the maximum depth by the breakdown voltage. Values calculated on the basis of breakdown fields given in Ref. 75 (p. 103).

true mapping mode to measure several tens of points per wafer, as each point takes only a few seconds. SIMS has poor mapping capabilities due to the slowness of the measurement.

2. THE FOUR-POINT PROBE

In the 4-pp a known current (I) is passed between two probes and the other two are used to measure the resulting voltage drop (V) (4-pp measurements are explained in ASTM standards; see Ref. 5) (Fig. 8). In the usual case for epitaxy, where four in-line probes are used to measure a very thin sample, the sheet resistance of the epitaxial layer is simply proportional to the ratio of voltage to current [64]; that is,

$$\rho_S = C \cdot \left(\frac{V}{I}\right). \tag{3}$$

The proportionality constant C is dependent on the probe geometry and also requires corrections—for edge effects, lateral sample dimensions, and thickness—which are automatically calculated by the instrument. The average resistivity, ρ, of the epitaxial layer can be calculated directly by multiplying the sheet resistance and layer thickness (t):

$$\rho = \rho_S \cdot t. \tag{4}$$

The 4-pp is best suited for the measurement of fairly thick, low-resistivity layers. Blunt probes and a small probe force must be used for the measurement of thin layers less than a few micrometers to avoid pressing the probe tips through the epitaxial layer into the substrate. The major limitation of the 4-pp for the resistivity measurement of epitaxial layers is that the current must be confined in the epitaxial layer by a pn junction (Fig. 9). Therefore, only p/N or n/P structures can be measured. The 4-pp can be used for p/P and n/N structures with the aid of a special test wafer of the appropriate type. However, if a P^+ substrate is replaced with an N test wafer (or N^+ with P), the effects of autodoping and dopant diffusion from

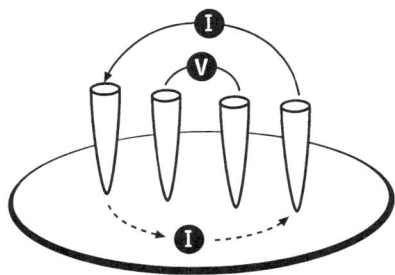

FIG. 8. Principle of four-point probe measurement.

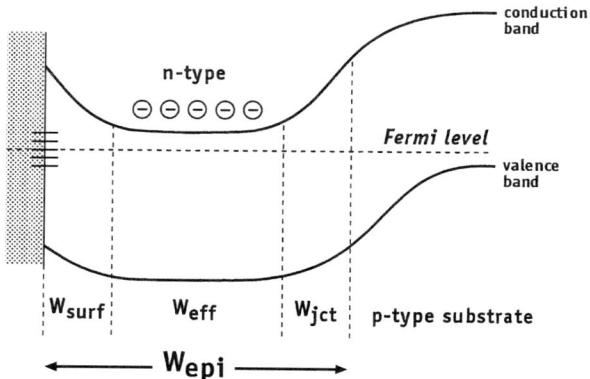

FIG. 9. A schematic energy band diagram of an n-type epitaxial layer on a p-type substrate. The effective electrical thickness of the layer is reduced by the depletion layers at the surface and the p/n junction.

the substrate into the epitaxial layer will not show correctly in the measurement, rendering the results unreliable.

If an accurate mapping of resistivity (and therefore carrier concentration) is required, thickness variations of the epitaxial layer should be taken into account by using the local thickness in Eq. (4). Because low-resistivity substrates are normally not used for 4-pp to avoid autodoping effects, the thickness distribution cannot be

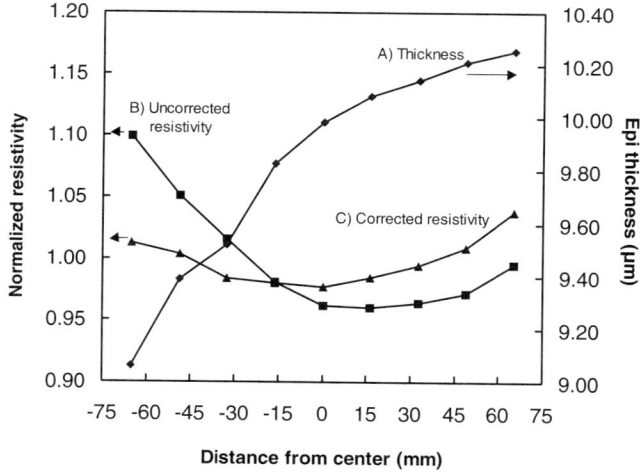

FIG. 10. Effect of thickness correction on the resistivity profile of a 150-mm epitaxial wafer measured by 4-pp. (A) Thickness profile of the epitaxial layer. (B) Resistivity profile calculated using a constant thickness to multiply the sheet resistance. (C) Resistivity profile obtained using the correct thicknesses from A.

determined directly from the resistivity test wafer. Single-wafer reactors generally produce extremely repeatable thickness profiles, and accurate resistivity profiles can be obtained by using the thickness distribution from a separate test sample to correct the resistivity, as shown in Fig. 10. However, in batch reactors the thickness profile can be quite skewed, making it difficult to get a valid resistivity distribution by the use of Eq. (4). In such cases it is more useful to use sheet resistance [Eq. (1)] or, preferably, CV measurement.

Another limitation for the use of Eq. (4) is the possibility that the electrical thickness may differ from the metallurgical thickness of the epitaxial layer, which is normally used for the conversion from sheet resistance to resistivity. This difference can be quite large for lightly doped epitaxial layers due to the depletion layers at the surface and at the pn junction (Fig. 9). The surface depletion width is strongly affected by the surface condition of the wafer, the maximum surface depletion occurring when the surface is in strong inversion. For instance, for a 1000-Ω-cm n-type layer on a P-type substrate the combined depletion layer thicknesses can vary from 12 up to 21 μm, depending on the surface depletion width (Fig. 11). As a result, the 4-pp measurements of high resistivity layers tend to be unstable. Even though it is possible to correct the sample thickness for the pn-junction depletion, the depth of the surface depletion layer is usually not known, causing considerable errors in the measurement. For this reason, the use of 4-pp should be limited to resistivities below 10 Ω-cm.

3. MERCURY PROBE CV

In mercury probe CV (Hg-CV) the test structure is a Schottky diode formed by pressing a capillary filled with mercury against the semiconductor surface [6, 32].

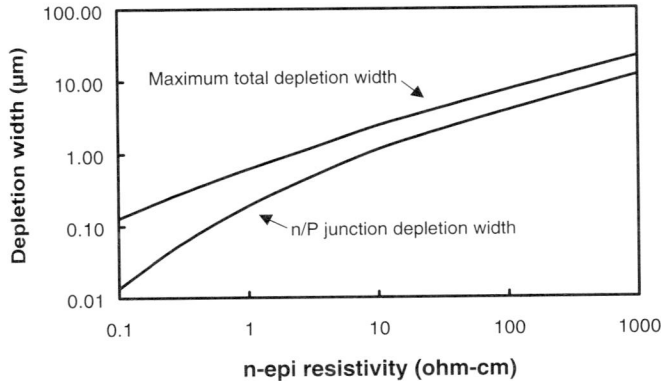

FIG. 11. Calculated width of depletion layers for an n-type epitaxial layer on a 10 Ω-cm P-type substrate as a function of the epitaxial layer resistivity.

The instrument measures the capacitance of the depletion region, and the concentration of ionized dopant atoms, N, at the depth W is calculated from the voltage dependence of the capacitance:

$$N(W) = \frac{\text{constant}}{A^2 \cdot [(d/dV)(1/C^2)]}. \tag{5}$$

The strong dependence of N on the junction area A in Eq. (5) is worth noting. This A^{-2} dependence requires the junction area to be determined and maintained very accurately. Depth profiling is achieved by changing the reverse bias voltage, V. The main advantage of Hg-CV is that the physics, capabilities, and limitations of the method are well understood. It allows quick, fairly noninvasive depth profiling combined with a limited mapping capability. The method also works with p/P+, n/N+, n/P, and p/N structures.

The equivalent circuit for Hg-CV sketched in Fig. 12 shows that, in addition to the quantity to be measured (Schottky junction capacitance C_D), there are two main parasitic elements that affect the measurement: the leakage conductance G_{LK} in parallel to the Schottky junction and the series resistance R_s due to the semiconductor bulk. Because G_{lk} and R_s cannot be determined accurately from the simple CV measurement, both factors must be kept sufficiently small to avoid

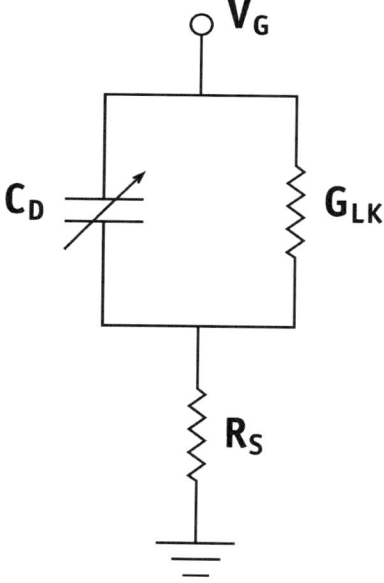

FIG. 12. The simplified equivalent circuit of the CV measurement. C_D is the voltage-dependent depletion capacitance; G_{lk}, the leakage conductance; and R_s, the series resistance.

excessive errors. The series resistance is increased if the epitaxial layer is thick and has a high resistivity, if the substrate resistivity is high, or if there is a pn junction at the substrate–epitaxial layer interface. The effect of the series resistance is reduced by a lower measurement frequency.

G_{lk} is affected by the surface condition and also by the bias voltage, so that low voltages usually have to be avoided, making it difficult to get reliable profiling near the wafer surface. Probing near the surface (to about 1 Debye length) is possible by using a thin thermal oxide on the wafer. The main culprit for problems in CV is leakage, which can be caused by various types of contamination such as mercury oxide and moisture. Moisture effects are prevented by using a nitrogen purge around the probe. The skin of Hg oxides is removed by doing a few probe down–up cycles prior to the first measurement. The condition of the wafer surface has a major effect on leakage currents. For this reason the correct surface treatment of the wafer is crucial. Organic contamination tends to create a negative surface charge, i.e., ionized acceptor states which drive the p-type surface into accumulation and reduce the surface barrier and increase leakage. The purpose of the surface treatment on p-type silicon is to achieve an acceptable barrier height by neutralizing the surface acceptor states. This is done with an HF dip or HF vapor treatment, which produce a passivated H-terminated surface. HF treatment is not stable and the properties of the surface will degrade after some hours. In fact, it is difficult to obtain stable measurements on p-type material with a resistivity below 3 Ω-cm. However, the long-term stability can be improved by using a suitable surfactant in the DI water rinse after the HF dip.

For n-type silicon the purpose of the surface treatment is to form a high-quality oxide on the wafer. The oxide can be formed with a number of treatments, for instance, chemically with hot H_2O_2 or $H_2O_2:H_2SO_4$, with furnace oxidation, and by plasma or ozone treatment. N-type material remains stable for a very long time, probably because most contamination increases the negative surface charge thus driving the surface into depletion. A 3-mm edge exclusion is minimum for the Hg-CV method due to the stray capacitances near wafer edges. The repeatability of CV is affected by the carrier concentration and becomes worse for low-resistivity samples.

4. Spreading Resistance Profilometry

Spreading resistance profilometry (SRP) has been extensively reviewed and is well covered by ASTM standards [7, 55]. The principle of the measurement is simple: a wedge-shaped sample is prepared by angle lapping a small test slice (typically about 2×5 mm^2) so that the bevel extends beyond the epitaxial layer–substrate interface and the resistance between two probes is measured at different points along the lapped surface (Fig. 13). The resistance between the two probes consists of three terms: spreading resistance, contact resistance, and the resistance

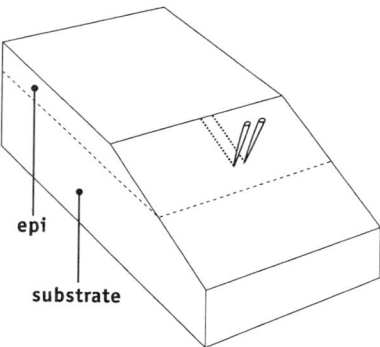

FIG. 13. Principle of two-point spreading resistance measurement.

of the semiconductor material. The effect of the semiconductor bulk is reduced by keeping the distance between the probes very small (typically 20–100 μm). The contact resistance contribution is kept constant by utilizing low voltages and by using carefully prepared probes with a constant probe load. The measured resistance is therefore proportional to the resistivity of the underlying material. The proportionality is calibrated using highly uniform calibration samples. Different sets of calibration samples are required for n- and p-type material for both (100) and (111) orientations. The thickness range can be set by measuring the bevel angle accurately.

The raw SRP resistance data are converted into a depth profile by using a suitable layer-by-layer correction algorithm [22, 28, 44, 56, 67]. To produce correct results, the correction calculation must start in a thick layer of constant resistivity. For this reason SRP profiling usually must extend all the way to the bulk silicon. A typical SRP profile is plotted in Fig. 14. Properly done, the SRP measurement can be surprisingly accurate and good agreement with profiles measured by SIMS has been reported [29]. For maximum accuracy, calibration measurements have to be done frequently. However, SRP is notoriously operator dependent and reproducibility is not good, especially when samples are measured at different locations. Large differences are found due to variations in sample preparation, bevel angle determination, data smoothing and correction algorithms, probe diameter, and load or probe diameter determination [30].

In spite of its limitations, SRP has become indispensable for the measurement of transition widths in n/N^+ and p/P^+ structures, autodoping profiles on patterned wafers, and layer thicknesses and resistivities in multilayer structures. SRP has a very wide dynamic range, being capable of measuring resistivities over nearly seven orders of magnitude (0.0005–2000 Ω-cm). The technique also has the capability to check the conductivity type of the material with the hot probe technique. When SRP is used for probing patterned wafers, the test stripe must be large enough

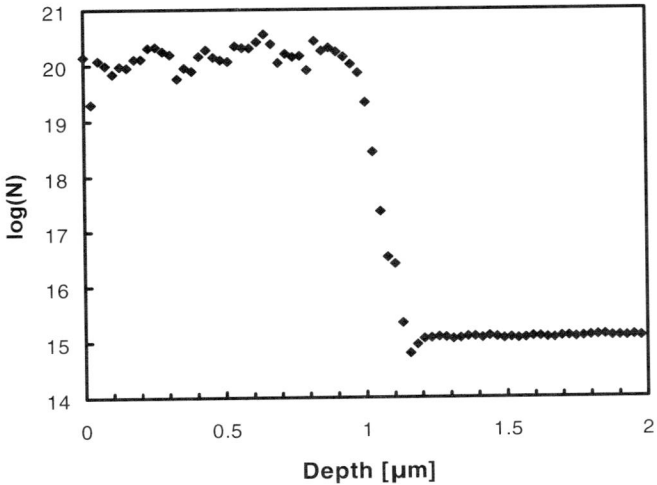

FIG. 14. An SRP profile of a heavily B-doped 1.0-μm epitaxial layer on a lightly doped substrate.

(minimum width = 3× probe spacing, and length sufficient for the bevel to reach bulk silicon).

5. NONCONTACT RESISTIVITY MEASUREMENTS: AIR-GAP CV, AIR-GAP SPV CV, AND AIR-GAP SPV

The invasiveness of Hg-CV, SRP, and 4-pp is a major cost factor for the processing of 200- and 300-mm wafers. A possible way to reduce the cost of samples for these techniques is to reduce the sampling frequency. Whereas such a reduction may be acceptable for depth profiling, it may not be possible for resistivity control. For this reason there has been growing interest in noncontact resistivity measurement techniques. Three commercial instruments based on different operating principles are currently available: air-gap CV, air-gap ac-SPV (SCP), and air-gap ac-SPV CV (Epimet) (Fig. 15).

Air-gap CV is the obvious noncontact extension of Hg-CV: the conducting probe is kept a short distance from the wafer surface to form a metal–insulator–semiconductor (MIS) diode. The measurement principle is similar to that of Hg-CV. The major weakness of the method is that the electrode has to be maintained exactly parallel and very close to the surface. In the commercially available instrument (SSM 3000 NanoGAP; Solid State Measurements, Inc.), the air-gap adjustment is based on capacitance measurement and is additionally controlled optically using the properties of evanescent waves. The adjustment is

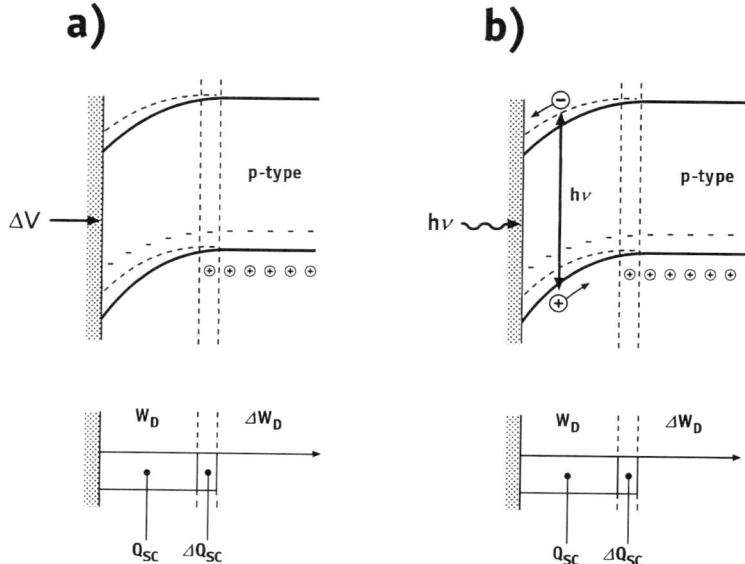

FIG. 15. Comparison of CV with SCP and SPV measurements. The modulation of the depletion layer width also causes a modulation of the surface space charge (ΔQ_{SC}). (a) In CV the depletion layer is modulated by a small change (ΔV) in the bias voltage. (b) In SPV and SCP the modulation is accomplished by minority carriers created by short light pulses ($h\nu$).

relatively slow but extremely accurate, being able to set the gap within 10 Å of the required value of 0.3–0.5 μm. In common with the conventional CV measurement, air-gap CV also requires surface treatment to reduce leakage currents. The edge exclusion for air-gap CV is also large (10 mm). The instrument has depth profiling capabilities, but the depth range is limited by the large voltage drop across the air gap, which limits the maximum bias voltage within the silicon to about 5 V. Throughput is limited by the slow electrode adjustment but the instrument has the appealing property of being similar to the familiar Hg-CV. The depletion capacitance limits measurements to resistivities above 1 Ω-cm.

Air-gap ac-SPV, the so-called surface charge profiler (SCP) (QC Solutions, Inc.) is based on an optical technique for the measurement of dopant concentration (the reader is referred to Danel [31] for a thorough review of the SCP technique). A low-intensity beam of light, modulated with frequency ω, is used to generate electron-hole pairs near the semiconductor surface. Because of the short wavelength of the excitation pulse, nearly all mobile charge is created within the surface depletion layer. The minority carriers drift to the surface, causing a small AC surface photovoltage (ΔV_s) which is detected by an electrode positioned above the wafer. From the real and imaginary parts of the ΔV_s, two parameters can be determined [48]—the surface recombination rate ($1/\tau_s$) and surface depletion width (W)—

$$\frac{1}{\tau_s} = -\omega \cdot \frac{\text{Re}\{\Delta V_s\}}{\text{Im}\{\Delta V_s\}} \qquad (6)$$

$$W = -C_{\text{opt}} \cdot \left(1 + \frac{1}{\omega^2 \tau_s^2}\right) \cdot \omega \cdot \text{Im}\{\Delta V_s\}, \qquad (7)$$

where C_{opt} is a constant. If the surface is in strong inversion, the concentration of ionized dopant atoms at the edge of the depletion layer can be calculated with the help of the depletion width W [75]. SCP is a very fast measurement, allowing full wafer mapping without any edge exclusion (Fig. 5). The measurement of the surface lifetime also produces information on the quality of the surface, for instance, surface contamination and damage. SCP is well suited for the measurement of high-resistivity layers and has been used successfully to measure epitaxial layers of over 1000 Ω-cm [80]. The low-resistivity limit of about 0.01 Ω-cm is set by the requirement that the semiconductor is nondegenerate. The disadvantage of the method is that depth profiling is not possible. Also, the wafer surface must be brought to strong inversion with correct surface treatment. However, *in situ* surface treatment is possible in the commercial instrument.

The principle of air-gap ac-SPV CV [the commercial instrument is called Epimet (Semi Test, Inc.)] is a combination of air-gap CV and SCP. In common with the air-gap CV, an electrode is brought very close to the wafer surface and a bias voltage connected between the electrode and the wafer. However, the voltage dependence of the capacitance is not used for the determination of the dopant concentration, but the sample is excited using modulated light and the depletion width is determined from the ac photocurrent signal [54]. The capability of the Epimet resembles closely that of the air-gap CV: it has a similar profiling range and also requires a large edge exclusion. The major drawback, as with air-gap CV, is the slow measurement, which makes mapping impractical. However, the method does not require as accurate an alignment of the electrode with the wafer as CV does, and a fast air-bearing technique is used to maintain the small separation. Epimet can also measure wafers without surface treatment, even though a relatively large minimum bias is required to minimize surface leakage for reliable results.

6. SECONDARY-ION MASS SPECTROMETRY

In secondary-ion mass spectrometry (SIMS) profiling is accomplished by sputtering a small area of the sample with an ion beam. The sputtering process causes the ejected atoms to become ionized. These secondary ions are detected using a suitable mass analyser. SIMS is a powerful profiling technique used widely for contamination and compositional analysis. SIMS is naturally destructive, and even

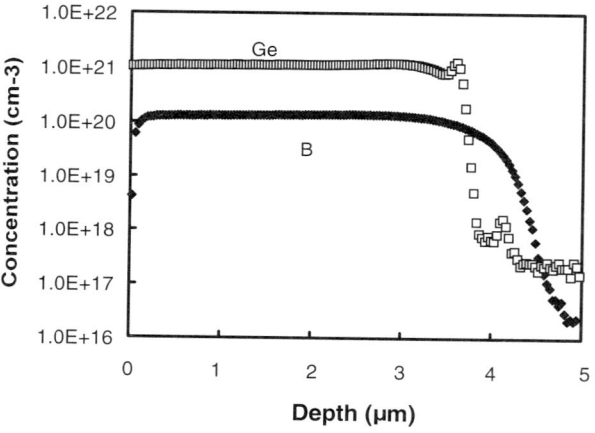

FIG. 16. The concentration profiles of boron and germanium in a heavily doped epitaxial layer measured by SIMS.

though tools capable of accepting whole wafers are available, it is more typical that a small sample piece is cleaved from the wafer to allow several samples to be loaded into the apparatus simultaneously. The ion beam technique requires an ultrahigh vacuum (UHV). SIMS is capable of detecting most elements. Its main limitation for the measurement of doping profiles is the relatively low sensitivity in the depth profiling mode, with detection limits of about 10^{14}–10^{15} cm^{-3} for the common dopants. An example of a SIMS profile of a heavily boron-doped epitaxial layer is plotted in Fig. 16. SIMS can be quantitative if suitable calibration samples are available.

IV. Thickness Measurements

1. STANDARD FTIR

The Fourier transform infrared reflectometer (FTIR) is based on a Michelson interferometer that contains one moving and one fixed mirror (Fig. 17). As sketched in Fig. 17b the main part of the infrared beam is reflected from the wafer surface and a small part from the epitaxial–layer substrate interface. When the length of the moving interferometer arm changes, constructive interference occurs at different positions for the beam reflected from the surface and for the beam reflected from the interface, giving the interferogram sketched in Fig. 17d. The spectrum has intense central peaks ("centerburst") caused by the reflection from the wafer surface and much weaker "sidebursts" caused by the reflection from the epitaxial–layer substrate interface. The thickness of the epitaxial layer is

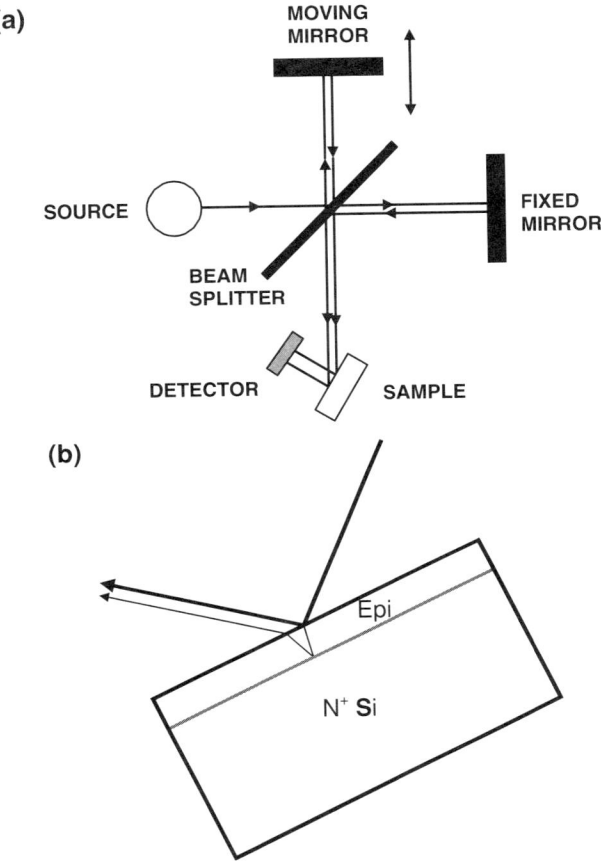

FIG. 17. Principles of the optical measurement of the epitaxial layer thickness. (a) The FTIR instrument is a Michelson interferometer with one moving mirror. The detector can be placed either in front of the sample (reflection mode depicted in the diagram) or behind the sample (transmission mode). (b) Light beams reflecting from the surface and the epitaxial layer–substrate interface travel different distances. (c) The interferogram is the graph of intensity vs moving mirror position. The intense peaks at the center (centerburst) of the interferogram are caused by the specular reflection from the surface. The minor peaks (sidebursts) come from the epitaxial layer–substrate interface. The epitaxial layer thickness is obtained directly from the distance of the sidebursts (n = refractive index in the epitaxial layer). (d) Reflectivity spectrum obtained from the interferogram by Fourier transformation. The epitaxial layer thickness can be calculated if the reflectance is corrected with the instrument function and a calculated reflectance spectrum is fitted to the measurement. (*Continues*)

obtained directly from the separation of the two sidebursts. Naturally, the second reflection can occur only if there is a sufficient difference between the refractive index of the epitaxial layer and that of the substrate. Usually this difference is achieved by using a conductive substrate with a resistivity of less than 0.025 Ω-cm.

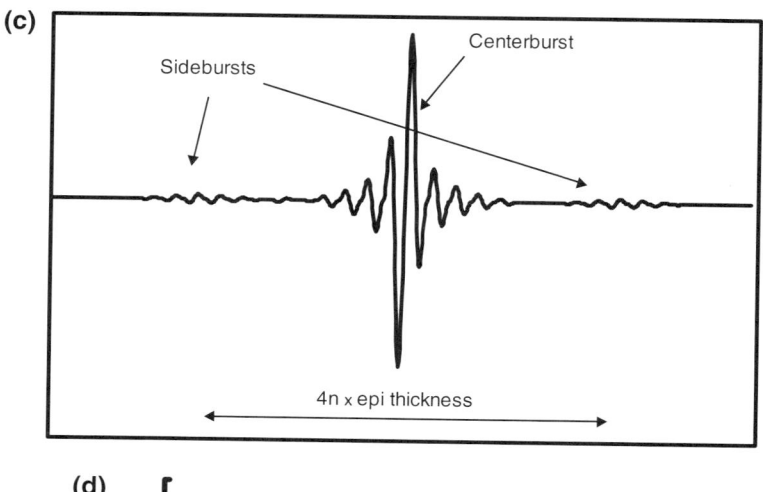

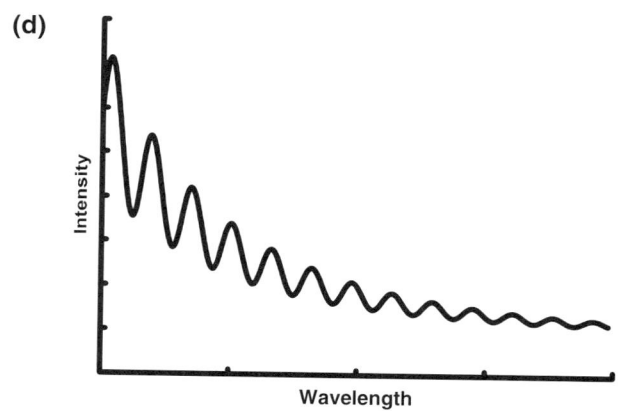

Fig. 17. (*Continued*)

FTIR is a reliable technique with good precision. However, the accuracy of the method is limited by the lack of suitable reference material. In addition, the FTIR suffers from a sample-dependent bias. Because of the finite bandwidth of the optical system, the sidebursts have a complicated shape consisting of several peaks, and the exact location of the sideburst cannot be defined unambiguously. For thin epitaxial layers the sidebursts become obscured by the more intense centerburst. To avoid this it is customary to remove the centerburst from the interferogram by subtracting a spectrum measured from a thick reference sample (usually more than three times thicker than the sample to be measured). To achieve a complete centerburst subtraction the epitaxial layer–substrate interface in both samples should be identical. Unfortunately, this is usually not the case and the subtracted interferogram contains spurious peaks which can, in extreme cases, lead to erroneous measurements of thin epitaxial layers.

2. MODEL-BASED FTIR

If the interferogram in Fig. 17c is Fourier transformed into a reflectance spectrum (Fig. 17d), and the spectrum is corrected with the instrument function to obtain the absolute reflectance, the thickness of the epitaxial layer can be obtained by fitting a calculated spectrum to the experimental result [86]. The original reason for avoiding this step was the increase in computation time required by the Fourier transformation and the fitting algorithm. Recently, however, commercial instruments utilizing this technique have become available. The fitting algorithm takes into account the refractive index of silicon, the resistivity of the substrate, and also the finite transition width between the epitaxial layer and the substrate. The advantages of the fitting method include [27, 84] the following.

- An epitaxial reference sample is not required because centerburst subtraction is not needed. Only a highly reflective internal reference is used to measure the instrument function.
- The measurement is highly stable, repeatable, and reproducible (Fig. 19).
- In addition to the epitaxial layer thickness, the free carrier concentration of the substrate and the thickness of the transition layer are obtained (Fig. 18). However, for carrier concentrations below about 2×10^{17} cm^{-3} the change in the refractive index is too small to be detected and the modeling of the transition region is limited to carrier concentrations above this value.

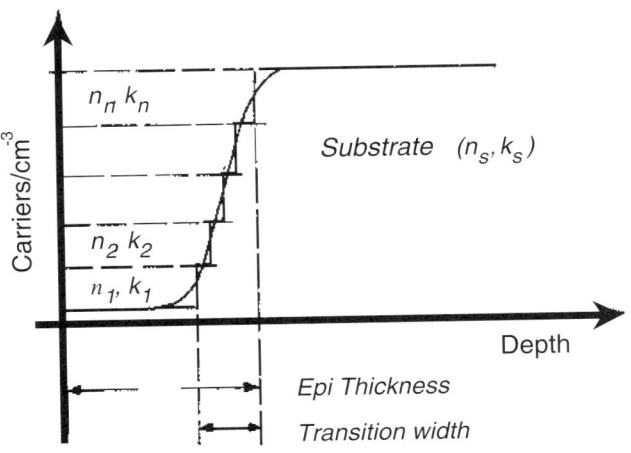

FIG. 18. Information obtainable from the reflectance curve. In the modeling the transition region is divided into several thin layers that have different carrier concentrations and refractive indices. (Diagram courtesy of G. Kneissl, ADE Corporation.)

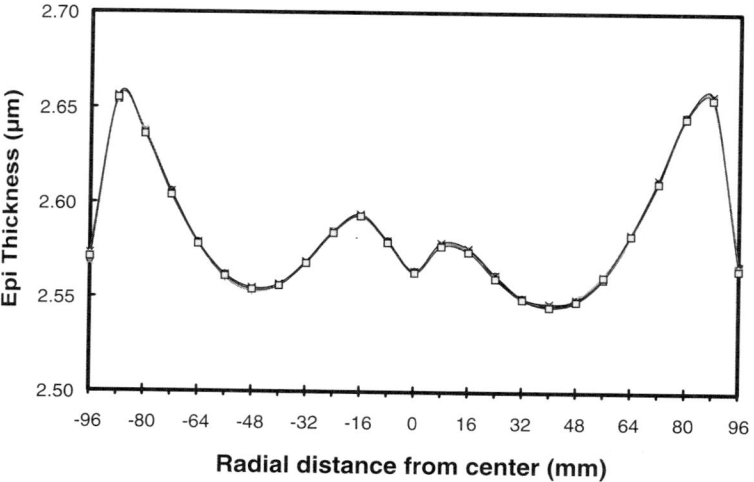

FIG. 19. Thickness profile of a 200-mm wafer grown in a single-wafer reactor. The profile was measured with a model-based instrument using 4-mm edge exclusion. The measurement was repeated 11 times within a 4-week time period to test the reproducibility. (Data courtesy of G. Kneissl, ADE Corporation.)

- A quality-of-fit parameter can be calculated to serve as an indicator of possible measurement problems.
- The measurement works well even for very thin epitaxial layers (0.5 μm).

3. THICKNESS MEASUREMENTS ON n/N AND p/P STRUCTURES

There are currently no nondestructive techniques available for the measurement of resistive substrates. If the resistivity of the substrate is above 0.05 Ω-cm, the reflection from the substrate–epitaxial layer interface is too weak for the FTIR technique to work. For these structures it is customary to use low-resistivity test wafers (either N^+ or P^+), making FTIR effectively a destructive technique. One way to solve this problem is to utilize a comparative technique: the thickness change of the wafer is determined from the difference in measurements made prior to and after epitaxial deposition.

The mean epitaxial layer thickness can be measured with an accurate scale from the weight change in the wafer. The precision of this technique is as good as with FTIR and the accuracy is better for thin epitaxial layers (below 10 μm). The major source of bias is caused by the difficulty of determining the surface area of the wafer accurately due to the edges.

Within-wafer variations can be obtained nondestructively with a capacitive wafer thickness measuring instrument by subtracting the thickness of the substrate from

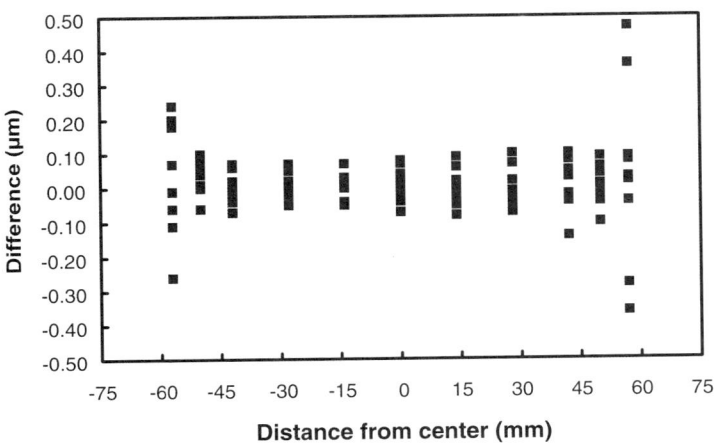

FIG. 20. Reproducibility of the capacitive thickness measurement tested with repeated measurement of eight 125-mm wafers.

the thickness of the epitaxial wafer. Commercial instruments with software supporting this capability are available, but no published data on the precision and accuracy exist. In a typical manner for a capacitive measurement the technique requires a large edge exclusion and the precision degrades substantially near the edges, where the shape of the wafer changes the fastest (Fig. 20). A precision of 0.05 μm is achievable with the capacitive measurement, making it a possible solution for the measurement of thick epitaxial layers.

4. RESISTIVITY AND THICKNESS MEASUREMENTS ON MULTILAYER STRUCTURES

The resistivity and thickness measurements on multilayer structures present some difficult problems for effective process control. The topmost epitaxial layer is accessible to measurement by some of the techniques outlined earlier. Deeper layers can sometimes be measured using test wafers which contain only the layers of interest. Such test structures can in principle be grown in a single-wafer reactor with little difficulty. However, the test structure entails a quality risk because it requires the use of a different recipe, with the possibility of different parameters being used for the test wafer and complete structure. For this reason, it is preferable to characterize the complete stack of epitaxial layers. Clearly, some kind of profiling is required, SRP usually being the preferred method.

The most frequently grown multilayer structures are for power applications, fairly thick, and generally grown on smaller wafers in a batch reactor. A typical

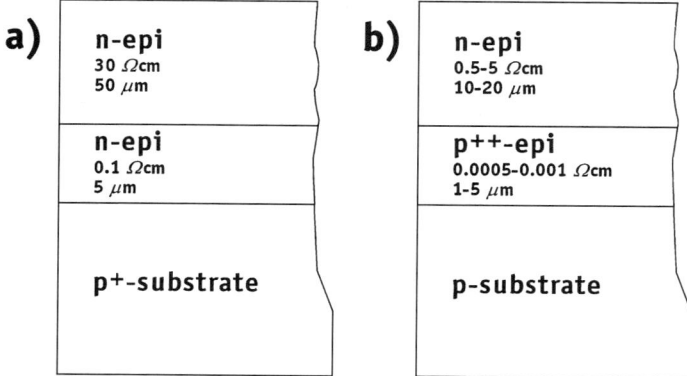

FIG. 21. Schematic diagram of (a) a typical IGBT structure and (b) a sensor structure with a p^{++} etch-stop layer.

power device structure is shown in Fig. 21a. The measurement of the IGBT structure can proceed as follows (Fig. 22).

1. Total thickness and thickness variation (layers 1 + 2) by FTIR.
2. Resistivity of layer 2 with Hg-CV.
3. Sheet resistance of the combined structure (layers 1 + 2) with 4-pp.
4. Resistivities and thicknesses of layers 2 and 1 by SRP profiling.

This scheme provides more than sufficient information, so either step 2, step 3, or the resistivity information from step 4 can be discarded.

Some sensor applications also require more than one epitaxial layer, for instance, a thin, heavily B-doped etch stop layer below the capping layer (Fig. 21b). This

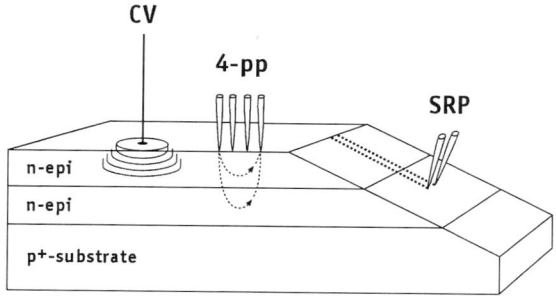

FIG. 22. Measurement strategies for resistivity control in a multilayer IGFET structure. The resistivity of the top, n-type layer can be measured with CV. 4-pp gives the combined sheet resistance of the two epitaxial layers. SRP can be used to determine the resistivities of both layers. SRP also gives layer thicknesses.

structure can be measured similarly to the IGBT structure, by a combination of 4-pp, FTIR, and SRP.

V. Contamination and Surface Quality

1. GENERAL

Many types of organic and inorganic (either metallic on nonmetallic) contamination in the form of atoms, molecules, or particles may be present on epitaxial silicon wafers. Some of the most common contamination and defect sources in the epitaxial process are as follows.

- Substrate: Metals, organic contamination, particles, defects
- Wafer handling: Particles
- Susceptor: Metals, particles
- Reactor chamber: Particles
- Process gases: Metals
- Process ambient: Particles, molecular contamination

Particles deposited after epitaxy can generally be removed with the usual postepitaxy wet cleaning. The vast majority of surface defects on epitaxial layers originate on the substrate surface, where, in addition to particles, also surface damage, molecular contamination, metal precipitates, and oxide or nitride islands can act as nucleation sites for defects. Therefore it is as important to control the quality of the substrate as the epitaxial wafer itself. Apart from the substrate, the most prevalent sources of particles in the epitaxial process environment are deposits accumulated on the quartz chamber, deposits in the exhaust line, and particles caused by wafer loading and handling mechanisms. Metal contamination is an issue because it can easily occur in the hot process environment using corrosive Cl-containing gases and can adversely affect crucial device properties in extremely small concentrations.

A large number of analytical techniques exist for contamination control and defect analysis. Essentially identical methods are used for controlling contamination in epitaxial and polished wafers. Visual inspection, microscopy, and scanning surface inspection equipment are used to detect particles, surface defects, and haze (microroughness). Molecular and atomic contamination on the wafer surface is analyzed with TXRF, ICP-MS, AAS, or SIMS. Bulk contamination is measured using minority carrier recombination (μ-PCD, SPV, or Elymat), SIMS, or DLTS. In addition, chemical etching techniques are also used to detect structural defects and surface contamination by certain metals. If necessary, contamination profiles of single defects can be measured by performing depth profiling with AES.

2. Surface Inspection: Particles, Defects, and Microroughness

a. Visual Inspection, Microscopy, and Slip Control

The different types of epitaxial defects (Fig. 23) are closely related to each other and almost invariably contain stacking faults which make the minimum lateral dimension of the defect larger than the thickness of the epitaxial layer [47]. Because of the large size, very few of these defects can be accepted on a wafer, making them a major cause of yield losses in epitaxy. As a result, the ability to locate and identify defects is very important to control the causes of defect formation.

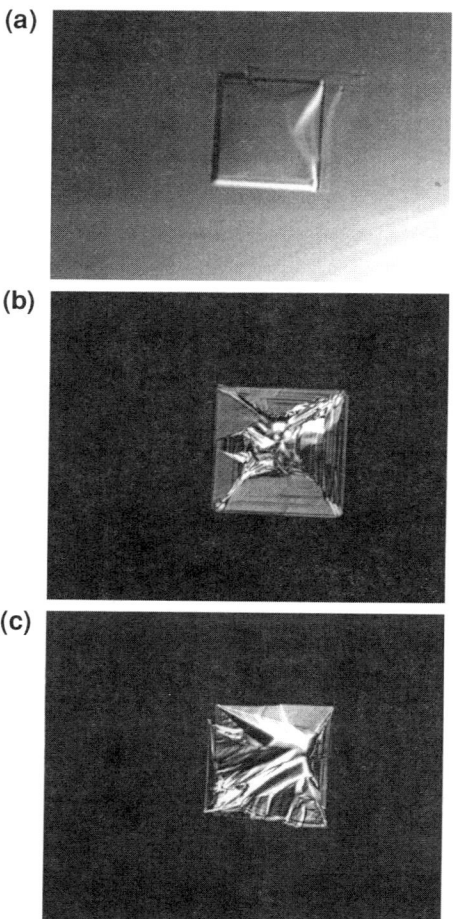

FIG. 23. Micrographs of common epitaxial surface defects: (a) stacking fault; (b) pyramid; (c) mound; (d) mound; (e) hillocks; (f) spike.

Visual inspection under fluorescent or bright light [18] has traditionally been the most important inspection because it allows a fast response to process problems. An experienced operator can detect large particles, mounds, spikes, haze, and other front or back surface anomalies quickly and relatively reliably. The high throughput combined with the low investment required have been the main attractions of visual inspection, which still remains in use for small wafers and thick epitaxial layers. However, the technique has some drawbacks: surface defects and particles cannot be classified reliably; the method is very much operator dependent and the consistency of rejection criteria is difficult to maintain. In addition, for larger wafers with more advanced specifications the size, density, and frequency of

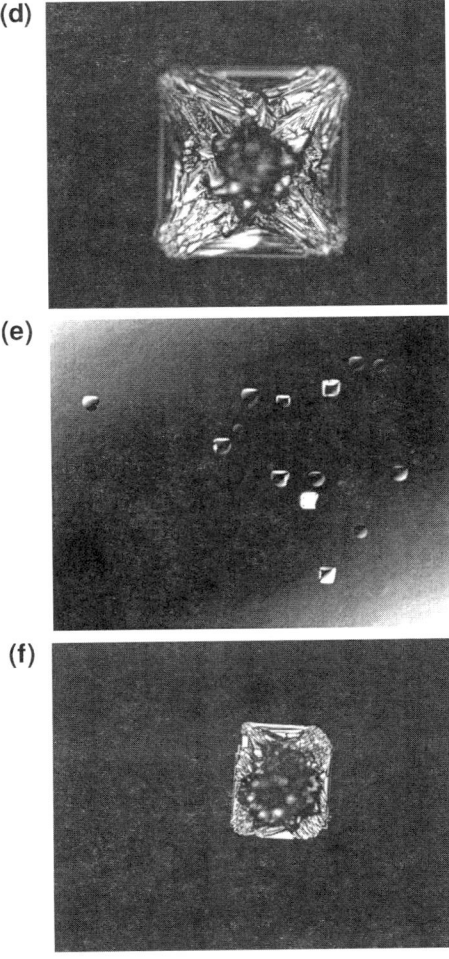

FIG. 23. (*Continued*)

defects are so small that the reliability of visual inspection becomes questionable. Defects can be identified with an optical microscope but manual inspection with the microscope is tedious and unreliable. However, scanning optical microscopes capable of classifying epitaxial defects automatically using optical image processing are available. These instruments are too slow for full-wafer scanning but can be used effectively when defect coordinates are transferred from the automated surface scanning equipment.

Very strong slip-lines can also be detected by visual inspection under fluorescent light. A much more sensitive method is to use a Nomarski interferece microscope at a fairly low magnification (100–200x), optimized so that the surface roughness has maximum visibility. With a suitable rotating sample stage, slip inspection can be made very rapidly and inexpensively from the wafer edges. However, the microscope inspection of central areas of the wafer is very slow. A much faster method for full wafer inspection is to use either a Magic Mirror (a trademark of Hologenix, Inc.) or a laser surface scanner.

b. Automated Surface Inspection and Scatterometry

As discussed earlier visual inspection does not detect or identify small defects reliably, especially when the epitaxial layer is thin. For this purpose automated optical scanning surface inspection systems (SSIS) used in the semiconductor industry for particle detection have been developed further to enable reliable identification of surface defects [33, 83]. In addition, SSIS can also provide defect site XY-coordinates which other defect review and inspection equipment can utilize to locate interesting defects for further analysis.

Scattering of light from the silicon surface is used to measure the number and size of particles and to identify surface defects. For particle measurement SSIS are calibrated using polystyrene latex spheres (PSL) [8]. The sizes of irregular particles are classified as PSL equivalents. Laser scatterometry also provides information on the microroughness of the surface [74]. A schematic of a particle scanner is shown in Fig. 24. Basically the scanner has a laser light source, a mechanism to scan the focused beam over the wafer surface, and collection optics and detectors. Many geometries are used for the incident light, the collection optics, and detectors. There are three main parts of scattered and reflected light that can be used to gain information on the surface quality: the directly reflected specular beam (called bright field), diffusely scattered light (dark field), and backscattered light. The parameters measured can include the intensity, polarization, and direction of reflected light. Either normal or oblique incidence and different polarizations (s, p, or circular polarization) can be used.

Surface structures such as steps or mounds cause slight changes in the intensity and direction of the specular beam which can be used to identify certain surface defects [33]. Surface defects and microroughness scatter light diffusely into the dark-field detector. It is also possible to collect light scattering into different angles

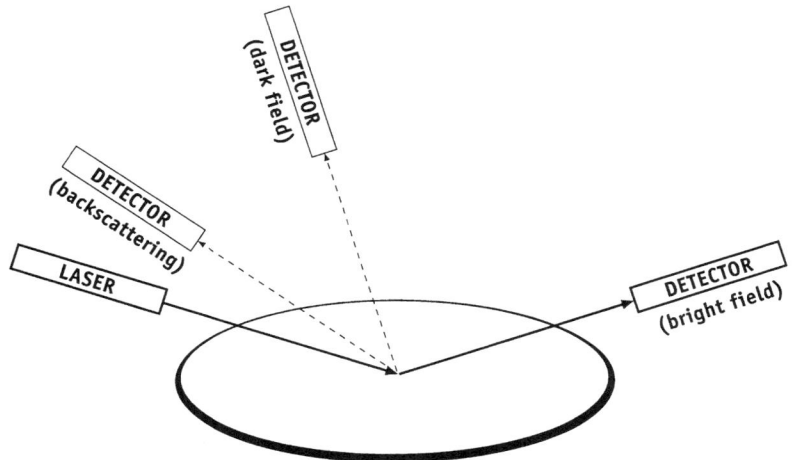

FIG. 24. Principle of laser scatterometry for the detection of surface defects.

to separate the intensity of light scattered into large angles from the small-angle scattering [83]. Subsurface defects scatter light efficiently toward the light source and this backscattered light can be used to identify such defects.

c. *Microroughness*

Microroughness can adversely affect the identification of small particles and have an impact on the quality of thin gate oxides. Microroughness can be measured by several methods: AFM or other types of scanning probe microscopy, optical profilometry based on phase-contrast interference, and laser scatterometry. Surface roughness is usually characterized by the power spectral density function (PSD), which describes the roughness power distribution as a function of the spatial frequency, and the root-mean-square (rms, or R_q) value, which is the standard deviation of topographical deviations *over a defined spatial frequency range*. A complete evaluation of the surface roughness requires PSD data to show how the roughness depends on the spatial frequency. It should be noted that surface microroughness is typically nonisotropic and PSD must be evaluated in two dimensions. R_q is essentially the integral of the PSD function [9]. The spatial frequencies depend on the measurement technique and typically range between 0.01 and 100 μm^{-1}. To compare different roughness measurements, the frequency ranges should be identical.

The most popular way to characterize surface roughness is scattering measurements with an SSIS due to their speed and convenience and the fact that the roughness information is obtained from the same measurement as used to count particles. The dark-field signal is used for roughness measurement. Because of the very low intensity this light is usually collected over a large area and the quantity usually calculated is the ratio between the dark-field and the incident beam intensities. If there are no particles or defects on the surface, the dark-field signal is dominated by the microroughness and is called *haze* because it is related to the visible haze seen with collimated bright light in visual inspection. The haze value given by the laser scanner is proportional to the square of the rms roughness value. The spatial frequency range depends on the design of the instrument. It is defined by the geometry of the optics (i.e., how far from the specular beam the dark-field collection begins) and the wavelength of the laser source. In practice the spatial frequency in haze measurements is generally limited between 0.2 and 3 μm^{-1}. The scattering intensity is also strongly dependent on the direction of the surface undulations relative to the plane of incidence, incident angle, source polarization, wavelength, and spot size. The effect of the nonisotropic nature of surface roughness can be minimized by using an instrument which is capable of determining the direction of maximum scattering. However, the haze values given by different laser scanners are usually not directly comparable. If quantitative data on the surface topography are required, the surface must be profiled with either an AFM or an optical profiler.

The AFM is usually used for the roughness measurement in the tapping mode to minimize possible damage to the wafer and probe. The lower spatial frequency range is simply the inverse of the scan length (L) and the upper frequency limit is set by the ratio of measurement points (N) to L:

$$\text{AFM spatial frequency range} = \frac{1}{L} \leftrightarrow \frac{N}{2 \cdot L}. \qquad (8)$$

If an AFM roughness measurement overlapping the spatial frequency range of laser scanner haze measurements is required, a relatively large scan area, $5 \times 5\ \mu m^2$, must be used to reach the lower spatial frequency limit of 0.2 μm^{-1}.

An optical profiler based on an interference microscope can also be used for measuring the PSD function. The spatial frequency range naturally depends on the optics, the magnification, and the wavelength of the light but it can typically be comparable to that achieved with the AFM: for instance, a range of 0.01–3.8 μm^{-1} is achieved with a commercial instrument (Wyko NT-2000) using red light and an objective with a 50× magnification. The correlation of different instruments is problematic but possible. Figure 25 shows that the correlation of the rms roughness values given by the AFM and interference microscope with the square root of the haze value can be quite good if correct spatial frequency ranges are used. One should bear in mind, though, that the spatial frequencies of different instruments may be biased and require an empirical correction to achieve the best correlation.

The surface roughness of epitaxial wafers is naturally strongly dependent on the surface quality of the starting substrate, the crystallographic orientation of the surface, and the processing parameters (temperature, deposit rate, etc.).

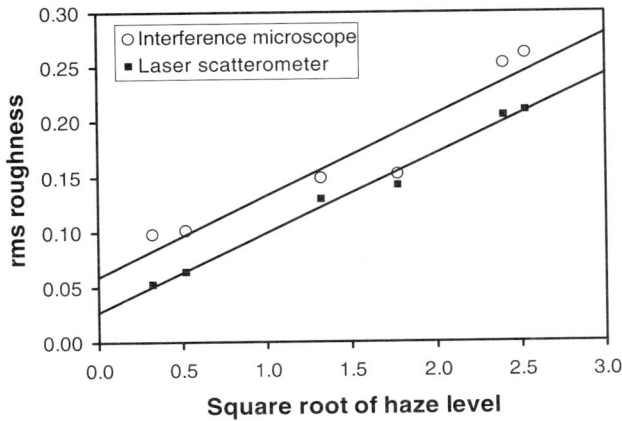

FIG. 25. R_a measured from polished wafers with different surface finishes using an AFM and an optical interference microscope correlated with the square root of the haze value measured with an SSIS from the same wafers. The best correlation is obtained by optimizing the nominal spatial frequency ranges: AFM, 0.4–2.2 μm^{-1}; interference microscope, 0.1–2.6 μm^{-1}; SSIS, 0.26–2.6 μm^{-1}. (Data courtesy of Olli Lehtonen.)

In general the haze level of epitaxial wafers tends to be higher than in polished substrates. This is due to increased roughness at high spatial frequencies (small steps), whereas roughness at lower spatial frequencies tends to decrease in the epitaxial process [74]. In Fig. 26 the strong anisotropy of surface roughness at high spatial frequencies due to the formation of atomic scale terraces is obvious.

3. Metal and Molecular Contamination

Epitaxy is a fairly violent process which subjects the wafer to very high temperatures for prolonged periods of time, and the probability of contamination is quite high. The possible metallic contaminants in epitaxial wafers are largely the same as in polished wafers. Metals with low solubilities and high diffusivities in silicon tend to be the most critical contaminants. In silicon epitaxy these include 3d transition metals and some noble metals: Fe, Cu, Ni, Ti, Mo, Zn, Cr, and Au [20, 41, 60, 81, 82]. Molecular contamination is mainly an issue for the polished substrate.

The techniques used for controlling contamination of epitaxial wafers are identical to those used with polished wafers. The most common techniques available include the following.

- Bulk metal contamination: Minority carrier recombination
                                                                          DLTS
                                                                         NAA

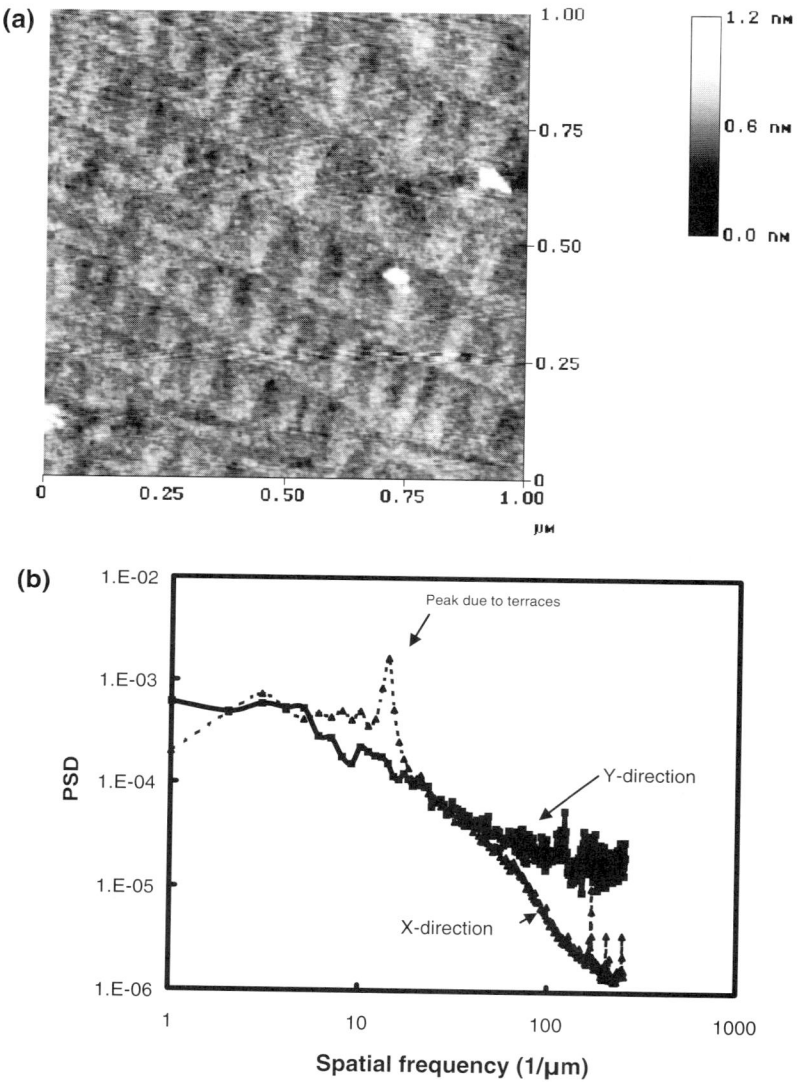

FIG. 26. AFM measurements of the (100) surface of an epitaxial layer grown in an atmospheric single-wafer reactor. (a) AFM image showing monolayer height terraces. (b) Microroughness characterization of the same sample. PSD curves show clearly the anisotropy of the surface. The peak at 15 μm^{-1} in the x-direction is due to the terraces. (Data courtesy of J. Ahopelto, VTT Microelectronics Center.)

- Surface metal contamination: TXRF
                                   ICP-MS
                                   Haze test (chemical etching)
                                   SIMS
                                   AAS
- Surface molecular contamination: SIMS
                                   AAS

a. Bulk Metal Contamination: Minority Carrier Recombination

The interaction of positive (holes) and negative charge carriers (electrons) in silicon can be understood on the basis of the energy band diagram (Fig. 27). (For good reviews on lifetime and diffusion length methods the reader is referred to Bullis and Huff [26], Schroder [65, 66], and ASTM standards [10].) A cloud of excess charge carriers can be created, for instance, with a pulse of light. The excess carriers diffuse through the silicon material before recombining. The characteristic distance for this process is called the diffusion length L, which depends on the average lifetime, τ, and diffusion constant D:

$$L = \sqrt{D \cdot \tau}. \tag{9}$$

Diffusion length and lifetime are proportional to each other and either parameter can be used to describe the recombination properties of the material. Silicon is an indirect semiconductor and the direct recombination of electrons and holes across

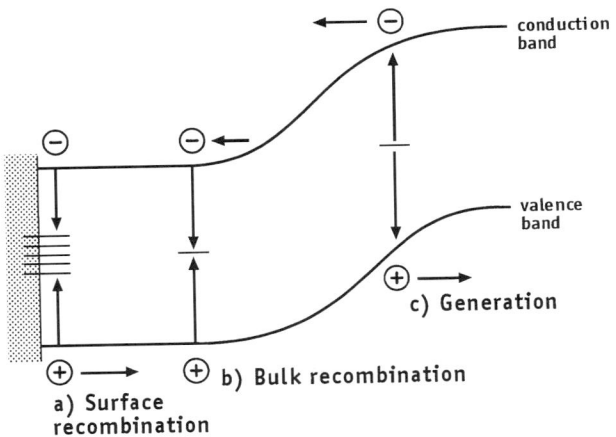

FIG. 27. Charge carrier recombination and generation mechanisms in a lightly doped semiconductor: (a) surface recombination; (b) recombination through a recombination center; (c) carrier generation in an electric field.

the forbidden gap is an unlikely event. Therefore, the lifetime of excess carriers in high-purity silicon is very long (in the millisecond range). The value of L and τ for contamination control comes from the fact that certain impurities or crystal defects can create energy levels deep inside the forbidden band gap. These levels, recombination centers, can increase the recombination rate and decrease the lifetime and diffusion length of carriers dramatically even when their density is very low. Also, recombination centers increase the probability of the reverse process, thermal creation of electron-hole pairs, especially when an internal electric field bends the energy bands. As a result both the recombination and generation of excess carriers can be used to detect certain impurities. It is much more common to use the recombination lifetime (or diffusion length) for contamination monitoring because the generation lifetime measurements are quite slow and require the formation of an internal electric field, usually with a MOS capacitor or diode structure [11]. The recombination lifetime is a function of the injection level η, which is defined as the ratio of the injected minority carrier concentration to the concentration of majority carriers [26]:

$$\tau = \frac{\tau_0 + \eta \cdot \tau_\infty}{1 + \eta}. \tag{10}$$

Therefore, the recombination lifetime (and diffusion length) will usually be dependent on the injection level, which must be identical if two measurements are to be compared.

The three main recombination techniques are all based on the creation of an excess carrier cloud in silicon by a short light pulse. The decay time of these excess carriers is then detected capacitively (SPV), from the photocurrent (Elymat), or with the aid of microwave reflection (μ-PCD). As shown in Fig. 27, carrier decay through bulk states is not the only possible recombination process. A clean silicon surface contains a very high density of energy levels, which provide an efficient path for *surface recombination.* The lifetime limited by an unpassivated surface is very short. For this reason it is usually necessary to passivate both wafer back and wafer front surfaces for the recombination measurement to work.

In μ-PCD the recombination lifetime is measured by detecting the transient in the microwave reflectance caused by the excess carriers. μ-PCD requires good surface passivation to reduce the surface recombination velocity. Commonly used passivation methods are thermal oxidation (often augmented by corona charging) [68], HF, and iodine–methanol [42]. An *in situ* passivation method based on the continuous corona charging of the native oxide is also available in a commercial instrument (Semilab, Inc.). The injection level in μ-PCD is typically medium to high and it is capable of measuring lifetimes from microseconds to several milliseconds.

In SPV very low-intensity light pulses are used to inject excess carriers into the bulk silicon. (For a good review of SPV see Ref. 52). The lifetime of the carriers depends on two competing processes: recombination in the bulk and diffusion to (and recombination on) the surface. The surface photovoltage transient, ΔV, is detected capacitively and can be used to determine the diffusion length if the absorption coefficient α is varied by using light of several wavelengths to inject the carriers,

$$\Delta V = \text{Constant} \cdot \Phi \cdot \left(1 - \frac{B(L)}{\alpha L}\right) \bigg/ \left(1 - \frac{1}{\alpha^2 L^2}\right), \quad (11)$$

where $B(L)$ is a function of L. SPV requires the surface to be in depletion or weak inversion for the minority carriers to be collected at the surface. Diffusion lengths larger than two to three times the wafer thickness cannot be measured. The main advantage of SPV is the very low injection level, which makes the measurements of different wafers directly comparable. SPV can also measure the surface charge, which is a sensitive indicator of surface damage.

In Elymat (GeMeTec, mbH), the wafer is placed in a HF solution which acts as an electrolyte and passivates the surface. As in SPV the minority carriers are collected at the wafer surface. The detection is done by measuring the photocurrent at the reverse-biased electrolyte–Si diode either at the front or the back surface. Elymat is capable of measuring diffusion lengths from about one-fifth up to four times the wafer thickness. Usable injection levels are similar to those in μ-PCD [34, 35, 53].

As noted before, the recombination lifetime is a function of the injection level, therefore it should be possible to identify contaminants from the injection level dependence [43]. This dependence has been reported for Fe, Ni, Co, W, Cr, Au, Ti, and oxygen precipitates [34, 43, 51, 63]. In practice, such identification does not tend to work due to several complications. Usually an insufficient range of injection levels is accessible. In addition, the actual injection level is unknown, for it is difficult to relate the intensity of the light pulse to actual excess carrier concentrations. Also, it is always possible that more than one-impurity levels are effective in the same sample.

Fortunately the most important contaminant, Fe, can be identified and even measured quantitatively by recombination techniques because Fe can exist in p-type silicon in two forms: as interstitial atoms (Fe_i) and in iron–boron pairs (Fe–B). At room temperature Fe–B is the stable form. However, Fe–B pairs can be broken either by heating of the sample to $>200°C$ or by illumination with intense light. As Fe–B and Fe_i have different effects on recombination, the presence of Fe is clearly revealed by the injection level dependence depicted in Fig. 28. In fact, it is even possible to obtain the concentration of Fe when the injection level is known (either very low or very high), a condition which is easily met by SPV as shown by Zoth and Bergholz [85]. Similar methods have been applied to Fe identification from generation lifetime measurements [61] and with μ-PCD [43].

Even lacking complete contamination identification capability, recombination techniques are invaluable for monitoring reactor cleanliness. In fact, a definite identification of the contaminant is normally unnecessary because the lifetime pattern is sufficient to reveal the source of the contamination. Figure 29a shows the most common type of contamination: iron diffusing out of a pinhole in the silicon carbide coating of the susceptor pocket causes a round area of low lifetime on the wafer. Iron is always present in any contamination caused by the susceptor.

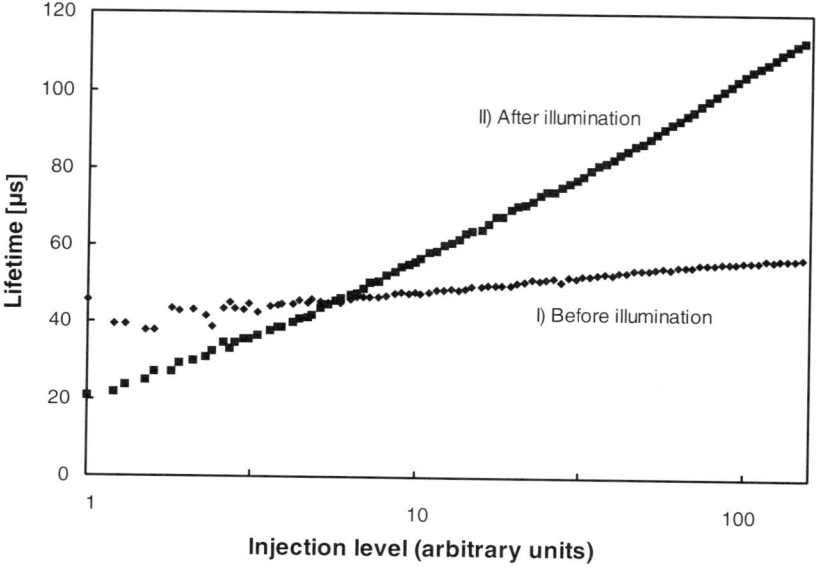

FIG. 28. Injection level dependence of recombination lifetime measured with μ-PCD. (I) Fe contamination in the form of Fe–B pairs: injection level has almost no influence on τ. (II) Injection level dependence on the same spot after illumination with bright light. Injection level typically has a strong influence on τ for Fe interstitials.

Leaks in gas lines are another fairly common occurrence and they can be detected easily from the typical uniform lifetime distribution (Fig. 29b).

The main weakness of recombination techniques for contamination control is the difficulty of obtaining valid data from samples with highly doped substrates (N^+ or P^+). Due to the high concentration of the majority of carriers, the recombination lifetime is necessarily very short (a few microseconds) in these structures and the lifetime measurement with μ-PCD requires special techniques [40, 62].

b. *Bulk Metal Contamination: DLTS and NAA*

Very low concentrations of transition metals in the bulk can be detected with recombination techniques. Apart from Fe, however, the impurities cannot be identified easily, nor can their concentrations be measured accurately. SIMS is capable of elemental analysis and depth profiling but detection limits are quite high. Bulk contamination can sometimes be identified quantitatively from transient capacitance measurements. The original technique is called deep-level capacitance transient spectroscopy (DLTS) [58]. Impurity levels within a space charge region are filled (for instance, with light or a current pulse) and the emission of majority carriers is detected by monitoring the change of capacitance. From the rate of change the activation energy, cross section, and concentration of the impurity levels can be

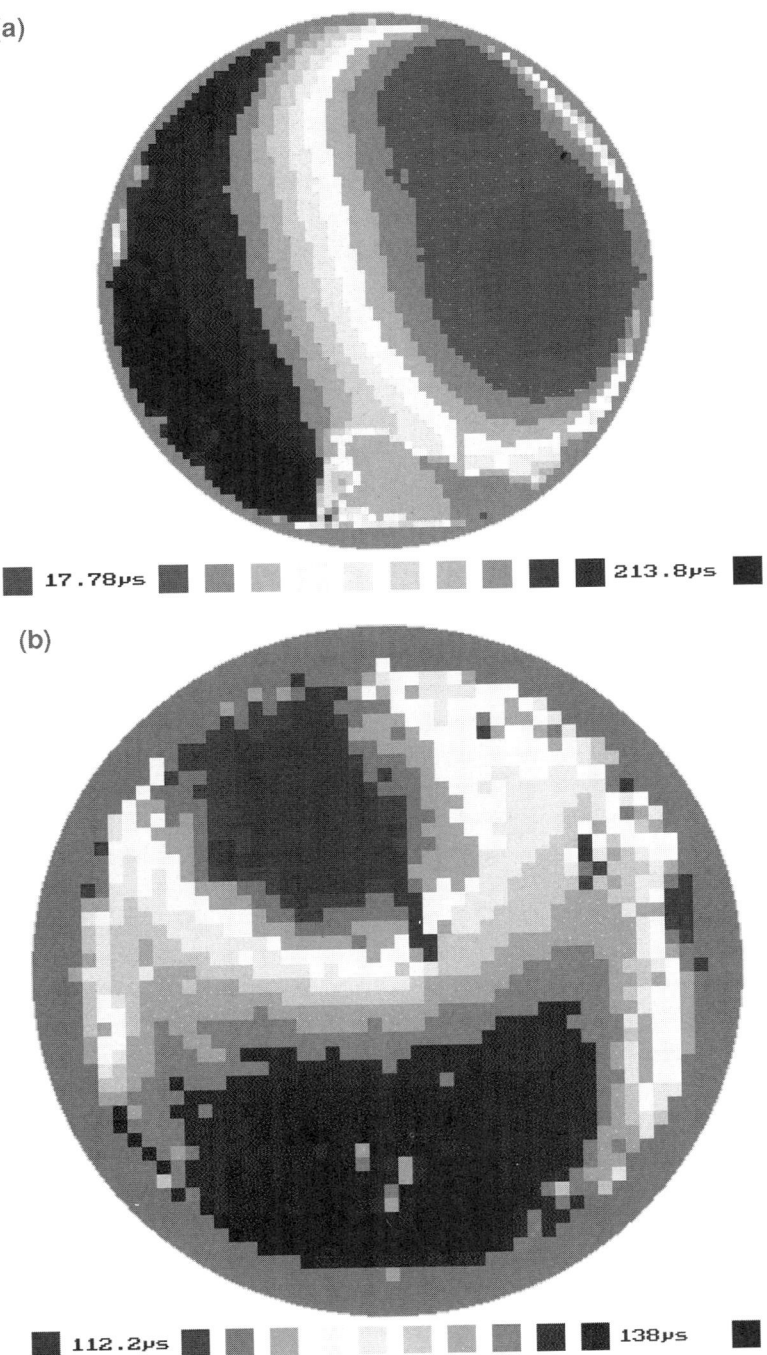

FIG. 29. (a) A μ-PCD lifetime map of a wafer showing iron contamination from a pin hole in the silicon carbide coating of the graphite susceptor. (b) A μ-PCD lifetime map of an epitaxial wafer displaying the characteristic uniformity caused by contaminated process gases. In this case the contamination was caused by a leaking bellows valve in the trichlorosilane line.

obtained and the contaminants identified on the basis of the activation energy and cross section. DLTS is quantitative and sensitive but requires the processing of a diode on the wafer. Good correlation has been demonstrated among DLTS, TXRF, and recombination lifetime measurements for low levels of iron contamination [3].

Neutron activation analysis (NAA) is a powerful quantitative technique for trace contamination analysis of the wafer surface, the epitaxial layer, and the bulk silicon [81]. NAA requires the samples to be irradiated with thermal neutrons and the concentrations of trace elements are determined by γ-ray spectroscopy. NAA is fairly expensive but invaluable when a definite identification of the trace elements is required [57].

c. Haze Test and Defect Delineation by Chemical Etching

Subsurface defects can be revealed by etching the substrate or epitaxial wafer chemically [12, 13]. Different chemical solutions can be used on either pure silicon or oxidized wafers to delineate structural defects, such as stacking faults, dislocations and slip [14, 15], or metallic contamination [16]. As the damage that can typically cause surface epitaxial defects is typically very shallow, it is generally advisable to use a slow etchant to probe a thin layer near the surface.

The solubilities in silicon of the fast-diffusing metals Ni and Cu are extremely low. When the wafer is cooled after epitaxy these metals have a strong tendency to move to the wafer surface, where they form precipitates, which can be revealed using a chemical technique developed by Graff [38]. In this so-called haze test the wafer is etched chemically and inspected visually. Metal precipitates are etched preferentially and form shallow pits which are visible as strong haze when the wafer is inspected under bright light. The haze test, even though nonquantitative, is quick and very useful because the recombination techniques can be ineffective in detecting Ni and Cu contamination.

d. Surface Metal Contamination: TXRF, ICP-MS, AAS, and SIMS

X-Ray radiation can excite impurity atoms resulting in fluorescence at characteristic energies which can be used to detect and identify the different elements. In TXRF this effect is used for surface-sensitive detection of contamination by letting a monochromatic X-ray beam strike the wafer surface at a very small angle so that it will undergo total reflection. At typical glancing angles of 1–3 mrad, the penetration depth of the X rays is only a few nanometers [23]. The range of elements that can be analyzed is dependent on the target used. With a W $L\beta$ radiation the elements from P to Zn, including S, Cl, K, Ca, Ti, Cr, Mn, Fe, Ni, Cu, and Mo, can be detected with great sensitivity (detection limits below 10^{10} cm^{-2} can be reached [21]). In addition, the distribution of contaminating elements on the wafer can be mapped with an area resolution of about 0.5 cm^2. Sensitivities over two orders of magnitude better are obtained using the vapor phase decomposition

(VPD) method to concentrate the impurities [36, 59]. In VPD the native oxide on the wafer is first dissolved with HF vapor and the impurities are collected with a small droplet of oxidizing solution, typically a dilute mixture of H_2O, HF, and H_2O_2. After the wafer surface has been scanned with the droplet, it is dried and the TXRF measurement is performed on the residue. Some mapping capability can be retained by using several droplets to collect impurities from different areas. Often it is sufficient to use just two droplets to check the center and the edge of the wafer. TXRF is calibrated by measuring a calibration sample with a known contamination level of a single reference element. Quantification for other elements is achieved by using sensitivity factors derived from the well-known relative fluorescence yields. The TXRF technique is well suited for the analysis of smooth and clean silicon wafers and the measurement techniques are well established [17, 71]. TXRF is an atmospheric measurement but it requires expensive X-ray equipment and the measurement is relatively slow. However, automated equipment for VPD preparation is available.

In ICP-MS the sample is vaporized, mixed with a carrier gas (Ar), and ionized in plasma. A small portion of the ions is fed through pressure reduction stages into the vacuum chamber, where the ions are identified with a suitable mass analyzer. Usually a quadrupole mass filter is employed, but time-of-flight (TOF) and magnetic sector instruments are also available. ICP-MS has a very high sensitivity and almost-complete elemental coverage. However, Ar atoms and polyatomic species (i.e., ArO and ArH) interfere with the detection of K, Ca, and Fe. By using cool plasma conditions or a special reaction cell interferences can be reduced and even Fe can be detected down to 1-ppt concentrations with quadrupole ICP-MS [76]. TOF ICP-MS has a lower sensitivity for most elements but can still detect the interesting elements down to 4- to 55-ppt concentrations [79]. The main advantage of the TOF technique is speed: analysis time is essentially independent of the number of elements analyzed. ICP-MS can provide quantitative concentration information when calibrated using solutions containing known concentrations of metals. However, the technique requires that the sample be vaporized or nebulized for ionization. Wafer surface analysis is possible by using the VPD technique to collect the impurities and vaporizing the VPD droplet [19, 37]. Surface mapping is naturally difficult with the ICP-MS.

In atomic absorption spectroscopy (AAS) a liquid sample, i.e., a droplet from VPD, is atomized, usually in a graphite furnace. The atomic composition of the sample is measured by detecting the characteristic absorption lines in light passed through the sample cell. AAS is a slow measurement but can achieve detection limits that are comparable to those in TXRF and SIMS [39].

SIMS can be used as a high-sensitivity surface analysis tool with either a magnetic sector or a TOS mass separator. The achievable detection limits are even lower than in TXRF, being in the range of 10^8–10^{10} cm^{-2} for the interesting contaminants [25]. SIMS can be used to analyze both metallic and organic contamination and is capable of detecting most elements with a high precision. Similarly to TXRF, SIMS is quantified with a known elemental standard from which other elements are calibrated by using relative sensitivity factors. Good correlation has been shown between TXRF and SIMS results for several metals [72]. Because SIMS is a vacuum technique, it is not ideal for the analysis of volatile species.

Typically the surface metal concentrations of epitaxial wafers are below the detection limits of TXRF, ICP-MS, and AAS, and these techniques are of more interest for the monitoring of substrates, i.e., the performance of wafer cleaning prior to epitaxy. The cleanliness of the epitaxy process can normally be monitored more conveniently with recombination techniques, with their powerful mapping capabilities. However, if postepitaxy *in situ* cleaning techniques in cluster tools become common, these surface analysis methods may become more important for the monitoring of epitaxial wafers.

e. Surface Molecular Contamination

Organic contamination is becoming steadily more critical in IC processing due to its prevalence in clean rooms and also because low thermal budget processing does not allow even relatively volatile compounds to vaporize from the wafer surface. In epitaxy it is known that organic molecules may cause surface defects due to the formation of silicon carbide islands [49]. It is to be expected that the defect problem will become more critical with the low-temperature epitaxy being developed for SiGe. The epitaxy process itself is not a major source of organics. However, chlorine-containing molecular contamination may be present on the wafers as they are removed from the reactor due to the fast cooldown from the processing temperature.

The most common methods used for analyzing molecular contaminants on wafer surfaces include the atmospheric techniques GC-MS, IMS (or MS-IMS), TD GC-MS, and FTIR and the vacuum techniques TOF-SIMS and XPS [73]. The contamination can either be desorbed from the surface for subsequent analysis (as in GC-MS and IMS) or be analyzed directly from the wafer surface (TOF-SIMS, XPS, and FTIR). The evaporation of volatile species has to be prevented by cooling the sample for vacuum measurement, making atmospheric techniques preferable in this respect [24].

VI. Future Developments and Conclusions

1. MEASUREMENTS ON AND CHARACTERIZATION REQUIREMENTS FOR 300-mm WAFERS

The International Technology Roadmap for Semiconductors provides a good basis for forecasting the near-future [46]. Single-wafer processing will expand to 300-mm wafers, but larger wafers will not become important for some time.

The forecasts for the most demanding epitaxial thickness requirements are not very radical and well within the capabilities of today's measuring equipment. It is likely that the traditional FTIR equipment will be replaced by the model-based instruments. If the use of P^- wafers for advanced logic applications becomes more common, thickness measurements on high-resistivity substrates will become necessary. Possible solutions can be found by measuring the thickness of the whole wafer either capacitively or optically as outlined in Section IV. The main issue for resistivity control will be noncontact measurements for which solutions already exist. Smaller edge exclusion limits down to 1 mm will be a difficult challenge for the capacitive techniques. The most difficult problems for thickness and resistivity characterization may come from silicon–germanium epitaxy and selective epitaxy. Silicon–germanium will have some special requirements due to the very thin layers and heterostructures. Improvements in the nondestructive means of controlling layer thicknesses and composition profiles can probably be found from ellipsometry and X-ray diffraction. It is likely that future IC technologies will involve more three-dimensional device structures with accurately controlled selective epitaxy, requiring three-dimensional compositional, thickness, and doping profiling techniques.

The contamination limits for critical surface metals near 10^{10} cm^{-2} are well within the present capabilities. Minority carrier recombination methods work well for metal contamination monitoring of high-resistivity substrates but more work will be required to improve the characterization of epitaxial layers on highly doped substrates. Surface contamination (either metallic or molecular) will not be a problem for epitaxial wafers unless *in situ* surface cleaning becomes common. The automated methods for surface defect classification have progressed well in recent years and no major improvements will be necessary. However, it is likely that the quality of the wafer backside will become more critical due to the impact backside particles and defects can have on wafer flatness and photolithography. Therefore, in the future, wafer backsurface inspection requirements will approach those of the front surface. Haze measurements may be sufficient for microroughness control if their correlation with PSD analysis can be improved, possibly with the help of haze standards. If the spatial frequency range of haze measurements is not sufficient, PSD analysis using optical interferometers will be needed.

2. Should We Monitor the Wafers or the Process?

Process monitoring through wafers is expensive. Even if nondestructive measurements are used, there is always a time delay before a process deviation or problem is identified and corrective action initiated. One possibility is to use *in situ* measurements to control the wafer and process, i.e., measure epitaxial wafer thickness, resistivity, and particle and defect levels. The main objection to this approach is cost. More might be gained by monitoring the process and equipment directly to maintain a more stable process and to create the ability to detect and correct problems before they cause defective products. For instance, process gas composition could be measured to detect leaks in the gas lines, or a leak-checking instrument could be installed in the gas panel. Oxygen or moisture monitoring of the wafer handler chamber and load locks could detect leaks in those parts of the reactor. The stability of the gas flow rates could be improved by using the real-time monitoring features of digital mass flow controllers and possibly by installing *in situ* calibration facilities or flow rate meters for critical controllers. The condition of the temperature control and heating system could easily be controlled by measuring the average power fed into each lamp zone in a lamp-heated reactor.

Acknowledgments

It was my pleasure to have productive conversations about characterization with people from several organizations. I would like to thank the following persons for their help: Robert J. Hillard and Stephen M. Ramey (Solid State Measurements, Inc.), Gerhart Kneissl (ADE Corporation), Marco Borgini (MEMC Electronic Materials), Adrien Danel (Euris), Sara Acerboni and Mario Cottini (ST Microelectronics), Tibor Pavelka (Semilab), Francois Tarif (Leti), Johannes Baumgartl (Infineon Technologies), Christo Bojkov (Texas Instruments), Anton Huber (GeMeTec), Doug Meyer (ASM), and Dennis F. Paul (Physical Electronics). I would also like to acknowledge the help of Steve Bell (Okmetic Inc.) and Maria Hokkanen and Mikko Hiltunen (Okmetic Oyj.) in collecting some of the material. Okmetic Oyj. provided financial assistance for this work. The provision of logistical support by Janice Airaksinen is also gratefully acknowledged, as is the permission by Tarja, Megan, Iain, and Ben Airaksinen to use the domestic computing facilities.

References

1. AIAG, *Measurement Systems Analysis (MSA) Reference Manual,* Automotive Industry Action Group, 1994.
2. American National Standard, *Sampling Procedures and Tables for Inspection by Attributes,* ANSI/ASQC Z1.4-1993, American Society for Quality, 1993.
3. Anttila, O. J., and M. V. Tilli, *J. Electrochem. Soc.* **139**, 1751 (1992).
4. ASTM, ASTM Standard F 723-97, *1998 Annual Book of ASTM Standards,* Sect. 10, American Society for Testing and Materials, West Conshohocken, PA, 1998.
5. ASTM, ASTM Standards F 374-94a and F 1529-97, *1998 Annual Book of ASTM Standards,* Sect. 10, American Society for Testing and Materials, West Conshohocken, PA, 1998.
6. ASTM, ASTM Standard F 1393-92, *1998 Annual Book of ASTM Standards,* Sect. 10, American Society for Testing and Materials, West Conshohocken, PA, 1998.

7. ASTM, ASTM Standards F 525-88, F 672-88, and F 674-92, *1998 Annual Book of ASTM Standards,* Sect. 10, American Society for Testing and Materials, West Conshohocken, PA, 1998.
8. ASTM, ASTM Standard F 1620-96, *1998 Annual Book of ASTM Standards,* Sect. 10, American Society for Testing and Materials, West Conshohocken, PA, 1998.
9. ASTM, ASTM Standard F 1811-97, *1998 Annual Book of ASTM Standards,* Sect. 10, American Society for Testing and Materials, West Conshohocken, PA, 1998.
10. ASTM, ASTM Standard F 1535-94, *1998 Annual Book of ASTM Standards,* Sect. 10, American Society for Testing and Materials, West Conshohocken, PA, 1998.
11. ASTM, ASTM Standard F 1388-92, *1998 Annual Book of ASTM Standards,* Sect. 10, American Society for Testing and Materials, West Conshohocken, PA, 1998.
12. ASTM, ASTM Standard F 47-94, *1998 Annual Book of ASTM Standards,* Sect. 10, American Society for Testing and Materials, West Conshohocken, PA, 1998.
13. ASTM, ASTM Standard F 80-94, *1998 Annual Book of ASTM Standards,* Sect. 10, American Society for Testing and Materials, West Conshohocken, PA, 1998.
14. ASTM, ASTM Standard F 1809-97, *1998 Annual Book of ASTM Standards,* Sect. 10, American Society for Testing and Materials, West Conshohocken, PA, 1998.
15. ASTM, ASTM Standard F 1727-97, *1998 Annual Book of ASTM Standards,* Sect. 10, American Society for Testing and Materials, West Conshohocken, PA, 1998.
16. ASTM, ASTM Standard F 1049-95, *1998 Annual Book of ASTM Standards,* Sect. 10, American Society for Testing and Materials, West Conshohocken, PA, 1998.
17. ASTM, ASTM Standard F 1526-95, *1998 Annual Book of ASTM Standards,* Sect. 10, American Society for Testing and Materials, West Conshohocken, PA, 1998.
18. ASTM, ASTM Standard F 523-88, *1998 Annual Book of ASTM Standards,* Sect. 10, American Society for Testing and Materials, West Conshohocken, PA, 1998.
19. Balazs, M. K., and J. Fucskó, in *Semiconductor Characterization: Present Status and Future Needs,* edited by W. M. Bullis, D. G. Seiler and A. C. Diebold, pp. 445–449, American Institute of Physics, Woodbury, NY, 1996.
20. Benton, J. L., D. C. Jacobson, B. Jackson, J. A. Johnson, T. Boone, D. J. Eaglesham, F. A. Stevie, and J. Becerro, *J. Electrochem. Soc.* **146**, 1929 (1999).
21. Bergholz, W., D. Landsmann, P. Schauberger, and B. Schoepperl, in *Crystalline Defects and Contamination: Their Impact and Control in Device Manufacturing,* edited by B. O. Kolbesen, P. Stallhofer, C. Claeys, and F. Tardif, p. 69, Electrochemical Society Proceedings Series, PV 93-15, Electrochemical Society, Pennington, NJ, 1993.
22. Berkowitz, H. L., and R. A. Lux, *J. Electrochem. Soc.* **128**, 1137 (1981).
23. Berneike, W., *Spectrochim. Acta* **48B**, 269 (1993).
24. Budde, K. J., in *Mater. Res. Soc. Proc.* Vol. 386, pp. 165–175, Materials Research Society, Pittsburgh, PA, 1995.
25. Budrevich, A., and J. Hunter, in *CP449, Characterization and Metrology for ULSI Technology: 1998 International Conference,* edited by D. G. Seiler, A. C. Diebold, W. M. Bullis, T. J. Schaffner, R. McDonald and E. J. Walters, pp. 169–181, American Institute of Physics, Woodbury, NJ, 1998.
26. Bullis, W. M., and H. R. Huff, *J. Electrochem. Soc.* **143**, 1399 (1996).
27. Charpenay, S., P. Rosenthal, G. Kneissl, C. Hoener Gondran, and H. Huff, *Solid State Technol.* **96**, 161–170 (1998).
28. Choo, S. C., M. S. Leong, and J. H. Sim, *Solid-State Electron.* **26**, 723 (1983).
29. Clapper, R. A., D. G. Schimmel, J. C. C. Tsai, F. S. Jabara, F. A. Stevie, and P. M. Kahora, *J. Electrochem. Soc.* **137**, 1877 (1990).
30. Clarysse, T., and W. Vandervorst, *J. Vac. Sci. Technol. B* **16**, 260 (1998).
31. Danel, A., *Caractérisation des Propriétés de Surface du Silicium par Analyse de Charges: Méthode SCP—Surface Charge Profiler,* Thése pour le titre de Docteur, L'Institut National Polytechnique de Grenoble, Grenoble, 1999.
32. DIN 50439, *Determination of the Dopant Concentration Profile of a Single Crystal Semiconductor Material by Means of the Capacitance Voltage Method and Mercury Contact,* Beuth Verlag GmbH.

33. Dou, L., D. Kesler, W. Bruno, C. Monjak, and J. Hunt, in *CP449, Characterization and Metrology for ULSI Technology: 1998 International Conference,* edited by D. G. Seiler, A. C. Diebold, W. M. Bullis, T. J. Schaffner, R. McDonald, and E. J. Walters, pp. 824–828, American Institute of Physics, Woodbury, NY, 1998.
34. Eichinger, P., in *Recombination Lifetime Measurements in Silicon,* ASTM STP 1340, edited by D. C. Gupta, F. R. Bacher and W. M. Hughes, pp. 101–111, American Society for Testing and Materials, 1998.
35. Eichinger, P., and M. Rommel, in *New Developments of the Elymat Technique,* Proc. Vol. 97-22, Electrochemical Society, Pennington, NJ, 1997.
36. Eichinger, P., in *Analytical Techniques for Semiconductor Materials and Process Characterization,* edited by B. O. Kolbesen, D. V. McCaughan, and W. Vandervorst, PV 90-11, pp. 227–237, Electrochemical Society, Pennington, NJ, 1990.
37. Fucsko, J., S. S. Tan, and M. K. Balasz, *J. Electrochem. Soc.* **140**, 1105 (1993).
38. Graff, K., in *Metal Impurities in Silicon-Device Fabrication,* edited by Springer Series in Materials Science, Vol. 24, Springer-Verlag, Heidelberg, 1995.
39. Gupta, P., Z. Pourmotamed, S. H. Tan, and R. McDonald, in *Semiconductor Characterization: Present Status and Future Needs,* edited by W. M. Bullis, D. G. Seiler, and A. C. Diebold, pp. 450–455, American Institute of Physics, Woodbury, NY, 1996.
40. Hashizume, H., S. Sumie, and Y. Nakai, in *Recombination Lifetime Measurements in Silicon,* ASTM STP 1340, edited by D. C. Gupta, F. R. Bacher, W. M. Hughes, pp. 47–58, American Society for Testing and Materials, West Conshohocken, PA, 1998.
41. Hayamizu, Y., T. Koide, and Y. Kitagawara, *27th Symposium on ULSI Ultra Clean Technology Proceedings,* p. 25, 1996.
42. Horányi, T., T. Pavelka, and P. Tüttö, *Appl. Surf. Sci.* **63**, 306 (1993).
43. Horányi, T. S., P. Tüttö, and Cs. Kovacsics, *J. Electrochem. Soc.* **143**, 216 (1996).
44. Hu, S. M., *J. Appl. Phys.* **53**, 1499 (1982).
45. Irwin, J. C., *Bell Syst. Tech. J.* **41**, 387 (1962).
46. ITRS, *The International Technology Roadmap for Semiconductors, 1999 Edition,* Semiconductor Industry Association, International SEMATECH, Austin, TX, 1999.
47. Iwabuchi, M., K. Mizushima, M. Mizuno, and Y. Kitagawara, *J. Electrochem. Soc.* **147**, 1199 (2000).
48. Kamieniecki, E., *J. Vac. Sci. Technol.* **20**, 811 (1982).
49. Kim, K.-B., P. Maillot, and A. E. Morgan, *J. Appl. Phys.* **67**, 2176 (1990).
50. Kimerling, L. C., J. Michel, H. M'saad, and G. J. Norga, in *Semiconductor Characterization: Present Status and Future Needs,* edited by W. M. Bullis, D. G. Seiler, and A. C. Diebold, American Institute of Physics, Woodbury, NY, 1996.
51. Kurita, K., and T. Shingyouji, in *Recombination Lifetime Measurements in Silicon,* ASTM STP 1340, edited by D. C. Gupta, F. R. Bacher, and W. M. Hughes, pp. 59–67, American Society for Testing, and Materials, West Conshohocken, PA, 1998.
52. Lagowski, J., P. Edelman, and V. Faifer, in *Recombination Lifetime Measurements in Silicon,* ASTM STP 1340, edited by D. C. Gupta, F. R. Bacher and W. M. Hughes, pp. 125–144, American Society for Testing and Materials, West Conshohocken, PA, 1998.
53. Lehmann, V., and H. Foell, *J. Electrochem. Soc.* **135**, 2831 (1988).
54. Liberman, S., in *In-Line Characterization Techniques for Performance and Yield Enhancement in Microelectronic Manufacturing,* SPIE Proceedings, edited by D. K. DeBusk and S. Ajuria, Vol. 3215, pp. 35–40, Bellingham, WA, 1997.
55. Mazur, R. G., and G. A. Gruber, in *Materials and Process Characterization of Ion Implantation,* edited by M. I. Current and C. B. Yarling, Ion Beam Press, 1997.
56. Mazur, R. G., *J. Vac. Sci. Technol. B* **10**, 397 (1992).
57. McGuire, S. C., T. Z. Hossain, A. J. Filo, C. C. Swanson, and J. P. Lavine, in *Semiconductor Characterization: Present Status and Future Needs,* edited by W. M. Bullis, D. G. Seiler, and A. C. Diebold, pp. 329–333, American Institute of Physics, Woodbury, NY, 1996.

58. Miller, G. L., D. V. Lang, and L. C. Kimerling, *Annu. Rev. Mater. Sci.* **7**, 377 (1977).
59. Neumann, C., and P. Eichinger, *Spectrochim. Acta* **46B**, 1369 (1991).
60. Nguyen, M. C., and A. Danel, *Electrochem Soc. Proc.* **90-13**, 432 (1998).
61. Obermeier, G., and D. Huber, *J. Appl. Phys.* **81**, 7345 (1997).
62. Ogita, Y.-I., N. Tate, H. Masumura, M. Miyazaki, and K. Yakushiji, in *Recombination Lifetime Measurements in Silicon*, ASTM STP 1340, edited by D. C. Gupta, F. R. Bacher, and W. M. Hughes, pp. 168–82, American Society for Testing and Materials, West Conshohocken, PA, 1998.
63. Pavelka, T., in *Recombination Lifetime Measurements in Silicon*, ASTM STP 1340, edited by D. C. Gupta, F. R. Bacher, and W. M. Hughes, pp. 206–216, American Society for Testing and Materials, West Conshohocken, PA, 1998.
64. Schroder, D. K., *Semiconductor Material and Device Characterization*, John Wiley & Sons, New York, 1990.
65. Schroder, D. K., in *Handbook of Semiconductor Silicon Technology*, edited by W. C. O'Mara, R. B. Herring, and L. P. Hunt, pp. 550–639, Noyes, Park Ridge, NJ, 1990.
66. Schroder, D. K., in *Recombination Lifetime Measurements in Silicon*, ASTM STP 1340, edited by D. C. Gupta, F. R. Bacher and W. M. Hughes, pp. 5–17, American Society for Testing and Materials, West Conshohocken, PA, 1998.
67. Schumann, P. A., Jr., and E. E. Gardner, *J. Electrochem. Soc.* **116**, 87 (1969).
68. Schöfthaler, M., R. Brendel, G. Langguth, and J. H. Werner, in *Proceedings of the 1st World Conference on Photovoltaic Energy Conversion*, Waikoloa, Hawaii, 5–9 Dec. 1995.
69. SEMI, Semi M2-1296, *Book of SEMI Standards 1997*, Semiconductor Equipment and Materials International, Mountainview, CA, 1997.
70. SEMI, Semi M11-0697, *Book of SEMI Standards 1997*, Semiconductor Equipment and Materials International, Mountainview, CA, 1997.
71. SEMI, Semi M33-0998, *Book of SEMI Standards 1997*, Semiconductor Equipment and Materials International, Mountainview, CA, 1997.
72. Smith, S. P., J. M. Metz, and V. K. F. Chia, in *1998, International Conference on Ion Implantation Technology*, Kyoto, 1998.
73. Smith, P. J., and P. M. Lindley, in *CP449, Characterization and Metrology for ULSI Technology: 1998 International Conference*, edited by D. G. Seiler, A. C. Diebold, W. M. Bullis, T. J. Schaffner, R. McDonald, and E. J. Walters, pp. 133–139, American Institute of Physics, Woodbury, NY, 1998.
74. Stover, J. C., in *Semiconductor Characterization: Present Status and Future Needs*, edited by W. M. Bullis, D. G. Seiler, and A. C. Diebold, pp. 399–412, American Institute of Physics, Woodbury, NY, 1996.
75. Sze, S. M., in *Physics of Semiconductor Devices*, 2nd ed., edited by, p. 369, John Wiley and Sons, New York, 1981.
76. Tanner, S. D., and I. Baranov, *Atom. Spectrosc.* **20**, 45 (1999).
77. Thurber, W. R., R. L. Mattis, Y. M. Liu, and J. J. Filliben, *J. Electrochem. Soc.* **127**, 1807 (1980).
78. Thurber, W. R., R. L. Mattis, Y. M. Liu, and J. J. Filliben, *J. Electrochem. Soc.* **127**, 2291, (1980).
79. Tian, X., H. Emteborg, and F. C. Adams, *J. Anal. Atom. Spectrom.* **14**, 1807 (1999).
80. Tower, J. P., E. Kamieniecki, M. C. Nguyen, and A. Danel, *SPIE Proceedings, Vol. 3884, In-Line Methods and Monitors for Process and Yield Improvement*, SPIE, Santa Clara, CA, Sept. 1999.
81. Van Dalsem, D. J., in *CP449, Characterization and Metrology for ULSI Technology: 1998 International Conference*, edited by D. G. Seiler, A. C. Diebold, W. M. Bullis, T. J. Schaffner, R. McDonald, and E. J. Walters, pp. 897–900, American Institute of Physics, Woodbury, NY, 1998.
82. Werkhoven, C. J., in *Aggregation Phenomena of Point Defects in Silicon*, edited by E. Sirtl, J. Goorissen and P. Wagner, p. 144, Electrochem. Soc. Conf. Proc. 83-4, Electrochemical Society, Pennington, NJ, 1983.

83. Williams, R., W. Chen, M. Akbulut, N. Khasgiwale, R. Persaud, and T. X. Tong, in *International Symposium on Semiconductor Manufacturing Conference Proceedings*, pp. 107–110, IEEE, Pittsburg, PA, 1999.
84. Zhang, W., M. Richter, P. Solomon, Y. Kostoulas, G. Kneissl, W. Aarts, and A. Waldhauer, *Solid State Technol.* **41**, S4–S14 (1998).
85. Zoth, G., and W. Bergholz, *J. Appl. Phys.* **67**, 6764 (1990).
86. Harbecke, B., B. Heinz, V. Offerman, and W. Theiss, in *Optical Characterization of Epitaxial Semiconductor Layers,* edited by G. Bauer and W. Richter, pp. 225–252, Springer Verlag, Heidelberg, Berlin, 1996.

Chapter 8

Epitaxy for Discrete and Power Devices

G. Beretta

ST MICROELECTRONICS
DSG EPITAXY CATANIA DEPARTMENT
CATANIA, ITALY

NOMENCLATURE	277
I. INTRODUCTION	278
1. *Epitaxy at High Voltages*	279
2. *Epitaxy at High Currents*	279
II. CONSIDERATIONS	280
1. *Breakdown Voltage in Epitaxial Junctions*	280
2. *Dopant Redistribution in the Epitaxial Process*	282
3. *Resistivity and Carrier Density*	283
III. SPECIFIC EPITAXY PROCESSES FOR POWER AND DISCRETE DEVICES	283
1. *Bipolar High-Voltage Transistors*	283
2. *Bipolar Medium-Voltage Mesa Transistors*	285
3. *Small-Signal Bipolar Devices*	285
4. *RF Power Devices*	287
5. *Medium- and High-Voltage Power-MOS*	287
6. *Low-Voltage Power-MOS*	288
7. *Insulated Gate Bipolar Transistors*	288
8. *Mixed Epitaxy Devices*	289
9. *Diodes and Thyristors*	290
IV. DEFECTS AND PROBLEMS	290
1. *Epi-Crown*	290
2. *Faceting*	291
3. *Back-Side Defects*	291
V. ACCESSORY CONSIDERATIONS	292

NOMENCLATURE

BV_{bceo}	Breakdown voltage, base–collector with emitter open
BV_{dss}	Drain–source breakdown voltage
h_{FE}	Forward transfer ratio (current gain)
Q_b	Total amount of majority carriers in the base region
R_{acc}	Accumulation contact resistance
R_{ch}	Channel resistance

R_{epi} Epitaxial layer resistance
R_{jfet} Junction FET resistance
V_{br} Voltage at breakdown
$V_{CE_{sat}}$ Collector–emitter voltage at saturation
$V_{DS_{on}}$ Drain–source voltage
V_f Voltage drop in forward bias

I. Introduction

During the last two decades, despite the small amount of literature, modern discrete and power device technologies have been found to be among the most challenging from various points of view. For example, a modern power-MOS (metal–oxide–semiconductor) device may have 1 cm^2 of active area, contain about 40 million cells whose dimension is of the order of 0.6 μm (within which a contact dimension of 0.25 μm must be made); and possess a total gate area well above the gate area of any VLSI device and a channel length of 0.18 μm with a total channel width of about 100 m. In this scenario, it will become apparent that epitaxial (epi) growth processes are one of the key steps in device manufacturing.

As a matter of fact, the main device features—such as the breakdown voltage in the interdiction region and $V_{DS_{on}}$ in MOS devices and, to some extent, the h_{FE} and $V_{CE_{sat}}$ in bipolar devices—are almost totally or mainly controlled by the epi layer resistivity and thickness and by minority carrier lifetimes in the epi silicon. This means that for any device listed in the data sheets there should be a precisely targeted epi process for which the process parameter distributions are exactly known and aimed at the design parameters for that particular device.

In addition, the high voltages and/or the high currents that must be handled by power devices require an epi thickness and resistivity that are somewhat extreme and not attainable with just any epi process and equipment, but make it necessary to apply specialized reactors and processing. This consideration led some manufacturers, mainly in the Far East, to give up on the epi approach and to address the development of high-voltage devices toward triple diffused technology, where the device collector is not made with epi, but floating-zone bulk silicon is used, and where the low-resistance back-side contact is achieved by heavy phosphorous diffusion, while the collector thickness is controlled by front-side lapping of the bulk substrate. This approach has led to a virtually unlimited capability in terms of collector electrical thickness and to extreme ruggedness of the devices. Unfortunately, this approach has the limitations of a lack of flexibility and of the necessity of handling very thin wafers (all the photolithography has to be done on very thin front-lapped wafers) with problematic mechanical yields.

The epi technology approach was initiated in the early 1970s by RCA, who, being aware that high-thickness/high-resistivity epi was not industrially applicable with the epi reactor models that were commercially available at that time, started to develop a specially designed reactor. The barrel design, along with induction heating, as we will see later in the discussion of the process, has been the key point; this design was further developed, starting in 1974, by SGS-ATES (which later become one of the divisons of ST Microelectronics) and by Preti Engineering (now LPE). Since then, many development steps have been successfully undertaken to improve these epi reactors: from the thirty-two 2-in. wafer model to the new fully engineered 2061S, with a double fourteen 6-in. wafer chamber, GEM–SECS interface, and operation through a network-connected Sun Sparc workstation.

1. EPITAXY AT HIGH VOLTAGES

   Process requirements for high-voltage epi are as follows.

   (a) *The ability to control very high resistivity, which means extremely low dopant concentrations.* Silicon resistivity is many orders of magnitude more sensitive to dopant concentration than any technique of chemical or physical analysis can detect. One cubic centimeter of silicon contains 5×10^{22} atoms: at a resistivity of 80 Ω-cm (typical for a 1000-V bipolar transistor collector), 5×10^{13} of them (i.e., 1 in a billion) are dopant atoms. This means that the epi process gases must contain (as order of magnitude) 1 dopant atom per 1 billion silicon atoms. But, to obtain the required crystal perfection, the silicon should not exceed 2% of the total process gas amount. The device specifications require a 2% tolerance in the control of resistivity, so that the 2% of the 2% 1 ppb must be controlled in the gas phase: this means approximately 1-μm precision over the entire circumference of the earth.
   (b) *The ability to grow up to 200-μm layers at a reasonable cost* (in terms of wafers per run, maintenance frequency, machine stress, etc.).
   (c) *The ability to achieve a minority carrier diffusion length in excess of 300 μm,* which, in terms of common contaminants such as iron, means concentrations about two order of magnitudes less than the already extremely low dopant concentration.

2. EPITAXY AT HIGH CURRENTS

   The process requirements for high-current epitaxy are as follows.

   (a) *The ability to control the epi resistivity also in the presence of heavily doped substrates acting as a dopant source.* The current that can be carried by a device is limited essentially by the h_{FE} in the case of a bipolar transistor and by the voltage drop, which is a function of the device area. It is highly desirable not to waste silicon by being excessive in the area when the device is designed. So the epi wafer must not add too many voltage drops, and it is often necessary to use very heavily doped substrates, which may act as a dopant source (autodoping) hindering the control of resistivity: this requires special precautions to screen dopant evaporation from the substrate.

(b) *Avoidance of effects that can adversely influence the* h_{FE} *of the device.* Some technique for monitoring the minority carrier lifetimes and knowledge of all the possible metallic contamination sources are necessary. In some cases (mesa technologies), not only the collector but also the base of the device is obtained by epi, and thus precise control of the base layer thickness and resistivity is necessary to keep the gain in current of the final device under control.

In the next section, we outline the epi process requirements for each class of power device and cover the main issues to be considered in designing a specific epi process recipe.

II. General Considerations

1. BREAKDOWN VOLTAGE IN EPITAXIAL JUNCTIONS

Each class of epi device has a differently named breakdown voltage (BV) parameter referring to the device regions: V_{br} for diodes, BV_{bceo} for bipolar transistors, and BV_{dss} for power-MOS. But all these parameters depend directly on the same epi parameters, i.e., the carrier density and thickness. In fact, in any of the above cases, we are dealing with a single (or, in some cases, double-)-sided step junction.

While increasing the reverse bias on the uniformly doped side of a step junction, the existing carriers are pushed away from the junction, leaving a depleted zone (space charge region). The width W of the space charge region depends on the carrier density N and the applied voltage U. The electrical field will decrease in the space charge region from the value $E(0)$ at the junction to 0 at the end of the depleted region, with a slope $dE/dW = f(N)$; in the presence of an epi layer (with a uniform carrier density), $N(W) =$ constant, so also $dE/dW =$ constant.

At $E = \nabla(U)$, the applied voltage U will be $\int E(W)dW$, i.e., the area of the triangular region in the E/W plot. The electrical field at the junction, $E(0)$, cannot increase indefinitely: when a critical value $E(cr)$ (which is an increasing function of the carrier density and, for lightly doped silicon, is of the order of about 20 V/μm) is reached, breakdown occurs by an avalanche multiplication mechanism (Fig. 1). The area in the E/W plot when $E(0) = E(cr)$ is the BV.

If there is not enough epi thickness to allow for widening of the space charge region, the breakdown will still occur at $E(0) = E(cr)$, but the area in the E/W plot will be limited by the fact that when the depletion reaches the heavily doped substrate, the electrical field E will suddenly drop to zero. The triangular plot will be cut to a trapezoid (we are neglecting the very small depletion extension in the substrate), thus limiting the BV by effect of the epi thickness (Fig. 2).

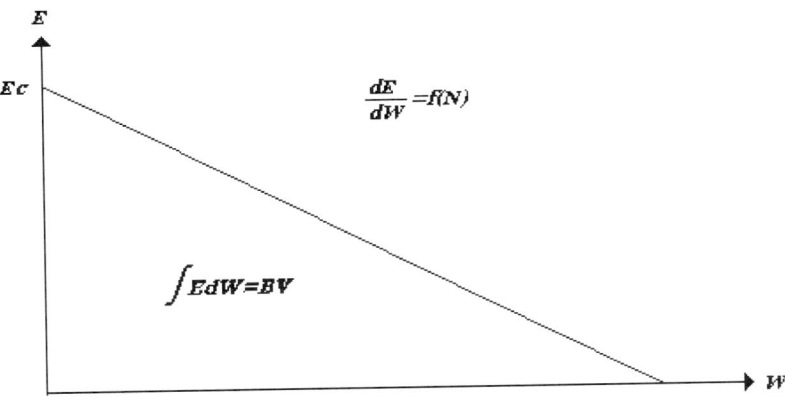

FIG. 1. Avalanche breakdown.

The above considerations about the BV of a step junction are valid for an ideal plane junction of infinite area. In the real case, at the edge of a planar junction, the critical value of E will be reached at a lower voltage, due to the curvature of the field at the edges. But for a fixed device layout, the controlling parameters for BV are the carrier density and the thickness of the epi layer.

The design of a discrete semiconductor device is always a compromise solution among many requirements. While there are infinite couples of values of epi thickness and resistivity which will result in the target BV, only certain of these couples will satisfy the other device requirements, so that the specification limits for epi parameters often must be very narrow.

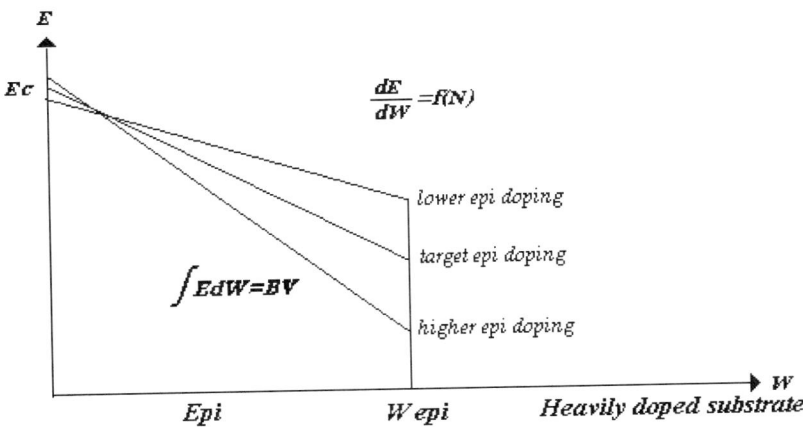

FIG. 2. Reach-through breakdown.

2. Dopant Redistribution in the Epitaxial Process

For high-current devices, in order not to add too many resistive voltage drops, a low substrate resistivity is needed: this means a very high substrate doping (very close to solid solubility). The dopant species (Sb, As, B) have precise vapor pressures, which, at the epi process temperature, are many orders of magnitude higher than the partial pressure of the related dopant gases used. This should mean that when such substrates, which act as a dopant solid source, are used, the control of epi resistivity is impossible. Fortunately, the growth rate is much higher than the diffusion speed of dopants, so that the grown silicon itself works as a screening against dopant evaporation from the substrate, in such a way that only the first part of the epi layer on top of the substrate (to the epi transition region) is affected.

Unwanted doping from the substrate (from solid to vapor and reincorporation into solid) at the very beginning of the epi layer is called "autodoping" and is often confused with out-diffusion (within a solid redistribution), which, in the epi process, due to the short duration, is often negligible and, in any case, much less important than autodoping. If there is no growth or no screening layer on the back side of a highly doped substrate, the autodoping effects cannot be screened and the substrate will still act as an unwanted doping source throughout the process. So it is necessary either to use prebacksealed substrates [usually with a chemical vapor deposition (CVD) silicon dioxide layer] or to allow for some simultaneous silicon growth on the substrate backside (*in situ* or mass-transfer backsealing). The latter technique requires the heat flow to be from the susceptor to the wafer (induction heating), and not vice versa (radiation heating).

It is very important to note that the autodoping model (Fig. 3), from both the substrate and the buried layers, cannot be implemented in any of the modeling software tools used in the design of processes/devices (such as Suprem). This is because it is a function of many local parameters, such as the heating efficiency, gas flows, shape and dimensions of the reaction chamber, position of the wafer, and number of wafers loaded, and some "exotic" parameters, such as the Reynolds, Prandtl, and Nusselt numbers.

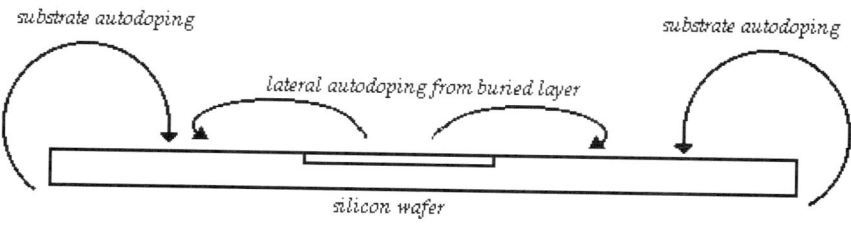

FIG. 3. Autodoping.

The best and probably the only safe approach to the simulation of processes in which autodoping can play an important role is to start with empirical data collected by spreading resistance profiling from some undoped epi growths on the real substrate. For high-voltage devices, with a high epi layer thickness, the shape of the transition region of a few microns between the substrate and the epi layer may not be significant [with the important exception of insulated gate bipolar transistors (IGBTs)], but it must be empirically determined and taken into account for low- or medium-voltage power devices.

3. Resistivity and Carrier Density

It is customary for epi layers to specify the resistivity, rather than the carrier density. This practice stems from the times when the four-point probe V/I was almost the only one available for doping evaluation. So now the designer of a transistor device still makes calculations and simulations by using carrier densities, then they are translated into resistivity to determine an epi specification. On the other hand, the epi engineer, who almost always uses an Hg-probe capacitance–voltage (CV), actually measures carrier densities and translates them into resistivity to compare them to the specification target and limits. This is often a source of misunderstandings, because quite a few correlations are available (Irwin's curve differs from the ASTM curve, which in turn differs slightly from the NBS curve, etc.). The safest practice when requiring epi wafers from different epi suppliers is to specify carrier density in place of resistivity and to make thorough use of correlation samples.

III. Specific Epitaxy Processes for Power and Discrete Devices

1. Bipolar High-Voltage Transistors

For both mesa and planar devices, a very high epi thickness (in excess of 150 μm) and resistivity (above 80 Ω-cm) are necessary. This poses many problems. To solve them it is common to use suitable epi reactors that are capable of high growth rates and of relatively long process times.

Induction heating reactors are mandatory for two reasons: (a) to prevent the wafers from sticking to the susceptor and (b) to allow *in situ* backsealing. The slight temperature gradient from the susceptor to the wafer causes the equilibrium to be more favorable for the deposition for the wafer and for the etch for the susceptor where they are in contact, so there is no encapsulation of the wafer and, if some silicon has been previously deposited on the susceptor, it is transferred onto the back of the wafers while they are processed (*in situ* backsealing) so hindering the unwanted doping effect caused by evaporation from the heavily doped substrates. This is particularly important when arsenic-doped substrates are used.

High-voltage bipolar devices have a wide base region (in forward operation, the base region may expand itself into the lightly doped collector), so it is important that the minority carriers are not subject to unwanted trapping during their transit across the base; i.e., their lifetime should be much longer than the transit time across the base, or (equivalently), their diffusion length should be much greater than the base width. Otherwise, a higher base current would be necessary to replace the unusable carriers in order to have the required collector current, i.e., the h_{FE} of the transistor would be low. So both the reactor and the facilities should be designed to avoid any trace of metallic contamination and some tools to measure and monitor diffusion lengths or minority carrier lifetimes are necessary.

The silicon source is trichlorosilane (TCS; $SiHCl_3$), because its lower decomposition activation energy with respect to tetrachloride allows lower process temperatures, with higher growth rates, better crystal quality (slip), and substantial process cost savings.

A process recipe may contain the following steps:

1. prepurge,
2. power-controlled temperature ramping,
3. final temperature-controlled ramping and temperature stabilization,
4. gas-etch (slight etch of the surface),
5. temperature stabilization at the deposition set point,
6. deposition of the first layer at a lower resistivity,
7. deposition of the second layer at a higher resistivity,
8. deposition of an eventual third layer of opposite conduction type (the base layer for mesa devices),
9. cooldown, and
10. postpurge.

After each deposition run, an etch-and-coat run to recondition the susceptor is necessary.

The measurements done on the wafers are as follows:

1. For planar structures, where the layers have the same conductivity type:
    a. Total epi thickness by Fourier-transform infrared spectroscopy (FTIR) (the single-layer thicknesses are assumed to be proportional to the respective growth times)
    b. Hg-probe CV carrier density for the outer layer (the inner layer resistivity is calculated, assuming it to be proportional to the resistivity of the outer layer according to the amount of doping)
    c. Surface photovoltage on an N^- test wafer for diffusion length assessment
2. For mesa structures, where the outer layer has the opposite conductivity type:
    a. Total epi thickness as above
    b. Four-point V/I for outer layer resistivity

c. Spreading resistance for inner layer resistivity
d. Epi junction photovoltage for diffusion length assessment

For productivity considerations and for morphological crystal reasons (edge-crowning, faceting), a total epi thickness in excess of 170 μm is unfeasible. Therefore, very high breakdown voltages are obtained by using a high epi resistivity, and these devices have a reach-through-type breakdown mechanism, so that their BV is controlled mostly by the epi thickness. The first high-doped epi layer has the purpose of increasing the device ruggedness by supplying a finite region where the stored energy of the base region can be dissipated. In consideration of edge-crowning effects related to a high epi thickness, substrates with a specially shaped edge are required (Fig. 4).

2. BIPOLAR MEDIUM-VOLTAGE MESA TRANSISTORS

This class of devices, also known as epi-base transistors, requires a double-layer epi of opposite conduction type because the outer epi layer will be the base of the transistor. So the Q_b, the total amount of carriers in the base region, must be, as much as possible, exactly known and controlled to have control of the h_{FE} of the device.

Unfortunately, while the sheet resistance can be easily measured by a four-point probe, the FTIR thickness measurement gives only the summed thickness of both the base and the collector layer, so knowledge of the external (base) layer thickness is desirable to apply the proper emitter diffusion for centering on the h_{FE} target. Thus it is necessary to apply, on a statistical basis (this being a destructive method), an angle-lapping and staining measurement of the junction depth. Both layers are grown in sequence in the same reactor run. It is obviously necessary that the reactor be equipped with double dopant gas lines and mass flow controller MFC sets, one for PH_3 (N-type) and one for B_2H_6 (P-type).

3. SMALL-SIGNAL BIPOLAR DEVICES

This class of devices, having low-thickness/low-resistivity epitaxy layers, does not require anything special besides an accurate and reliable method and instrument to measure the carrier density in the epi layer. Modern commercial Hg-probe CV meters are very suitable. The challenging part of the manufacturing is not in epi, but in diffusion, where extremely thin base widths must be obtained and where the switching speed must sometimes be controlled by very tricky gold-doping techniques. The epi growth process is straightforward, and several runs can be done between one susceptor etch and another.

In some cases PNP transistors have to be produced, therefore, P-type epitaxy on P^+ substrates is required. This can pose some boron autodoping problems, which can be overcome by using low process temperatures, *in situ* or,

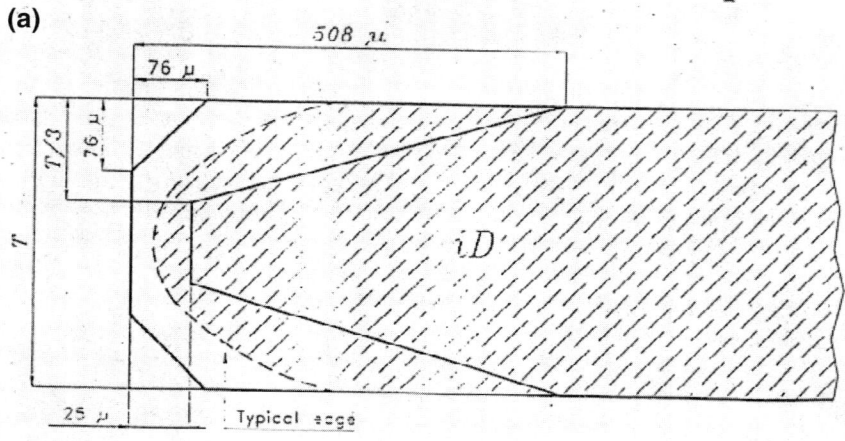
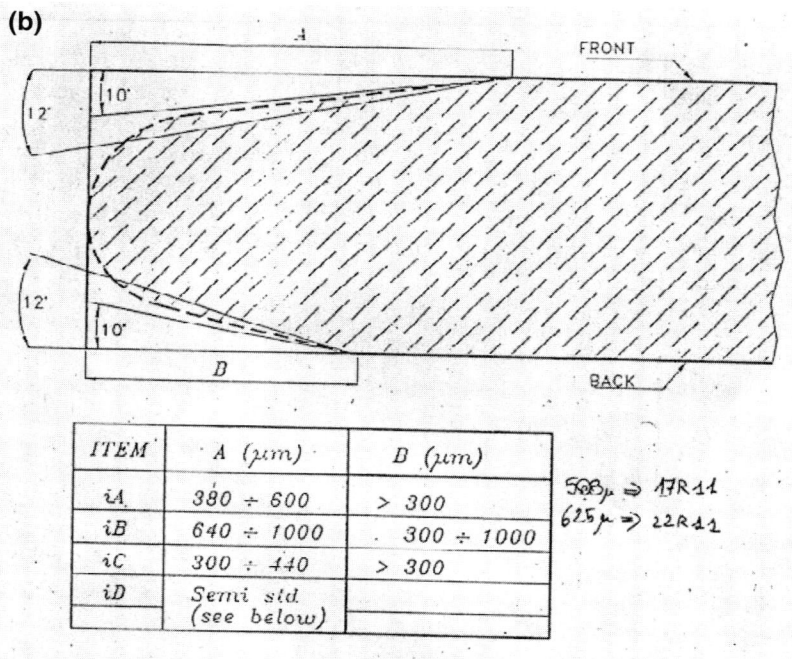

FIG. 4. Wafer edge shape: (a) round; (b) elliptical.

preferably, oxide backsealing, and (in the case of *in situ* backsealing) susceptor etching and coating between each run.

4. RF POWER DEVICES

These devices are rather similar to low-voltage power-MOS, from the point of view of their structure, as well as concerning their epi layer specifications. To obtain the minimum power loss at high frequencies, the resistivity of the epi layer must be kept as low as possible compatibly with the required BV and the thickness should be such as to allow room for no more and no less than the depletion width at breakdown. The measurements here are also the thickness by FTIR and carrier density by Hg-probe CV. Also in this case, no special epi process requirement, besides strict and accurate control of the epi resistivity, is necessary.

5. MEDIUM- AND HIGH-VOLTAGE POWER-MOS

The main point of concern in the design of a power-MOS structure is the $V_{DS_{on}}$, i.e., the voltage drop across the device when in the conduction state. This can be thought of as the effect of many resistances in series, the sum of which is the so-called R_{on}.

The components of the R_{on} are as follows:

R_{acc}: Accumulation contact resistance
R_{ch}: Channel resistance
R_{jfet}: Junction FET Resistance
R_{epi}: Epitaxial layer resistance

While in low-voltage power-MOS the R_{epi} is about 20% of the total R_{on}, for medium- to high-voltage devices the R_{epi} can represent about 80% of the R_{on}. Thus, the use of sophisticated metrology and process control techniques is mandatory, to be able to control the process at the minimum resistivity required to attain the desired breakdown voltage, because any increase in resistivity is reflected in the device $V_{DS_{on}}$. A narrow distribution of epi parameters allows designers to target precisely the specification for epi thickness and resistivity in order to reach the lowest possible $V_{DS_{on}}$ for the desired BV_{dss}.

Some tricks in profiling the carrier concentration of the epi layer can result in some $V_{DS_{on}}$ improvement, such as graded doping in the drain region and higher doping in the JFET region, as long as they do not adversely affect the process control by increasing the difficulty of process measurements.

Of course, also the substrate should not add unwanted series resistances or voltage drops, so arsenic-doped substrates (which can reach a resistivity lower than 4×10^{-3} Ω-cm) are used. The autodoping problems posed by the high vapor pressure of the arsenic dopant are overcome by *in situ* backsealing techniques, so that the use of induction-heated reactors is preferred.

6. Low-Voltage Power-MOS

For all power-MOS devices, the compromise between BV_{dss} and R_{on} is very strict, so the product can range only around a very limited tolerance. For a low voltage, the residual drain depth under the deep body region is small compared with the total epi thickness, so any eventual epi thickness nonuniformity impacts completely on this thin region: to meet the required tight BV tolerances it is mandatory to have an epi thickness as uniform as possible. Therefore, single-wafer or "minibatch" epi is almost always necessary.

Also in this case, the use of very heavily arsenic-doped substrates is necessary to attain $V_{DS_{on}}$ values as low as possible, but the lower epi resistivity requirements allow for LTO backsealed substrates instead of *in situ* backsealing.

From the point of view of crystal quality, MOS devices are, for obvious reasons, much more sensitive than bipolar devices; while bipolar power transistors are almost insensitive to defects such as stacking faults, a single stacking fault may short circuit the channel of a MOS device.

A special note must be made here about the substrate-induced defects. The body–source junction and, in general, any active region of the device, in the case of low-voltage devices, are very close to the Czochralski (CZ) substrate. It is well known that oxygen is the main impurity in CZ material, and to avoid undesired defects effects from oxygen precipitation in the substrate bulk, about 20 μm of "denuded zone" (i.e., oxygen-depleted zone) is necessary. In devices that use an epi layer, the oxygen-free epi layer itself generally constitutes the denuded zone. But in low-voltage power-MOS, the epi thickness ranges from 3 to 10 μm, according to the required BV, so that the oxygen precipitates can occur not far enough from the active regions or, due to diffusion of oxygen from the substrate into the epi silicon, inside the active part of the device, causing a very high presence of defects. For this reason, the choice of the oxygen concentration in the substrate should be made in accordance with its compatibility with the thermal processes after epi growth.

The oxygen concentration on highly doped substrates cannot be measured on wafers by a non-destructive technique such as IR absorption, because of the opacity due to the high concentration of free carriers, so some statistical process control should be applied by the substrate supplier using destructive and time-consuming measurements on some of the ingots (gas fusion analysis).

7. Insulated Gate Bipolar Transistors

IGBT devices overcome the $V_{DS_{on}}$ problem of the power-MOS, especially in the field of very high current densities, by offering, together with the very high input impedance of the MOS devices, a $V_{CE_{sat}}$ (as in bipolar devices) instead of a $V_{DS_{on}}$, i.e., a voltage drop in conduction state much more independent from the output current.

Two layers on a P^+ substrate constitute the epi structure of an IGBT: the first layer ("buffer") is N^+ and the second is N^-. The buffer layer must be as highly doped as possible, to assure an efficient recombination of minority carriers; this requires that the reactor be specially equipped with a subsidiary dopant gas panel designed for high dopant concentrations.

The epi recipe must be designed to take into account not only the heavy autodoping from the boron-doped substrate, but also the possible out-diffusion from the substrate toward the lightly doped N^- layer. In fact, if even a small amount of boron dopant reached the N^- region at any moment during epi growth or the subsequent thermal processes, the overall carrier concentration profile could be severely altered, leading to compensation or to a permanent type inversion of the deeper region of the N^- layer (the appearance of a so-called "phantom layer").

Very efficient backsealing techniques must be adopted, such as *in situ* backseal with many microns of silicon, or special LTO backseal, or epi prebacksealed substrates. In addition, the buffer layer should be thick enough not to be passed over by the diffusion of the boron substrate (which is to be considered an "infinite source") during any subsequent thermal process.

The metrology issues for IGBT epi are the same as for the power-MOS devices for the N^- layer and the carrier concentration and, despite the underlying P^+/N^+ junction, can be measured well by mercury-probe CV.

The doping of the N^+ layer can be easily measured and monitored by four-point V/I: the resistance from the upper N^- layer in the V/I measurement model is parallel that to of the N^+ layer but being some order of magnitude higher, does not affect the measurement, so that the measured V/I is practically the same as it would be measured directly on the N^+ layer.

During the development of the process, the spreading resistance must be used to evaluate and verify the amount of boron autodoping and out-diffusion. In a well-stabilized and controlled epi process, the use of spreading resistance can be limited to statistical checks instead.

8. MIXED EPITAXY DEVICES

In some cases some circuitry can be integrated in the same chip together with a power device. This poses various supplemental challenges, because it requires growing high-resistivity layers on patterned wafers, where both P^+ and N^+ buried layers coexist. This implies contrasting process requirements so limiting as to reduce the process-conditions window to almost-zero.

A classic process for integrating circuitry with high-voltage power devices implies an initial epi process that is the same as for a planar bipolar or a power-MOS device; after that, a P well is implanted and diffused; inside it, the N^+ buried layer for the integrated circuitry is formed and the wafer is processed again in the epi reactor for deposition of the collector layer of the integrated circuit; at the same time this last layer will complete the power device epi structure. This can result in difficult process control when a very high resistivity is required, because of the unwanted doping effects of the P-well boron and of the antimony (or, even worse, arsenic) buried layer on the epi silicon.

The presence of P^+ necessitates a low temperature to reduce autodoping, while to avoid pattern shifts and/or to reduce arsenic autodoping, a high-temperature process is necessary. So designing a proper process is not easy, because it includes a compromise process temperature and the choice of a "capping" technique for masking the autodoping effects; such capping techniques may be differently aimed according to the structures, the types, and the extensions of the buried layers.

Wherever pattern distortion control is the main issue, high temperatures, close to 1200°C, must be used. In that case the preferred silicon source is $SiCl_4$ (silicon tetrachloride) instead of $SiHCl_3$, because the higher activation energy of decomposition of the tetrachloride limits the unwanted depositions of silicon on the quartz walls of the reaction chamber that occur when $SiHCl_3$ is used.

9. Diodes and Thyristors

Generally speaking, the epi process for diodes does not differ very much from the process for transistors. For a high current and low voltage, the main issue is control of the V_f, which means tighter control of the resistivity and thickness in terms of epi parameters: the requirement for a low forward voltage drop also means here, as in power-MOS, setting the process targets very close to the minimum necessary to assure the required V_{br} and, consequently, good and well-calibrated metrology equipment (also here Hg-probe CV and FTIR) and good process control tools (SPC).

The epi process is also suitable for making thyristors, and in this case, also, the information reported for high-voltage bipolar transistors is applicable.

IV. Defects and Problems

As for high epi thickness and resistivity, some additional problems, of which it is necessary to be aware, may arise.

1. Epi-Crown

The epi growth rate is governed by the thickness of the laminar boundary layer. The wafer edge represents a surface discontinuity and the fluid motion of gas is more turbulent at any discontinuity (i.e., the boundary layer is thinner and the growth rate is locally higher). For a high epi thickness, the edge of the wafer may grow higher than the wafer surface. This can make it impossible to put the wafer surface in contact with the photo mask when contact or close-proximity photolithography is applied. For an epi thickness of less than 150 μm, using specially edge-shaped substrates can minimize the epi-crown (Fig. 5).

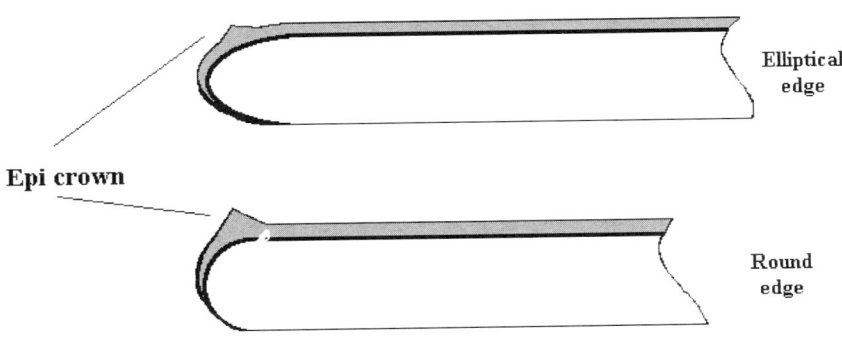

FIG. 5. Wafer edge shape and epi-crown.

2. FACETING

Epitaxial growth occurs by lateral additions of silicon atoms to the <111> plane. This means that any <111> plane existing on the wafer will increase this area while silicon is growing on the wafer. On <100> wafers, there are four points corresponding to the <111> orientation. When a thick epi layer is grown, four small facets will appear on the wafer's edge that can extend even above the wafer's surface plane, similarly to an epi-crown.

For the same reason, it is not possible to grow an epi layer oriented exactly <111>, but some degrees of "off-orientation" are needed (generally 4°). Anyway, due to edge rounding, there is always a point on the wafer edge where the exact <111> plane is exposed. During growth, that point will extend, forming a rather large facet corresponding to a true <111> plane and forming, with respect to the wafer surface, an angle that is obviously equal to the off-orientation. This facet is called the "orientation flat" and can represent a point of brittleness causing some loss in mechanical yield, correlated with the epi thickness.

3. BACK-SIDE DEFECTS

When *in situ* backsealing is applied, the susceptor quality plays an important role. The silicon carbide coating must be in the perfect state, because any out-gassing or "pinhole" will cause inhomogeneity and localized roughness, or "nodules," on the silicon backseal layer. This will adversely affect the workability of the wafers in photolithography, either because the back-side asperities could not allow the vacuum to hold in the photolithography chucks or because any back asperity, if vacuum is applied, is reported by elastic deformation to the front surface of the wafer.

The rate of mass transfer of silicon from the susceptor to the wafer back side during the *in situ* backsealing is not the same at any point: it is significantly higher

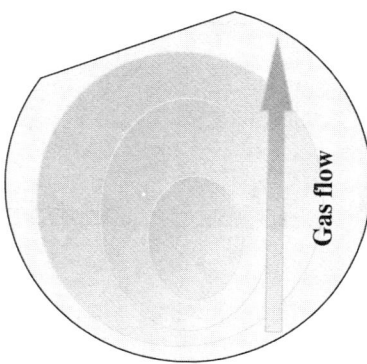

FIG. 6. "Goose egg" effect due to back transfer of silicon.

at the periphery and lower at the center, due to the diffusion path of chloride ions, acting as transport agents, from the periphery to the center. So if the process time is not long enough for a complete transfer, some silicon can remain on the susceptor in correspondence with the center of the wafer, while the mass transfer is already complete in the periphery. This results in an abnormal TTV that also can severely disturb the photolithography processes (Fig. 6).

V. Accessory Considerations

In epi in power devices, very large quantities of chlorosilanes and hydrogen chloride (for etching the silicon deposited on the susceptor) are employed, with chemical yields of the reactions rather low, so that the process effluents carry large amounts of hydrogen chloride and chlorosilanes. This poses special problems concerning environmental protection, and no epi facility should be started before having solved them. In the absence of adequate exhaust treatments, in a short time the whole facility would be seriously damaged by external corrosion, not to mention the ecological damage. So a highly efficient and specialized system for exhaust gas scrubbing is mandatory, and no generic acid fume treatment will work, because the hydrolysis product is silica, which will quickly clog any simple absorption column.

So the system must be specially designed with a preabatement of silica, followed by a multistage absorption column. The efficiency of the abatement must be as high as possible, considering that the effects of chloride corrosion are cumulative, and even with a 99% efficiency 1% of HCl would still be emitted (e.g., 10 tons of HCl used per year is 100 kg of emission in 1 year!). Obviously 99% efficiency is inadequate and at least 99.99% efficiency is necessary. And a system of sensors for continuous monitoring of scrubbers efficiency is also highly desirable.

Another issue, when high amounts of silicon have to be frequently etched from the susceptor as in power device epi, is the handling of anhydrous hydrogen chloride. The properties of liquid anhydrous HCl in terms of material compatibility are not as well known as those of gaseous HCl or of HCl solutions and are very different from them. What should be absolutely avoided is the condensation of liquid HCl in any part of the delivery system; and condensation is very probable (because of the Joule–Thompson effect) at the pressure regulator, which, unluckily, is the part most likely to be damaged by liquid HCl. To avoid this, the storage temperature of the HCl bottles should be kept significantly lower than the ambient temperature (this will lower the pressure and reduce the amount of pressure drop at the regulator), and if possible, the pressure regulator should be supplied with an amount of heat at least equal to the adiabatic expansion heat of the delivered HCl.

CHAPTER 9

Epitaxy on Patterned Wafers

S. Acerboni

ST MICROELECTRONICS
CFM-AG1 DEPARTMENT
AGRATE BRIANZA, ITALY

I. INTRODUCTION	295
1. *What Is a Pattern on an Epitaxial Wafer?*	296
II. DEVICE REQUIREMENTS AND PROCESS COMPLEXITY	300
1. *Bipolar, Bipolar/C-MOS, and Mixed Technologies*	300
2. *General Comparison between Bare and Patterned Epitaxial Wafers*	303
3. *Operations and Epitaxial Equipment*	306
III. SURFACE PREPARATION: PRE-EPITAXIAL CLEANING	306
IV. GEOMETRICAL PATTERN INTEGRITY	308
1. *Pattern Shift*	309
2. *Pattern Distortion and Washout*	312
3. *Evaluation of Geometrical Pattern Integrity*	313
V. DOPING PATTERN INTEGRITY	316
1. *The Autodoping Problem*	316
2. *Doping Profile in The Presence of Autodoping*	318
3. *Autodoping Characterization*	321
4. *Buried Layer Parameters Affecting Autodoping*	322
5. *Recipe Parameters Affecting Autodoping*	324
VI. CRYSTAL DEFECTIVITY	327
1. *Thermal Treatments Before Epitaxial Growth*	329
2. *Lattice Damage and Recovery After Implantation*	332
3. *Dislocations, Slip-Lines, Epitaxial Stacking Faults, and Others*	335
VII. CONCLUSION: THE BEST EPITAXIAL RECIPE?	343
BIBLIOGRAPHY	343

I. Introduction

When we talk about the epitaxy (epi) process, we mean a really complicated miracle of chemistry, something that we compare with the building of the Egyptian pyramids. Epitaxy on bare single-crystal wafers is a glamourous process, but epi on complicated structures, such as a patterned, multiimplanted wafer, is a real challenge for the process engineer. When I was introduced to this work, it was incredible to think of all the requirements necessary to realize a good-quality epitaxial (epi) layer on complicated structures, which may induce a bad doping profile, or dopant compensation, or a high defectivity. Moreover, a good-quality (epi) layer must be grown in a repetitive and reliable manner.

This chapter describes some typical production problems I encounter every day in processing thousands of patterned epi wafers through their final characterization. Using a photographic reportage approach, I report typical problems and solutions (sometimes learning from production disasters) using daily data from production lines. My intention is to provide a general "atlas" of epi on patterned wafers.

1. What Is a Pattern on an Epitaxial Wafer?

First, what is a pattern? Why do we have to grow an epi layer on a patterned, structured silicon wafer? How do we obtain it? To answer these questions we start with the general requirements for the fabrication of silicon devices.

Silicon substrates used to manufacture bipolar integrated circuits, bipolar and C-MOS, and general mixed technologies are usually patterned (by masking). Then they are implanted and diffused prior to epi to create an array of heavily doped regions (the subcollector and isolation regions). Under these areas vertical transistors are built after epi layer deposition just into the epi silicon layer.

Figure 1 shows that the pattern can be easily inspected with a Nomarski optical microscope. Using different gray shades, we can make two situations evident at the surface, where the big square or rectangular regions are depressions

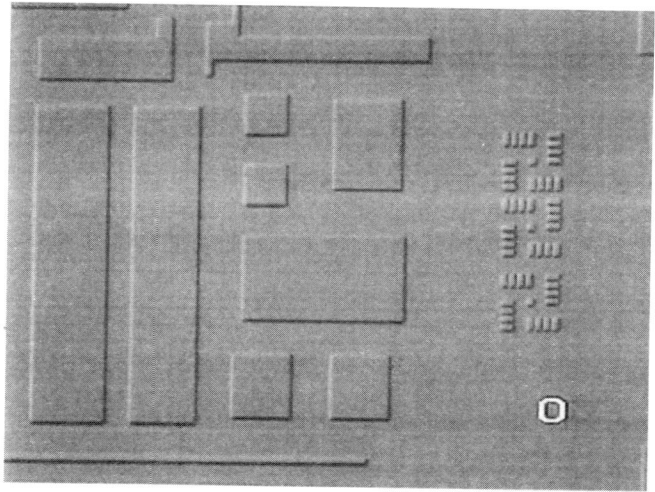

Fig. 1. Top view of a patterned silicon surface before epi growth. The square and rectangular shapes are depressed areas (about 2000 Å) compared to the surface plane. Using the Nomarski effect (or interference-contrast mode) makes the depressed areas evident. Original magnification, ×200.

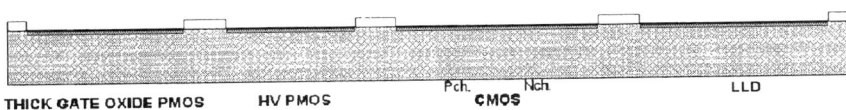

FIG. 2. Cross section showing the first operations on silicon to realize patterned wafers for generic mixed technology: from starting material, substrate oxidation, buried layer mask and oxide etch, and, finally, buried layer implantation.

(1000–1500 Å) under the surface. Of course, this pattern structure is similar to a stepped surface with wells in the silicon substrate in the proximity of implanted and diffused zones for buried layers. This is usually done by thermal oxidation of the substrate (Fig. 2).

The oxide is then patterned using masking and etching operations, and the opened areas are implanted with arsenic, antimony, or phosphorus to obtain N-doped regions and with boron to obtain P-doped regions. Then a high-temperature thermal diffusion step takes place to distribute the dopant (Fig. 3) called "drive-in." During the drive-in of the implanted dopant, some oxygen is used to grow a thin oxide on the exposed silicon surface. This helps to seal the surface to reduce dopant evaporation from the surface during drive-in, to accelerate drive-in, and to produce a surface step depressing the patterned area 50 to 1500 Å below the surface of the substrate. The longer the diffusion/oxidation step lasts, the deeper the depression is. Silicon consumption during oxidation makes the area above the

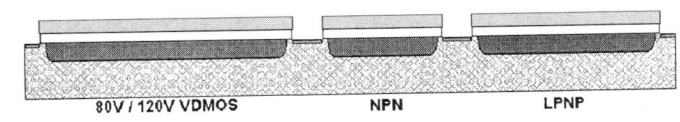
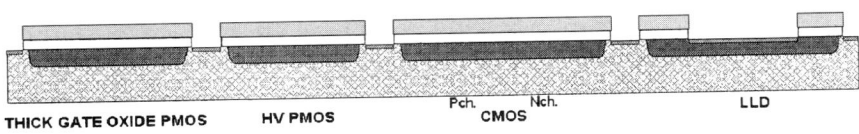

FIG. 3. Cross section (like Fig. 2) after the first buried layer diffusion (see brown wells) and after a second buried layer mask, typically P^+, and oxide etch. The yellow regions are the photoresist to define the "screen" oxide for the second buried layer mask.

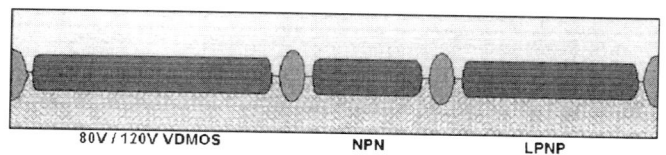
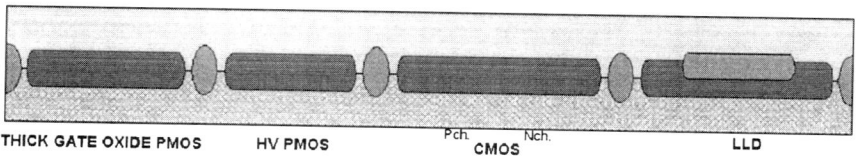

FIG. 4. Cross section of generic mixed technology after epi growth (pink region). The diffusion of both buried layers into the epi layer is evident.

buried layer areas depressed with respect to the original surface. This occurs only at local oxidation areas during the diffusion operation.

After dopant diffusion, when the surface masking oxide (sometimes Si_3N_4 is used as the masking layer) is removed, the pattern of the diffused area can be inspected using interference contrast optical microscopy, which produces light/dark contrast lines at small surface steps as shown in Fig. 1. These locally preferentially doped areas will be buried layer regions after epi growth (see Fig. 5). The surface steps, previously generated in the buried layer diffusion steps, can still be seen as surface steps on the epi layer (Fig. 4).

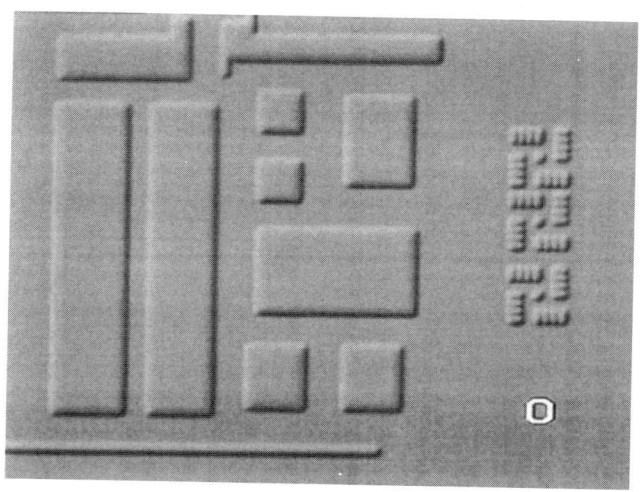

FIG. 5. Top view of the same patterned silicon surface as in Fig. 1, after 10-μm epi growth. The difference from Fig. 1 is evident in terms of border definition induced by epi growth. Original magnification, ×200.

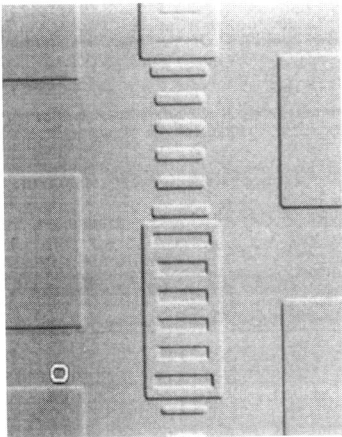

FIG. 6. Typical alignment marks for stepper lithography equipment, before epi growth. The vertical profile of the walls after epi growth is critical for alignment equipments. Original magnification, ×200.

The presence of this pattern is the crucial point. We cannot avoid these particular structures. The problem is to reproduce these surface steps properly after epi growth in terms of proportioning, positioning, and sharp shape. These characteristics constitute the "pattern integrity." We have to preserve this pattern integrity on particular structures called "alignment marks" (see Fig. 6), which are used as a reference to align the postepi layer mask over the mask used before epi (last buried layer mask).

The alignment mark is usually built onto the silicon substrate by masking the first thick oxide (4000 Å) using the first mask after the first oxidation. Shapes and dimensions are functions of epi thickness and alignment equipment requirements. Sometimes, the alignment mark is made directly into the substrate at the beginning of the process, by masking and etching the silicon substrate itself. This technique is called "zero mask" and it can be a good solution when the bulk device does not require a starting oxidation step.

Now we know what the pattern structure is and how it is built in the silicon substrate. But how do we recognize and evaluate the pattern before and after epi growth? The standard procedure is to use an *optical microscope,* which is one of the most important analytical tools for monitoring the epi process to understand what is occurring in real time during the manufacturing sequence. In semiconductor wafer processing applications, the most important features of the optical microscope are the resolution, magnification, and mechanical stability and a wide field of view; in patterned wafer applications the *bright field* and *dark field,* important is the *Nomarski interference contrast* capabilities. The *bright-field* image is formed by reflection of the light received by the sample as shown in Fig. 7a. Figure 7b shows a *dark-field* image. The light is directed onto the wafer at angles outside of the

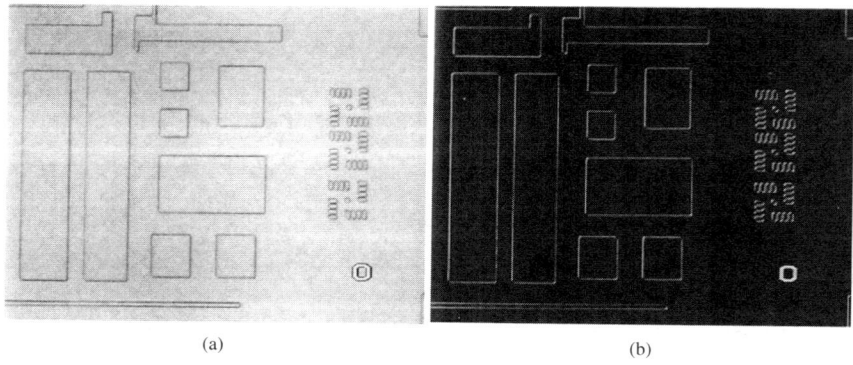

FIG. 7. Bright-field (a) and dark-field (b) effect of Fig. 1, without Nomarski contrast.

cone that the objective encompasses, so that the light strikes the wafer obliquely. Only the light that is reflected or refracted by some features on the wafer surface is collected by the optics. The sample appears with a black background, while the features that reflect or refract the light appear bright. This means that dark-field illumination enhances the visibility of details normally washed out by bright-field illumination: small details below the resolution limit. This makes the technique useful for scanning a wide viewing field to find particles, scratches, dislocations, and general defects on the surface.

Under illumination in the Nomarski interference contrast mode, features on a substrate which are at different relative elevations appear as different shades of gray (see Figs. 1–3 and 5). Such a contrast is achieved by splitting the primary illuminating beam into two beams. The two beams strike the stepped surface of the wafer a short distance apart. Then they are reflected back into the microscope and reconstructed. The optical path lengths of the two beams will be different if one of them encounters a step or a change in the index of refraction, caused by a phase boundary not encountered by the other beam. These differences in optical path length produce a microscopic image that contains a contrast effect (interference contrast effect).

II. Device Requirements and Process Complexity

1. Bipolar, Bipolar/C-MOS, and Mixed Technologies

Now we know that the pattern is a buried layer array under the epi layer. A further step is to understand the historical evolution that introduced the patterned wafer approach. Starting from discrete and power devices, silicon epitaxy was developed to enhance the electrical performance of discrete bipolar transistors. The technology change from junction transistors to diffused planar structures in the early 1960s produced a need for a material structure not achieved by blanket diffusion of dopant from the surface. The breakdown voltage (BV) of discrete transistors was limited by the field avalanche BV of the substrate material.

The use of higher-resistivity substrates produces a higher BV despite the poor performance caused by the increased collector series resistance. The structure needed was a thin, lightly doped, single-crystal layer of high protection placed over a more heavily doped substrate. In this structure the more heavily doped substrate material reduces the collector series resistance, while the base collector BV is governed by the lighter doping in the near-surface region. We can achieve this type of material structure with the epi deposition of a lightly doped P or N layer on an N^+ or P^+ substrate.

Breakthroughs using epi layers are the accurate control of doping levels plus other advantages deriving from a low oxygen and low carbon content. Nowadays, multilayer epi structures are also used for discrete high-voltage or high-power products.

The development of planar bipolar integrated circuits introduced the requirement to build devices on the same substrate to obtain electrical isolation. The use of an epi layer having opposite doping with respect to the substrate met part of that requirement. The completion of the device isolation was obtained by the diffusion of isolation regions through the epi layer to contact the substrate between active areas. A good compromise between epi thickness requirements (e.g., $10-15$ μm) and layout geometrical constraints is "top–bottom isolation," in which the bottom is implanted before epi, and the top is implanted after epi growth and is connected with the bottom by thermal diffusion. In bipolar devices the junction breakdown (BV) is a very important parameter, which increases with epi resistivity and thickness.

In planar bipolar circuits, it is common to employ a heavily doped implanted and diffused region under the transistor, something like the subcollector area under the epi layer, usually called the buried layer (BL); this structure is used to lower the lateral series resistance between the collector area below the emitter and the collector contact.

In mixed technology bipolar/C-MOS/D-MOS (BCD), the N^+ buried layer also acts as a collector structure for vertical D-MOS (see Fig. 8) and the P^+ buried layer acts as a bottom isolation region to be connected with the P well after epi to be the isolation structure (see Fig. 9).

General benefits of bipolar devices are the simplified isolation, independently controlled dopant profile, higher switching speeds, and improved high voltage with linearity characteristics. Benefits of MOS ICs are the lower diffused line capacitance, better diffused line charge retention, better control of spurious charge, improved DRAM performance, and C-MOS latch-up protection.

The necessity of mixed bipolar/C-MOS, so-called BiCMOS, technology has recently emerged as a key step in solving the traditional pure C-MOS process limitations. For example, especially in circuits with a very high frequency and speed, it is important to solve problems such as noise margin, drive capabilities, and

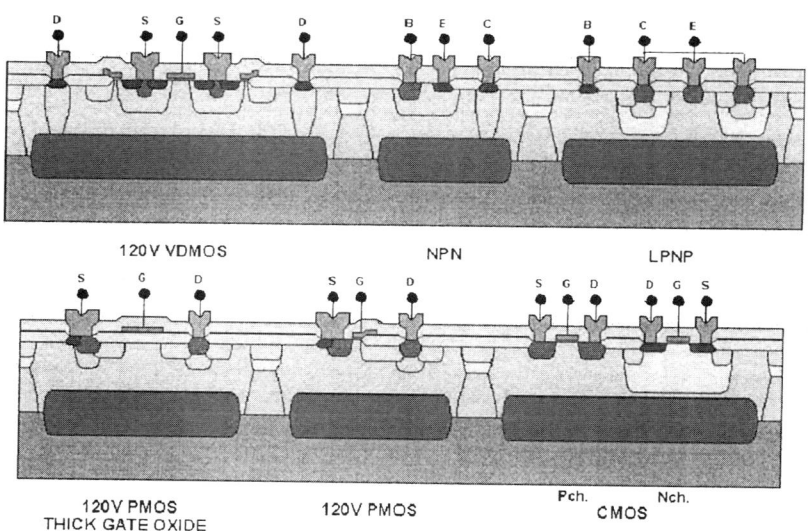

FIG. 8. A first-generation bipolar/C-MOS/D-MOS technology cross section with several devices realized in the epi layer (orange; such as a "device tank").

latch-up. The merging of a few steps in a pure C-MOS process, which represents the fundamental baseline of the new BiCMOS process, permits better performance just by adding a few mask levels (five or six levels are added in more recent process versions). The improvements in circuit performance, not only electrical but from a reliability standpoint, balance the increased costs of that new process. The crucial step in transforming a pure C-MOS into a BiCMOS process is epi growth.

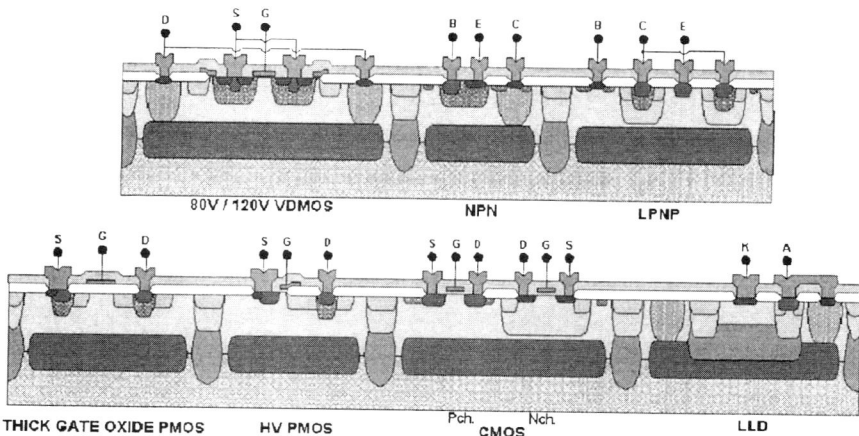

FIG. 9. A second-generation bipolar/C-MOS/D-MOS technology with N^+ and P^+ buried layers (brown and violet wells), both implanted before epi growth. The epi layer is the pink area.

We usually need an N epi layer, for three reasons.

(a) To drive the BVs of bipolar transistors to satisfy design requirements. We usually fix the BV for the standard n–p–n device using the epi layer as the collector.
(b) To eliminate MOS latch-up problems with the aid of an N-well layer implanted and diffused inside the epi.
(c) To maintain a good current drive in the transistor, with increased doping compared with the pure N-well pocket situation, as we normally have in a pure C-MOS flow.

Usually, for all the previous requirements the resistivity specification of the epi layer should be about 1 Ω-cm. The epi process can introduce limiting factors in packing density for the devices inside a chip. This is due to the necessity to isolate the epi pockets by lateral P diffusions and to avoid the introduction of parasitic capacitances due to the junctions toward the substrate, which act as a speed degradation factor.

A number of particular features are normally required for epi for BiCMOS processes. We can summarize the main features as follows.

(a) The epi layer must be very thin (less than 1.5 μm for high-speed BiCMOS).
(b) The epi resistivity must be very uniform throughout the layer. This is very difficult due to the preimplanted P^+ and N^+ buried layer autodoping effects. A direct consequence of this problem is the low-pressure/low-temperature growth conditions normally adopted in BiCMOS processes.
(c) The wafer flatness after epi growth must be maintained as tightly as possible to assure masking capabilities for processes with a minimum size lower than 0.35 μm. This is not possible if the back of the wafers is contaminated at even a few points by silicon/particles.
(d) The control of particles on the front of the wafers must be tight to avoid the presence of defects, much tighter than in a pure C-MOS process. Therefore, the defect density must be very low (a few defects, with dimensions of about 0.15 μm, on the whole wafer surface).

2. General Comparison between Bare and Patterned Epitaxial Wafers

As reported previously, the history of the wafer does not begin just before the epi process; there is an earlier history, with many thermal steps, mask operations, and implant operations. All operations can impact the defectivity, and we analyze this later.

We focus on the pattern presence because the doped regions are both a discontinuity in the geometrical silicon surface and a local discontinuity of the dopant uniformity. This can influence the doping profile of the epi layer. In this section we introduce the main characteristics of the epi process on patterned wafers, including an epi recipe declinated on the presence of the pattern. We illustrate the deep impact of the pattern structure, for its presence but also for the role of different operations that contribute to its realization on the silicon wafer.

The basic steps in an epi recipe are the temperature ramp-up, the pre-epi hydrogen bake, the silicon deposition, and, finally, the temperature cooldown. The chamber and susceptor etchings are managed in a separate run in batch reactors or within the growth recipe in single-wafer reactors.

a. *Temperature Ramp-Up in H_2*

This is performed after the room-temperature nitrogen purge of the process chamber and it is different for single-wafer and batch reactors.

1. In batch-type reactors using pyrometer temperature readings, after wafer loading there is a double hydrogen ramp: the first is from room temperature to 900°C; the second is from 900 to 1150–1200°C (depending on the silicon source). For 150-mm wafers, the temperature rate is 20–25°C/min.
2. In single-wafer reactors the wafer is loaded at 700–900°C (depending on the loading system), in hydrogen, and the ramp-up occurs up to 1100–1200°C at a very high rate: 5°C/s (for 150-mm wafers).

b. *Hydrogen Pre-epi Bake*

This step is used to remove silicon oxide before epi growth by H_2 reduction. The temperature must be the highest possible to reduce the surface defectivity and to anneal residual implant damage. During the bake step the previously locally introduced dopant (buried layer) can evaporate from the surface and/or diffuse without control throughout the wafer, outside the pregenerated pattern. These effects are different for different dopant species. This step must be the best compromise in terms of *temperature, pressure,* and *step length* to avoid mixing among the locally doped buried layers and the future epi layer (autodoping problem). This is the most critical of all the epi process steps.

c. *Silicon Deposition*

This is the same as for bare wafers, but a higher temperature is used to preserve the geometrical pattern integrity and with the same constraints as for the pre-epi bake step. During the deposition step, the epi film reproduces the pattern morphology layer by layer. When the deposition is completed, the pattern is "translated" at the surface. Depending on the pattern characteristics, the deposition step can be single or double (or more). In the latter case, the first step occurs at a lower growth rate and temperature to preserve the pattern integrity for boron-doped buried layers and at a lower growth rate and higher temperature for arsenic-doped buried layers.

If both boron- and arsenic-doped buried layers are present, a compromise between the two situations must be found.

The peculiarities of epi on patterned wafers versus bare wafers are summarized in Table I. From the epi recipe description and Table I, we can highlight four macro areas that impact the epi process on patterned wafers and on which we must concentrate our attention to create a good-quality epi layer:

TABLE I

EPITAXY ON BARE VERSUS PATTERNED WAFERS

Epi on bare wafers	Epi on patterned wafers
1. *Applications:* Discrete ICs, power MOS (epi, 10–100 μm), C-MOS (3–7 μm)	1. *Applications:* Integrated bipolars, BiCMOS, HFCMOS, BCD (bipolar/C-MOS/D-MOS)
The epi layer is lightly doped over a heavily doped substrate.	The epi layer is lightly doped over a "stepped" surface with several implanted and diffused areas which are heavily doped with different types and species of dopants.
2. *Pre-epi surface treatment* • Standard RCA (optional) + scrubber (optional) *in situ* etch or high-temperature H$_2$ bake in the epi reactor before deposition	2. *Pre-epi surface treatments* • *HF to remove pre-epi oxide* (the oxide is grown during several buried layer diffusion steps before epi growth) • Standard RCA + scrubber • Often without *in situ* etch: to be avoided for crystal defects on implanted structures and for low-energy implantation (there is a risk of totally removing the buried layer structures) • H$_2$ high-temperature bake, paying attention to autodoping problems
3. *Productivity:* High Very high growth rate, the limit being only the mono/poly transition	3. *Productivity:* Low Lower growth rate for pattern integrity and autodoping issues
4. *Defectivity:* Low Less prone to defects coming from previous steps	4. *Defectivity:* High Prone to formation of defects introduced during the several pre-epi process steps in high-temperature furnaces (warpage and dislocation problems, ESFs, slip-lines) and during implant operations (dislocations and general lattice damage). Defects also come from masking and etching operations. The risk of scratches is higher due to the several handling steps.
5. *Low-temperature epi* Less risk of slip-lines and wall deposition	5. *High temperature epi* For pattern integrity and to rebuild the single crystal after implant damage not completely recovered by pre-epi anneal. There is a risk of silicon deposition on the walls of the reactor chamber.
6. *Atmospheric-pressure epi*	6. *Reduced-Pressure epi* For pattern integrity in terms of doping (arsenic autodoping) and geometrical issues

- Wafer surface preparation
- Pattern integrity preservation in terms of geometrical issues
- Pattern integrity preservation in terms of doping
- Defect minimization, when introduced before and during the epi process

3. Operations and Epitaxial Equipment

The equipment dedicated to epi growth on patterned wafers is the same as that normally used for epi on bare wafers. The pattern presence requires several "loop" operations before epi, as follows:

> Layer deposition or thermal growth → MaskingMask etching
> → Dopant implanting → Wafer cleaning
> → Crystal annealing and dopant thermal diffusion

Each loop will generate a specific buried layer in the silicon substrate whose total area on the wafer is delimited by the mask.

Regarding epi growth, a standard approach is to use barrel reactors for high-thickness epi wafers, up to 150 mm in diameter, and single-wafer reactors for low-thickness wafers, up to 200 mm in diameter, depending on the epi layer requirements. Generally speaking, for manufacturing epi Fabs, the choice depends on several parameters, the most important being the reactor productivity and the device sensitivity to the formation of defects. There is not one "ideal epi reactor" for all processes; if one is better for doping control, another is better for defectivity.

The use of epi reactors for patterned wafers started in the 1970s with linear, horizontal, RF-heated pancake reactors. With the increase in wafer diameter from 100 to 125 mm in the beginning of the 1980s, linear reactors showed some weaknesses, as they were sensitive to the high defect presence on the epi layer. The problem was the front-to-back temperature gradient for RF-heated reactors and the center-to-edge temperature gradient for lamp-heated reactors. The lamp-heated barrel reactor was normally used for C-MOS technologies without patterns until the advent of 125- and 150-mm diameter wafers.

Our starting experience was with epi growth on masked wafers using two types of reactors: barrel reactors with lamps and barrel reactors with RF coils. We moved to the single-wafer concept at the beginning of the 1990s. We think that the 200-mm-diameter wafer is the limit for batch reactor systems; the future for patterned wafers leans toward single-wafer reactor generation.

III. Surface Preparation: Pre-epitaxial Cleaning

When a patterned wafer is incoming for epi growth, it usually presents a thick silicon oxide layer generated during the previous thermal steps (implanted dopant

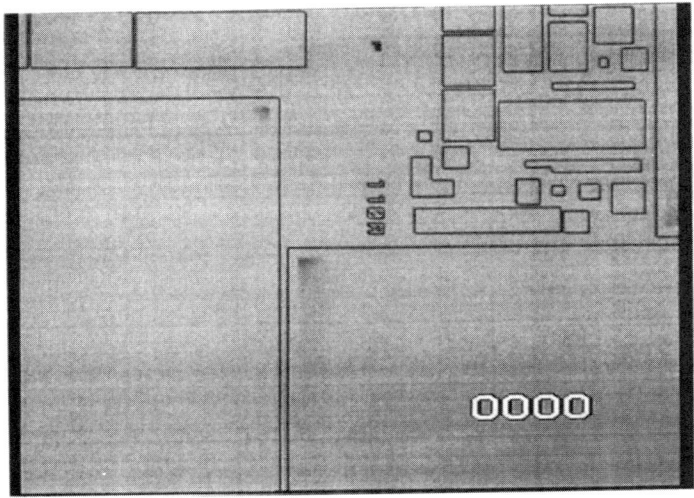

FIG. 10. Local staining during HF oxide removal. The angle stained point can be the source of epi defects after epi growth.

annealing and diffusion as well as oxidation to produce an oxygen-free zone near the surface, called the denuded zone). To grow a perfect monocrystalline layer we must start the epi process with a completely oxide-free surface. Diluted or concentrated HF (40%) etching solutions are used to obtain a totally hydrophobic (i.e., oxide-free) stepped surface. But some residual oxide coming from the oxidant reagents that compose the cleaning solutions will be removed during the pre-epi bake by H_2 reduction. During the wet cleaning, the HF can locally overetch and "stain" the pattern over doped regions, as shown in Fig. 10.

The pre-epi wafer cleaning sequence is similar to a prefurnace cleaning sequence, sometimes including a wafer scrubber or megasonic treatments, to remove particulate, organic, and inorganic contamination (such as heavy metals and sodium) from the front and back surfaces of the wafer. The cleaning sequence is based on an alkaline step with a diluted NH_4OH/H_2O_2 (called the SC1 solution) mixture, followed by an acid step with a diluted HCl/H_2O_2 (called the SC2 solution) solution. The SC1 step can be associated with megasonics. It is responsible primarily for particle and organic residual removal: the H_2O_2 oxidizes the surface and the ammonia removes the "dirt" oxide, which includes the particles present on the surface. The SC2 step is responsible primarily for metallic contaminant removal, by complexing the metallic impurity with Cl^- ions.

Often such cleaning procedures are not sufficient, and an additional *in situ anhydrous HCl* (1% or less HCl in H_2) *etch* at a temperature higher than 1100°C is required to remove any residual contaminants from the surface. Care must be taken to prevent the removal of a significant thickness of the silicon, previously locally doped, substrate. As the etch rate is elevated at high temperatures (200–600 Å/min, depending on the equipment used), the risk is of removing the buried layer regions completely, mainly for low-energy implanted dopants with a concentration peak near the surface.

It is mandatory to use high-quality/high-purity HCl, but this is a general requirement for all gases in the epi process. In the past it was believed that, for good bipolar device quality, epitaxial layers could be prepared only with an *in situ* HCl etch or in the presence of some Cl^- species during the deposition. However, a recent study, also supported by our experience, has shown that excellent device-quality epi material can be prepared on wafers that have received only a pre-epi cleaning.

Our experience with "HF last"—with a fast dip in HF immediately before a low-temperature-bake epi growth—has shown that using SiH_4 gas for deposition gave bad results in terms of process defects without a 1050–1100°C bake before a low-temperature (900°C) deposition.

It is sometimes standard procedure, immediately before epi growth, to perform a chemical–mechanical treatment using scrubber technology, with DI water, on the preclean wafer, sometimes adding diluted acid or a tensioactif solution. This pre-epi scrubber treatment can assure a high quality regarding defectivity after epi film deposition when pre-epi cleaning is not enough.

Finally, many problems can be seen on patterned structures after epi growth, such as incomplete oxide removal, N^+ or P^+ doped region staining with concentrated HF, and NH_4OH overetch of the overdoped areas during the SC1 cleaning step. All of them can induce general roughness or crystallographic defectivity (epi stacking fault generation) or, in the worst case, polysilicon.

IV. Geometrical Pattern Integrity

After the pre-epi cleaning step the epi process is performed using the general sequence described in Section II.2. The first issue is to preserve geometrically the pattern characteristics created on the substrate before the deposition of the epi layer. While theoretically the epi layer will exactly reproduce on the top surface the same geometry as underneath, in reality several problems can arise, connected with the surface of the substrate and/or the epi recipe. How to control and minimize them to optimize the mask alignment after epi growth is a main issue for all epi process engineers.

During epi growth, while the steps at the edge of the patterned areas generally propagate upward, lateral shifts of the geometry from the original position do occur. These shifts lead to shift and distortion of the pattern on the top of the epi layer. When the left and right parallel edges of the pattern geometry (for example, a square depression of a buried layer) shift in the same direction by an equal amount,

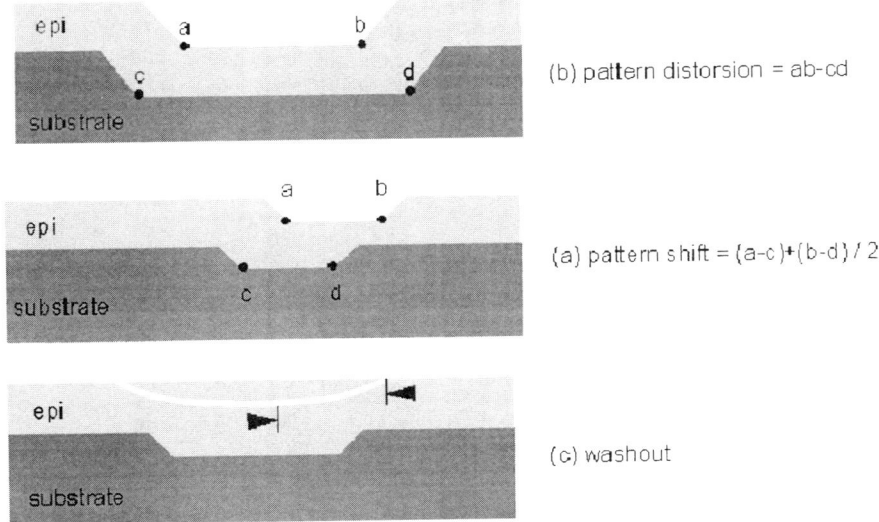

FIG. 11. Schematic drawings illustrating pattern distortion (a), pattern shift (b), and washout (c).

a simple shift is produced, called a *pattern shift* effect. As shown in Fig. 11, we can define the pattern shift value by

$$\text{Shift} = (a - c) + (b - d)/2. \quad (1)$$

However, the atomic structure of the left and right sides of the depression is different and may lead to changes in the values on the left and right sides of the depression. This difference leads to a change in the spacing between the sides of the depression (see Fig. 11). This size change is known as *pattern distortion*. Extremely hard pattern distortion, with flattening of the steps, is known as *washout*. Different studies have performed.

Generally the degree of shift is substrate connected, while distortion and washout are influenced by the process parameters (deposition temperature, pressure, and growth rate), by the silicon source (SiH_4, SiH_2Cl_2, $SiHCl_3$, and $SiCl_4$), and by the reactor itself (heating and process chamber type).

1. PATTERN SHIFT

It is normal experience to correlate the direction of pattern shift with the crystallography of the substrate wafer. For (100)-oriented wafers the edges of rectangular buried layers are generally oriented parallel to the surface traces of (111) planes. The four edges of a rectangular depression are equivalent and move together. Figure 12 shows a pattern shift effect after chemical etching from the top view, with evident misplacement in the upper direction. Therefore, for a wafer accurately

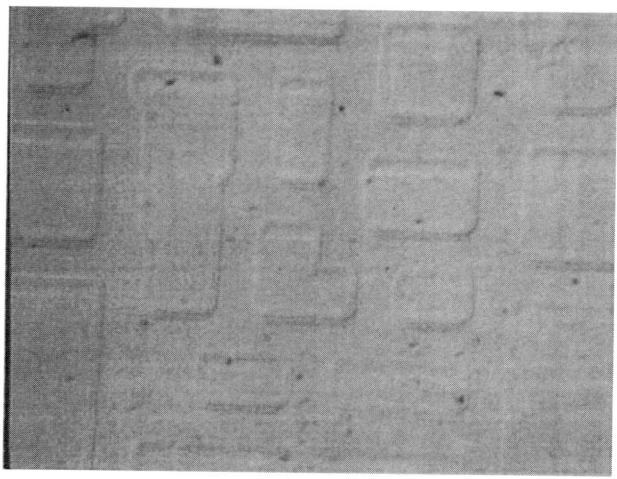

FIG. 12. Pattern shift effect image after chemical etching, with evident misplacement in the upper direction (view from top).

cut on the (100) plane, the shift will be small and the pattern may only expand or contract, with distortion and washout connected only with the epi recipe. Although the pattern shift is generally small for accurately oriented (100) substrates, it can be significant if the substrate is slightly misoriented or if epi growth is done at low rates and low temperatures.

For the (111) orientation, pattern shift is parallel to the wafer's major flat in a direction opposite the approximately 3° tilt off the exact (111) plane. For (100) substrates, the edges of the pattern parallel to the wafer's major flat move in opposite directions, contributing a size change (distortion) but little or no shift.

Pattern shift and distortion are lowest for silane-based epi growth chemistry and increase with increasing chlorine in the family of chlorosilanes: lowest for SiH_2Cl_2, for moderate for $SiHCl_3$, and highest for $SiCl_4$. Pattern shift is reduced in higher-temperature growth. Use of reduced-pressure epi growth can decrease pattern shift for (111) substrates. On (100) silicon substrates, the pattern shift is higher for higher epi thicknesses.

When problems arise from (100) substrate misalignment, due to an error during the ingot cutting operation, a 1° tilt off the (100) orientation is enough to create misalignment from the pre-epi pattern to the postepi mask, and the whole process is seriously compromised, mainly for high-thickness epi layers (more than 8–10 μm).

For (111) oriented wafers, the shift is a consequence of imposing square or rectangular depressions on a surface having basically threefold symmetry. Two edges of a rectangular pattern parallel to the wafer flat lie parallel to a mirror-image symmetry plane and will move by a similar amount but in opposite directions. This provides a distortion or size change for the two sides of a rectangle parallel to the

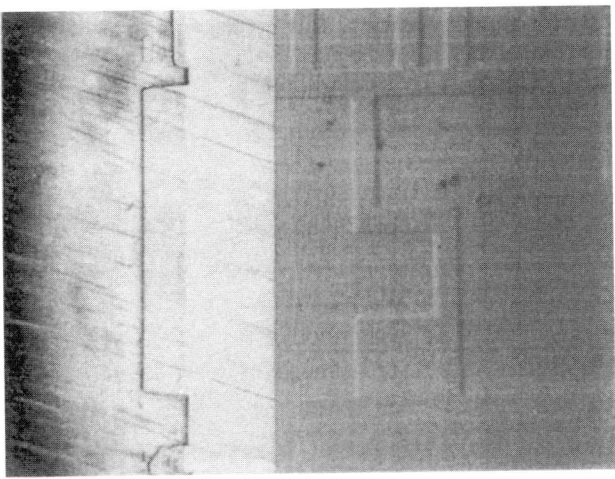

FIG. 13. Pattern shift effect in a (111) substrate on a lapping section with evident buried layer misalignment (top-left side of the image).

wafer flat. However, for the two sides perpendicular to the flat the sidewalls edges will be different. Under atmospheric-pressure epi growth the two edges will shift by similar amounts in the same direction, leading to a shift. Figure 13, shows a pattern shift effect on a (111) substrate on a lapping section with evident buried layer misalignment (top-left side of the image). As we know, epi growth on (111) Si masked substrates can shift the geometry of the device; this effect can lead to misalignment of subsequent masking steps and, if severe, to shorting of normally isolated regions in the device.

By definition, pattern shift is a displacement of a surface feature relative to the underlying buried layer. Two quantities allow us to measure it.

Absolute pattern shift:

$$\text{APS} = (a - c) + (b - d)/2 \tag{2}$$

Relative pattern shift:

$$\text{RPS} = \text{APS}/t \tag{3}$$

where t is the epi layer thickness.

Figure 14 shows a cross section (lapped and stained) of an epi pocket with buried layer asymmetric growth into the epi layer, due to wafer misalignment from (100). This problem, not easily detectable during the epi process, can cause wafer scraps at parametric testing of the BV parameter of one lateral PNP transistor for several months. This is due to the punch-through of the transistor collector with the wafer substrate grounded during the measurement. A deeper analysis revealed that the P-well top isolation region was misaligned by about 4 μm in the direction parallel to the flat.

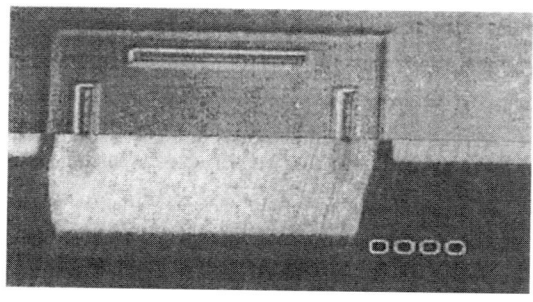

FIG. 14. Epitaxial pattern shift in a buried layer on a misaligned substrate (top); good epi on a good substrate (bottom).

2. Pattern Distortion and Washout

As stated, pattern distortion is lowest for silane-based epi growth chemistry and increases with increasing chlorine in the family of chlorosilanes: lowest for SiH_2Cl_2, moderate for $SiHCl_3$, and highest for $SiCl_4$. Washout is a strongly increasing function of Cl^- atoms at a fixed temperature, but only weakly increasing if we use the chlorosilane compounds in the optimal reaction range temperature for each (850–950°C for $SiCl_4$, 1000–1100°C for SiH_2Cl_2, 1150–1200°C for $SiHCl_3$, and higher than 1190°C for $SiCl_4$). Figure 15 shows one example of a washout effect on an alignment mark in the deposition temperature range 1150–1200°C. Effectively the faceting of the edges is reduced in higher-temperature growth.

Lowering the growth temperature to 1150°C, or below that for $SiCl_4$, leads to large distortion and, in some cases, to flattening of the steps (washout), which obliterates the pattern for the following mask operation. The washout problem is directly connected with the beginning of the deposition step (0.2- to 0.6-μm 0.6 epi), thus it is crucial to control the temperature at that moment and to avoid starting the deposition at a temperature lower than required by the recipe.

We can lower the effect of washout by reducing the chamber pressure, with results such as those shown in Figs. 16–18. In a single-wafer reactor we correlated the amount of washout at atmospheric pressure and 1180°C with that at 60-Torr

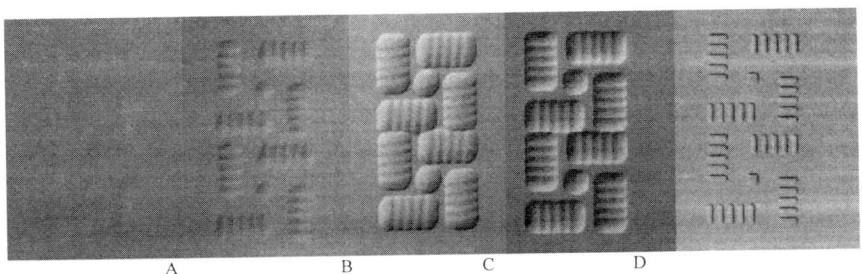

FIG. 15. Effects of washout degradation in the 1150–1200°C temperature range (A, 1150°C; B, 1165°C; C, 1180°C; D, 1200°C) using a $SiHCl_3$ atmospheric pressure recipe, in a lamp-heated single-wafer reactor.

pressure and 1050°C. At atmospheric pressure, lower growth rates decrease distortion and prevent washout.

To summarize, as shown in Table II, we reduce both pattern distortion and washout using the highest deposition temperature, a lower pressure, and the lowest growth rate. The productivity issues in patterned wafer manufacturing force the process engineer to find a compromise among these parameters.

3. EVALUATION OF GEOMETRICAL PATTERN INTEGRITY

As illustrated in Figs. 16–18, we can use different methods to recognize and evaluate these pattern reproducibility problems.

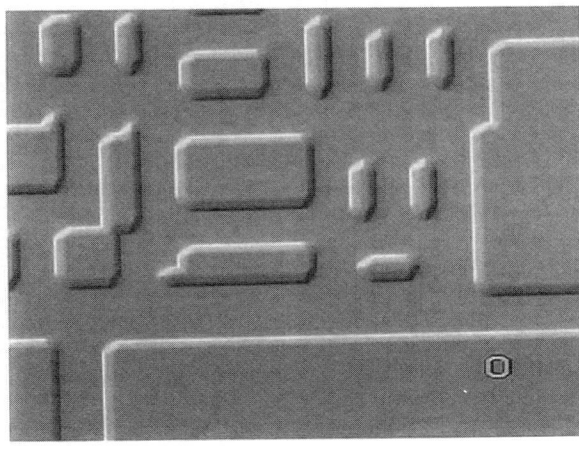

FIG. 16. Washout with pattern edge deformation.

FIG. 17. The same device as in Fig. 16 with (1150°C atm) a washout effect.

a. Chemical Etch

This consists of removing the epi layer by chemical etch with a 6% solution of HF (40%) in HNO_3, until two boundaries of the buried layer island appear. These boundaries are visible under a Nomarski microscope, and from their distance it is possible to compute the amount of pattern shift. This method is very easy to use—even if useful only in the case of pattern shift—but it is sensitive

FIG. 18. The same device as in Figs. 16 and 17 without (1150°C, reduced pressure) a washout effect.

TABLE II

Correlation of Pattern Shift, Pattern Distortion, and Washout with Process Parameters[a]

Parameter	Parameter behavior	Pattern shift	Pattern distortion	Washout
Slice orientation	On (100)	—	—	—
	On (111)	++	+	—
Pressure	Increasing	+	+	++
Temperature	Increasing	——	——	——
Growth rate	Increasing	—	+	+
Cl atoms	Increasing	+	+	++

[a] (+) More shift from the good case; (−) less shift from the good case.

to etching time and interferential contrast adjustment. Both these factors decrease the accuracy of the method (see Fig. 13).

b. Lapping and Staining

The lapping and staining method consists of the lapping of a big buried layer island along a direction perpendicular to the main flat. It is useful to work with a large lapping angle (5–10°). Then the lapped surface is P stained (see Chapter 7 of this book). If the lapping direction is perpendicular to the main flat direction, good accuracy and reproducibility of the method are assured. We evaluate the pattern shift entity from the buried layer misadjustment with the optical microscope (see Fig. 14).

c. Profilometer Measurements on Alignment Targets

While the Nomarski microscope can supply a synthetic top view of the pattern, the *profilometer* tool can rebuild the shape and height of pattern islands and windows in the silicon, before epi growth, as a *pattern cross section,* and it allows a comparison with the pattern profile after epi growth, as in Fig. 19. This method consists in a profiler tip (0.1–0.2 μm) that rebuilds the vertical and lateral profiles of both alignment marks and buried layer island shapes. It is a technique to be associated with lapping and staining to complete the overall image of pattern integrity loss. Using a capacitive tip, the profilometer method allows a construction analysis of the pattern profile, as a vertical map with vertical resolution up to 0.008 Å.

Pattern distortion and washout are important for the fine-tuning and qualification of the epi recipe, measuring the alignment mark's height and lateral size and the flattening of the pattern wells in the silicon. Pattern shift is monitored by the profilometer measuring the left-to-right asymmetry of the alignment mark.

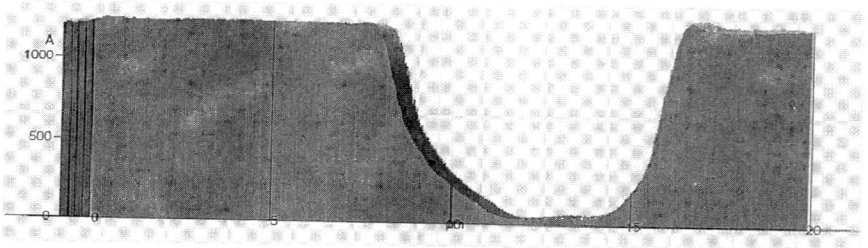

FIG. 19. Profilometer measurement: well profile of the alignment mark (with a pattern shift effect).

V. Doping Pattern Integrity

In this section we focus on doping problems arising when unintentional doping, coming mainly from patterned structures, influences the general epi layer electrical characteristics. The epi autodoping problem in the presence of buried doped regions is described for different and competitive dopant types. Finally, how to control and minimize autodoping and out-diffusion phenomena is discussed.

1. The Autodoping Problem

While most VLSI applications require the epi deposition of a lightly doped epi layer (10^{14}–10^{17} atoms/cm^3) on a heavily doped substrate (10^{19}–10^{21} atoms/cm^3), the patterned wafer consists of isolated regions of heavily doped buried layers implanted and diffused in a lightly doped substrate. For C-MOS device structures uniformly doped substrates are used, while in most bipolar, BiCMOS, and mixed bipolar/C-MOS/D-MOS (BCD) applications buried layers are used. During epi film deposition, in addition to intentionally added dopants, unintentional doping may come from the front or back side of the silicon substrate and from the patterned layers, doped and diffused under the epi layer. In batch reactors all wafers contribute to enhancing unintentional doping. This unintentional doping is called *autodoping*.

Autodoping may also come from dopant released from the reactor walls as a result of prior deposition cycles. This effect is called "reactor memory." Intentional epi dopants are substitutional impurities in the silicon crystal lattice with more (n-type) or fewer (p-type) valence electrons than silicon (which has four valence electrons), and the concentration of these impurities is used to control the resistivity of the epi film; we define the autodoping effect as the incorporation of unwanted dopants that originate from the wafers themselves.

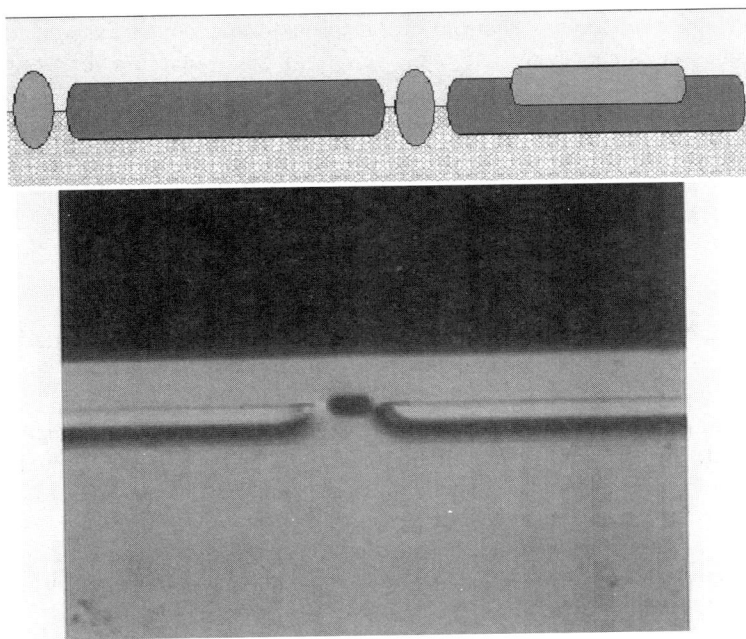

FIG. 20. Schematic image (top) and photograph of a cross section of an N^+ buried layer (lateral black areas) with a middle boron P^+ buried layer (black oval in the middle) under a 10-μm epi layer cross section. The boron vertical diffusion into the epi layer is evident. Dopant moves laterally and vertically in the silicon (or nitride or oxide). It has typical units of square microns per hour. Original magnification, ×500.

Regarding physical sources, all the autodoping phenomena can be classified into two categories:

(a) *Macroautodoping:* Dopant coming from the wafer surfaces (front, side, back).
(b) *Microautodoping:* Dopant coming from N^+ and/or P^+ buried layer islands that migrate to another location on the same patterned wafer.

This second contribution is critical because it leads to different competitive autodopant species occurring at the same time, with behavior opposite to that in autodoping reducing trials. Figure 20 shows a cross section of a patterned wafer after epi growth: an N^+ buried layer (lateral black areas) with a middle boron P^+ buried layer (black oval in the middle) under a 10-μm epi layer cross section (500×). Vertical diffusion of boron into the epi layer is clearly seen. Dopant moves laterally and vertically in the silicon.

We can also summarize autodoping following the geometrical distribution in the wafer.

(a) *Vertical autodoping* is the upward incorporation of dopant evaporated from the wafer and from patterned regions of the wafer into the grown epi layer.
(b) *Lateral autodoping* is the lateral migration by evaporation and incorporation of dopants from heavily doped regions on the wafer to lightly doped regions. This phenomenon causes an undesirable dopant concentration inside the epi layer and laterally connects different buried layers.

Two mechanisms can cause the following autodoping phenomena.

(a) Solid-state diffusion of implanted and diffused dopant species under the epi layer at elevated temperatures, known as *out-diffusion*.
(b) *Reincorporation of dopants* that have evaporated from the substrate and buried layer during the pre-epi bake process. The process sequence is solid-state diffusion of the impurity to the surface and evaporation from the surface in the gas phase, depending on the temperature and epi recipe parameters; mass transport in the gas phase (by convection and/or gas-phase diffusion, depending on the chamber geometry and fluodynamics considerations); and adsorption and incorporation into the growing layer, depending on the film growth rate.

2. Doping Profile in The Presence of Autodoping

When the epi doping profile is measured for undoped or lightly doped film over a heavily doped substrate, it shows a transition region instead of the expected ideal abrupt junction, a stepped change in dopant concentration. Figure 21 shows the ideal epi cross section in the presence of both N^+ and P^+ buried layers; the dopant is "confined" in each different region.

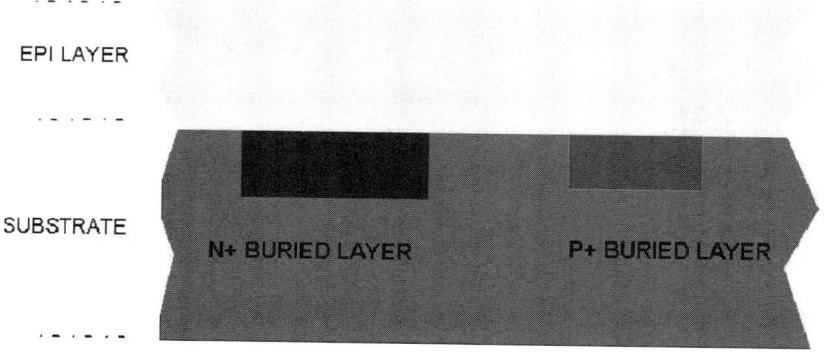

FIG. 21. Ideal epi cross section with multiple buried layers (N^+ and P^+).

The real transition behavior may be understood in terms of at least three components:

(1) the substrate/epi layer interface,
(2) the out-diffusion region (gas-phase autodoping region), and
(3) the contribution from other surfaces in the reactor.

a. The Substrate/Epitaxial Layer Interface

This component is due to the solid-state diffusion of the substrate dopant into the growing film during the epi growth process or during high-temperature diffusion processes after epi growth, such as P-well or N-well diffusion in C-MOS technology.

Rice, Grove, Price, and Goldman introduced different models for the epi profile near the interface, over a uniformly doped substrate; it generally follows a complementary error function expected for a step change in concentration after redistribution by solid-state diffusion:

$$C_{epi}(x) = C_{BL}/2 \left[1 - \text{erf}\left(x/2(Dt)^{\frac{1}{2}}\right) \right] \quad (4)$$

where C_{BL} is the initial buried layer dopant concentration at the interface, $C_{epi}(x)$ is the concentration of dopant in the epi layer, D is the diffusion coefficient of the buried layer impurity, and t is the deposition time. Formula (4) assumes that the epi growth rate far exceeds the dopant diffusion rate from the buried layer, which is generally the case for routine epi depositions in semiconductor manufacturing.

Because of the lower diffusion coefficient of antimony in silicon compared with phosphorus or arsenic (see Table III), heavily doped antimony substrates are frequently used as monitor wafers for thickness uniformity and growth rate assessment.

The presence of differents type of buried layers (P^+ and N^+) on patterned wafers under a uniform epi layer complicates the simple model described previously, as shown in Fig. 22. At fixed dopant species (for example, boron and arsenic) the "weight" of the different buried layers is more or less important, depending on the

TABLE III

SOME BURIED LAYER PARAMETERS IMPORTANT FOR AUTODOPING

Dopant species	Vaporization (kcal/g-atoms)	Atomic number	Diffusion coefficient	
			D_H (eV)	D_0 (cm^2/s)
Boron	126	11	3.69	10.5
Phosphorus	2.97	31	3.69	10.5
Arsenic	6.62	75	3.56	0.32
Antimony	46.6	121	3.95	5.6

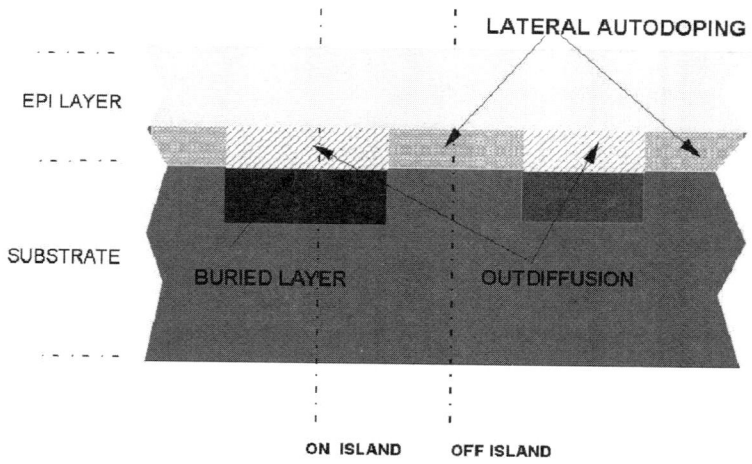

FIG. 22. Cross section of an epi profile with multiple buried layers.

percentage area, energy, and dose of implanted islands. If we monitor the epi doping profile on one buried island, the result is affected by the typical out-diffusion of this buried layer; between two different buried layers the situation is mixed and is affected mainly by lateral autodoping.

b. The Out-Diffusion Region (Gas-Phase Autodoping Region)

The second component of autodoping, the gas-phase autodoping region or vapor-phase autodoping, is responsible for lateral autodoping. A model has been suggested in which dopant atoms evaporate from the surface, mix with the gases above the substrate, and then reincorporate in the growing layer at a rate dominated by their concentration in the gas stream. It has also been assumed that the atoms are reincorporated at a rate proportional to the surface dopant concentration and to the epi growth rate. Thus the autodoping is proportional to the silicon atoms available for the reaction. A model based on trapping of surface absorbed dopant species by the growing epi layer has also been proposed. Since the gas stream carries away most of the evaporated dopant, these three models predicted an exponential decrease in the autodoping contribution with epi layer thickness away from the interface with the substrate.

c. The Contribution from Other Surfaces in the Reactor

The third component of autodoping summarizes the contribution from other surfaces in the reactor. In a batch reactor this tail can also contain the effect of dopant released from the back side of the wafer or from other wafers. This source generally decreases with time and with thicker epi layers.

3. Autodoping Characterization

The four-point probe and C-V measurements are the standard tools for epi doping profile monitoring. Vertical and lateral autodoping must be characterized by monitoring the epi layer–substrate and epi layer–buried layer interfaces instead. The *spreading resistance probe* (SRP) is the correct tool for this. The two-point probe spreading resistance technique has been developed for diffusion profile measurements. For a two-point probe set, the total spreading resistance is given by

$$\text{SRP} = \rho/2a \tag{5}$$

where ρ is the average resistivity near the probe parts, at the probe radius.

This technique employs a cleaved and beveled sample. A two-point probe is used to measure the resistivity sequentially at a number of points, starting at the epi layer surface and continuing through the epi layer/buried layer or substrate interface. By correlating the change in resistivity at each point with the carrier concentration, the doping concentration and profile are calculated. This technique is very sensitive to local impurity concentration variations (it has a high spatial resolution), to sample surface, and to probe condition.

While the C-V method works well with doping concentrations in the range of 10^{14}–10^{17} atoms/cm^3, for higher doping concentrations the SRP technique is used. Measuring a wide range of doping concentrations (8×10^{14} to 10^{20} atoms/cm^3), it suffers from poor accuracy and complex data reduction. When the SRP profile is taken through the epi layer into the buried layer region, the vertical autodoping profile is measured as shown in Fig. 23. When the profile is taken close to the buried layer region the lateral autodoping profile is

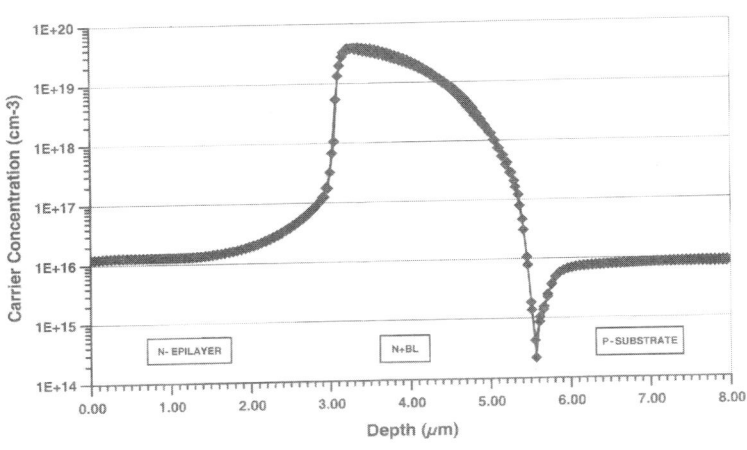

FIG. 23. SRP dopant concentration profile: N$^-$ epi/P$^+$ iso/N$^-$ iso/P$^-$ substrate.

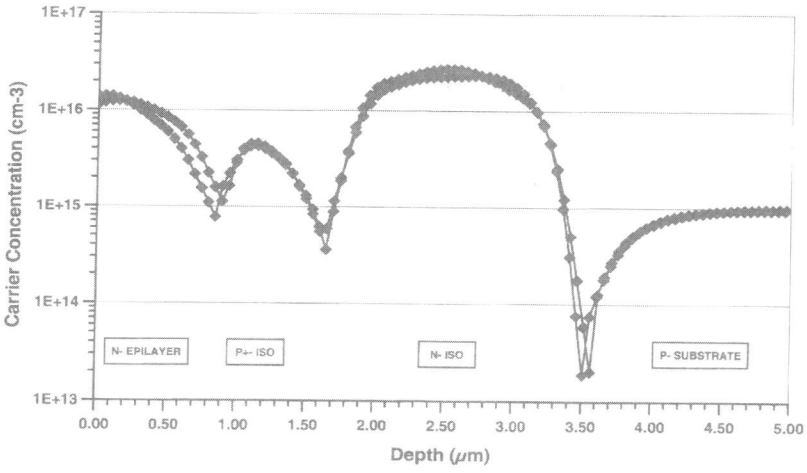

FIG. 24. SRP multiple buried layer structure after epi, lapped and stained.

measured. This technique is used mainly as a comparative method for autodoping qualification of new epi recipes.

Out-diffusion and autodoping effects (both vertical and lateral) are manifest as an increased thickness of the *transition layer* at the interface between the implanted/diffused regions and the *epi flat* region, where the resistivity is a constant parameter versus the thickness. The transition region is shown schematically in Fig. 24.

Close to the patterned substrate, solid-state diffusion from diffused islands and the substrate is the predominant effect, causing a broader transition profile. The rapid growth of the film relative to the motion of diffused species limits this effect. Thereafter, the transition zone is controlled by the dopant incorporation from the vapor phase. When the atoms evaporated from the substrate exceed those intentionally introduced into the reactor, an autodoping tail develops. Since the high-concentration buried layers quickly become covered with more lightly doped material, the autodoping effect ends and the desired concentration and resistivity are reached.

The extent of the autodoping tail depends on the deposition temperature, silicon source, growth rate, reactor geometry, pressure, and epi recipe Alternatively, in addition to the SRP method, the lapping–staining method can be used (Fig. 25), to recognize and evaluate the shape and profile of the transition interface between the buried layer region and the beginning of the "flat doped" epi layer.

4. BURIED LAYER PARAMETERS AFFECTING AUTODOPING

Most researchers agree that the *prebake process conditions* are the most influential parameters that determine the level of autodoping, as they represent the

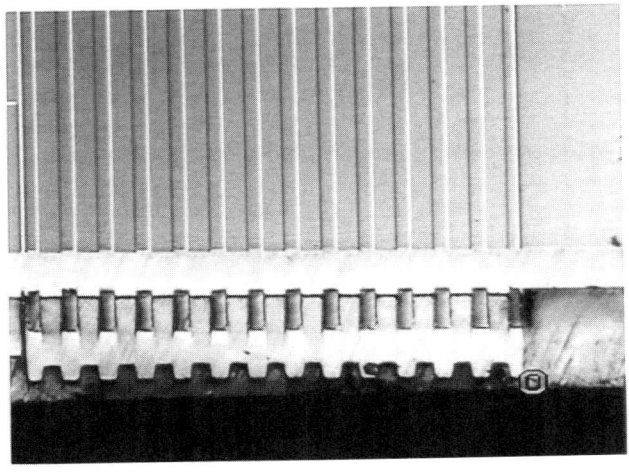

FIG. 25. Stained cross section with a comb-shaped buried layer, with lateral autodoping.

process *interface* between the wafer pre-epi history and the physical deposition of epi film.

Table III summarizes some buried layer-related parameters that are important for autodoping, such as the *dopant diffusivity* and *vapor pressure* of the dopant species. Autodoping increases with increasing vapor pressure (or decreasing heats of vaporization) and diffusion rates of the buried layer dopant species. For example, antimony causes the least autodoping, followed by arsine, boron, and phosphorus. Optimal processing conditions vary by different dopant species, not by type N or P. Boron and phosphorus follow similar trends, while arsine behaves very dissimilarly.

Autodoping is also very dependent on the dopant concentration at the surface and it becomes worse and worse with higher surface concentrations. Therefore autodoping depends on the implant energy and dose and the thermal diffusion process, which leads to a buried layer profile like that before epi deposition. Table IV

TABLE IV

DOPANT PEAK CONCENTRATION DISTANCES FROM THE SURFACE CALCULATED FOR DIFFERENT IMPLANTATION CONDITIONS

Dopant species	Energy (keV)	Doses (ion/cm^2)	R_p (Å)
Sb	80	2.5×10^{15}	449
B	150	2.5×10^{14}	4861
B	150	2.0×10^{14}	5679
As	60	4.0×10^{15}	443
As	80	2.5×10^{15}	563

lists some examples of dopant peak concentration distances from the surface, calculated for different implantation conditions.

Concerning lateral autodoping, many studies were conducted by Srinivasan (1978, 1980), who developed a model for autodoping effects by considering evaporation from a point source. The evaporated dopant was considered to be moving by diffusion but influenced by gas flows near the surface. The contours of dopant equiconcentration, which would be circular (two-dimensional model) under zero stream flow, are distorted to elongated ellipses in the flow direction.

If we consider a typical patterned wafer, we can extrapolate the Srinivasan model to the combined effects of multiple point sources, represented by different buried layer regions in the device and the different devices on the slice. In the case of small, widely spaced buried layers, the autodoping concentration will vary laterally around each buried layer point source and extend over relatively long distances (1–6 mm calculated in the upstream and downstream directions, respectively). As the spacing of the autodoping sources is reduced, the lateral effects overlap, and beyond 500 local sources per wafer, the autodoping concentration varies less than 5% in the space between local source areas. The model also predicted a square-root dependency of the mean autodoping level on the buried layer density. This was confirmed by subsequent experiments.

Pre-epi thermal treatments can strongly affect autodoping. Thermal oxide on silicon preferentially absorbs boron, and the concentration near the single crystal is reduced. The opposite is true for arsenic and phosphorus. When possible, preepi doping for arsenic and phosphorus should be accomplished by higher-temperature dry oxidations rather than lower-temperature wet oxidations; the opposite applies for boron.

5. Recipe Parameters Affecting Autodoping

In Section II.2 and Table I, we summarized the critical recipe parameters for patterned wafers. Here we analyze them in detail.

a. *Effect of Pre-epi Bake Parameters on Autodoping and Methods to Reduce It*

In our experience, arsenic autodoping is a quite flat function of the *bake time* and it is quite independent of the H_2 main flow. On the other hand, it is strongly dependent on the temperature and pressure. The reduced pressure effect is really effective (mainly under 100 Torr) to obtain a sharp transition and minimize lateral autodoping. A reduced pressure associated with a high-temperature bake can create a surface region depleted of arsenic. If the pre-epi bake is followed by a high-H_2 flow purge, epi film deposition can start with minimized autodoping effects. Whereas for antimony the bake is not so critical, for boron autodoping we have effects opposite to those of arsenic. To avoid diffusion in the future growing layer, the bake is shorter and at the lowest temperature and atmospheric pressure.

b. Effect of H_2 Flow Rate During Deposition

Lateral autodoping decreases at increasing gas flows. At a fixed silicon source flow, a higher H_2 flow is related to a lower growth rate. At a low rate (less than 0.8 μm/min), the arsenic autodoping peak concentration increases at an increasing growth rate, and this effect is directly proportional to the decreasing pressure during the deposition step.

c. Effects of Cl Atoms

HCl vapor etching before deposition can also lower arsenic lateral autodoping, proportionally with the time at which the HCl etch occurs. We think that the explanation is the removal of the surface filled by arsenic atoms. Considering the effect of the number of chlorine atoms in chlorosilane Si sources, vertical autodoping is independent of the type of chlorosilane, but near the surface the epi layer is more concentrated with a lower number of Cl atoms, at the same process temperature. Regarding lateral autodoping of boron and arsenic, at a fixed temperature, mainly for boron there is less autodoping at a higher Cl atom concentration—the autodoping peak changes one order of magnitude from SiH_4 and $SiCl_4$; for arsenic this is not as critical.

Autodoping imposes a limit on the minimum thickness and doping level for an epi layer and therefore must be minimized. Generally speaking, the best system for minimizing the autodoping is to control the *input parameters* of the reactor: the temperature, pressure, and time of different steps in the epi growth recipe. A general technique for reducing the out-diffusion component for phosphorus or arsenic relies on the *evaporation* of substrate dopant from the wafer during a high-temperature bake just prior to the deposition step. Longer or higher-temperature bake cycles deplete the surface and subsurface concentrations prior to the commencement of deposition. During deposition, the substrate dopant out-diffuses to fill the depletion region, minimizing the dopant encroachment or compensation coming from the sum of out-diffused and intentionally added dopants.

Another technique is the use of a *cap layer*. Increased velocity of gas flow will make the boundary layer thinner and will promote the escape of dopant coming from the buried layer implanted region into the exhaust stream of the reactor. We can increase the gas velocity by reducing the pressure in the reactor chamber or by using a high gas flow. Regarding pressure reduction, it is known that for a thin epi layer deposited on a high-concentration. As implanted buried layer, decreasing the reactor pressure reduces the transition width but maintains the same H_2 flow. Increasing the velocity of the gases passing over the substrate under reduced pressure conditions is also effective for suppressing lateral autodoping.

The advent of single-wafer reactors totally eliminated the autodoping contribution coming from other wafers in the same batch.

Several techniques are available to process engineers for reducing autodoping effects, which are summarized below.

(a) Prevent Boron autodoping. The epi deposition process should be performed at the lowest deposition temperature providing good-quality films and a reasonable growth rate. Note that this technique is effective only for reducing boron autodoping since arsenic autodoping increases at lower temperatures.

(b) Use substrate and buried layer dopants that have low vapor pressures and low diffusivities; for example, Sb should be selected instead of As (high vapor pressure) or P (high diffusivity).

(c) Use reduced-pressure (subatmospheric) reactors to minimize lateral autodoping. A reduction in lateral autodoping is achieved since the diffusivity of gas molecules at reduced pressure is increased. This allows the dopant atoms rapidly to reach the main gas flow (mainly H_2) and be swept out of the reactor into the exhaust line (also, exhaust in RP). This method works well for P and As but not for B.

(d) Use a sequence such as cap → purge → grow, in which a very thin layer (0.1–0.2 μm) of undoped silicon is grown to seal the As coming from the buried layer.

(e) Use a low-temperature purge after the HCl etch to make sure that the dopant evolved during the etching is swept out of the system. It is also important to use an appropriate etch time to avoid a memory effect from the past growth run.

(f) Reduce the pressure. This increases the gas velocity and cuts the average gas residence time by a factor of about 10, which improves the abruptness of junctions between layers and their uniformity. This also influences film properties related to pattern shift and autodoping.

(g) For As-doped buried layers, perform a two-step deposition at a high temperature and low growth rate to deposit a 0.5-μm-thick cap.

(h) Use a boron counter-doping cap for As-doped buried layers.

(i) Reduce the load size if using batch systems.

(j) Replace As with Sb as the n-dopant species for a dopant concentration of less than 10^{19} at/cm^3. This is because Sb exhibits much less autodoping due to its low rate of evaporation and diffusion.

(k) Dilute SiH_4 with HCl during the deposition step.

(l) Reduce the surface density of heavily doped diffused regions.

(m) Use rapid thermal processing.

TABLE V

SUMMARY OF TECHNIQUES FOR REDUCING AUTODOPING EFFECTS

Recipe parameter	Parameter behavior	Autodoping[a]			
		Boron	Phosphorus	Arsenic	Antimony
Bake temperature	Increasing	++	+	− −	−
Bake time	Increasing	++	+	− −	−
In situ etch	Increasing time	−	−	−	−
Pressure	Decreasing	+	+	−	−
Main H_2 flow	Increasing	−	−	−	−
Growth rate	Increasing	++	−	+	+
Deposition temperature	Increasing	+	−	−	−
Cl atoms in chlorosilane formula	Increasing Cl atoms	−	−	−	−

[a] (+) Worse; (−) better.

(n) Use a low temperature, low growth rate, and reduced pressure cap (950°C, 0.25 μm/min, 80 Torr) for wafers with both P^+ and N^+ diffused buried layers.

(o) Decrease the As concentration to less than 10^{18} atoms/cm^3 if possible.

Table V summarizes all the techniques described above.

VI. Crystal Defectivity

In talking about the crystallographic epi layer defectivity on patterned wafers, the first noteworthy concept is that the crystalline quality of the layer cannot surpass the quality of the substrate.

As described previously (see also Table I), the history of the patterned wafer before the epi process can seriously affect the final quality of the epi layer. During epi deposition, process-induced lattice defects can also occur. The pattern integrity requirements restrict the useful window for recipe input parameters on the reactor (higher temperature, lower pressure and growth rate). Mainly a higher temperature is worse for defectivity.

Thus, general epi defectivity on patterned wafers results from several factors.

(a) Defects previously present on the wafer, due to implantation or high-temperature diffusion processes involving the bulk silicon properties. These defects become apparent after epi deposition. They are dislocations, epi stacking faults, and roughness (called "orange peel").

(b) Surface defects (particles and organic residuals), introduced mainly by cleaning operations, as shown in Fig. 26, prefurnaces, resist removal, pre-epi

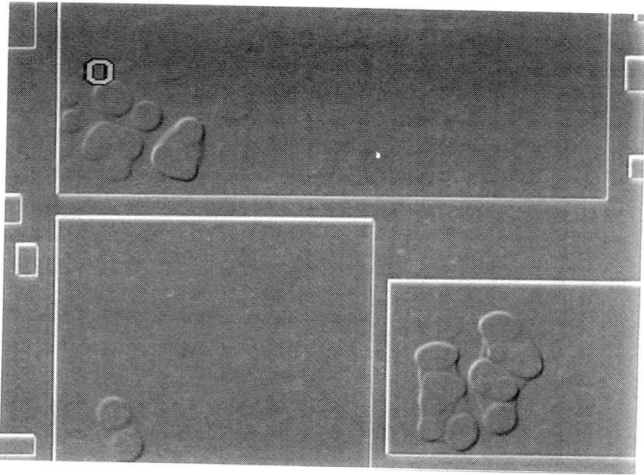

FIG. 26. Defectivity induced during a preepi oxidation step by a bad drying treatment.

cleaning, or dry etches for mask removal (see Fig. 27). They are very difficult to detect visually. The net result is to leave at the pattern surface microscopic steps that become defect sources and are detected only after epi growth by a "lens effect" as shown in Fig. 28.

(c) Line or area crystallographic defects due to from the epi process itself, such as slip-lines (see Fig. 29) and wafer warpage. Atmospheric deposition on

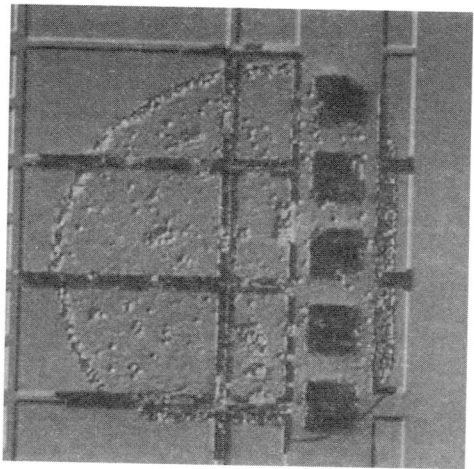

FIG. 27. Defectivity induced during preepi dry etches.

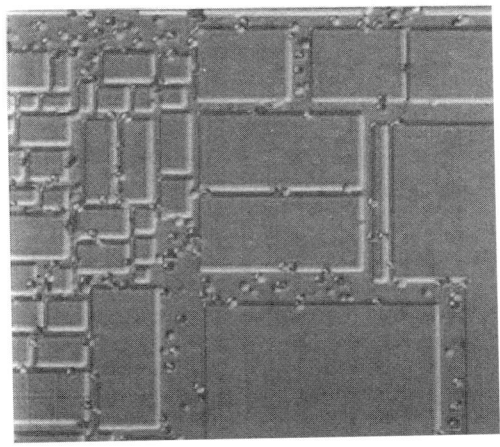

FIG. 28. Little bubbles induced during preepi dry etches: the epi layer deposition makes them more detectable, by the "lens effect."

patterned wafers requires a temperature higher than 1100–1180°C (DCS or TCS source, respectively) for pattern integrity.
(d) Defects coming from contamination of the reactor itself, such as pyramids, growth hillocks, and haze.

1. THERMAL TREATMENTS BEFORE EPITAXIAL GROWTH

During pre-epi pattern building, several thermal treatments can be applied:

(a) oxidations, to create a silicon oxide masking layer or to remove the oxygen from the surface and create a precipitate-free region near the active device (denuded zone treatment),
(b) annealing treatments to recover the implant damage, and
(c) diffusion treatments to diffuse the implanted dopant and to create dopant wells for several device requirements.

During diffusion treatment an O_2 atmosphere is used to create a surface oxide layer trapping the dopant inside the silicon, for a very low dopant diffusivity in the oxide. The same silicon oxide will be used as the dopant mask.

A nitridation process is also used to create a masking layer: Si_3N_4 deposition at 500–600°C. During thermal treatments at temperatures higher than 1000–1100°C the center-to-edge wafer thermal gradient can induce dislocation arrays called "slip-lines." If the thermal stress is not relaxed, general deformation of the wafer, called "warpage," can occur. In the case of nitride deposition, the stress is due to

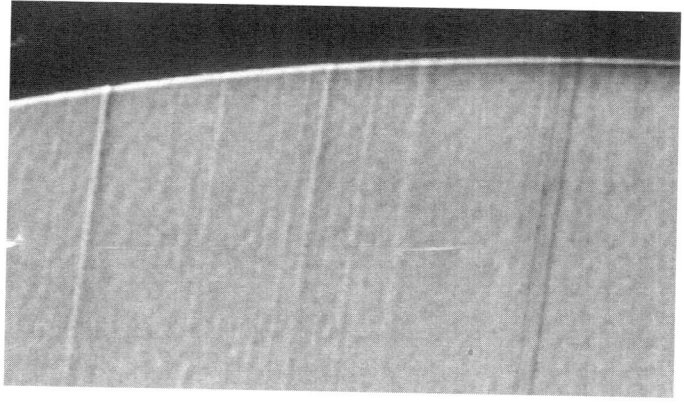

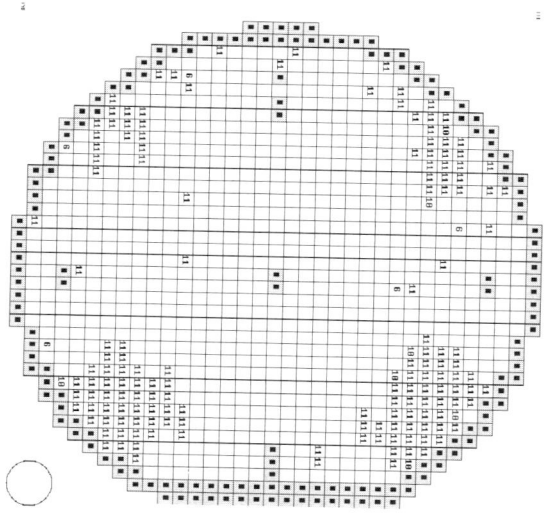

Fig. 29. Slip-lines on a wafer edge inspected with a Hologenix tool (top) and a relative yield map at the final testing of the same wafer, showing (black dots) the region scrapped for leakage, due to slip-line presence. The "ears-like" position, due to crystallographic preferential slip-planes for the (100) oriented wafer, is evident.

the different lattice constants between Si_3N_4 and silicon substrate, mainly at the pattern edges and corners, where the stress is more localized.

Why are slip-lines produced? The most important parameter connected with slip-lines for high-temperature thermal treatments is the *silicon yield strength* (also called the *critical shear stress*), which is the maximum stress the material can withstand without irreversible damage. An estimate of this critical value, S_c, for silicon is

$$\log S_c = 3.61 - 0.00385 T \tag{6}$$

where the temperature T is in degrees Celsius and S_c is in units of 10^8 dyne/cm^2.

Dislocation and slip-lines can be generated by the thermal process if care is not taken to minimize temperature gradients across a wafer during heating and cooling. Both when closely stacked wafers are pulled from a furnace and the wafer borders cool faster than the wafer centers and during epi growth there is a center-to-edge thermal gradient. The *thermal stress* S_{th} produced by the center-to edge temperature difference DT can be calculated as

$$S_{th} = a\, E \Delta T \tag{7}$$

where a is the coefficient of thermal expansion ($2.62 \cdot 10^{-6}\,°C^{-1}$ for silicon) and E is the Young's modulus [$1.067 \cdot 10^{12}$ dyne/cm^2 for (100) silicon].

As an example we can consider a process set at 600°C. Using Eq. (6) the critical shear stress becomes $S_c = 2 \times 10^9$ dyne/cm^2. Then, according to Eq. (7), the limit thermal stress ΔT would have to exceed 714°C, therefore, reasonably, dislocations are not thermally generated at 600°C.

Considering instead a temperature of 1200°C, which is the normal deposition temperature for patterned wafers, a $\Delta T = 10°C$ is calculated. This is enough to produce plastic deformation with slip-line formation.

a. Denuded Zone (DZ) Treatment

One technique is to create an oxygen-free zone near the surface of the silicon wafer, to avoid thermal donor generation and the presence of precipitates in the active device's junction zone. In particular cases (e.g., vertical high-voltage D-MOS applications), the presence of precipitates can seriously affect the device behavior. Oxygen precipitates can also affect the device performance as aggregation nuclei of metallic contaminations. These bulk precipitates can generate extended defects, such as sinkers for heavy metal contaminations, with a cleaner surface as the final result (intrinsic gettering).

The thermal treatment to create a DZ consists of a high-temperature step (1100°C) to diffuse and evaporate the oxygen from the surface, followed by a low-temperature step (700–800°C) to nucleate oxygen precipitates, far from the active surface. The result is shown in Figs. 30 and 31.

We can use different thermal cycles to create a good DZ.

(1) Hi–Lo: The denuding step is before the nucleation step.
(2) Lo–Hi: The nucleation step is before the denuding step.

In the second case, a higher bulk defect density can result.

The required specification for DZ thickness is different for MOS and bipolar technologies, ranging from 10 to 50 μm below the surface, respectively.

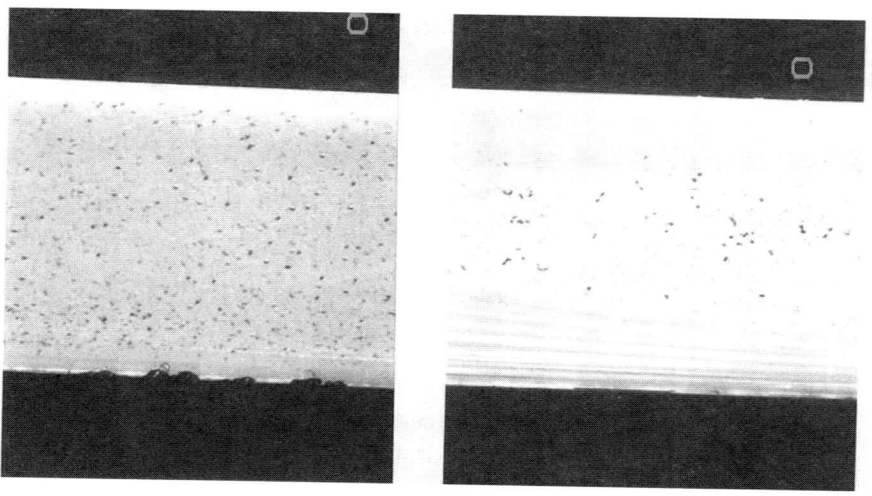

FIG. 30. Wafer cross section after epi growth, with narrow (left) and wide (right) denuded zones.

2. Lattice Damage and Recovery After Implantation

On patterned wafers epi growth is performed after two or more dopant deposition or implantation operations, depending on the device layout. In the history of patterned wafers, dopant deposition operation was replaced by the introduction of locally doped regions by implantation. The implantation process allows a

FIG. 31. Wafer cross section of the denuded zone (lower) and buried layers after epi growth in a defect-free surface region.

well-defined dopant location and concentration but leaves a damaged lattice by point defects (vacancy and interstitials) and extended defects (dislocations), until total amorphization of the silicon layer occurs. These effects are proportional to the implanted dose (measured as the ion concentration per square centimeter) and to the acceleration energy (orders of kilo-electron volts up to mega-electron volts).

To put an implanted dopant ion in a substitutional position (to electrically activate it) and to rebuild the crystal lattice integrity, annealing is usually performed in the 500–1000°C temperature range. If this annealing is the last operation before epi growth, it becomes critical to match the effect of the annealing step in the furnace with the effect of the annealing step in the reactor, performed during the hydrogen ramp-up and H_2 pre-epi bake process steps. The final crystal quality is the result of this compromise. Note that batch reactors generally anneal better (due to longer ramp-up steps at lower heating rates) than single-wafer reactors, which behave similarly to RTP systems (e.g., in single-wafer systems ramping from 900 to 1200°C takes only 1 min, while in batch systems it takes 10–15 min). Figure 32 shows a bad annealing result.

The crystal damage and the recovery efficiency also depend on the implanted species, the dose, and the implantation energy. For boron implantation at a dose higher than 10^{15} ions/cm^2, the damage can also be unrecoverable with annealing; the result is a high density of dislocations localized in the implanted regions, well detectable after epi growth (see Figs. 33 and 34).

Annealing temperatures that move implanted dopants into substitutional sites to make them electrically active may not eliminate all the damage-induced defects. The number of displaced atoms is hundreds to thousands times higher than the number of implanted dopant atoms. This requires annealing processes designed to recover crystallinity and minimize residual defects. Isochronal annealing behavior

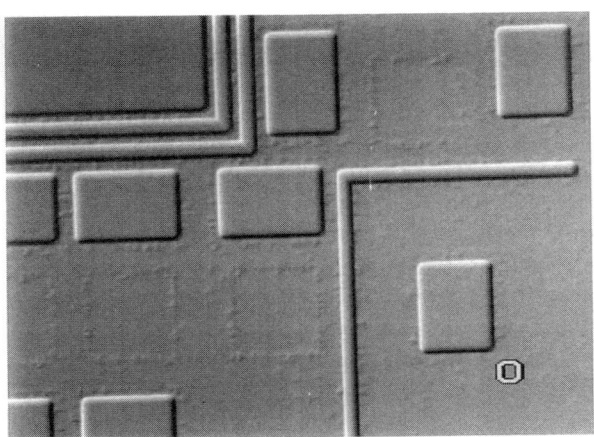

FIG. 32. Postepitaxial stacking faults (small square shapes in square frames), induced by bad annealing of preepi implantation. Original magnification, ×100.

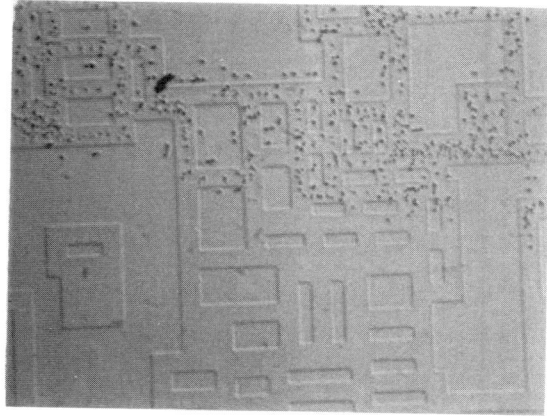

FIG. 33. Dislocations after epi growth on a sample without preepi annealing. The upper zone is implanted; the lower zone is not implanted. The dislocations are only in the implanted area.

for silicon implanted with low, medium, and high doses of boron (same times at each temperature) was analyzed. High doses can displace 250 times as many silicon atoms as a low dose. Higher displacement concentrations have a higher probability of forming defect clusters, while isolated point defects are annealed by ion beam heating or removed by low annealing temperatures (400–500°C). At annealing temperatures between 500 and 600°C, dislocations are observed. An agglomeration of vacancies or interstitials can lead to dislocation formation when the localized stress exceeds the yield strength of the silicon material.

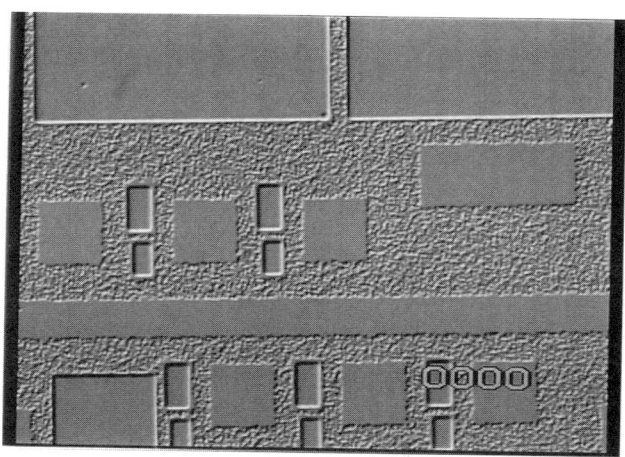

FIG. 34. Roughness after Secco chemical etch of heavily boron-implanted region before epi growth on an unannealed sample: the radiation damage in the crystal is evident.

3. DISLOCATIONS, SLIP-LINES, EPITAXIAL STACKING FAULTS, AND OTHERS

Common types of crystal defects resulting after epi deposition are dislocation and epi stacking faults. These defects are also generated on polished nonpatterned wafers but the origins can be very different.

a. *Dislocations*

Dislocations are generated in epi films by several mechanisms.

(a) Propagation into the growing film of a dislocation line coming from the substrate that reaches the wafer surface. Dislocation-free material must be used to growth a dislocation-free epi layer.
(b) The existence of a large lattice misfit between a lightly doped epi film and a heavily doped buried layer (misfit dislocation).
(c) Thermal generation by stress that exceeds the yield strength of the silicon, resulting in a slip-line. Slip-lines can be produced before epi during the furnace steps (e.g., during diffusion of a Sb buried layer), just at the contact point of the wafer with the furnace boat (mainly if the boat presents some irregularities or is dirty as shown in Fig. 35), and tend to grow with the epi layer itself.
(d) Slip-lines can occur during the epi deposition itself.

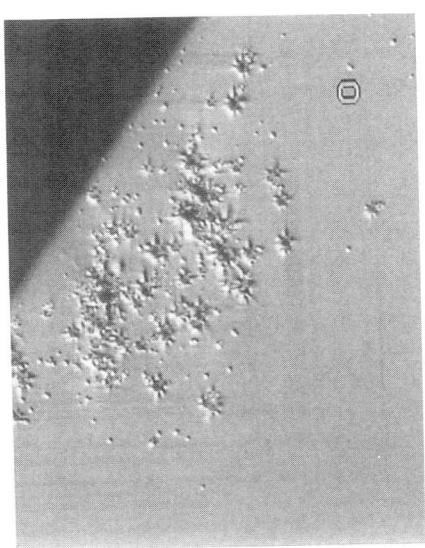

FIG. 35. Defects on a wafer edge induced by a dirty furnace boat before epi.

Higher temperatures are set for pattern integrity, so thermal stress comes from front-to-back temperature gradients of the wafer during the deposition process or center-to-edge gradient temperatures, depending on the reactor chamber geometry. In the case of RF-heated barrel reactors, the source of thermal stress is induced by the first effect, the heating coming only from the back of the wafer, the graphite susceptor. A particular compromise was tested on an RF batch system using specially shaped gold-plated bell jars or gold-plated induction coils for slip-line quality improvement. In the case of lamp-heated chambers, this front-to-back effect is minimized (front and back lamp heating in single-wafer equipment), but the center-to-side thermal gradient remains critical. Our experience with lamp-heated single wafer reactors is that, using an atmospheric recipe at 1180–1200°C to preserve the geometrical pattern integrity, imperfections of susceptor pockets become sources for defects.

Regarding slip-lines generated previously during oxidations, their initial point is on the wafer edge near the cassette contact point and they can propagate only toward the center of the wafer: for (111) silicon wafers the typical shape of the slip pattern, coming from the edge, is star-like, for (100) silicon wafers the slip pattern is "four ears"-like. If the typical high-temperature (>1000°C) thermal history before epi has two or three thermal steps, also the smallest slip-lines, a few millimeters long, can become, after epi growth, during MOS well diffusion (two or three steps up to 1000°C), slip-lines 2–3 cm long from the edge (the "ears" situation as in the map in Fig. 29).

The presence of thermal stress on an epi wafer, not associated with slip-line release, induces a general wafer deformation, defined as "wafer warpage," deviation from the center plane of a wafer (or wafer centerline). Warp is a bulk property. Standard experience is that epi growth flattens patterned wafers: before the epi process standard values of warpage on 150-mm wafers are in the range of 40–30 μm, depending on the thermal history; after the epi process the medium value decreases to about 20 μm.

b. Epitaxial Stacking Faults

An *epitaxial stacking fault* (ESF) is a deviation from the normal stacking sequence of atoms in a silicon crystal during epi growth (see Fig. 36). If it ends within the interior of the crystal, it will terminate at a partial dislocation. These defects are dangerous for leakage of the final device because, like dislocations, they can set a preferential "wrong" way for carriers. On (111) silicon wafers, ESFs can appear as either closed or open equilateral triangles, separated or intersecting, or as a partial triangle, with the edge length $L_{ESF(111)}$ related to the epi thickness TH_{epi} by the relation

$$L_{ESF(111)} = 1.62 TH_{epi} \qquad (8)$$

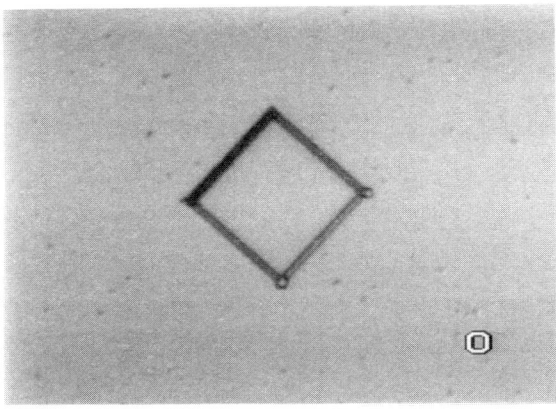

FIG. 36. Epitaxial stacking fault on (100) material, after Secco etch.

On (100) silicon wafers, they appear as either closed squares or triangular portions of squares, and the edge length $L_{ESF(100)}$ is related to the epi thickness TH_{epi} by the relation

$$L_{SF}(100) = 1.4 TH_{epi} \tag{9}$$

There is a correlation between the stacking fault shape and the epi recipe (silicon source and chamber pressure), type of buried layer, and reactor. Our last point is that recipes using a reduced pressure and DCS source at a low thickness produce mainly partial triangular-shaped stacking faults on (100), while recipes using atmospheric-pressure TCS, in both barrel and single-wafer reactors, produce mainly square-shaped faults.

The nucleation of stacking faults in the epi layer is always at the epi layer/substrate interface and is believed to result from two distinct causes:

1. microscopic surface steps on the substrate and
2. impurities either on the substrate or within the reactor itself.

If we accept the atomistic model that explains epi growth as little sideways steps moving in two-dimensional directions along the $\langle 110 \rangle$ axis, "if" an obstacle is encountered on the substrate a fault in the stacking sequence is generated. The critical point is the obstacle encountered on the substrate: the presence of slip bands in the substrate or particles on the substrate can also nucleate ESFs.

On patterned wafers the working steps before epi can also affect the final quality of the crystal layer, typically

(a) incomplete oxide removal by the preepi cleaning (see Fig. 37),
(b) incomplete nitride removal by dry etches,
(c) preepi scratches (see Fig. 38), and
(d) process chamber leakage during epi growth.

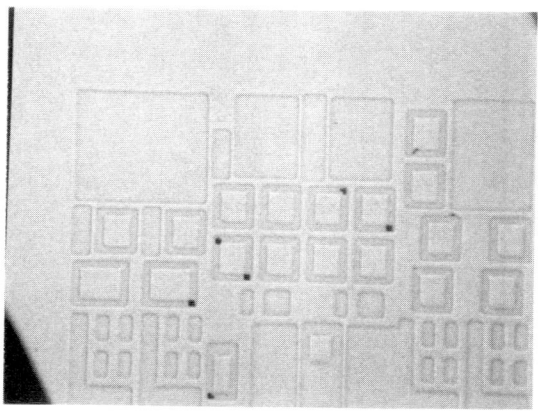

Fig. 37. Epitaxial stacking faults at the corners of buried layer regions, due to incomplete oxide removal before epi, associated with local stress.

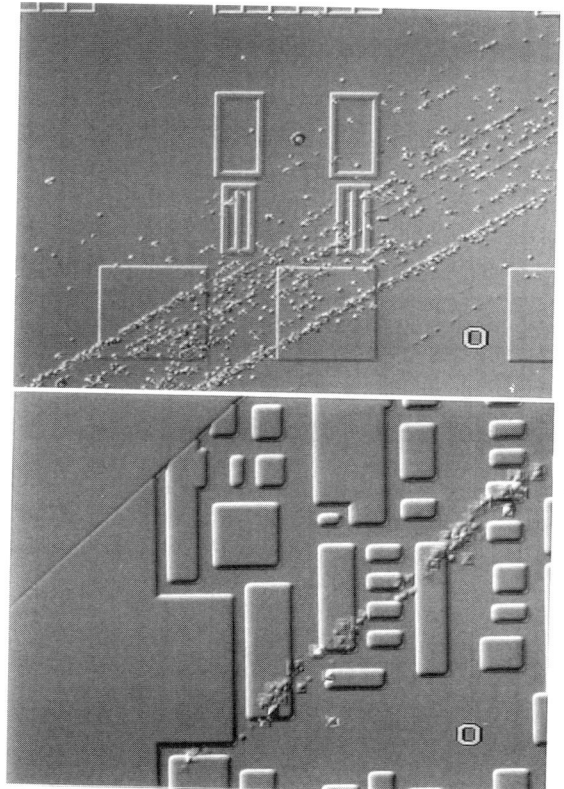

Fig. 38. Epitaxial stacking fault aggregates at different magnifications: both were generated by preepi scratches.

9 EPITAXY ON PATTERNED WAFERS   **339**

FIG. 39. Example of a stacking fault with inglobated material.

Scrubber treatments can generally solve only ESF problems induced by pre-epi residuals. Other examples of ESFs are illustrated in Figs. 39 and 40.

c. *Other Defects*

Other defects connected with the pattern presence are as follows.

(a) *Orange peel:* A roughness characterized by numerous rounded irregularities like those on an orange skin which can be observed under fluorescent light

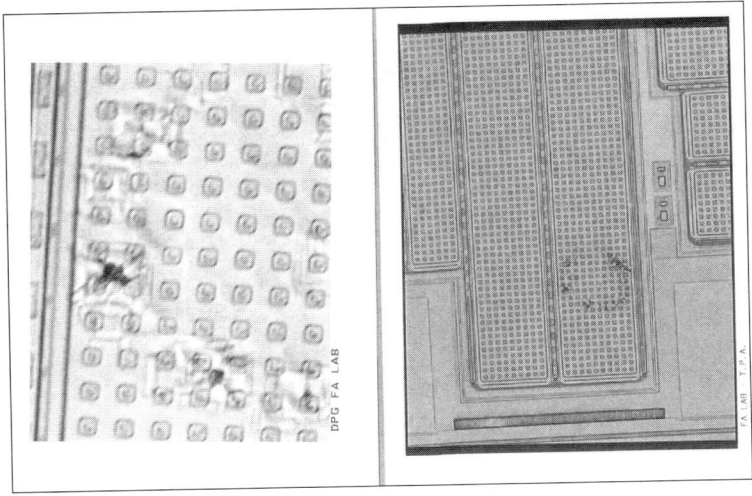

FIG. 40. Two magnifications for the same stacking fault aggregate on power MOS cells, detected in a finished device. The root cause was probably a preepi scratch (see the circular shape).

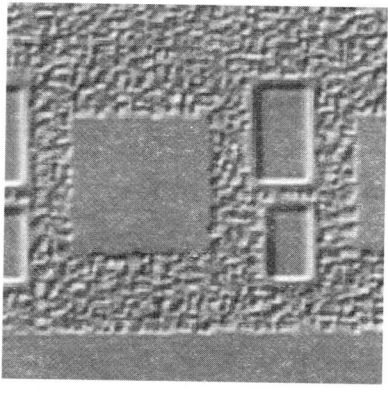

FIG. 41. Example of orange peel in boron-implanted regions, induced by the presence of humidity in the gas lines of the epi reactor.

but usually not under collimated light. This is normally seen (Fig. 41) on heavily implanted boron and phosphorus buried layers.
(b) *Pyramids:* Very easy to detect using an optical microscope (see Fig. 42) or light scattering methods, many layers after epi growth (Fig. 43). After chemical etch an associated stacking fault is visible as a square or triangular black border with dislocation at the corners.

FIG. 42. Example of a large epi pyramid.

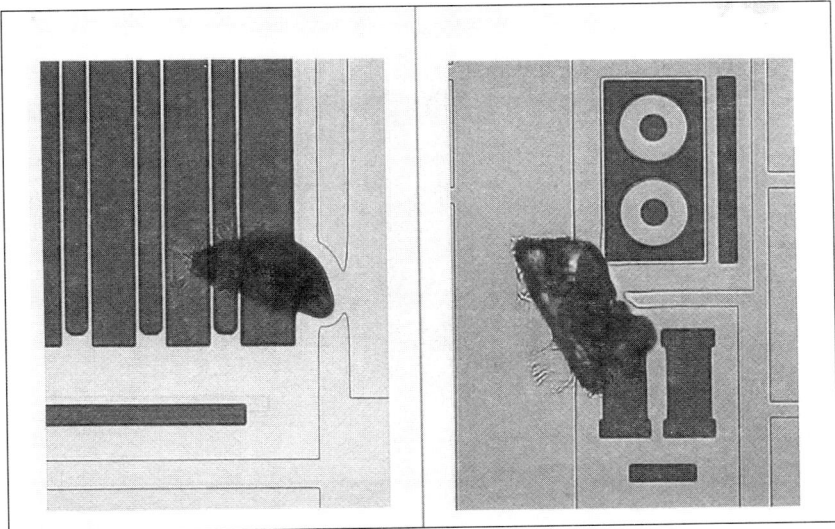

FIG. 43. Pyramids detected in some operations after epi growth.

d. Defects Characterization

The standard defect characterization method is to inspect them with an optical microscope. A dark-field setting is used to inspect dislocations, slip-lines, and ESFs (holes at the surface) better, and a Nomarski contrast setting to inspect growth hillocks and pyramids (peaks on the patterned surface).

The slip-finder (an optical tool with inspection under fluorescent light, which makes slip-lines and edge-chipping evident) is used mainly to inspect with a higher sensitivity, similar to that of chemical etch methods. In the semiconductor market other fast inspection instruments using photoluminescence methods or measuring material-localized stress with depolarized light (Fig. 44) are available.

Epitaxial stacking faults are detected using an optical microscope with Nomarski interference contrast, which resolves surface features 3–5 nm in deep, far beyond what can be achieved by the ordinary optical microscope. ASTM suggests a "field inspection," a statistical estimation of ESF density, with a limitation for epi thicknesses <3 μm. Light-scattering automatic tools are in the early stages for automatic inspection of ESFs on patterned epi wafers.

Chemical etch-based methods can be used to delineate defects prior to microscopic examination, by etch pit evaluation. The etch pit can be defined as a sharply defined depression in the immediate region of a stressed or defective crystal lattice, which has resulted from preferential etching. For silicon epi the most frequently used etchant is the Wright etch, a mixture of HF, HNO_3, CrO_2 $Cu(NO_3)_2$,

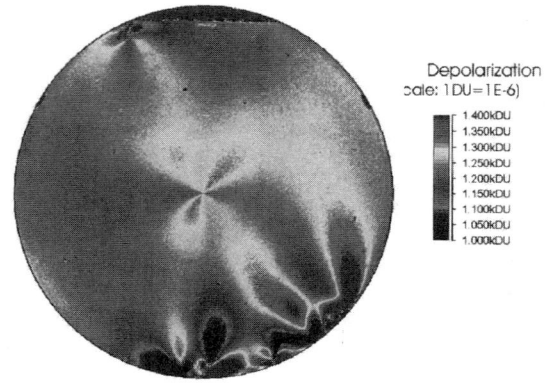

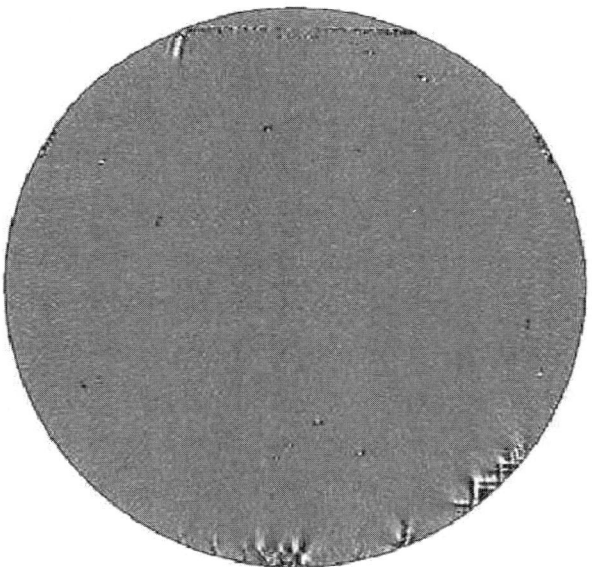

Fig. 44. New method of inspection of epi thermal stress on a wafer: (top) image of stress fields and (bottom) image of slip-lines detected with the depolarized light method.

and acetic acid in deionized water. Other solutions, Secco and Sirtl, are often used to delineate dislocations and pits.

VII. Conclusion: The Best Epitaxial Recipe?

"Higher temperatures are good for pattern integrity, but not so good for autodoping and slip-lines." When process engineers try to find the best epi recipe for patterned wafers, they must also consider the reliability of the equipment and productivity issues. Therefore, the epi recipe for patterned wafers is a difficult equilibrium, like a symphony, in which all the instruments must play well together!

ACKNOWLEDGMENTS

I would like to thank Nicola Alba, Vittorio Scaravaggi, Michael Golding, Francesco Sindaco, and Anna Sabatini for consulting and photographic material. Special thanks go to Rosangela Radaelli and to Maurizio Pozzi for special measurements and support and to Mario Cottini, who "inspired" this work.

BIBLIOGRAPHY

Chang, H. R., in *Proceedings of Symposium on Defects in Silicon*, edited by W. M. Bullis and L. C. Kimerling, pp. 549–557, Electrochemical Society, Pennington, NJ, 19XX.
Drum, C. M., and C. A. Clark, *J. Electrochem. Soc.* **115**, 664 (1968).
Herring, R. B., in *Handbook of Semiconductor Silicon Technology*, edited by W. C. O'Mara, R. B. Herring, and L. P. Hunt, Chap. 5, Noyes, Park Ridge, NJ, 19XX.
Leroy, B., and C. Plougovel, *J. Electrochem. Soc.* **127**, 961 (1980).
Middlemann, S., and A. K. Hochberg, in *Process Engineering Analysis in Semiconductor Device Fabrication*, Chap. 13, McGraw–Hill International, New York, 19XX.
Pearce, C. W., in *VLSI Technology*, 2nd ed. edited by S. M. Sze, Chap. 2, McGraw–Hill International, New York, 19XX.
Prussin, S., *J. Appl. Phys.* **32**, 1876 (1961).
Shepard, W. H., *J. Electrochem. Soc.* **115**, 652 (1968).
Srinivasen, G. R., *J. Electrochem. Soc.* **125**, 146 (1978).
Srinivasen, G. R., *J. Electrochem. Soc.* **127**, 1334 (1980).
ST Microelectronics internal comunications (Agrate Brianza, Italy).
Tung, S. K., *J. Electrochem. Soc.* **112**, 436 (1965).
Weeks, S. P., *Solid State Technol.* **24**, 111 (1981).
Wolf, S., and R. N. Tauber, in *Silicon Processing for the VLSI Era, Vol. 1. Process Technology*, 2nd ed. Chap. 7, Lattice Press, 19XX.

CHAPTER 10

Si-Based Alloys: SiGe and SiGe:C

D. J. Meyer[1]

ASM EPITAXY
PHOENIX, ARIZONA

I. INTRODUCTION	345
II. APPLICATIONS OF Si ALLOYS	349
1. Bipolar Applications of $Si_{1-x}Ge_x$	349
2. CMOS Applications of $Si_{1-x}Ge_x$	353
3. Optoelectronics: Superlattices for Photodetectors and Emissive Devices	358
4. Alternative Substrate Material: Pseudo-Substrates for GaAs Solar Cells	358
5. Other Si Alloys	360
6. The Role of Carbon in $Si_{1-x}Ge_x:C$ Alloys	362
III. LOW-TEMPERATURE SURFACE PREPARATION	362
1. Native Oxide Removal by H_2 Bake	363
2. Native Oxide Removal by a Liquid-Phase HF Process	364
3. Native Oxide Removal by Vapor-Phase HF Processes	365
4. Other Potential Vapor-Phase Native Oxide Removal Processes	368
IV. PROCESS CHEMISTRY FOR Si ALLOY DEPOSITION	370
1. Influences of Temperature, Pressure, and Gas Composition	371
2. Nonselective Mixed Poly/Epi Deposition	376
3. Selective Deposition of $Si_{1-x}Ge_x$	378
4. Inclusion of C in $Si_{1-x}Ge_x$	378
5. Oxygen Control in $Si_{1-x}Ge_x$ Alloys	382
V. METROLOGY OF $Si_{1-x}Ge_x$ LAYERS	386
1. SIMS	386
2. Auger	387
3. Ellipsometry	388
4. X-ray Diffraction	388
5. The Four-Point Probe	388
VI. PRODUCTION ROBUST $Si_{1-x}Ge_x$ PROCESSING	389
1. Composing a High-Throughput Process	389
VII. SUMMARY	393
REFERENCES	394

I. Introduction

The epitaxial (epi) deposition of Si alloys differs substantially from that of traditional production Si homoepitaxy in that the process is almost always carried

[1]Current address: ATMI, Phoenix, Arizona.

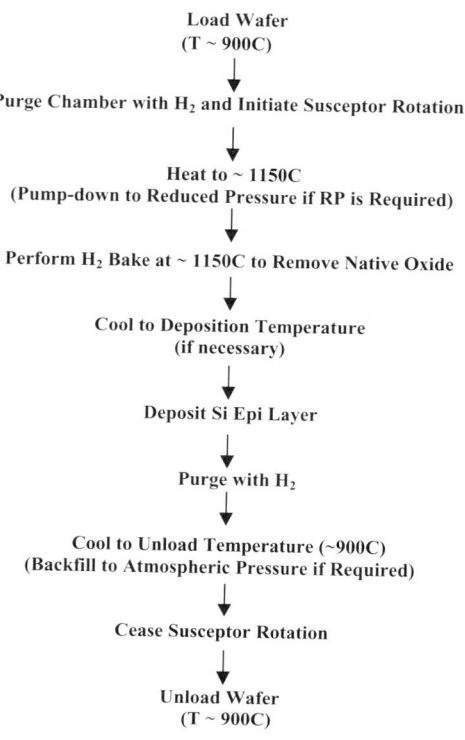

FIG. 1. "Typical" production Si epitaxy process recipe.

out at a low temperature. The typical production Si epitaxy (epi) recipe, outlined in Fig. 1, uses a high-temperature H_2 bake to remove the native oxide on the substrate prior to epi deposition. If this native oxide is not removed, the deposited film will be polycrystalline, or if the temperature is sufficiently low and a chlorosilane is used as the Si precursor, little or no deposition may occur.

An outline for a $Si_{1-x}Ge_x$ epi process, where the base of a bipolar transistor is to be grown, is presented in Fig. 2. The process integration sequence for a bipolar CMOS (BiCMOS) integrated circuit utilizing a $Si_{1-x}Ge_x$ base in some of the bipolar elements, requires that a substantial amount of the CMOS transistor fabrication be completed prior to epitaxially depositing the $Si_{1-x}Ge_x$ layer. This places limitations on the thermal budget for the epi process, which usually prevents the use of a high-temperature H_2 bake for removal of the native oxide.

The lattice constant of Si at room temperature is about 5.43 Å, while that of Ge is ~5.66 Å. Thus, as one might expect, the native lattice constant of a $Si_{1-x}Ge_x$ alloy will be somewhere between these two values and dependent on the Ge content of the alloy. Most $Si_{1-x}Ge_x$ applications require that the $Si_{1-x}Ge_x$ layer be deposited as a pseudomorphic layer on top of a Si layer. The term pseudomorphic

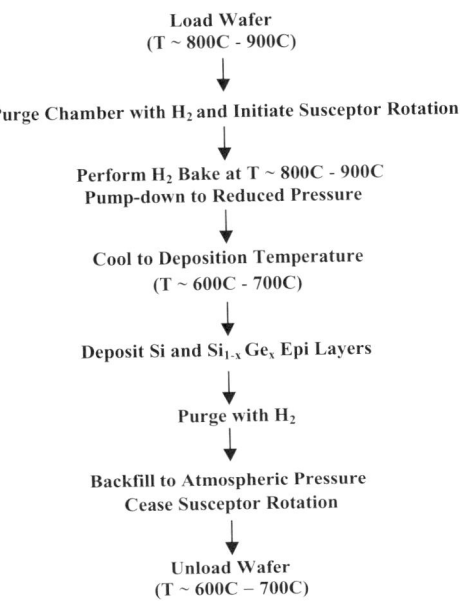

FIG. 2. "Typical" $Si_{1-x}Ge_x$ epitaxy process recipe.

refers to the fact that the lattice constant of the $Si_{1-x}Ge_x$ layer is taking on the lattice constant of the Si layer beneath it. The layer is also referred to as a strained layer since the pseudomorphic $Si_{1-x}Ge_x$ layer is in compressive strain in the plane of the Si substrate. There is, however, nothing to constrain the vertical growth of the $Si_{1-x}Ge_x$ layer above the Si substrate. Perpendicular to the plane of the Si substrate, the $Si_{1-x}Ge_x$ layer lattice constant is larger than its native lattice constant. An illustration of both pseudomorphic and relaxed $Si_{1-x}Ge_x$ films on Si is provided in Fig. 3.

The pseudomorphic $Si_{1-x}Ge_x$ layer is metastable. The thickness and Ge content of the layer determine the temperature at which the layer can be deposited and the thermal budget that the layer can tolerate after it has been deposited. Figure 4 presents a compilation of critical thickness data as functions of deposition temperature and Ge content of the layer. The critical thickness is defined as the maximum $Si_{1-x}Ge_x$ layer thickness that can be grown at a specified temperature and Ge content while maintaining a strained layer [1]. Growing the layer at a higher temperature will result in relaxation of the layer. That is, the strain in the $Si_{1-x}Ge_x$ layer will be relieved by the layer taking on the native lattice constant of this film. The result is a misfit dislocation network, as shown in Fig. 5, with threading dislocations growing vertically from the intersection of each misfit dislocation line. PN junctions violated by these threading dislocations will exhibit substantial leakage currents and, therefore, be rendered essentially useless.

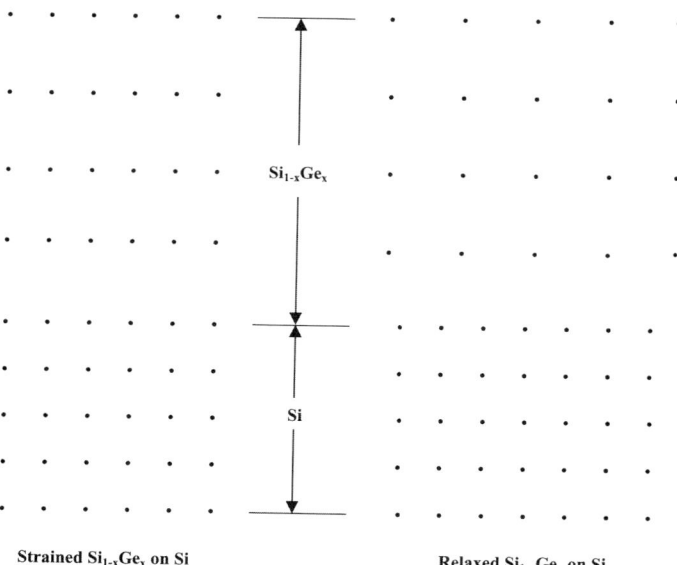

FIG. 3. Pseudomorphic (strained) and relaxed $Si_{1-x}Ge_x$ on Si.

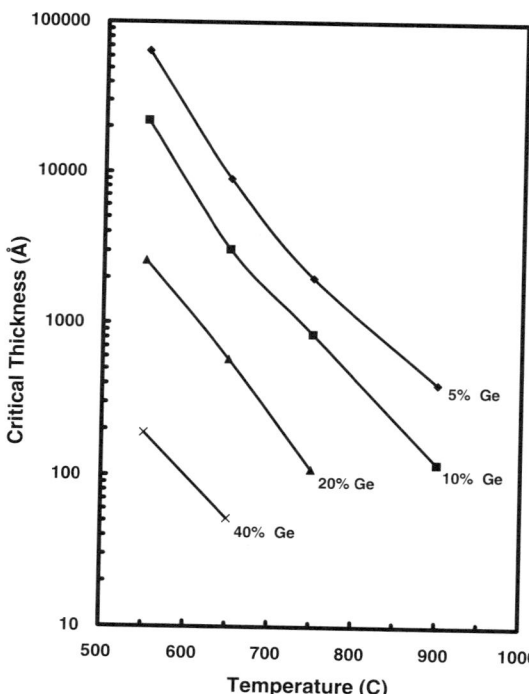

FIG. 4. Critical thickness of $Si_{1-x}Ge_x$ on Si as a function of temperature and Ge content.

FIG. 5. Nomarski micrograph of a relaxed $Si_{1-x}Ge_x$ layer on Si showing misfit dislocations.

Sounds a bit forbidding, does it? Well don't be too concerned. Most of these nasty issues have been worked out to a reasonable degree of functionality. So let's move on to the applications section, where we talk about bipolar, CMOS, optoelectronics, III–V's and so on. It's almost unbelievable to consider that the use of Si and Ge to form a heterojunction device was first proposed back in 1948 by none other than William Shockley [2]. Apparently, good things come to those who wait.

II. Applications of Si Alloys

Applications exist for $Si_{1-x}Ge_x$ materials in bipolar, CMOS, and optoelectronics segments of the Si device markets. There is also an application for $Si_{1-x}Ge_x$ materials as an inexpensive substrate for GaAs solar cells. The relevance of $Si_{1-x}Ge_x$ materials for these markets is discussed in this section.

1. BIPOLAR APPLICATIONS OF $Si_{1-x}Ge_x$

The need for power-efficient radiofrequency (RF) output devices and low-noise RF input devices, which can be integrated with Si CMOS and bipolar technology, has driven the development of Si alloy processing and device technology. At the time that I author this chapter (roughly, 11 PM or so on a Thursday in April 2000, since it is due at the publisher on Friday) the primary application of Si alloys is that of the epitaxially deposited $Si_{1-x}Ge_x$ or $Si_{1-x}Ge_x$:C base of an NPN bipolar transistor used in a BiCMOS process flow for communications devices.

Why should we be interested in $Si_{1-x}Ge_x$ devices when the III–V community can offer clever InGaPHBTs and other very high-speed devices? The answer lies in the term "monolithic integration." The ability to include $Si_{1-x}Ge_x$ devices in a monolithic integrated circuit along with Si CMOS devices will allow us to build VLSI and ULSI integrated circuits on a single die at, ultimately, a very low cost. The ability to monolithically integrate III–V devices with Si, at an adequate level of capability and complexity to meet the needs of the rapidly expanding communications industry, has never been robustly demonstrated. Further, the cost of making ULSI III–V integrated circuits would be astoundingly high compared to that of Si. Thus, the compatibility of $Si_{1-x}Ge_x$ materials with Si for IC manufacturing has provided the motivation.

There are, fundamentally, two types of $Si_{1-x}Ge_x$ bipolar transistor structures presently under consideration. The heterojunction bipolar transistor (HBT), shown in Fig. 6, and the drift field transistor (DFT), shown in Fig. 7. Each structure is discussed below, but please be aware that the industry has somewhat contorted the term HBT to refer to both of these device structures regardless of whether or not the emitter/base and collector/base junctions actually

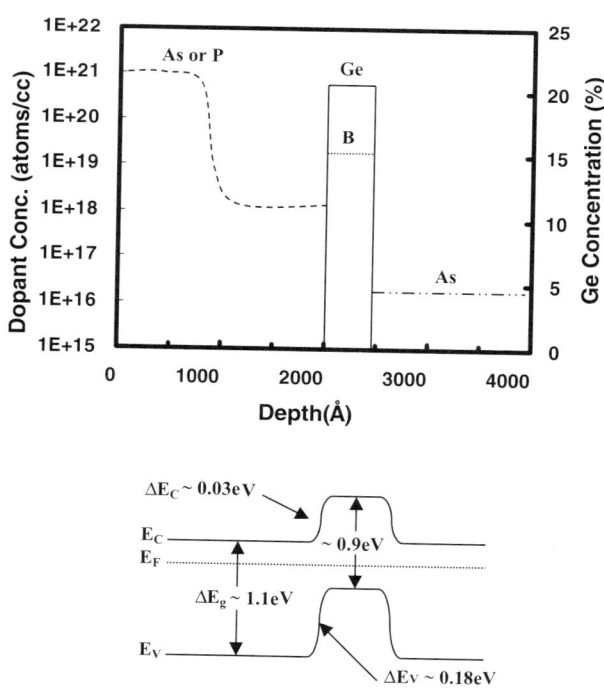

FIG. 6. $Si_{1-x}Ge_x$ HBT depth profile and band structure.

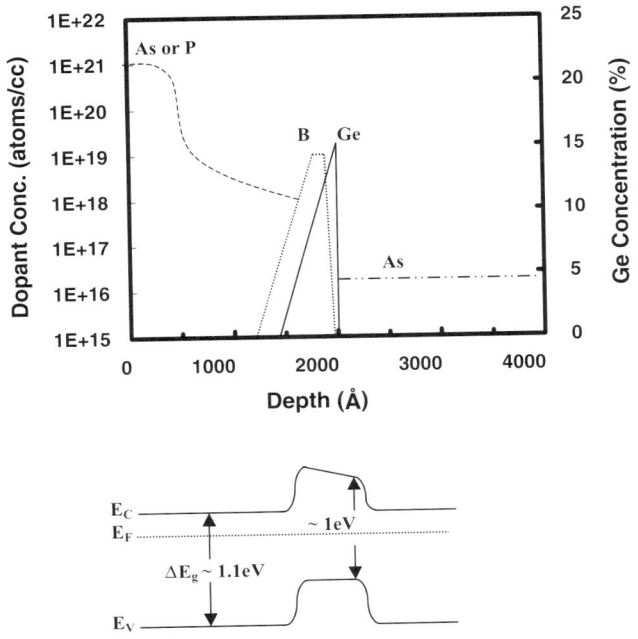

FIG. 7. $Si_{1-x}Ge_x$ DFT depth profile and band structure.

occur at different materials interfaces. So, to avoid too much confusion, keep an open mind when you read through the literature.

a. The Heterojunction Bipolar Transistor

The $Si_{1-x}Ge_x$ heterojunction bipolar transistor, or HBT, is shown in Fig. 6 along with a simplistic diagram of the band structure for this device. The HBT utilizes a box Ge profile with the collector/base and emitter/base junctions located at the $Si_{1-x}Ge_x$ to Si interfaces. During the deposition of the base, the B doping profile is often inserted inside the $Si_{1-x}Ge_x$ region by 50 Å or so to accommodate the diffusion of B caused by additional thermal processing of the wafer. The base width of the HBT is commonly in the range of 500 to 600 Å, which is about 30 to 50% of the base width of a Si bipolar junction transistor (BJT). The frequency at which the gain of a bipolar transistor becomes unity, f_T, alternately referred to as the gain–bandwidth product or cutoff frequency, can be roughly summarized as [3]

$$f_T = \{2\pi \left(W_B^2/\eta D_h + R_C C_{CB} + R_B C_{EB} + \text{other } RC \text{ terms}\right)\}^{-1}. \quad (1)$$

Obviously the f_T of the device is strongly impacted by the base width, W_B, and, therefore, decreasing the base width will increase the f_T.

If the base doping level for the HBT were held at the same level as that of the BJT, the Gummel number in the HBT base, calculated by integrating the base doping profile over the base width, would decrease, due to the use of such a narrow base. Since the base Gummel number largely controls the gain of the transistor, this would be an undesirable effect. This situation would also cause an increase in the base resistance, increasing the time constants of any terms in the f_T equation which include this parameter and, therefore, decreasing the f_T value. Fortunately, the solubility of B at low temperatures in $Si_{1-x}Ge_x$ is much higher than that of B in Si [4–6], which allows the base doping level to be increased substantially over that used in a BJT. It is common to use B doping levels in the base of an HBT which are well in excess of $5 \cdot 10^{18}$ cm^{-3} and sometimes as high as $1 \cdot 10^{20}$ cm^{-3}. Thus, the base Gummel number for a $Si_{1-x}Ge_x$ HBT can be larger than that of a Si BJT and the base resistance for the HBT can be lower than that of a Si BJT.

An additional benefit for the "true" $Si_{1-x}Ge_x$ HBT structure is attributed to the valence band discontinuity which exists between the emitter and base of this transistor. As shown in Fig. 6, most of the change in bandgap occurs in the valence band, and only a small change occurs in the conduction band. This "offset" of the valence bands provides a barrier to the reverse injection of holes from the base of the NPN HBT into the emitter of the transistor, which in turn reduces the base current. The minority carrier injection efficiency, γ, varies with the E_g change [3]

$$\gamma \sim e^{\Delta E_g / kT}. \tag{2}$$

The change in the bandgap, caused primarily by the change in the position of the valence bands between the emitter and the base relative to the Fermi level, therefore causes an improvement in the efficiency of injecting electrons (the minority carrier in the P-doped base) across the base into the collector. This is a very desirable characteristic for a power device since the minority carrier injection efficiency will decrease as the device temperature increases, thus decreasing the transistor gain. The "true" $Si_{1-x}Ge_x$ HBT therefore provides some resistance to thermal runaway based on the intrinsic physics of its design.

b. The Drift Field Enhanced Transistor

The $Si_{1-x}Ge_x$ drift field enhanced transistor (DFT) shown in Fig. 7, is built by using a graded Ge content across the base. For an NPN transistor the Ge concentration is higher near the collector than it is near the emitter, thus providing an electric field which improves the "drift" component of electron transport from the emitter to the collector. This electric field serves to sweep the electron across the base more quickly, thereby reducing the transit time across the base, which in turn reduces the recombination current in the base. The DFT is similar to the HBT in that it also uses a base width which is approximately 30 to 50% that of a Si BJT. Therefore, like the HBT, the f_T for the DFT is also improved by the relationship between the base width and the gain–bandwidth product.

The most common DFT designs grade the Ge content from ~15% near the collector to, effectively, 0% in the region near the emitter/base junction. In such cases improvements in the minority carrier injection efficiency of the device, due to reduced hole injection from the base to the emitter, are not realized. Other DFT designs are hybrids of the HBT and DFT designs where the Ge content of the layer is graded from ~15% near the collector to ~5% near the emitter. The abrupt change of Ge content near the emitter provides some reduction in reverse hole injection similar to that of a "true" HBT.

Perhaps the most significant benefit of the DFT over the HBT is the use of a lower peak Ge content and a lower average Ge content in the film. This provides a metastable base region which is tolerant to higher temperatures for longer times than would be possible for the HBT structure. Indeed, the thermal cycle required to drive the As or P in a "poly-emitter" structure can often dictate the choice of a DFT or HBT.

2. CMOS Applications of $Si_{1-x}Ge_x$

Although, at this time, the industry appears to be enamored with bipolar applications of $Si_{1-x}Ge_x$, there are indeed some interesting CMOS opportunities for this materials group. The elevated source/drain structure and the possibility of using $Si_{1-x}Ge_x$ as a gate electrode are the most near-term applications. The subject of enhanced mobility channel materials is fascinating, but realistically, probably a bit farther off in the future.

a. The Elevated Source/Drain Structure

There are, of course, applications of Si alloys which are amenable to CMOS device structures. The elevated source/drain, shown in Fig. 8, uses a selective deposition of Si or $Si_{1-x}Ge_x$ to enhance or even replace the use of silicides as contacts to the source, drain, and gate electrodes [7, 8]. The films deposited on the source and drain regions are single-crystal and the film deposited on the gate is polycrystalline. In the early 1980s, when this structure was proposed, it was envisioned that a selective Si layer, deposited at a temperature of 800 to 850°C, would be an enabling technology for achieving submicron critical device dimensions. The successful implementation of salicides (self-aligned silicides), initially $TiSi_2$ and more recently $CoSi_2$, provided a more cost-efficient solution for achieving low-resistance contacts and, as a result, selective epi source/drain elevation processes did not become a mainstream process technology.

Interest in the elevated source/drain structure has, however, reemerged for sub-1500 Å devices, but the bar has been raised with respect to process temperature. These small devices have thermal budgets, dictated primarily by B diffusion from the source/drain regions into the channel area and from the gate electrode into

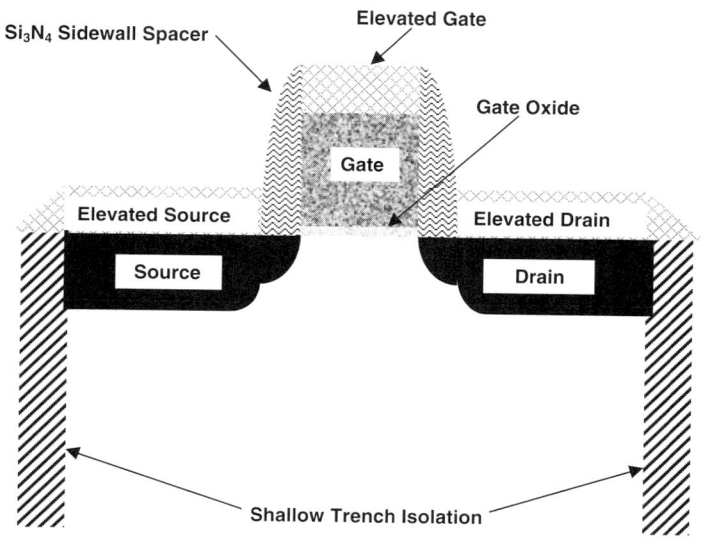

FIG. 8. Elevated source/drain structure.

the gate dielectric, which require process temperatures to be <800°C. Figure 9 provides a comparison of the deposition rates for selective Si and $Si_{1-x}Ge_x$. Clearly, the productivity of an epi reactor growing a 500 Å selective $Si_{1-x}Ge_x$ film at a temperature of, say, 700°C, will be much higher than that of a selective Si deposition at the same temperature. Also, the thermal budget of the selective $Si_{1-x}Ge_x$ process will be substantially less than that of the Si process due to the shorter deposition time. And, finally, the higher solubility of B in $Si_{1-x}Ge_x$ compared to Si, by about two orders of magnitude at 700°C, will provide a lower contact resistance for the $Si_{1-x}Ge_x$ film compared to Si on the PMOS device.

Unfortunately, the selective $Si_{1-x}Ge_x$ process does not appear to provide the same degree of dopant solubility benefits for the NMOS transistor. An increase of only a factor of 3 or so in the P doping level is achieved for $Si_{1-x}Ge_x$ films compared to Si at 700°C [8]. Further, the P in these low-temperature $Si_{1-x}Ge_x$ films appears to be <50% electrically active. Thus, the PMOS transistor is provided the largest advantage from the $Si_{1-x}Ge_x$ elevated source/drain process, while the NMOS transistor receives a more moderate benefit.

b. $Si_{1-x}Ge_x$ Gate Electrodes

Please permit me to make a mild digression from the world of epi into the land of polycrystalline and amorphous $Si_{1-x}Ge_x$ films. The gate electrode, shown in Fig. 8, is clearly not single crystal and therefore not an epi layer. However, the same

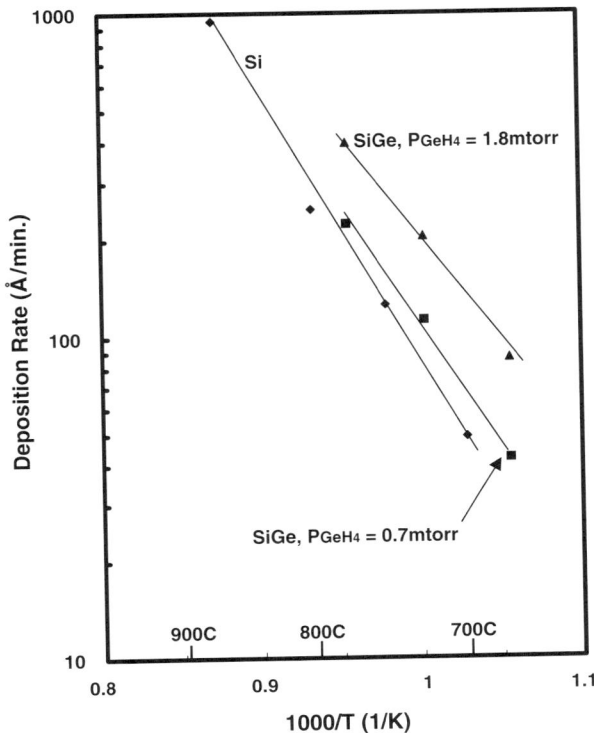

FIG. 9. Arrhenius plot comparing selective Si and SiGe deposition rates.

equipment that can deposit epitaxially aligned films is fully capable of depositing polycrystalline or amorphous films on dielectric surfaces, if the proper precursors are chosen (i.e., SiH_4 or Si_2H_6 instead of a chlorosilane) [9]. Further, the interest in this application for $Si_{1-x}Ge_x$ materials is such that it cannot, in good conscience, be ignored just because this book is targeted at the epi audience.

The present CMOS technology of depositing amorphous Si, performing the self-aligned gate stack etch process, and implanting this gate electrode p-type for PMOS devices and n-type for NMOS devices brings with it some inherent limitations.

- Poly-depletion effects in the gate electrode (a depletion region which extends through the gate dielectric into the gate electrode) decrease the capacitive coupling between the gate electrode and the channel.
- Relatively large implant doses are needed in the channel region of the PMOS device to achieve acceptable threshold voltage levels.
- Diffusion of B from the gate electrode into the gate dielectric increases the gate leakage current and gate resistance.

- The need for two dopant implant steps, one for the PMOS gate and one for the NMOS gate, requires multiple photomask operations.

A poly or amorphous $Si_{1-x}Ge_x$ layer, to be used as the gate electrode [10], provides the following potential benefits.

- The work function of $Si_{1-x}Ge_x$ allows a substantial reduction of the implant dose necessary to achieve the required threshold voltage for the PMOS device. This, however, comes at an expense to the NMOS device, where some increase in the threshold voltage implant dose must occur.
- The higher solubility of B in $Si_{1-x}Ge_x$ provides a lower gate resistance. This also helps to reduce the poly-depletion effect.
- The slower diffusion of B in $Si_{1-x}Ge_x$ than in Si may help to reduce the amount of B which diffuses into the gate dielectric. At the time this chapter was authored a firm conclusion had not been reached on this issue. The arguments, pro and con, are as follows.
  (a) Pro: Decreased diffusion in each grain makes less B available to diffuse to the grain boundaries and through the individual grains to the gate dielectric.
  (b) Con: Diffusion in polycrystalline material is dominated by the grain boundaries. Therefore, diffusion of dopants within the individual grains is not a significant issue.
- The "within-the-bandgap" position of $Si_{1-x}Ge_x$ may allow a single dopant species to be used for both NMOS and PMOS devices. Clearly, B would be the species of choice due to its higher solubility compared to P or As.
  (a) If a single dopant species can be used, the film can be doped *in situ*, thus eliminating any implantation of the gate electrode and, potentially, two photomasking steps.
  (b) If the film can be doped *in situ*, there may be less dopant segregated at the grain boundaries, compared to doping via implantation, and this could have a positive impact on B diffusion into the gate dielectric.

Initially, Ge contents of up to 100% in the $Si_{1-x}Ge_x$ film were investigated for gate electrodes [11]. More recent work indicates that the Q_{ss} increases by a factor of ~ 3 for films containing 50% Ge, compared with a conventional Si gate electrode or films with 30% Ge. In general, the PMOS device receives increasing benefits as the Ge content of the film is increased [11]. However, it appears that, due to Q_{ss} concerns, the maximum acceptable Ge content for these films will likely be somewhere between 30 and 50%.

c. Enhanced Mobility Channels

The deposition of multilayer stacks of Si and $Si_{1-x}Ge_x$ for enhanced mobility channels provides an opportunity for improved switching times for the PMOS device, if a strained $Si_{1-x}Ge_x$ channel is used, and both the PMOS and the NMOS devices, if a strained Si channel is used. This is due to an increase in the mobility of the hole in $Si_{1-x}Ge_x$ materials, whereas the mobility of the electron is substantially unaffected. For strained Si channels, the mobilities of both the hole and the electron can be improved substantially [12].

Oxides containing Ge are not well characterized. Therefore, it is unlikely that the thermal oxidation of a $Si_{1-x}Ge_x$ layer would be undertaken to form a gate dielectric. Indeed, the oxidation of $Si_{1-x}Ge_x$ at high temperatures ($T > 700°C$ or so) tends to be composed entirely of SiO_2, with the Ge being "pushed down" to form a very Ge-rich material immediately under the oxide [13]. $Si_{1-x}Ge_x$ films oxidized at low temperatures have been found to contain Ge in the oxide, due, presumably, to insufficient mobility of the Ge atom at such low temperatures. The lack of understanding regarding the dielectric breakdown fields, relative dielectric constant, and Q_{ss} for Ge-containing oxides, as well as the Q_{ss} associated with a high Ge content at the gate dielectric/channel interface, will most likely preclude the use of $Si_{1-x}Ge_x$ strained layers as channel regions in depletion-mode CMOS devices.

Besides, why improve only one device type (the PMOS device) when you can have it all? The strained Si channel should provide an improvement in the mobility of both PMOS and NMOS devices. So, as Walt Benzing would say, "Do the last experiment first." And, being a big fan of the late and very great Dr. Walter, I certainly agree.

To have a strained Si layer there must be a layer underneath whose lattice constant is a mismatch to that of Si. This is, of course, where the $Si_{1-x}Ge_x$ layer makes its entrance. A relaxed $Si_{1-x}Ge_x$ structure, similar to that shown in Fig. 10, is first deposited. Then a Si layer is deposited on the $Si_{1-x}Ge_x$ structure, with the Si layer taking on the lattice constant of the $Si_{1-x}Ge_x$ layer. In this application Si is therefore the pseudomorph.

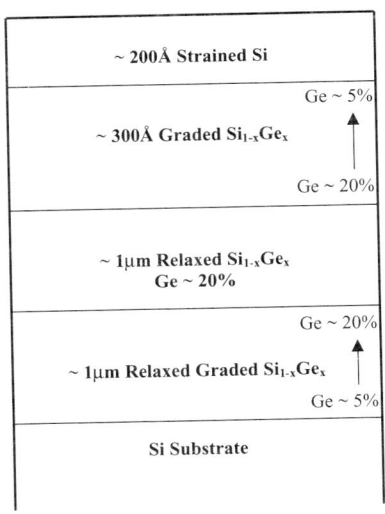

FIG. 10. Strained Si enhanced mobility channel on relaxed $Si_{1-x}Ge_x$.

But what about the crystal defects? If we are to grow relaxed $Si_{1-x}Ge_x$ on Si, then there should be a misfit dislocation network and threading defects propagating to the top of the layer. These defects will wreak havoc on the source/drain-to-well junctions and just about everything else they touch. So something must be done to prevent these defects from intersecting sensitive regions of the device structure.

The approach to this problem consists of growing an "optimized buffer layer" which forces the defects to occur as dislocation loops instead of line defects [14]. Such a structure is claimed to position these defects at the edge of the wafer, or at least near enough to the edge of the wafer, such that most of the wafer area is substantially defect free and amenable to high-performance CMOS device manufacturing. The successful growth of such a structure depends strongly on the Ge composition profile of each layer, the overall thickness of each layer, and, likely, a number of other issues not openly shared by the researchers. The secrets surrounding the growth of these layers are closely guarded by those companies who are developing this capability. I'm sworn to silence. I also openly admit that I don't know all the answers to this one.

3. OPTOELECTRONICS: SUPERLATTICES FOR PHOTODETECTORS AND EMISSIVE DEVICES

$Si/Si_{1-x}Ge_x$ superlattices can be grown that provide two-dimensionally constrained electron and/or hole gases that can, in turn, be used as the basis of photodetector and photoemissive devices [15]. One goal of such devices is to provide a low-cost repeater network for fiber-optic cables operating in the 1.3-μm wavelength region of the optical spectrum. Due to the economy of scale commonly realized in large-volume Si device manufacturing, it is conjectured that Si- and $Si_{1-x}Ge_x$-based repeaters can be made at a much lower cost than III–V-based devices. Further, high levels of integration for Si and $Si_{1-x}Ge_x$ with other Si devices, within a monolithic integrated circuit, can provide additional benefits, such as error detection/correction and self-diagnostics, which are not easily included in a cost-effective manner in III–V technologies.

Figure 11 shows a cross-sectional scanning electron micrograph (SEM) of a $Si/Si_{1-x}Ge_x$ superlattice deposited at 625 and 700°C at a reduced pressure in a production single-wafer epi system. A photoluminesence spectrum of this device is shown in Fig. 12. Figure 13 shows the targeted compositional structure of the superlattice.

4. ALTERNATIVE SUBSTRATE MATERIAL: PSEUDO-SUBSTRATES FOR GaAs SOLAR CELLS

The use of GaAs solar cells for providing electrical power to satellites provides yet another opportunity for $Si_{1-x}Ge_x$ materials [16]. Obviously, a high power output

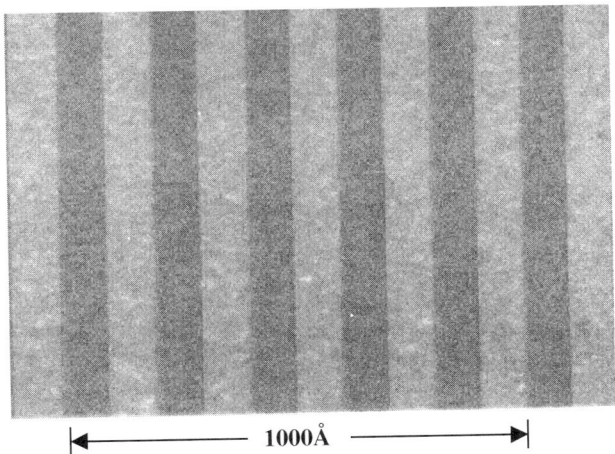

FIG. 11. Scanning electron micrograph of a $Si_{1-x}Ge_x/Si$ superlattice.

but a very low weight is desired for the solar panels in space-based applications. Since GaAs provides a much higher conversion efficiency for solar radiation than does Si, GaAs solar cells are highly desired for the satellite power supply. To minimize the weight, much of the substrate of the solar cell is lapped or polished away after the device has been built. This leaves a very thin, lightweight layer which comprises the solar cell.

GaAs substrates are extremely expensive, with a 150 mm GaAs substrate costing more than U.S. $500. On the other hand, a 200 mm Si substrate sells for less than U.S. $80. Thus, it would be very cost-effective if a Si substrate could be used. Additionally, Si substrates are much less brittle than GaAs and the technology for lapping and polishing Si is very mature.

The lattice constant for Si is about 5.43 Å, compared to 5.65 Å for GaAs and 5.66 Å for Ge, at room temperature. If we can grow a relaxed $Si_{1-x}Ge_x$ structure on Si, and grade it up to about a 96% Ge content, we should be able to provide a nearly perfect lattice match to GaAs, as illustrated in Fig. 14. The cost of manufacturing the pseudo-substrate, based on a prime-quality 200 mm Si substrate as the starting material, would likely be less than U.S. $150. Thus, on the basis of substrate surface area ($/cm^2), the cost of a 200 mm $Si_{1-x}Ge_x$ pseudo-substrate has the potential of being less than 10% the cost of a prime 150-mm GaAs substrate.

It should be pointed out, however, that due to minority carrier lifetime issues, it is unlikely that such a pseudo-substrate technology would be applicable to III–V-based transistors [17]. This is due to the requirement for a very low threading dislocation density to achieve proper device performance, and thus far, threading dislocation densities of $<10^5$ cm^{-2} have yet to be demonstrated in

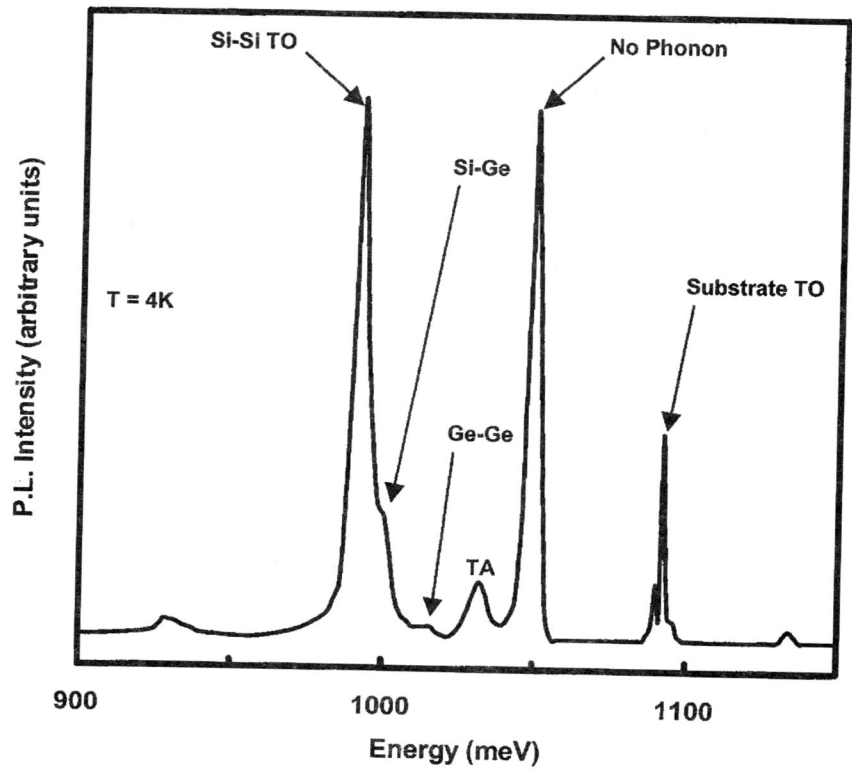

FIG. 12. Photoluminescence spectra of a $Si_{1-x}Ge_x/Si$ superlattice.

pseudo-substrate material structures. Photovoltaic devices, however, are capable of operating with higher dislocation densities and, therefore, are amenable to this technology.

5. Other Si Alloys

There are, of course, other elements which can be alloyed with Si to make semiconducting materials with various characteristics. $Si_{1-x}Sn_x$, $Si_{1-x-y}Sn_xC_y$, $Si_{1-x-y}Ge_xSn_y$, and $Mg_2Si_xGe_ySn_{1-x-y}$ are examples of such alloys [18–29]. These materials are most commonly deposited using molecular beam techniques and are in an amorphous state, as opposed to monocrystalline material. As such, they are not truly epi depositions since they do not take on the crystal structure of the material upon which they are deposited. The difficulty in achieving a

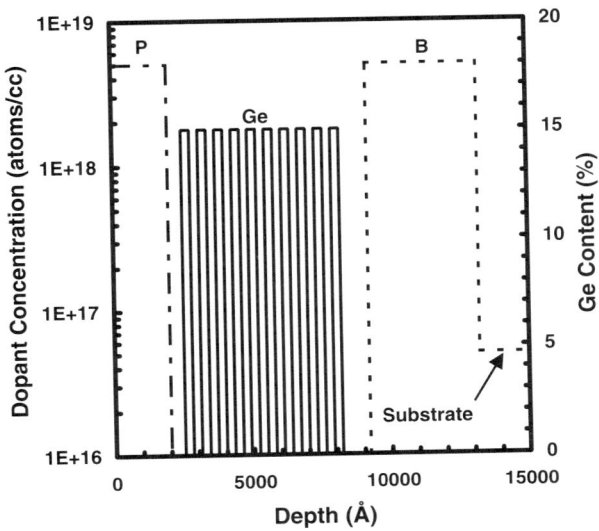

FIG. 13. Targeted Si/Si$_{1-x}$Ge$_x$ superlattice structure.

monocrystalline structure appears to be due to the very large lattice constant of Sn, about 6.5 Å, compared to the lattice constants of Si and Ge (5.43 and 5.65 Å, respectively).

The primary interest in these materials is for group IV optoelectronics applications. However, at this time, these materials are not being embraced by the semiconductor industry manufacturing community. Thus, I have provided a list of

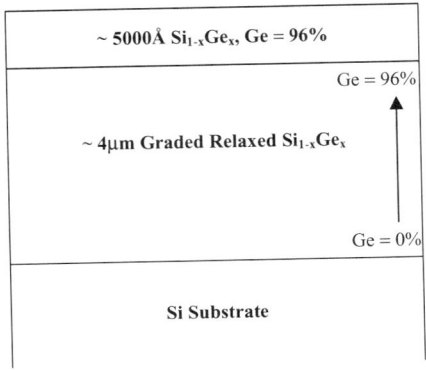

FIG. 14. Si$_{1-x}$Ge$_x$/Si pseudo-substrate for GaAs epitaxy.

6. THE ROLE OF CARBON IN $Si_{1-x}Ge_x$:C ALLOYS

I have written quite a bit regarding $Si_{1-x}Ge_x$ alloys without much discussion of the role that C plays in this technology, other than an oblique reference to it in the preceding section. There are, at this time, two roles that C can play in these alloys. The primary role is to inhibit B diffusion in $Si_{1-x}Ge_x$ materials. It has been found that the addition of small amounts of C, i.e., ~0.5%, greatly reduces the diffusion of B. Thus, very narrow high B concentration spikes can be grown, and these spikes can substantially keep their shape in the face of additional thermal processing, if some C is included in the $Si_{1-x}Ge_x$ layer.

A second application for C in $Si_{1-x}Ge_x$ is for strain compensation. Since C is a very small atom, its addition to Si serves to decrease the lattice constant. This effect is opposite that of adding Ge to Si, and if approximately 1 atom of C is added for each 9 atoms of Ge, the lattice constant of the resulting $Si_{1-10y}Ge_{9y}C_y$ layer can be made to match the lattice constant of Si. Initially, this appears to offer an opportunity to eliminate the metastability of the $Si_{1-x}Ge_x$ base region in an HBT or DFT structure, thus enabling the layer to withstand higher postdeposition thermal budgets. Unfortunately the strain which exists in the pseudomorphic structure plays a major role in altering the bandgap of the semiconductor. Thus, elimination of the strain by including sufficient C in the layer, is counterproductive to the goals sought by using $Si_{1-x}Ge_x$ in HBT and DFT devices. The use of small amounts of C in $Si_{1-x}Ge_x$ has, however, very little impact on the strain in the crystal and has a very large impact on suppressing B diffusion. Thus, $Si_{1-x}Ge_x$:C appears to have a very bright future in bipolar applications where narrow B profiles are required.

III. Low-Temperature Surface Preparation

The essence of epitaxy is that the growth of one material on a substrate requires that the growing material acquire the lattice constant of the substrate. That is the hard fast truth, but often we bend this rule a bit. Bending this rule is most common in the case of heteroepitaxy, where one material is grown on a substrate composed of a different material. Some examples of this are Si on sapphire (Al_2O_3), GaAs on Ge or on $Si_{0.04}Ge_{0.96}$ (the pseudo-substrate application discussed previously), relaxed Si on $Si_{1-x}Ge_x$, and relaxed $Si_{1-x}Ge_x$ on Si, just to name a few. In many of these aforementioned heteroepitaxy applications, misfit dislocations occur. These misfits allow the slightly different lattice constant of the growing material to form a substantially single-crystal material, albeit with a large crystal defect component.

In any event, it is clear to see that if we desire to grow monocrystalline material on a substrate, we are not likely to succeed if the substrate is covered with an amorphous film; the relevant example being the native oxide which exists on a Si substrate. This oxide must, therefore, be removed prior to deposition to achieve an epitaxial (i.e., monocrystalline) film.

1. NATIVE OXIDE REMOVAL BY H_2 BAKE

Most production Si epi processes occur at high temperatures ($T > 1000°C$), which allow the native oxide resident on the surface of the wafer to be removed by interaction with H_2. This high-temperature H_2 bake is very desirable for preparing the Si surface since, in addition to removing native oxide, it also can assist in annealing crystal damage, providing a region near the Si surface which is denuded in oxygen, and causing a desirable surface reconstruction to occur, which assists in the growth of a high-quality epi layer. There are two chemical reactions which are primarily responsible for the reduction of native oxide at high temperatures:

$$SiO_2(s) + H_2(v) \rightleftarrows SiO(v) + H_2O(v), \qquad (3)$$

and

$$SiO_2(s) + Si(s) \rightleftarrows 2SiO(v). \qquad (4)$$

Both of these reactions require relatively high temperatures to proceed at rates adequate for the removal of ~30 Å of native oxide from the surface of a Si substrate, as shown in Fig. 15. The reaction in Eq. (4) is initially a solid-phase reaction but is usually not significant at high temperatures. This is because the SiO, trapped at the Si/SiO_2 interface due to its inability to diffuse through Si or SiO_2, causes the reaction to self-terminate.

The first reaction, where water vapor is released, is obviously very sensitive to the presence of moisture in the reaction chamber since any H_2O in the gas phase will tend to shift the reaction away from reduction of the oxide and toward oxidation of the Si. Indeed, if the H_2O partial pressure is sufficiently high, the Si surface will oxidize via the reaction

$$Si(s) + 2H_2O(v) \rightleftarrows SiO_2(s) + 2H_2(v). \qquad (5)$$

The sensitivity of a bare Si surface to oxygen and moisture in the gas phase was studied, most famously, by Ghidini and Smith [30]. They found that at, ~1100°C, O_2 and H_2O partial pressures of, nominally, 1 Torr and 1 mTorr could be tolerated before the onset of oxidation. At a temperature of 900°C the maximum tolerable O_2 and H_2O partial pressures were found to be approximately 100 and 2 μTorr, respectively (some extrapolation of the H_2O data is necessary to extract the value at 900°C). It is therefore quite clear that if we are to process at low temperatures (typically 600 to 750°C for $Si_{1-x}Ge_x$ deposition), only very small amounts of O_2 and H_2O can be tolerated in the reaction chamber.

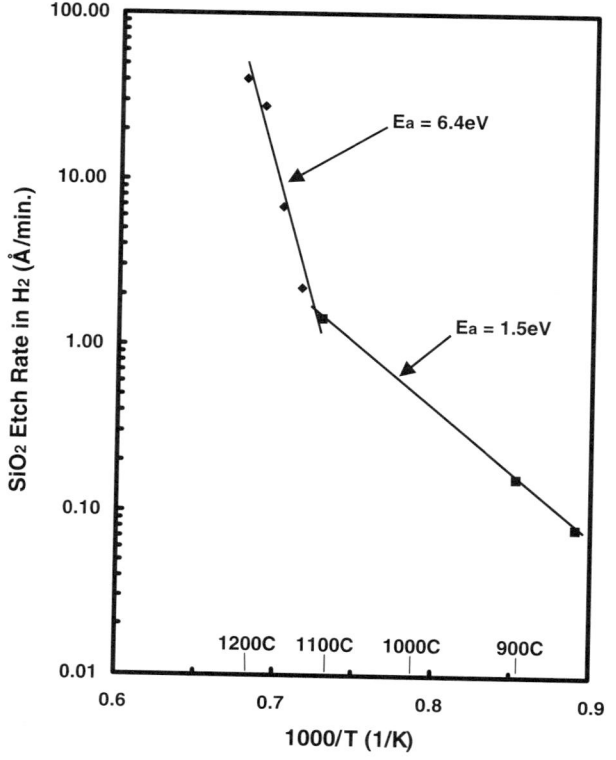

FIG. 15. Reduction of SiO_2 in H_2 as a function of temperature.

But the first significant issue we must face prior to deposition is the removal of the native oxide on the Si surface upon which we wish to grow. The process integration strategy for $Si_{1-x}Ge_x$ in DFT and HBT devices requires that much of the CMOS device structure be in place prior to the deposition of the $Si_{1-x}Ge_x$ base of the bipolar transistors. This puts a severe thermal budget limitation on the process, which precludes the use of a high-temperature H_2 bake for the removal of native oxide. Therefore, some chemical pretreatment of the wafer surface must first be done to enable a lower-temperature H_2 bake to be used effectively.

2. NATIVE OXIDE REMOVAL BY A LIQUID-PHASE HF PROCESS

The removal of SiO_2 by a liquid-phase HF process, often referred to as an "HF dip," is a very common process in Si processing. The SiO_2 is etched via the reaction

$$SiO_2(s) + 6HF(l) \leftrightarrows H_2(v) + SiF_6(l) + 2H_2O(l), \qquad (6)$$

with both HF and HF_2^- being active etchant species [31]. The HF dip effectively removes all of the SiO_2 from the surface of the wafer and results in the wafer surface being hydrogen terminated [32]. At room temperature, hydrogen termination of the Si surface is reputed to protect the surface from oxidation for a period of 1 to 2 h after removal from the HF bath. This is very important since it allows the wafers to be transported to the reactor after processing in a wet sink.

But, of course, the whole story is really not that simple. In a manufacturing environment we can't be dipping wafers in an HF bath and running around the clean-room with them. Safety concerns require that the wafers be rinsed prior to transportation to the next processing tool, and rinsing, with deionized water has been observed to greatly reduce the stability of the hydrogen surface passivation. Thus, it is common for wafers, which have received an HF dip and a deionized water rinse, to require some amount of H_2 bake, at a low temperature, to achieve an epi/substrate interface completely free of oxygen. This H_2 bake also provides the benefit of an epi/substrate interface free of carbon, which is not the case for wafers processed solely at low temperature. The source of the C at the epi/substrate interface is believed to be adsorption of hydrocarbons from the clean-room air onto the wafer surface during transportation of the wafer from the wet sink to the epi process tool.

An HF dip procedure commonly used to prepare the wafer surface is as follows.

- Immerse wafers in 100:1 DI-H_2O:reagent HF for 2 min at 40°C. Etch rate, ~100 Å/min.
- Rinse wafers, *in situ*, in the overflowing weir bath for 15 min at room temperature. Rinse water: TOC (total oxidizable C), <3 ppb; dissolved O_2, <50 ppb.
- Hot N_2 dry in spin rinser/dryer for 5 min.
- Transfer the wafers into the reactor load-lock and pump-down and/or purge with inert gas.

Following this HF dip procedure an H_2 bake is performed at 900 to 800°C for 1 to 5 min prior to low-temperature deposition. Figure 16 compares the H_2 bake time and the equivalent $D \cdot t$ product (diffusion coefficient multiplied by time, for B) of the high-temperature H_2 bake and the H_2 bake used after the HF dip. This $D \cdot t$ product is reduced by approximately three orders of magnitude when the HF dip is used to remove the native oxide.

3. NATIVE OXIDE REMOVAL BY VAPOR-PHASE HF PROCESSES

The liquid-phase HF process is simple and inexpensive but suffers from the need to transport wafers in an uncontrolled environment from the wet sink to the epi process tool. A low-temperature vapor-phase native oxide removal tool, which could be clustered to the epi process tool and thereby provide a controlled (i.e., inert) environment through which wafers can be transferred to the process chamber, is highly desired.

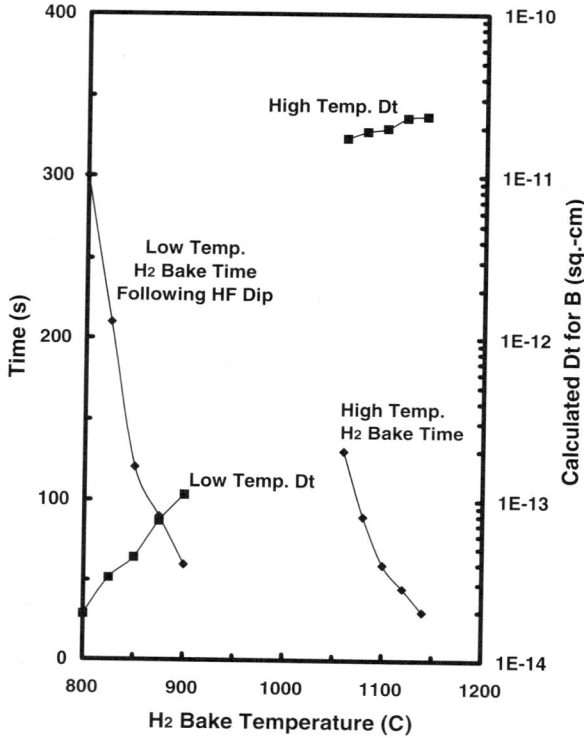

FIG. 16. Comparison of H_2 bake time and boron $D \cdot t$ product for high-temperature H_2 bake without HF dip and low-temperature H_2 bake with HF dip prior to bake.

Vapor-phase chemical etching of native oxide, other than the use of reactive ion technology, can be accomplished using anhydrous HF [33]. The etching of SiO_2 in the gas phase takes place through the reaction

$$SiO_2 + HF \xrightarrow{\text{initiator}} SiF_4 + H_2O. \tag{7}$$

The initiator for the reaction, e.g., H_2O, is a material which assists in the adsorption of the HF onto the surface of the wafer. Thus, the etch rate for an anhydrous HF-based process increases rapidly with time due to the generation of H_2O as a reaction product, which effectively catalyzes the reaction. This uncontrolled etch rate is a major detraction to the use of only anhydrous HF as an etch medium. A second detracting issue is that hydrocarbons on the surface of the oxide block the adsorption sites required for the vapor-phase etch species. Thus, it is possible to etch only a portion of the wafer's surface, or none of the surface at all, depending on the amount of hydrocarbon contamination of the surface. Finally, large disparities in etch rate between different types of oxides exist. Thermal oxide has been found to etch the slowest, plasma-assisted oxides from silane etch somewhat faster, and undensified plasma TEOS oxides etch the fastest.

The etch rate of the reaction can be controlled by first introducing an excess of H_2O vapor into the etch chamber prior to introducing HF [34], which makes the contribution of H_2O from the reaction products negligible. However, the process still remains sensitive to hydrocarbons blocking surface adsorption sites. Disparities in the etch rates of different types of oxides can be reduced by increasing the H_2O:HF partial pressure ratio.

The use of CH_3OH as an initiator, introduced in excess prior to the introduction of HF, solves both issues of controllable etch rate and surface hydrocarbon contamination [35]. Yet another process chemistry uses CH_3COOH as the initiator for the reaction [36]. Once again, disparities in the etch rates of different types of oxides can be reduced by increasing the initiator:HF partial pressure ratio.

The etch rate of thermal oxide as a function of HF charge and CH_3COOH partial pressure is shown in Fig. 17. Figure 18 compares the etch rate of a clean thermal oxide with that of an oxide intentionally contaminated with approximately

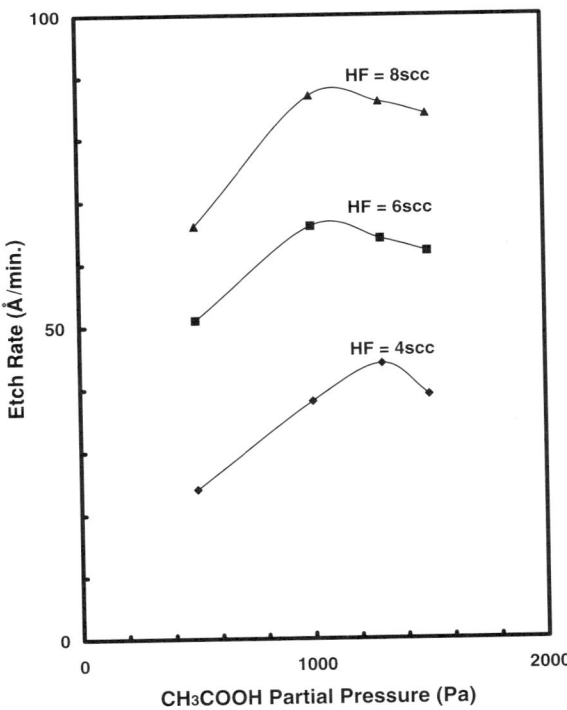

FIG. 17. Thermal oxide etch rate as a function of HF charge and CH_3COOH partial pressure for the vapor-phase etch process.

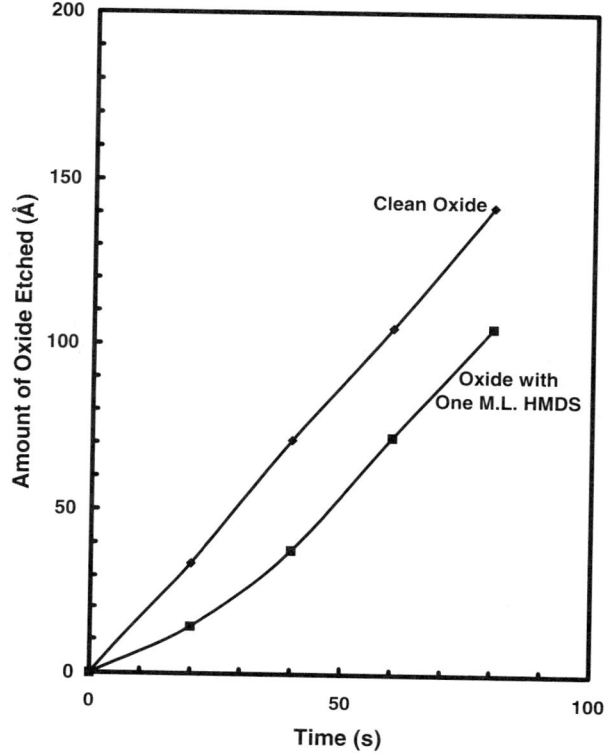

FIG. 18. Etch rate of clean thermal oxide and thermal oxide contaminated with HMDS for the HF/CH$_3$COOH vapor-phase etch process.

one monolayer of hexamethyldisilazane. A reduced etch rate is observed in the first 20 s, after which the etch rate becomes equivalent to that of the clean oxide surface. A comparison of the etch rate of undensified TEOS to that of thermal oxide is shown in Fig. 19. Increasing the CH$_3$COOH:HF ratio greatly reduces the disparity between the etch rates of these oxides.

Thus, there appear to be low-temperature, vapor-phase, clusterable oxide etch techniques which can be used to prepare the wafer surface for epi. The H$_2$ bake requirements for such clustered processes, if any, remain to be defined.

4. OTHER POTENTIAL VAPOR-PHASE NATIVE OXIDE REMOVAL PROCESSES

Although HF-based processes have historically been the "methods of choice" for native oxide removal at low temperatures, it should, in principle, be possible to use a reactive ion etch technology to affect the same results. One concern with such a process would be that the energetic ions may damage the Si surface. Thus, it

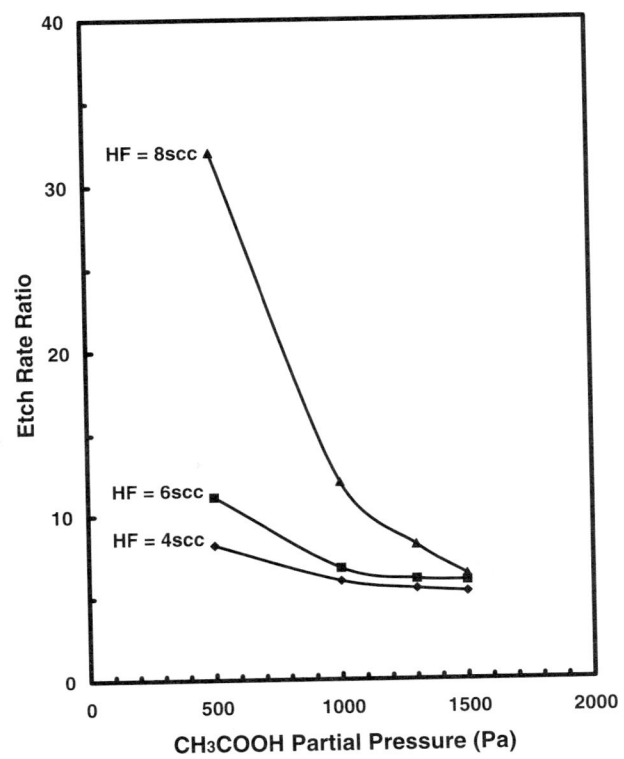

Fig. 19. Thermal oxide:TEOS etch rate ratio as a function of CH_3COOH partial pressure demonstrating etch selectivity for the HF/CH_3COOH vapor-phase etch process.

is likely that the plasma would need to be generated remotely and that the excited radicals would then need to find their way to the wafer surface to interact with the oxide.

Process gases such as NF_3, C_2F_6, CF_4, and F_2 are known to be effective oxide etch chemistries in RIE systems. Whether they would be acceptable for removing such a thin oxide layer as native oxide without damaging the Si surface is not immediately obvious.

One apparently obvious etch chemistry choice would be the use of a remotely generated hydrogen plasma. Hydrogen radicals, which have been given an "effective high temperature" due to RF excitation, may be capable of removing the native oxide in a reaction similar to that of a high-temperature H_2 bake and yet have little interaction with the exposed Si once the native oxide has been removed. Such a process for native oxide removal has yet to find acceptance in the manufacturing environment. It may, however, offer promise for a simple low-temperature native oxide removal process which is not only capable of being clustered to the epi reactor, but may be implemented *in situ* with the epi process chamber.

IV. Process Chemistry for Si Alloy Deposition

The deposition of $Si_{1-x}Ge_x$ alloys typically uses either SiH_2Cl_2, SiH_4, or Si_2H_6 as the Si precursor but has almost always centered around the use of GeH_4 as the Ge precursor. This may be due largely to the availability of GeH_4 as a high-pressure gas, but it does seem curious that $GeCl_4$ has not received much attention. For mixed deposition (sometimes termed "differential deposition"), where polycrystalline material is deposited on oxide and epi on the single-crystal Si surface, the use of GeH_4 seems clearly justified since the presence of Cl inhibits nucleation on the oxide. However, in selective depositions the additional Cl contributed from a $GeCl_4$ precursor would at least minimize the need for added HCl and might eliminate adding HCl completely.

It is likely that GeH_4 exists as the dominant Ge precursor for, basically, historical reasons. Still, the expense of GeH_4 is significant and an alternate less costly Ge precursor would be quite desirable, as this technology moves to full-scale manufacturing.

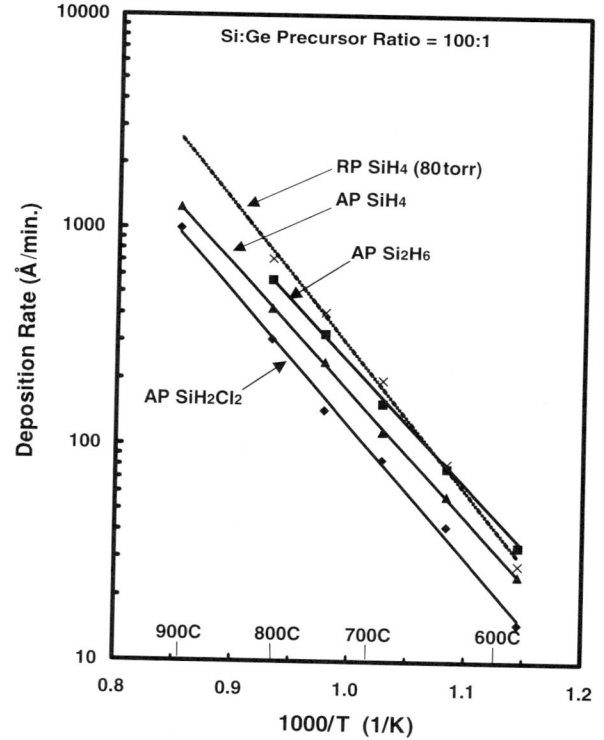

FIG. 20. Arrhenius plot for $Si_{1-x}Ge_x$ epitaxy from SiH_2Cl_2, SiH_4, and Si_2H_6 comparing Si precursors and atmospheric-pressure versus reduced-pressure behavior.

1. INFLUENCES OF TEMPERATURE, PRESSURE, AND GAS COMPOSITION

An Arrhenius plot for the deposition of $Si_{1-x}Ge_x$ from several Si precursors at atmospheric pressure, and SiH_4 at a reduced pressure, is shown in Fig. 20. The deposition occurs in an apparently reaction rate-limited regime over the temperatures of interest (600–700°C). Also, the deposition rate increases for a fixed GeH_4 partial pressure as the Si precursor is changed from SiH_2Cl_2 to SiH_4 to Si_2H_6. Figure 20 also shows that the reduced-pressure process demonstrates higher deposition rates than the atmospheric-pressure process for the same Si:Ge precursor partial pressure ratio. The Arrhenius plot in Fig. 21 demonstrates that decreasing the Si:Ge precursor ratio, by increasing the GeH_4 partial pressure, results in a decrease in the apparent activation energy of the process. This effect is explained in Fig. 22, where the deposition of Ge from GeH_4 is shown to be mass transport limited at temperatures above ~550°C, while the deposition of Si from either SiH_2Cl_2, SiH_4, or Si_2H_6 is reaction rate limited for temperatures <750°C. Thus, the apparent

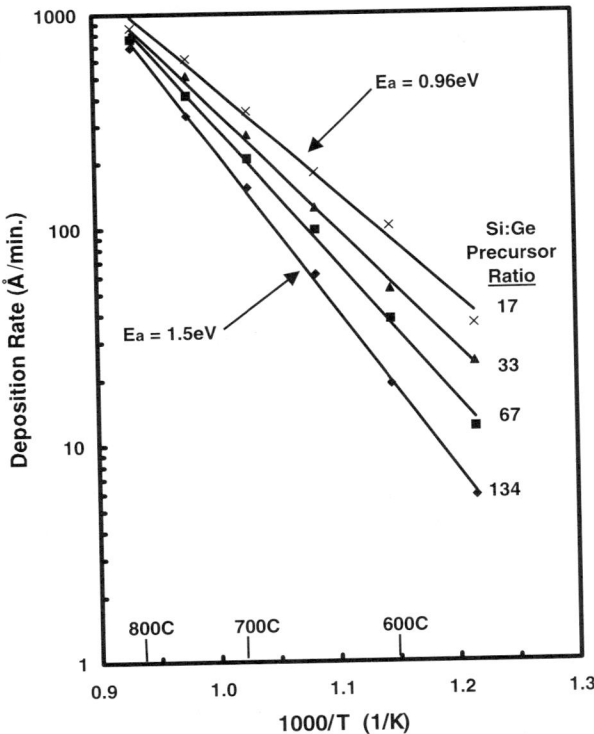

FIG. 21. Arrhenius plot for $Si_{1-x}Ge_x$ epitaxy demonstrating the impact of Si:Ge precursor ratio on deposition rate and effective activation energy.

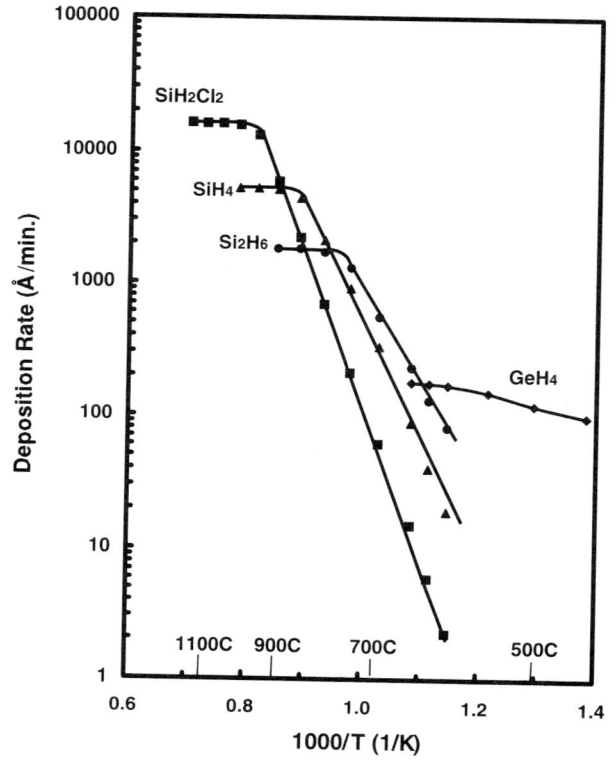

FIG. 22. Arrhenius plot comparing Si homoepitaxy from SiH_2Cl_2, SiH_4, and Si_2H_6 with poly-Ge deposition on poly-Si from GeH_4.

activation energy observed for $Si_{1-x}Ge_x$ deposition appears to be a weighted average of the activation energy for the Si precursor and the essentially zero activation energy associated with the mass transport limited deposition of Ge from GeH_4.

Increasing the temperature while holding the Si and Ge precursor partial pressures constant causes a decrease in the Ge content of the deposited alloy, as shown in Fig. 23. Also, fixing the GeH_4 partial pressure and Si precursor partial pressure but changing the Si precursor from SiH_2Cl_2 to SiH_4 causes a decrease in the Ge content of the alloy. Operation at a reduced pressure results in a slightly lower Ge content alloy at the same Si:Ge precursor ratio as shown in both Fig. 23 (100:1 ratio) and Fig. 24 (17:1 ratio). The impact of Si precursor choice and GeH_4 partial pressure on the Ge content of the alloy is demonstrated in Fig. 25. The Ge content initially changes very quickly as the GeH_4 partial pressure is increased but then slows at higher GeH_4 partial pressures. Figure 26 shows that decreasing the Si precursor partial pressure while holding the GeH_4 partial pressure constant also results in an increase in Ge content of the alloy. The gas-phase Si:Ge precursor

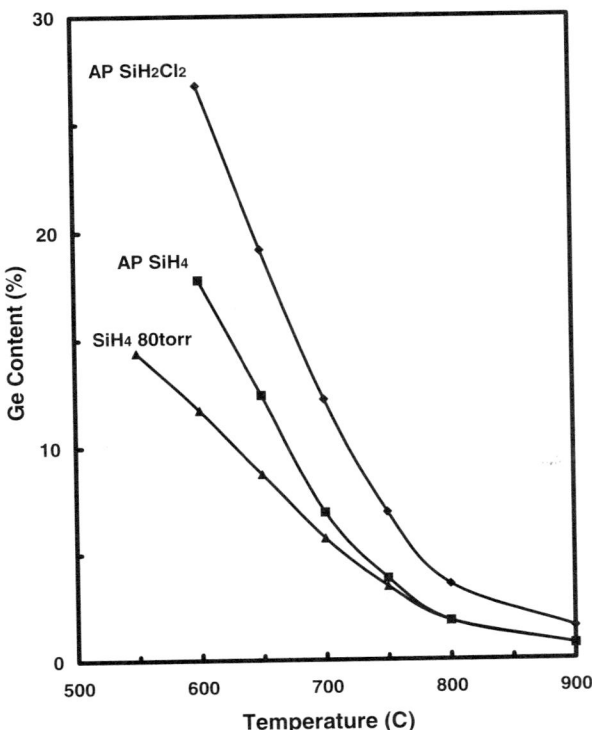

FIG. 23. Effect of temperature on the Ge content of $Si_{1-x}Ge_x$ epi layers deposited from SiH_2Cl_2, SiH_4, and Si_2H_6.

ratio is therefore the dominant factor controlling the Ge content of the alloy at a fixed temperature.

At this point it is worth making a comment on the choice of processing pressure for $Si_{1-x}Ge_x$ deposition. There is, fundamentally, no requirement for using a subatmospheric-pressure process to deposit oxygen-free $Si_{1-x}Ge_x$ alloys. Many people find this point very difficult to believe, probably due to molecular beam epitaxy (MBE) [37] and ultrahigh-vacuum chemical vapor deposition (UHV/CVD) [38] having been the earliest successful $Si_{1-x}Ge_x$ processing technologies. In actual fact, numerous investigators have grown $Si_{1-x}Ge_x$ HBT structures at atmospheric pressure with oxygen levels, at the epi/substrate interface and in the $Si_{1-x}Ge_x$ layers, below the limit of detection by secondary-ion mass spectroscopy (SIMS) (roughly $<5 \cdot 10^{17}$ cm^{-3}) [39–42]. No caveats, no exclusions, no excuses, no tricks, and *no oxygen!* So why is reduced pressure the most commonly used mode for this process? Reduced pressure allows us to achieve deposition rates for polycrystalline material and epi, which are nearly equal, as shown in Fig. 27. The importance of this issue is discussed in the next section.

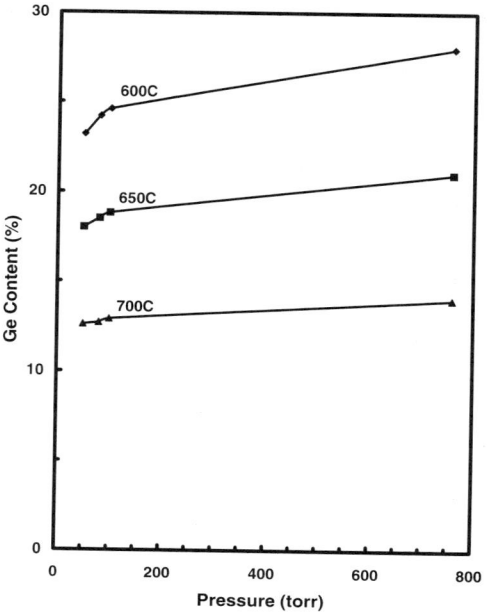

FIG. 24. The impact of pressure on the Ge content of $Si_{1-x}Ge_x$ epi layers at 600, 650, and 700°C.

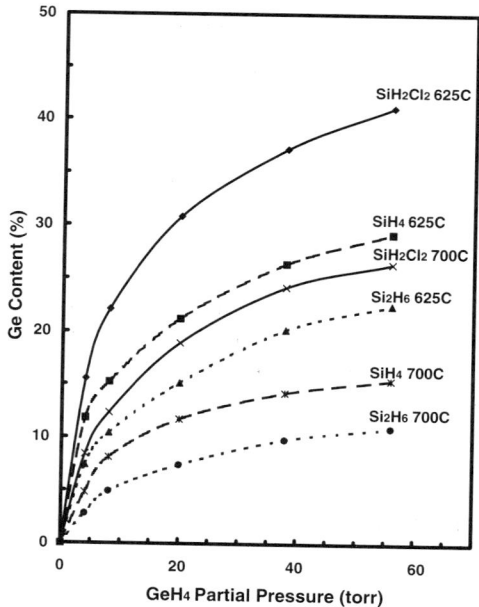

FIG. 25. The effect of GeH_4 partial pressure on the Ge content of $Si_{1-x}Ge_x$ epi layers deposited from SiH_2Cl_2, SiH_4, and Si_2H_6 at 625 and 700°C.

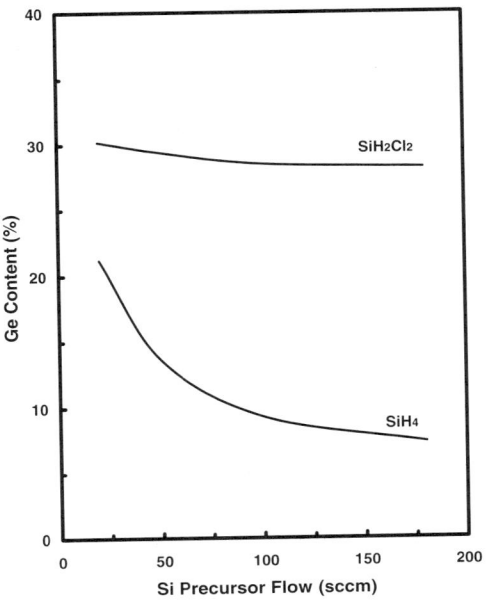

FIG. 26. Increasing the Si precursor flow results in a decrease in Ge content of the $Si_{1-x}Ge_x$ alloy. The effect is found to be much stronger for SiH_4 than for SiH_2Cl_2.

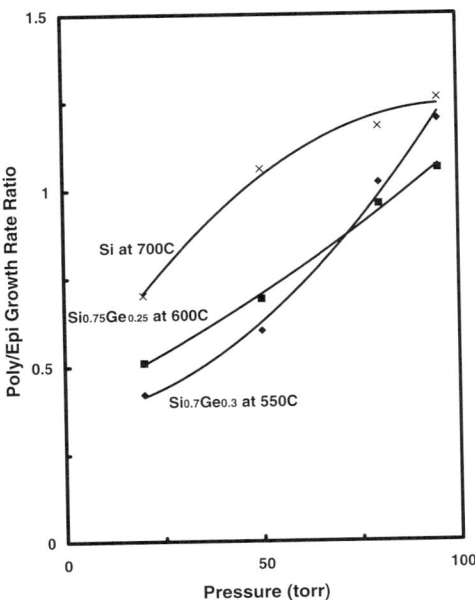

FIG. 27. A comparison of Si and $Si_{1-x}Ge_x$ polycrystalline and epitaxy growth rates as functions of temperature and pressure.

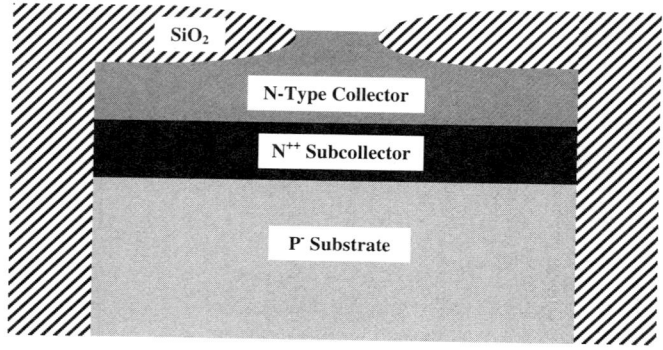

FIG. 28. The double-poly transistor structure just prior to mixed poly/epi deposition of the base.

2. Nonselective Mixed Poly/Epi Deposition

The most common transistor structures used for $Si_{1-x}Ge_x$ HBT and DFT devices use a "double-poly" design, where the first polycrystalline layer acts as the contact handle to the base of the transistor and the second poly layer forms the poly-emitter. The $Si_{1-x}Ge_x$ base of the transistor is commonly deposited over a structure similar to that in Fig. 28. A technique known as "mixed deposition" (also referred to as "differential deposition") is used, whereby polycrystalline material is grown on top of the oxide and epi is grown on the single-crystal Si region. Since most oxides of Ge are volatile, it is not possible to grow a continuous poly-$Si_{1-x}Ge_x$ film directly on top of the oxide regions. Thus, a polysilicon layer must first be deposited on top of this oxide.

One method of achieving the poly-on-oxide structure requires the *ex situ* deposition of an ~500-Å polysilicon layer (referred to as the "poly seal") on top of the oxide in a low-pressure CVD (LP-CVD) furnace. Openings are then etched in the poly-oxide and the wafer is transferred to the CVD reactor for deposition of the $Si_{1-x}Ge_x$ layer. This technique is most often used for UHV-CVD due to the very long times associated with nucleating poly-Si or α-Si on oxide at very low pressures of operation (1 to 10 mTorr).

The use of a production single-wafer Si epi reactor allows the poly-seal layer to be grown *in situ* and much thinner than the *ex situ* approach. By using operating pressures of ~100 Torr, the poly nucleates very quickly on the oxide and, at 700°C, a 100-Å silicon layer (typically referred to as the "Si seed") can be deposited in ~30 s. The layer is polycrystalline over the oxide and monocrystalline (epi) over the single-crystal region. The $Si_{1-x}Ge_x$ layer for the base is then grown immediately following the Si seed and an ~300-Å Si cap layer is then deposited on the $Si_{1-x}Ge_x$ layer to stabilize the structure and host the emitter of the transistor. This entire process sequence is grown *in situ* in a single process recipe. The resulting structure is shown in Fig. 29. The final transistor structure, including contacts, is diagrammed in Fig. 30.

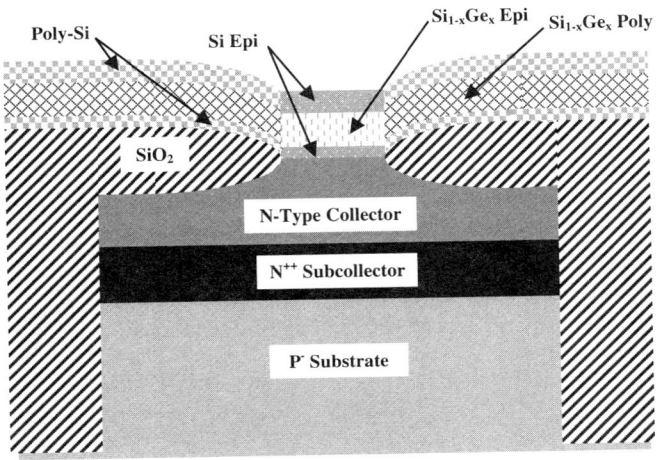

FIG. 29. The double-poly transistor after deposition of the Si cap, $Si_{1-x}Ge_x$, and Si seed layers. The film thicknesses are grossly exaggerated for clarity.

It is important that the polycrystalline portion of the structure be at least as thick as the single-crystal region since this is a critical component of the electrical contact to the base. If the polycrystalline region is too thin, the electrical resistance between the base and the contact will be high, and this will negatively impact the transistor speed. The upper limit to the thickness of the poly region compared to the epi region is not immediately clear. It is not uncommon for device designers to

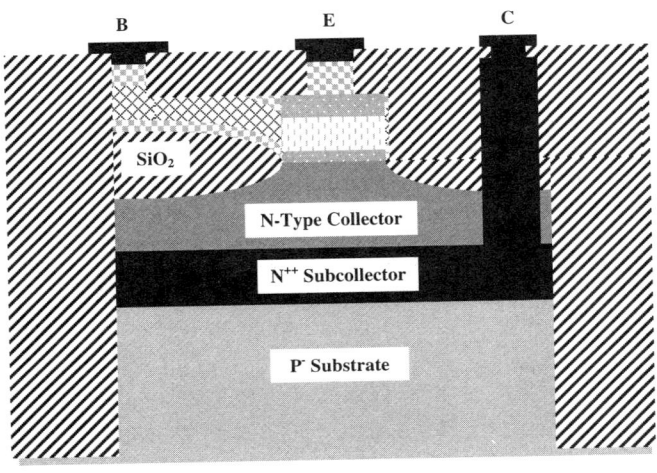

FIG. 30. The completed double-poly $Si_{1-x}Ge_x$ transistor (with exaggerated film thicknesses).

request that the poly layer thickness be equal to, or even up to 50% thicker than, the epi region layer thickness.

Chemical loading effects can present a challenge to monitoring the process without using product substrates. The size of the single-crystal Si openings and the amount of exposed Si on the wafer have an impact on the Ge content, doping level, and deposition rate of the growing film. A bare Si wafer or a wafer 100% covered with oxide is useful for obtaining an approximation to the desired structure. However, the actual patterned wafer may have film thicknesses and compositions which are significantly different. Thus, in a manufacturing environment where we are interested in film variations of the order of 1%, it is desirable to use monitor substrates which have an exposed Si area and Si openings similar in size to those of the product wafer.

3. Selective Deposition of $Si_{1-x}Ge_x$

It is also possible to deposit $Si_{1-x}Ge_x$ using selective epi. In this process deposition occurs only on the exposed monocrystalline and polycrystalline Si regions, and not on the oxide or nitride. The selective deposition process is relevant both to the base of a bipolar transistor and to the elevated source/drain structure for CMOS devices.

Pseudo-Arrhenius plots for selective Si and selective $Si_{1-x}Ge_x$ are shown in Fig. 31. These plots do not truly following the Arrhenius rule since the HCl flow required to achieve selectivity is different at each temperature. In general, as the temperature is decreased, less HCl is required to achieve selectivity, as shown in Fig. 32 for Si.

The case for $Si_{1-x}Ge_x$ is more complicated. Since adding GeH_4 increases the deposition rate, more HCl is required at a given temperature for $Si_{1-x}Ge_x$ than would be required for Si alone. It is also worth noting that since the base of these NPN transistors is doped *in situ* with B, and B increases the deposition rate slightly at low temperatures, the amount of HCl required for B doped films is slightly larger than that required for intrinsic films.

In the selective epi approach there is no need to balance polycrystalline versus epi deposition rates. However, loading effects, which cause smaller exposed Si regions to grow faster than larger regions, are very strong and are controlled by pressure. As shown in Fig. 33, pressures of <50 and <30 Torr are necessary to eliminate loading effects in selective Si and $Si_{1-x}Ge_x$ processes, respectively.

4. Inclusion of C in $Si_{1-x}Ge_x$

The principal interest in adding C to $Si_{1-x}Ge_x$ is to reduce the amount of B diffusion in the base region of the transistor. It has been found that B diffuses more slowly in $Si_{1-x}Ge_x$ than in Si and that the addition of 0.2 to 0.5% C, substitutionally

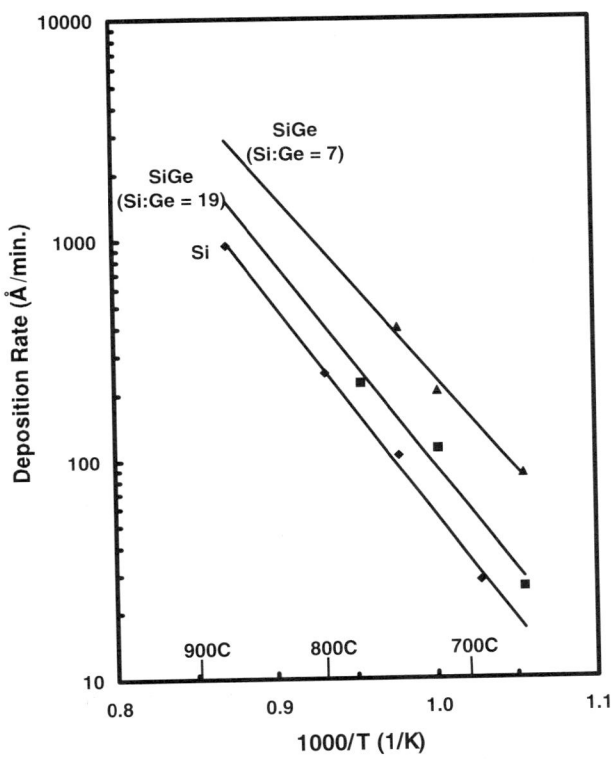

FIG. 31. Pseudo-Arrhenius plot for the selective deposition of Si and $Si_{1-x}Ge_x$ epitaxy at two Si:Ge precursor ratios.

in the $Si_{1-x}Ge_x$ lattice, reduces the diffusion of B even further [43]. The most common precursor used for including C in $Si_{1-x}Ge_x$ is $SiCH_6$. Other precursors, such as C_2H_4, also are effective C sources [44]. Simple alkanes such as CH_4, C_2H_6, and C_3H_8 do not decompose adequately at the temperatures of interest and, therefore, are not useful C precursors.

It is fortunate that only a small amount of C is necessary to impede the diffusion of B since C concentrations in excess of 3% or so can result in the film becoming amorphous [44]. Figure 34 compares the total C in a $Si_{0.8}Ge_{0.2}$:C film, as measured by SIMS, with the substitutional C component of the films, as measured by X-ray diffraction. As the process temperature increases, the C content of the film will increase for a fixed $SiCH_6$ partial pressure. The lower-temperature process, however, provides a higher substitutional C concentration than the higher-temperature process. Thus, lower temperatures (e.g., <625°C) are preferred for $Si_{1-x}Ge_x$:C processes, to achieve the highest possible substitutional C concentration and, at the same time, minimize the total C content of the film.

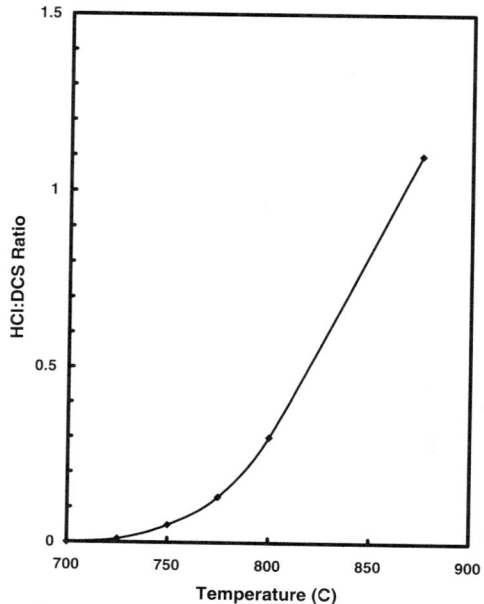

FIG. 32. The required HCl:SiH$_2$Cl$_2$ (DCS) gas-phase ratio for selective Si at temperatures of 700 to 875°C.

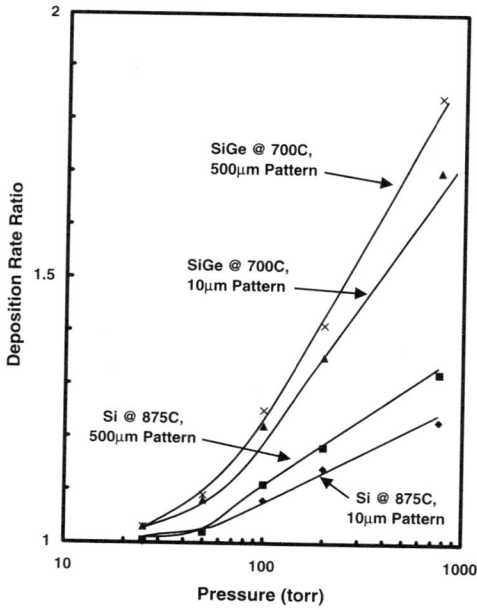

FIG. 33. Microloading effects for the selective deposition of Si and Si$_{1-x}$Ge$_x$ epitaxy. The deposition rate ratio is defined as the deposition rate of a 10 × 10- or 500 × 500-μm pattern divided by that of a 1 × 1-μm pattern.

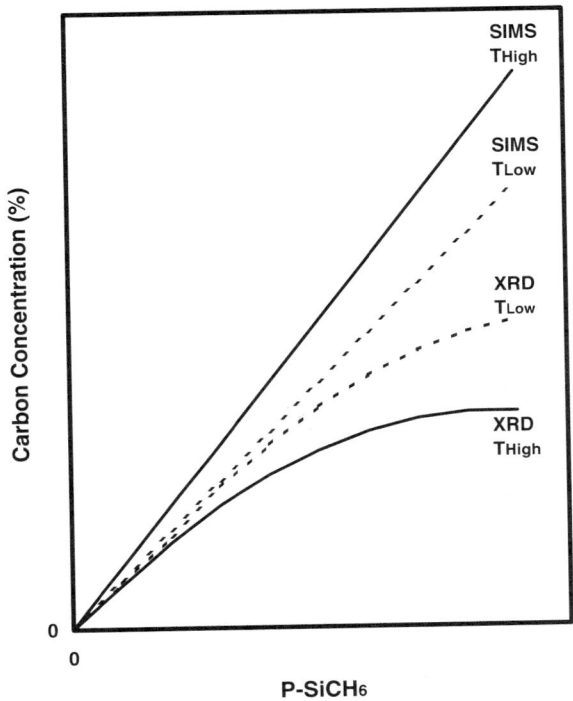

FIG. 34. A comparison of the C content in a $Si_{1-x}Ge_x$:C epi layer as measured by SIMS (total C content) and by XRD (substitutional C content) as a function of temperature and $SiCH_6$ partial pressure.

As one might expect, increasing the $SiCH_6$ partial pressure results in an increase in the C content of the film, as shown in Fig. 35. This increase in $SiCH_6$ partial pressure does not have any significant impact on the deposition rate or Ge content of the film over the range studied. Increasing the total pressure results in a lower C content in the film.

Figure 36 shows that increasing the GeH_4 partial pressure, while holding the $SiCH_6$ partial pressure constant, results in an increase in the C content of the film. The C content of the film will decrease when all of the precursors are held at a constant partial pressure and the temperature is increased. Therefore, if the Ge content of the film is to be held constant, increasing the temperature will result in an increase in the C content. This is caused by the need to increase the GeH_4 partial pressure to compensate for the effect of increased temperature on Ge content (Ge content decreases with increasing temperature).

The pressure dependence for Ge and C incorporation, and the linking between Ge content and C content in the film, also provides an interesting situation regarding process pressure changes. If the process pressure is increased, the C content of the film will decrease and the Ge content will increase. It therefore becomes

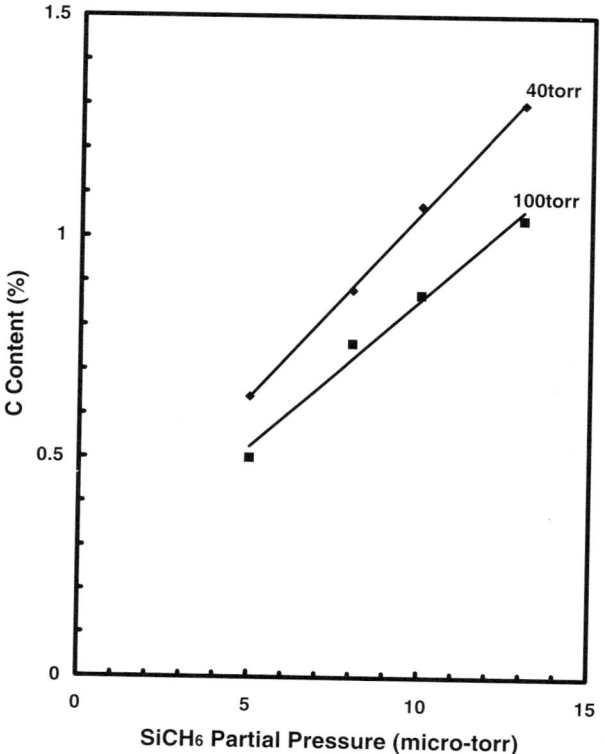

FIG. 35. The direct impact of $SiCH_6$ partial pressure on total C content of a $Si_{1-x}Ge_x$:C epi layer at 40 and 100 Torr.

necessary to decrease the partial pressure of GeH_4 in the system to hold the Ge content of the film constant, which results in an additional decrease in the C content of the film. Thus, a substantial increase in the $SiCH_6$ partial pressure is necessary to counteract the impact of increasing pressure when the Ge content of the film is to be held constant.

5. Oxygen Control in $Si_{1-x}Ge_x$ Alloys

The use of load-locked, inert gas-purged (or vacuum-based) wafer transfer systems coupled to the epi reactor process chamber has enabled the growth of O-free $Si_{1-x}Ge_x$ alloys. Previous generations of production epi reactors, in which wafers were loaded and unloaded from the process chamber in the laboratory ambient, precluded their ability to grow $Si_{1-x}Ge_x$ materials with acceptable O levels.

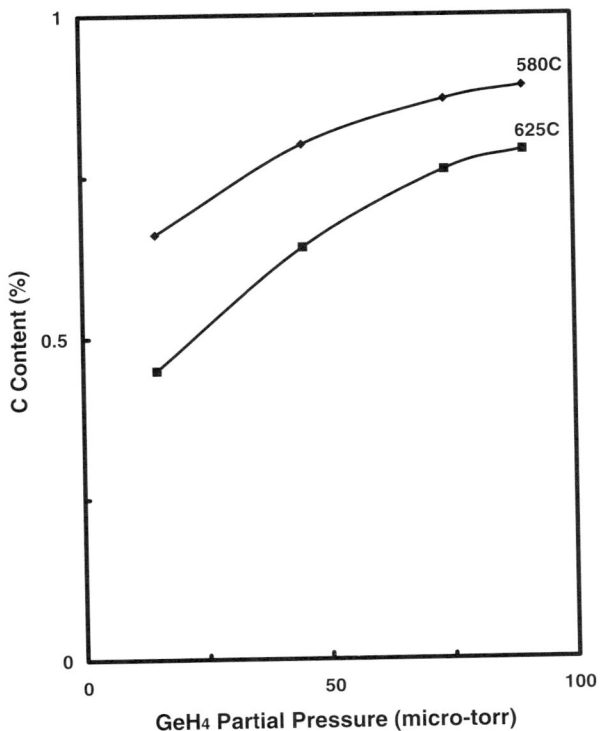

FIG. 36. The impact of GeH$_4$ partial pressure on total C content of a Si$_{1-x}$Ge$_x$:C epi layer at 580 and 625°C.

Oxygen has a very high affinity for incorporation in Si$_{1-x}$Ge$_x$ alloys. A low-temperature Si epi process may show no oxygen in the deposited layer, while a Si$_{0.9}$Ge$_{0.1}$ film, deposited at the same temperature (e.g., ~700°C), may show an oxygen content $>1 \times 10^{19}$cm^{-3}. It is common for a newcomer to the technology to suspect that the Ge precursor is responsible for the source of O atoms, since it was not present when the Si layer was deposited. Although this is a possibility, it is most often not the case.

The SIMS profile in Fig. 37 demonstrates that the Si layer between each Si$_{1-x}$Ge$_x$ layer is essentially O-free, while the Si$_{1-x}$Ge$_x$ layers deposited at 725 and 700°C show clear evidence of O. After the installation of a purifier in the H$_2$ supply the O is eliminated from Si$_{1-x}$Ge$_x$ layers, as shown in Fig. 38, down to a deposition temperature of at least 625°C. No conclusion can be drawn from Fig. 38 regarding the absence or presence of O in the film deposited at 600°C since the O from the surface of the wafer (native oxide) obscures the signal from this layer. It is, however, completely realistic to achieve O-free Si$_{1-x}$Ge$_x$ films at deposition temperatures down to at least 550°C using atmospheric-pressure or reduced-pressure CVD [40].

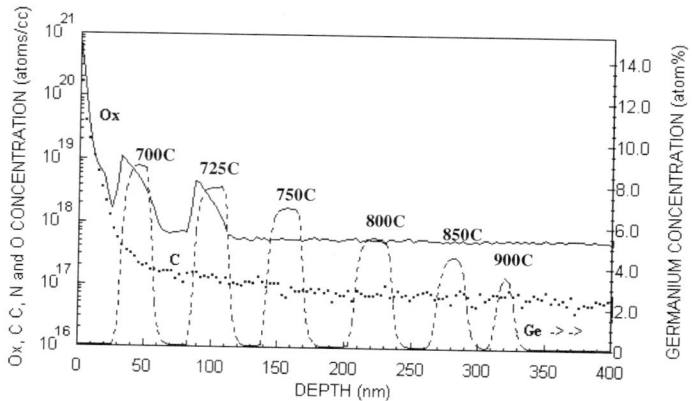

Fig. 37. SIMS of oxygen contamination in low-temperature $Si_{1-x}Ge_x$ epi layers. Note that the higher-temperature $Si_{1-x}Ge_x$ layers do not show any oxygen. Futher, oxygen is not found in the Si epi or at the epi/substrate interface. The H_2 was from a liquid source and has an O_2 and H_2O content of approximately 200 ppb.

It's time for a word about the term "oxygen-free." Every SIMS tool has a background level of oxygen, and as shown in Figs. 37 and 38, this background level is not zero (a very difficult level to achieve on a logarithmic scale). Thus, when the claim is made that a layer is free of oxygen, I am stating that there is no observable O feature above the background of the SIMS system. But we cannot use the SIMS background as a place to hide. Therefore, it is essential that the SIMS have a background level of 1×10^{18} cm^{-3} or less for the measurement to be useful (as a reference point, please note that the interstitial O content of a CZ Si wafer is

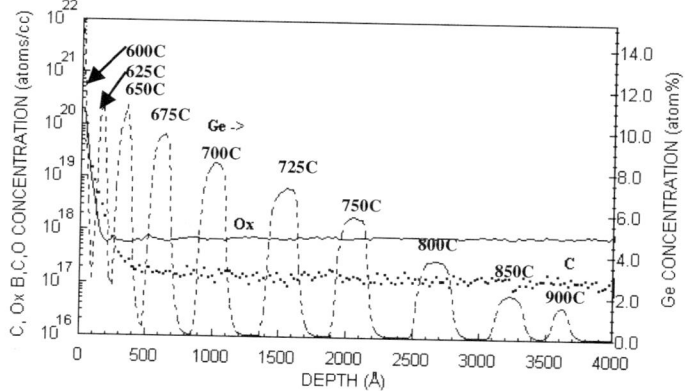

Fig. 38. SIMS of oxygen-free $Si_{1-x}Ge_x$ after installation of a purifier in the H_2 delivery system (estimated O_2 and H_2O content in the H_2, <10 ppb).

about 20 ppm atomic, which corresponds to an O level of $\sim 1 \times 10^{18} \mathrm{cm}^{-3}$). One more issue regarding SIMS, which is rather obvious but bears mentioning: If you are looking for O in a film, O cannot be used as the sputtering agent. Thus, Cs is commonly used for sputtering the film.

What about the sources of O which contribute to the contamination in the $Si_{1-x}Ge_x$ layer? From a species perspective, H_2O is the biggest player due to its higher oxidation potential compared to O_2. Also, any leak from the laboratory ambient which would contain O_2 would also likely contain H_2O, due simply to the relative humidity of the air.

Nobody likes to acknowledge the existence of leaks, but, in actual fact of course, everything leaks. The real questions are, "How much?" and "Are the leaks significant?" So leaks in the reaction chamber or the associated gas delivery system are potential sources of O from the outside world. There is also the issue of leaks in the wafer transfer system, which allow O_2 and H_2O to enter the reactor when the wafer is loaded. This can contaminate the $Si_{1-x}Ge_x$ layer but it often provides an additional signal by which it can be diagnosed: an O peak at the epi/substrate interface. Often, oxygen (or water vapor) which was introduced with the wafer has a long enough residence time in the reactor that the low-temperature H_2 bake actually oxidizes the surface slightly. Thus, an O peak at the epi/substrate interface is an indication of either an insufficient H_2 bake, an ineffective chemical removal of native oxide, or oxygen entering the reaction chamber during the wafer transfer.

When the reaction chamber is assembled following a maintenance activity, it is usually completely contaminated with H_2O on the walls. Therefore, some period of time is necessary to "bake-out" the chamber and remove this contamination. Fortunately, production Si epi reactors are capable of temperatures which easily exceed 1150°C, so this bake-out can occur in a relatively short time (1 or 2 h). However, the gas delivery system immediately upstream of the reaction chamber, the wafer handling chamber, and the load-locks do not reach such high temperatures. Therefore, if these regions are exposed to an ambient with a high humidity during the reactor maintenance procedure, a significant "dry-down" period may be necessitated.

The SIMS in Fig. 38 illustrates a very effective procedure by which the quality of the epi system can be evaluated regarding O content. The growth of $Si_{1-x}Ge_x$ layers, each separated by a Si layer, from 900 down to about 600°C, on a single wafer, can show at which temperature O becomes a problem. As the temperature decreases, the affinity for the film to take up O increases. Also, as the temperature decreases, the Ge content of the film increases, for a fixed GeH_4 partial pressure, and therefore the affinity of the $Si_{1-x}Ge_x$ film for O increases even further. This is, therefore, a very sensitive test for the presence of O sources in the epi reactor.

Purification of the H_2, N_2, and HCl in the epi reactor is a practical necessity for achieving a low O content in $Si_{1-x}Ge_x$ films. Since H_2 is used as the carrier gas in great excess, any O in this gas can dominate the O content of the reactor. N_2 is typically the "inert" gas used in the wafer transfer chamber and to purge the process chamber after a process cycle. Thus, since N_2 is introduced to the process chamber before and after the process, purifying the N_2 is strongly recommended. HCl is used to etch the process chamber, so, once again, since it is introduced to the chamber between each wafer processed, purification is strongly recommended. In the case of the selective deposition of the $Si_{1-x}Ge_x$ layer, HCl is used to control selectivity and is therefore present during the deposition process. Obviously, this is a further motivation to purify the HCl.

It is also possible to purify the SiH_4 and GeH_4 sources. In my experience, pure component sources such as SiH_4 and SiH_2Cl_2 rarely have contamination problems. However, gas mixtures, such as 2% GeH_4 in H_2, 2% SiH_4 in H_2, and 1% $SiCH_6$ in H_2, can become contaminated during the mixing process at the gas vendor. Thus, purification may be desirable. It is not reasonable to attempt to purify very dilute gas sources (e.g., 100 ppm B_2H_6 in H_2) since the dilute nature of these gases makes it virtually impossible to saturate the purifier, and therefore none of the active component (e.g., B_2H_6) will exit the purifier.

The use of cartridge-type purifiers, sometimes referred to as resin-based purifiers, is very common [45–48]. Palladium alloy diffuser technology can be applied to the H_2 source if desired, but cartridge purifiers presently dominate the production $Si_{1-x}Ge_x$ market.

V. Metrology of $Si_{1-x}Ge_x$ Layers

The metrology of $Si_{1-x}Ge_x$ layers has not been a pretty subject. In this section, SIMS, Scanning Auger, Ellipsometry, X-ray diffraction and four-point probe are discussed for evaluating $Si_{1-x}Ge_x$ epi layers.

1. SIMS

Historically, SIMS has been the basis for most measurements but it is not generally considered a "line measurement tool" for a fabrication facility. The operation of SIMS, which requires meticulous calibration methodology as well as a great amount of skill on the part of the operator to interpret the data, has limited the use of this tool in manufacturing environments. Further, sample preparation and turnaround times are long, and, therefore, these measurements are expensive (several hours and hundreds of dollars per data point).

SIMS, however, provides benefits which are unequaled by other measurement tools: primarily, the capability to provide depth profile information for Ge content as well as B, P, As, C, and O down to a limit of detection in the $\sim 1 \times 10^{16} cm^{-3}$ concentration range (dependent on the atom, ion yield, background contamination level, etc.). This allows the measurement of dopant profiles, the Ge profile, and the C profile and detection of C and O contaminants with a single measurement tool. As mentioned in Section IV.5, Cs, rather than O, must be used as the sputtering agent when O is to be detected in the films, to prevent masking of the O signal by the sputtering beam. SIMS is the only tool capable of elucidating O and C at growth interfaces as well as detecting O in the deposited layers at relatively low concentrations ($<10^{20}$ cm^{-3}, if one can consider that a low concentration). SIMS also possesses some negative attributes, as listed below.

- It is not possible to measure low As concentrations in the presence of Ge since the ^{74}GeH ion fragment confounds the measurement of the ^{75}As isotope.
- SIMS is an "atom counter" (actually a mass-to-charge ratio counter) and cannot distinguish between a substitutional As, B, P, or C atom and its interstitial counterpart.
- A phenomenon known as "knock-on" occurs when Cs is used as the sputtering agent, which causes narrow B profiles to be artificially elongated. Therefore, O sputtering is often necessary to get a more accurate picture of the B profile.
- The relative uncertainty for measuring Ge content via SIMS is ±5% for Ge concentrations <20% and becomes much larger as the Ge content increases.
- The relative uncertainty for measuring As, B, P, and C is ±15% for concentrations about 10× the detection limit (i.e., concentrations >10^{17} cm^{-3}).
- The Cs sputtering rate for $Si_{0.85}Ge_{0.15}$ is approximately 10% faster than that of Si (the O sputtering rate for $Si_{0.85}Ge_{0.15}$ is only about 3% faster than Si). The sputtering rate is a function of Ge content and, therefore, changes during sputtering of graded $Si_{1-x}Ge_x$ layers.
- Most SIMS plots do not account for the change in sputtering rate and are basically plots of ion counts versus sputtering time. Therefore, the depth profile observed in the plot is not the true metallurgical profile.

When we consider that present technology single-wafer epi reactors are capable of providing thickness and compositional uniformities in the ±1 to ±2% range, it becomes immediately obvious that SIMS cannot be used for film uniformity measurements. The SIMS technique is considered a destructive technique, even if a full wafer can be handled by the instrument, since the sputtering of the surface causes irreparable local damage.

Since SIMS requires calibration standards some method must be chosen as a reference for atomic concentrations. Ion implantation is the common method of making most standards for dopants in Si, but for Ge, another opportunity exists: Rutherford backscattering (RBS). RBS is a standardless technique by which the Ge content and thickness of Si on $Si_{1-x}Ge_x$ or $Si_{1-x}Ge_x$ on Si can be measured. This technique is not usually a full-wafer measurement technique and is therefore considered destructive. It is useful for measuring Ge contents >1% and establishing calibration standards for SIMS.

2. AUGER

Scanning Auger uses Ar as the sputtering agent and is very useful for the measurement of Ge concentrations in the range of 5 to 100% with a relative uncertainty of ~1%. Therefore, it has application for enhanced mobility channels and pseudosubstrates where the Ge content is graded to levels at or near 100%. Scanning

Auger provides depth profiling information but does not have an adequate limit of detection for measuring dopants or O and C contamination. Like SIMS, this technique is considered destructive.

3. ELLIPSOMETRY

Spectroscopic ellipsometry has demonstrated the capability for measuring Ge content as well as layer thicknesses for structures such as Si on $Si_{1-x}Ge_x$ and $Si_{1-x}Ge_x$ on Si [49, 50]. Single-wavelength ellipsometry has shown some promise [51] but it appears that multiwavelength techniques hold a greater promise for this application. Ellipsometry is an alluring technique since it has a history as a line measurement tool in semiconductor manufacturing. It does not have the capability to measure dopants or contaminants such as O and C, but its ability to measure thickness and Ge content, in a time- and cost-efficient manner, is a significant benefit. The measurement of box Ge profiles is generally straightforward. However, graded layers provide a bigger challenge in data interpretation. The technique is optically based and considered to be nondestructive.

4. X-RAY DIFFRACTION

Double-crystal X-ray diffraction (XRD) is a technique which essentially measures the crystal lattice constant using a grazing angle of incidence X-ray beam [52–54]. This technique has been demonstrated to be capable of measuring Ge content and film thickness with a precision of $\sim 2\%$ $1 - \sigma$ for Si on $Si_{1-x}Ge_x$ or $Si_{1-x}Ge_x$ on Si. The measurement is relatively fast, requiring approximately 30 min to compute a nine-point on-wafer map. XRD is also capable of measuring substitutional C concentrations in Si and $Si_{1-x}Ge_x$ and determining whether a layer is relaxed or pseudomorphic. Although XRD is not historically a line measurement tool, the availability of cassette-to-cassette full-wafer measurement tools will likely provide the necessary convenience for this technology to enter the manufacturing market [54]. Once again, making measurements with box Ge profiles is much easier than with graded Ge composition layers. This technique is not capable of measuring trace levels of components, such as dopants or C and O contaminantion. XRD is considered to be nondestructive measurement.

Other tools such as the Optiprobe by Thermawave [55] and the Nanospec 8000 [56] have claimed to provide the ability to measure Ge content and layer thicknesses by optical spectroscopy techniques. Like ellipsometry and XRD, neither of these tools is capable of measuring dopants or C and O contamination. Since these techniques are optically based, they are considered to be nondestructive.

5. THE FOUR-POINT PROBE

As archaic as it may seem, the four-point probe is still the best way to measure dopant concentration, particularly B in the doped base of an HBT or DFT. Recent advances in probe head technology now allow large-area probes, which minimally penetrate the Si layer, to measure reliably the R_s of 500-Å and even thinner films. The relative uncertainty of four-point probe technology can approach $\pm 0.1\%$ when used with care and, therefore, makes this tool ideal for monitoring on-wafer and wafer-to-wafer dopant uniformity and repeatability.

VI. Production Robust $Si_{1-x}Ge_x$ Processing

So much has been written that it is easy to get lost in the information. Thus, the question begged by this section is, "How do we make a useful process out of all of the aforementioned gibberish?" Toward that end I am going to distill all of the previously discussed "stuff" and show you exactly how to grow an HBT and DFT, using the mixed poly/epi process, on a commercial load-locked single-wafer epi reactor. It's up to you to choose the right brand of epi tool.

1. Composing a High-Throughput Process

Achieving a relatively high-throughput $Si_{1-x}Ge_x$ epi process in a single-wafer process tool requires more effort than simply composing a process recipe [57]. Assuming that your device designers and process integration architects have done their jobs and provided you with a set of specifications to which to grow the epi layers, we must first come up with a chemical clean to remove the native oxide. The HF dip, outlined in Section III.2, may not appear terribly elegant, but it works. So the 100:1 HF dip at an elevated temperature, followed by a very clean cool water rinse and hot nitrogen dry, will be chosen for the low-temperature removal of native oxide.

After the wafers emerge from the hot N_2 dry, they should be placed in the load-lock of the epi reactor as soon as practically possible. Waiting 5 or 10 min is not a problem; 30 min is acceptable; but after an hour, you are begging for trouble. Much of this queue time sensitivity seems to be driven by the actual purity of the "ultrapure" rinse water. The higher the dissolved O and TOC level of this water, the less queue time can be tolerated. Thus, it appears that the O and TOC level of the rinse water has a significant impact on the degree of hydrogen passivation of the wafer surface.

A vacuum-capable load-lock is highly desired to remove the air, with its attendant impurities, from this area once the wafers have been placed inside. Perform at least two, and preferably three, pump-down/backfill sequences in the load-lock, backfilling with purified N_2 or a true inert gas. If a vacuum-capable load-lock is not present, simply purging the load-lock with purified N_2, for 30 min to an hour, can often provide an adequately clean environment for the process (this is true for the Epsilon One series of single-wafer epi reactors and I presume that it will work for other brands as well).

The wafer is then transported, in an ultraclean environment, which is substantially free of O_2 and H_2O, into the epi process chamber. Transferring the wafer at atmospheric pressure is perfectly acceptable as long as the environment is adequately clean (e.g., O_2 and H_2O levels, <200 ppb).

The epi process is outlined in Fig. 39. The actual temperature for the H_2 bake will have a substantial impact on the productivity of the process. At 900°C only about 1 min is necessary to prepare the surface (defined as no measurable O or C dose at the epi/substrate interface by SIMS), while at 800°C 5 min is required. Some decreases in these times and temperatures are certainly possible, but as a bad example, a 20-min H_2 bake at 700°C will commonly leave an O dose

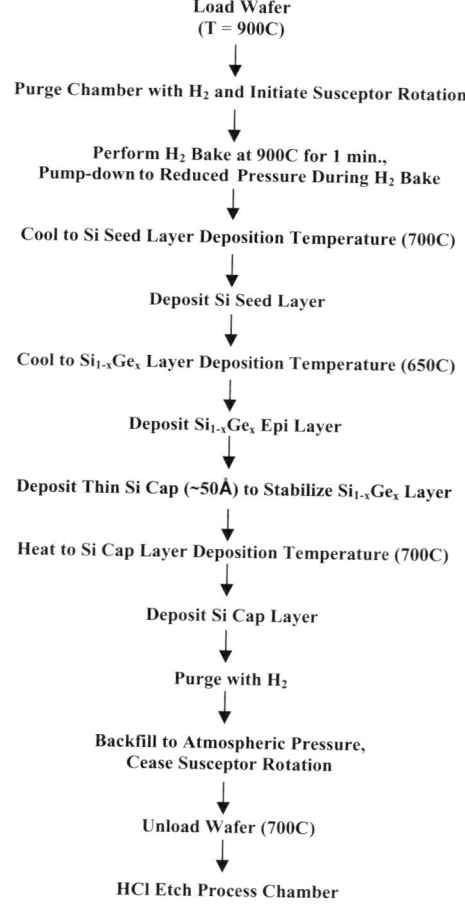

FIG. 39. Process recipe outline for a reduced-pressure $Si_{1-x}Ge_x$ HBT mixed poly/epi deposition process. The process includes a Si seed layer, a doped $Si_{1-x}Ge_x$ layer, and a Si cap layer.

of $\sim 10^{14}$ cm^{-2} at the epi/substrate interface. Thus, it is important in the process integration strategy that sufficient thermal budget be allowed for an H$_2$ bake in the 800 to 900°C range if the process is to have a reasonable throughput and be free of O and C at the first epi interface.

To extract the most productivity from a single-wafer process it is desirable to grow at the highest possible temperature to achieve high growth rates. The temperature for deposition of the Si seed layer is usually chosen to be close to that of the Si$_{1-x}$Ge$_x$ layer. If this layer is thin (e.g., ~ 100 Å), there is no need to use high temperatures. The temperature for deposition of the Si$_{1-x}$Ge$_x$ layer is chosen based on the critical thickness limitation and is therefore a function of the Ge content and thickness of the Si$_{1-x}$Ge$_x$ layer. The Si cap layer temperature should, in theory, be the same as the Si$_{1-x}$Ge$_x$ layer deposition temperature. In practice, however, depositing the cap layer at a temperature 50°C or so higher than that of the Si$_{1-x}$Ge$_x$ layer usually does not present a problem. A very typical deposition temperature for the Si seed and cap layers is 700°C. If the Ge content of the layer is less than 15%, is it likely that the Si$_{1-x}$Ge$_x$ layer can also be grown at 700°C. This provides a great benefit for the productivity of the process since heating, cooling, and stabilization of the temperature between the different deposition temperatures are eliminated. At Ge concentrations of between 15 and 25%, a deposition temperature of 650°C is normally used for the Si$_{1-x}$Ge$_x$ layer. Obviously, higher Ge content layers and greater thicknesses will require lower deposition temperatures for the Si$_{1-x}$Ge$_x$ layer and the Si cap layer.

The maximum Si precursor flow rate that can be used for deposition of the Si seed or cap layer is a function of the temperature. At 700°C a SiH$_4$ flow of 200 sccm, with a 20-slm H$_2$ carrier gas flow, will yield a completely specular surface. At 400 sccm the surface will have a grainy appearance, indicating that three-dimensional growth is occurring and that polycrystalline grain boundaries are present in the region which is supposed to be single-crystal. Maximizing the Si precursor flow rate allows the deposition rate to be maximized and therefore improves the process throughput.

The choice of pressure is not overly complex. Usually 40 to 100 Torr is acceptable. Very low pressures can sometimes lead to unacceptable roughness of the poly over the dielectric regions. Exceptionally high pressures (e.g., atmospheric pressure) can cause large disparities between the poly and the epi growth rates.

Rather than babble on, step by step, through the entire process sequence, Fig. 40 presents an epi recipe from the Epsilon One, for the mixed deposition of an HBT base. The structure grown is as follows.

Top: 300-Å undoped Si cap layer
Middle: 500-Å Si$_{0.8}$Ge$_{0.2}$, B-doped (10^{19} cm^{-3})
Bottom: 100-Å Si seed layer

The entire process requires just over 7 min, which suggests a throughput of about 8.5 wafers/h (any wafer size: 3 in. to 300 mm), exclusive of chamber cleaning.

Step #:	1	2	3	4	5	6	7
Step Name:	Load	Begin Rotation	H2-Bake	Cool-Down	Deposit Si Seed	Cool-Down	Deposit SiGe
Duration:	15	5	60	70	25	30	95
Token:	LOAD		RP_SLOW				
Temp.:	850	900	900	700	700	650	650
N2H2:	20H	20H	20H	20H	20H	20H	20H
HCl:	0V	0V	0V	0V	0V	0V	0V
P-Dope-SRC:	0	0	180	180	180	180	180
P-Dope-INJ:	0	0	180V	180V	180V	180	180
P-Dope-DIL:	0	0	4	4	4	4	4
SiH4:	0	0	0	100	100	20	20
1.5% GeH4:	0	0	0	0	0	115	115
DEP/VENT:	VENT	VENT	VENT	VENT	DEPOSIT	VENT	DEPOSIT
Rotation:	0	35R	35	35	35	35	35
Pressure:	ATM	ATM	100	100	100	100	100

Step #:	9	10	11	12	13	14	15
Step Name:	Purge	Deposit Thin Si Cap	Heat-up	Deposit Si Cap	Purge	Stop Rotation	Unload
Duration:	10	120	30	60	10	30	15
Token:						BACKFILL	UNLOAD
Temp.:	650	650	700	700	700	700	650
N2H2:	20H	20H	20H	20H	20H	20H	20H
HCl:	0V	0V	0V	0V	0V	0V	0V
P-Dope-SRC:	0	0	0	0	0	0	0
P-Dope-INJ:	0	0	0	0	0	0	0
P-Dope-DIL:	0	0	0	0	0	0	0
SiH4:	50	50	200	200	0	0	0
1.5% GeH4:	0	0	0	0	0	0	0
DEP/VENT:	VENT	DEPOSIT	VENT	DEPOSIT	VENT	VENT	VENT
Rotation:	35	35	35	35	35	0	0
Pressure:	100	100	100	100	100	ATM	ATM

FIG. 40. A complete process recipe for the deposition of a $Si_{1-x}Ge_x$ HBT structure on the Epsilon One single-wafer epi system.

The recipe can be modified for the growth of a linearly graded Ge profile used in a DFT by breaking up the $Si_{1-x}Ge_x$ deposition step into 5 to 10 segments and ramping the GeH_4 flow rate at each step. Since the deposition rate decreases with decreasing GeH_4 flow rate, the deposition time in each successive step must be increased to have an equal thickness in each step. Figure 41 presents the $Si_{1-x}Ge_x$ deposition sequence for a layer graded from 20% Ge at the base/collector junction to 0% at the emitter/base junction. The longer deposition times associated with the DFT, compared to the HBT, result in a decreased throughput for this process.

One of the strong opportunities provided by single-wafer processing is that the reaction chamber is largely ignorant of the diameter of the wafer being processed. Thus, it is quite reasonable to develop a process on 100- or 150-mm-diameter wafers. The productivity, on a device count basis, is then increased by simply changing to a larger wafer size as the process transitions from development to manufacturing.

To be sure, there are issues such as chemical loading effects and on-wafer temperature uniformity requirements which will require attention during a wafer

Step #:	7a	7b	7c	7d	7e	7f	7g
	Begin	SiGe	SiGe	SiGe	SiGe	SiGe	SiGe
Step Name:	SiGe	Ramp 1	Ramp 2	Ramp 3	Ramp 4	Ramp 5	Ramp 6
Duration:	5	2	3	30	40	55	70
Token:							
Temp.:	650	650	650	650	650	650	650
N2H2:	20H	20H	20H	20H	20H	20H	20H
HCl:	0V	0V	0V	0V	0V	0V	0V
P-Dope-SRC:	180	180	180	180	180	180	180
P-Dope-INJ:	180	180	180	180	180	180	180
P-Dope-DIL:	4	4	4	4	4	4	4
SiH4:	20	20	20	20	20	20	20
1.5% GeH4:	115	113R	95R	70R	45R	20R	10R
DEP/VENT:	DEPOSIT	DEPOSIT	DEPOSIT	DEPOSIT	DEPOSIT	DEPOSIT	DEPOSIT
Rotation:	35	35	35	35	35	35	35
Pressure:	100	100	100	100	100	100	100

Step #:	7h	7i
	SiGe	SiGe
Step Name:	Ramp 7	Ramp 8
Duration:	75	100
Token:		
Temp.:	650	650
N2H2:	20H	20H
HCl:	0V	0V
P-Dope-SRC:	180	180
P-Dope-INJ:	180	180
P-Dope-DIL:	4	4
SiH4:	20	20
1.5% GeH4:	5R	1R
DEP/VENT:	DEPOSIT	DEPOSIT
Rotation:	35	35
Pressure:	100	100

FIG. 41. Deposition sequence to replace step 7 in Fig. 46 if a DFT (graded Ge content layer) structure is used rather than an HBT.

size change. However, the very nature of a single-wafer process tool completely eliminates the issue of "within-a-batch" performance by transferring this issue to a "run-to-run" repeatability requirement. For epi the primary process variables are temperature, gas flow rates, and pressure. The ability to repeat these conditions accurately over thousands of runs has been demonstrated over the last 10+ years to be exceptional.

VII. Summary

The primary differences between the deposition of traditional Si epi and that of the alloys $Si_{1-x}Ge_x$ and $Si_{1-x}Ge_x{:}C$ are the requirements to process at low temperatures and take into consideration the different lattice constants of the materials. Tremendous benefits for bipolar technologies have been demonstrated with HBT and DFT structures which are capable of integration with Si CMOS devices. Applications for enhanced mobility channels in both NMOS and PMOS devices, as well as the elevated source/drain structure, offer promise for the continued development of higher-speed CMOS devices. Optoelectronics applications using $Si_{1-x}Ge_x$ superlattices appear to be feasible and lower-cost pseudo-substrates for GaAs solar cells may indeed be possible.

The deposition of $Si_{1-x}Ge_x$ and $Si_{1-x}Ge_x$:C materials can be conveniently carried out in production load-locked single-wafer epi reactors. The throughputs of these processes are highly dependent upon the details of the film specification as well as the thermal budget constraints, but can often exceed seven wafers per hour, regardless of the wafer size.

A complete epi process recipe for a $Si_{1-x}Ge_x$ HBT has been provided as a reference point for process development. Modifications to this recipe for the growth of a graded Ge content DFT have been presented.

No more excuses, it's time to grow the films.

References

1. People, R., and J. C. Bean, *Appl. Phys. Lett.* **47**(3), 322 (1985).
2. Shockley, W., U.S. Patent 2,569,345; application date, June 26, 1948.
3. Wolf, S., *Silicon Processing for the VLSI Era, Vol. 2. Process Integration,* Lattice Press, 1990.
4. Borisenko, V. E., and S. G. Yukin, *Phys. Status Solidi A* **101**(1) (1987).
5. Lippert, G., H. J. Olsen, and D. Kruger, *Strained Layer Epitaxy—Materials, Processing and Device Applications,* MRS Symposium Proceedings, Vol. 379, 1995.
6. Meyerson, B. S., F. K. LeGoues, T. N. Nguyen, and D. L. Harame, *Appl. Phys. Lett.* **50**, 113 (1987).
7. Osburn, C. M., M. Y. Tsai, S. Roberts, C. J. Lucchese, and C. Y. Ling, in *VLSI Science and Technology,* edited by C. J. Dell'Oca and W. M. Bullis, Electrochemical Society, Pennington, NJ, 1982.
8. Meyer, D. J., and J. Italiano, Selective Si and $Si_{1-x}Ge_x$ for elevated source/drain structures, 193rd Meeting of the Electrochemical Society, 1998.
9. Meyer, D. J., and M. R. Hawkins, The Deposition of *in Situ* Doped Polysilicon in a Single Wafer Reactor, 180th Meeting of the Electrochemical Society, 1991.
10. King, T. J., J. R. Pfiester, J. D. Shott, J. P. McVittie, and K. C. Saraswat, *IEDM-90,* p. 253, 1990.
11. Bensahel, D., Y. Campidelli, C. Hernandex, F. Martin, I. Sagnes, and D. J. Meyer, *Solid State Technol.* Suppl. 5 (1998).
12. Catalano, A., and M. S. Turnoy, *Strategic Report: Silicon Germanium,* Technology Assessment Group, Boulder, CO, Nov. 1999.
13. Jiang, H., and R. Elliman, *IEEE Trans. Elec. Dev.* **43**(1), 97 (1994).
14. Currie, M. T., S. B. Samavedam, T. A. Langdo, C. W. Leitz, and E. A. Fitzgerald, *J. Vac. Sci. Technol.* **72**(14), 1718 (1998).
15. Fukatsu, F., N. Usami, Y. Kato, H. Sunamura, Y. Shiraki, H. Oku, T. Ohnishi, Y. Ohmori, and K. Okumura, *J. Crystal Growth* **136**, 315 (1994).
16. *8th E.C. Photovoltaic Conf.,* p. 1522.
17. *J. Mater. Res.,* **6**, 376.
18. Por, A., A. Audouard, N. Malfoufi, G. Marchal, and M. Gerl, *J. Non-Cryst. Solids* **97/98** (1987).
19. Verke, C., J. F. Rochette, and J. P. Rebouillat, *J. Phys.* **C4**(10), 42 (1981).
20. Vergant, M., M. Piecuch, J.-F. Geny, C. Courey, G. Marchal, and M. Gerl, *J. Phys.* **C8**(12), 46 (1985).
21. Morrison, I., and M. Jaros, *Supperlatt. Microstruct.* **2**(4) (1986).

22. Soref, R., and C. Perry, *J. Appl. Phys.* **69**, 1 (1991).
23. Girginoudi, D., N. Georgoulas, and A. Thanailakis, *J. Appl. Phys.* **66**, 1 (1989).
24. Nicolaou, M. C., *J. Non-Cryst. Solids* **97/98** (1987).
25. Zdetsis, A. D., D. Girginoudi, G. Kiriakidis, Z. Hatzopoulos, A. Thanailakis, and A. Christou, *J. Non-Cryst. Solids* **97/98** (1987).
26. Demichelis, F., G. Kaniadakis, A. Tagliaferro, and E. Tresso, *Phys. Rev. B* **40**, 3 (1989).
27. Girginoudi, D., N. Georgoulas, N. Thanailakis, A. Zdetsis, G. Kiriakidis, and A. Christou, *Mater. Res. Soc. Symp. Proc.*, p. 118, 1988.
28. Johnson, K., and N. Ashcroft, *Phys. Rev. B* **54**(20) (1996).
29. Kobayashi, N., H. Katsumata, M. Hasegawa, N. Hayashi, Y. Makita, H. Shibata, and S. Uekusa, *Proc. 7th Intl. Symp. Adv. Nuclear Energy Res.*, JAERI-Conf 97-003, 1996.
30. Ghidini, G., and F. W. Smith, *J. Electrochem. Soc.* **131**, 2924 (1984).
31. Verhaverbeke, et al., *MRS Symp. Proc.* **315**(1993).
32. Singer, P. H., *Semiconduct. Int.* **Dec.**, (1992).
33. Vermuelen, W., *Electrochem. Soc. Extend. Abstr. Oct.* (1993).
34. Watanabe, H., H. Kitajima, I. Honma, and H. Ono, *J. Electrochem. Soc.* **142**, 4 (1995).
35. Caymax, M., JESSI E233 Report for ADICT, Jan.–Sept. 1996.
36. Sprey, H., A. B. Storm, J. W. Maes, E. H. A. Granneman, and M. Hendriks, 1st International Epsilon Users' Group Meeting on SiGe and Low Temperature Epi, Amsterdam, June 1998.
37. Kasper, E., *Appl. Phys.* **8**, 199 (1975).
38. Meyerson, B. S., *Appl. Phys. Lett.* **48**, 797 (1986).
39. de Boer, W. B., and D. J. Meyer, *Appl. Phys. Lett.* **58**(12), 1286 (1991).
40. Sedgewick, T. O., V. P. Kesan, P. D. Agnello, D. A. Grutzmacher, D. Nguyen-Ngoc, S. S. Iyer, and D. Meyer, *Proceedings of the IEDM Device Technology Meeting*, 1991.
41. Agnello, P. D., T. O. Sedgwick, T. S. Kuan, G. Scilla, D. Meyer, and A. P. Ferro, *Proceedings of the American Vacuum Society Meeting*, Toronto, Canada, Oct. 1990.
42. Meyer, D. J., and T. I. Kamins, *Thin Solid Films* **222**, 30 (1992).
43. Salm, C., D. T. van Veen, D. J. Gravesteijn, J. Holleman, and P. H. Woerlee, *J. Electrochem. Soc.* **144**(10), 3665 (1997).
44. Atzmon, Z., A. E. Blair, E. J. Jaquez, J. W. Mayer, D. Chandrasekhar, D. J. Sith, R. L. Hervig, and McD. Robinson, *Appl. Phys. Lett.* **65**(20), 2559 (1994).
45. Matheson Semi-Gas Systems, 625 Wool Creek Drive, San Jose, CA 95112; www.mathesongas.com.
46. Aeronex, Inc., 6975 Flanders Drive, San Diego, CA 92121, www.aeronex.com.
47. Millipore Microelectronics Divisions, 80 Ashby Road, Bedford MA 01730; www.millipore.com.
48. ATMI, Danbury, CT; www.atmi.com.
49. Pickering, C., R. T. Carline, D. J. Robbins, W. Y. Leong, S. J. Barnett, A. D. Pitt, and A. G. Cullis, *J. Appl. Phys.* **73**(1), 239 (1993).
50. SOPRA, 26 rue Pierre-Joigneaux, F 92270 Boise-Colombes, France.
51. Kamins, T. I., *Electron. Lett.* **27**, 451 (1991).
52. Bede Scientific Instruments Ltd., Bowburn South Industrial Estate, Unit 13D, Bowburn, County, Durham DH6 5AD, UK; www.bede.co.uk.
53. Rigaku, 199 Rosewook Drive, Danvers, MA 01923, www.rigaku.com.
54. Philips DCD2-H, Philips Analytical, 125 West. Gemini Drive, Suite E20, Tempe, AZ 85283.
55. Therma-wave, 1250 Reliance Way, Fremont, CA 94539; www.thermawave.com.
56. Nanometrics, Inc., 310 De Guigne Drive, Sunnyvale, CA 94086; www.nanometrics.com.
57. Meyer, D. J., and J. Italiano, Productive single wafer $Si_{1-x}Ge_x$ processing, 196th Meeting of the Electrochemical Society, Oct. 1999.

CHAPTER 11

Silicon Epitaxy: New Applications

D. Dutartre

ST MICROELECTRONICS
CENTRAL R&D
CROLLES, FRANCE

I. INTRODUCTION	397
II. EQUIPMENT: STATE OF THE ART	398
1. Molecular Beam Epitaxy	399
2. Chemical Beam Epitaxy	400
3. Plasma-Enhanced Chemical Vapor Deposition	401
4. Ultrahigh-Vacuum Chemical Vapor Deposition	402
5. Rapid Thermal Chemical Vapor Deposition	404
III. EPITAXY PROCESSES	406
1. New Challenges	406
2. Growth Regimes	407
3. Process Details	415
IV. NEW APPLICATIONS	422
1. Low-Cost Epitaxial Wafers	422
2. SOI Wafers	424
3. Bases of RF Bipolar Transistors	426
4. Elevated Sources and Drains in CMOS	435
5. Channel Engineering in CMOS	438
6. New CMOS Architectures	446
7. Other Applications	451
V. CONCLUSION	453
REFERENCES	455

I. Introduction

It is of interest at this time [about four decades after the first reports on Si epitaxy (epi) [1]] to reflect upon the progress and to look at our task in the future. Si epi is now of major importance for the manufacturing of a number of devices and circuits, such as epitaxial (epi) substrates for complementary metal-oxide-semiconductor (CMOS) and dynamic random access memory (DRAM), buried layers of bipolar and bipolar CMOS technologies, and power bipolar transistors. For these applications, epi growth is performed on full-sheet silicon wafers by chemical vapor deposition (CVD) using well-established processes. High temperatures ($>1000°C$) are used both to clean the Si surface and to achieve high deposition rates. The quality of the epi has to be very high, but epi characteristics (thickness, doping level, etc.) very often do not determine the main device characteristics [threshold voltage of MOS transistors, time retention of memory cells, characteristics of radiofrequency (RF) bipolar transistors, etc.]. All these fields are addressed in previous chapters.

During the last decade, huge progress was made in low-temperature epi, and today it is possible to grow high-quality Si or Si-based films with a reduced thermal budget ($<900°C$). This facilitates the growth of very thin films and the abrupt junctions that are required by small device dimensions. In addition, if the thermal budget is low enough, we will be able to make the epi deposition late in the device fabrication sequence. This enables an increase in the number of possible device architectures, opening new fields to silicon epi.

This chapter focuses on the new developments in Si or Si-based epi depositions and their possible applications in various technologies. Section II is devoted to state-of-the-art equipment. The most important epi techniques are reviewed, and the specific advantages and limitations of each are indicated. Section III deals with new epi processes. It is presented in three successive parts: (1) New Challenges, (2) Growth Regimes, and (3) Process Details. This organization should help the reader to relate application, process, and apparatus together and to understand further the possible evolution of epi technology. Low-temperature nonselective epi is usually obtained with silane and selective deposition within the $SiH_2Cl_2/HCl/H_2$ system. These two systems are considered, and as surface reactions are known to play a dominant role at low temperatures, the present understanding of the decomposition/behavior of the various reactants is reviewed. Section IV gives a selection of possible applications of epi in future technologies, which have currently led to a significant level of research, namely, low-cost epi wafers, silicon-on-insulator (SOI) wafers, bases of RF bipolar transistors, elevated sources and drains, canal engineering in CMOS, strained Si layers for CMOS, and nonconventional CMOS architectures. For each application, we endeavor to list the following: motivations, film requirements, specific epi challenges, process solution, and results. The list is not exhaustive, and even more numerous applications can be found in the literature, especially when considering applications based on strained IV–IV heterostructures. However, these are based on epi deposition with similar issues, growth regimes, and processes.

II. Equipment: State of the Art

In this section, we review the different types of epi reactors: for molecular beam epitaxy (MBE), chemical beam epitaxy (CBE), plasma-enhanced CVD (PE-CVD), ultrahigh-vacuum CVD (UHV-CVD), and rapid thermal CVD (RT-CVD). This review is neither exhaustive nor very detailed; however, we outline the advantages and limitations of the techniques mentioned. At a slight risk, we provide further details on systems which are more suitable for industry and, therefore, have more future. So today's RT-CVD systems, which have been developed to an industrial level and are used in laboratories as well in factories, are detailed more than MBE, which is no longer a reference technique for Si and Si-based epi. In this way, the reader will receive simple information and become more familiar with the characteristics specific to each technique. Indeed, the purpose of this part is to provide the reader with some background on epi reactors, to facilitate and enable a deeper understanding of the parts that follow.

1. MOLECULAR BEAM EPITAXY

MBE is described first because this technique is conceptually simple and is used for a number of materials. It consists of evaporating atomic species on a somewhat heated substrate. This technique is based on two main requirements: the mean free path of atoms has to exceed the chamber size (1 m corresponds to approximately 10^{-5} Torr) and the evaporated species must not reflect on the different internal surfaces.

Modern silicon MBE, which is currently run in UHV chambers with an ultimate vacuum around 10^{-10} Torr, easily meets these two conditions. To obtain such low base pressures, the stainless-steel chamber is baked and is evacuated by high-capacity cryogenics and, eventually, turbomolecular pumps, and samples are introduced through load-locks [2]. Under these vacuum conditions, it is acknowledged that silicon epi with a very high crystalline quality is achieved using a deposition rate of 2–4 Å/s and temperatures above 450°C.

Because of their high evaporation temperatures, silicon and germanium are evaporated using electron beam evaporators. Once evaporated, the Si and Ge atoms have a sticking coefficient of 1 at all of the usual growth temperatures. Consequently, control of the atom fluxes by mechanical shutters, inserted between the sources and the wafer, allows easy and perfect control of the deposition in terms of composition and thickness. The unity sticking coefficient also means that deposition kinetics from room temperature to above 900°C is virtually constant. Thus, MBE is an unbeatable technique whenever a very low temperature is required. For example, MBE enabled the growth of very short-period Si/Ge superlattices, which were thought to be highly desirable due to their expected direct bandgap and optical properties. In this case, a very low thermal budget was used to avoid Si/Ge interdiffusion, and atomic layer control was demonstrated.

Unfortunately, dopants generally exhibit a much more complex behavior, making the control of the epi characteristics much more difficult. The dopants that are easily evaporated can reevaporate from the wafer surface; others will undergo surface segregation. N-type doping is very difficult: tricks such as very low-energy implantation have to be used to achieve high doping levels, and an abrupt doping decrease is almost impossible.

In addition, in terms of any industrial application, MBE suffers from other important drawbacks. First, MBE very often exhibits metal contamination, almost certainly originating from the wafer heater and/or from the different heated sources.

To overcome this problem, great care has to be taken in the design of the system as well as in its operation. Consequently, only a minority of the groups working on MBE was actually able to limit this contamination [3]. Second, because of its excessive sensitivity to the residual base pressure, the MBE technique requires a considerable number of precautions and efforts to produce high-quality films (extended chamber bake-outs, wafer degassing, etc.), which are very time-consuming. Third, MBE has another inherent weakness: particle control. As stated previously, MBE uses species (atoms) that deposit very efficiently on walls. In situations of intensive use, very large (>100-μm) thicknesses of silicon will build up on internal surfaces and peel off in the form of particles. Regular maintenance of the system, including venting and cleaning, should prevent this problem but will have a negative effect on productivity. Note that, even if the substrate is placed front side downward, it has been shown that the atom beam is able to drive particles back up to the growing layer [4]. The author believes that a number of such particles will be created even more rapidly near the silicon source.

In conclusion, despite its interesting performances at very low temperatures ($<500°$C), silicon MBE has been discarded from any recent industrial development, undoubtedly because of the inherent weaknesses mentioned previously.

2. Chemical Beam Epitaxy

CBE is very similar to MBE except that its sources are gas injectors instead of solid sources. This technique, which is also referred to as gas-source MBE (GS-MBE), was first developed by Japanese groups [5]. After that, variants were developed, differing in terms of chamber geometry, heating system, and pressure operating range. Today, the utilization of CBE is not standard. However, we believe that a brief description of this technique is worthwhile here for the following three reasons: CBE has several unique deposition characteristics, it is a very interesting tool for understanding the complex mechanisms of CVD reactions, and little consideration is given to it in a number of reviews.

If we take the example of the Epineat system, which is commercially available from the Riber Company [6], the schematics of a CBE system typically consists of three modules: the process module, the gas control box, and the transfer module. The transfer module is a complete UHV system, which isolates the process chamber from any atmospheric contamination. The gas distribution system is designed to ultrapurity standards and allows precise gas injection onto the wafer with very short (<1-s) switching transients. The process chamber is a 200-mm single-wafer chamber designed to combine the advantages of rapid thermal processing (RTP) and UHV-CVD. The wafer is heated through a quartz window that is cooled, and the external halogen-lamp heater permits processes over a large temperature range, $400°$–$1200°$C. The chamber is evacuated by a combination of a high-capacity molecular pump roughed by a mechanical pump, and processes can be run in a very large pressure range: from almost 10^{-10} to 10 Torr. Thus, CBE combines some of the advantages of MBE,

- UHV capability,
- cold walls,
- an external heater, and
- a molecular beam in the very low-pressure domain,

together with some of CVD,

- cleanliness of gases,
- flexibility of the gas distribution, and
- flexibility in terms of the process (temperature and pressure).

As a consequence, this technique has some very specific characteristics. For example, the process, where "cold" molecules impinge on a hot wafer surface, is unique in all CVD techniques. Perhaps this has led to the publication of some interesting processes, such as selective epi growth (SEG) using silane or disilane chemistries without any chlorine atoms [7]. On the other hand, with a very low operating pressure and cold walls, CBE is favorable to the installation of efficient gas and surface analysis tools. This technique can therefore be considered the most interesting growth method for scientific purposes. However, CBE also combines the difficulties of MBE with those of CVD:

- UHV difficulties (metal seals),
- sensitivity to residual gases (especially at very low pressures),
- heating uniformity/control together with highly temperature-activated processes, and
- no *in situ* chamber clean (very critical with "through-the-window" heating).

Up to now, these difficulties, especially the latter two, have slowed down the development of this technique for industrial applications.

3. PLASMA-ENHANCED CHEMICAL VAPOR DEPOSITION

Because silicon epi by standard CVD is a thermally activated process, it has some limitations in terms of the thermal budget. Thus, several approaches to reducing the process temperature have been investigated. One approach to improving a given CVD process uses plasma to provide energy to the active gases. A PE-CVD that is routinely used in deposition has also been adapted to epi processes.

Conventional PE-CVD was studied by Donahue and Reif [8] using the system they describe. The technique enabled silicon epi between 650 and 800°C with highly enhanced kinetics. More recently, PE-CVD using low-kinetic energy particles or remote plasma has been used to obtain silicon epi at lower temperatures, and epi deposition has been demonstrated at 150°C [9]. In both cases, plasma was used to preclean the surface with argon and/or hydrogen, as well as to increase the growth kinetics. Although PE-CVD enabled very low temperatures to be used, this technique was not ideal, as a high defect density (above 10^4 cm^{-2}) was reported, together with high contamination levels (about 5×10^{19} and 1×10^{18} cm^{-3} for O and C, respectively). This contamination, which occurred unless ultraclean conditions were used, may be due to the chemical activity of the plasma, which is active on the wafer surface and on all the reactor parts as well. For the same reason, we can also speculate that the metallic contamination of the films was above the standards of microelectronics.

Because of these difficulties and the progress made by other simpler CVD techniques, little effort was given to developing PE-CVD for epi applications, and today it has not been developed to an industrial level.

On the other hand, hydrogen-plasma cleaning, which is believed to occur through a chemical effect of atomic hydrogen without any sputtering, should occur without the previous deleterious effects of plasma. In the author's opinion, this could probably be integrated in equipment and used *in situ*, prior to standard epi by CVD, for applications where the thermal budget has to be reduced.

4. ULTRAHIGH-VACUUM CHEMICAL VAPOR DEPOSITION

The UHV-CVD technique, first reported by Meyerson [10] in 1985, is a hot-wall, multiwafer deposition technique. The reactor geometry is very similar to that of the low-pressure (LP)-CVD reactor used for polycrystalline silicon deposition. The wafers are closely placed on a boat, which is introduced in a quartz tube, heated by a resistive external furnace. The epi reactor, as described in Ref. 10, also incorporates some specific features to meet the very strict requirements in terms of vacuum tightness and purity: UHV sealing techniques are used, a residual gas analyzer is installed for easier diagnosis of reactor cleanliness, a turbo pump is used to purge and pump the reaction chamber down to a pressure below 10^{-9} Torr, and wafers are introduced using a separately pumped load-lock. In this way, no traces of carbon or hydrocarbon are detected, and water vapor as well as oxygen is present at a partial pressure below 10^{-10} Torr.

Conventional silicon epi by CVD was limited to a high-temperature domain ($>950°$C). This occurred because epi growth was disrupted at lower temperatures by the presence of SiO_2 on the silicon surface, as a consequence of the small amount of water vapor in the epi reactor. Indeed, quantitative investigations of the $Si/SiO_2/H_2O$ system kinetics by Ghiddini and Smith [11] have shown that an oxide-free Si surface can be maintained at a given temperature only if the water partial pressure is less than a critical value. This value is very sensitive to temperature, with an apparent activation energy of about 3 eV. As such, extrapolation of Ghiddini and Smith's data indicates that if the temperature is decreased from 900 to 700°C, the water partial pressure has to be reduced from 10^{-5} to 10^{-8} Torr. In the author's opinion, these extrapolations would not be totally valid, especially in the CVD process, where other kinetic limitations have to be considered. However, such extremely low values explain the extreme difficulty of low-temperature silicon epi.

Thus, the UHV-CVD technique is capable of extending the epi process to lower temperatures, provided that the previously mentioned stringent conditions apply. Silicon surface preparation is a very critical aspect of this technique. Wafers are etched in diluted HF just prior to being introduced into the load chamber, where they are supposed to be pumped at less than 10^{-6} Torr. They are then inserted in the process chamber. This procedure is adequate to prepare a hydrogen-passivated surface, preserving the initial growth interface during different wafer transfers and purges, so that growth can start under clean conditions without recontamination by C and O. It should be noted that it is important for the wafer not be heated above 400–500°C before being introduced to the ultraclean environment. Growth is usually carried out using $SiH_4/(H_2)/(GeH_4)$ chemistry in a 1- to 10-m Torr pressure range. In this quasi-molecular regime, the gas residence time is very short, and molecule collisions are rare. Therefore, it can be argued that the chemical kinetics is significantly reduced and films can be deposited with a good uniformity, even at relatively high (800°C) temperatures [10]. Moreover, this technique is very often used in the 500–600°C range, where growth is hydrogen desorption limited. In this case, gas depletion is very low, and deposition kinetics, which is determined by a single parameter (temperature), is independent of reactor geometry. This is why the UHV-CVD technique is used successfully for batch deposition. This furnace batch configuration has important advantages in terms of productivity, uniformity, and repeatability.

It has been argued that such a process raises important issues [3]. First, because of the significant level of activation energy (2 eV), this process requires extraordinarily accurate temperature control, both across the reactor and from run to run. However, this difficulty is similar to that of polysilicon batch processing, and has been overcome for a long time, as this technique routinely provides very well-controlled polysilicon films in industry. Second, the SiH_4 reactants will produce deposition on the hot quartz walls of the reactor, where those deposits should eventually produce the same particulates as seen in MBE. Here again, this concern has also been addressed in polysilicon batch processing. Also, assuming that 20 μm of deposit can be tolerated on the tube, and that a single run corresponds to a 100-nm-thick deposition on 50 wafers, 10,000 wafers could be processed between wet cleans.

The absence of a hydrogen bake before deposition may be a more serious difficulty, as a very well-controlled preclean is mandatory. Even if the HF-last treatment enables epi growth without bake, significant C and/or O interfacial contamination is possible. In addition, as the silicon surface is very reactive after the HF-last preparation, very small particulates may adhere during transfer and generate crystalline defects during epi. Unfortunately, to the best of the author's knowledge, no clear data concerning interface contamination and film defectivity obtained in UHV-CVD have been published. A further weakness arises from the technique's poor level of flexibility in terms of the process. For example, selective epi with the chlorine chemistry has never been reported, undoubtedly because of the equipment's incompatibility with HCl. In addition, the use of dopant atoms may present some difficulties, such as memory effects and incorporation efficiency and trailing transitions for n-type doping.

In conclusion, UHV-CVD, which has been extensively developed by IBM researchers, seems to be mature enough today for industry. This technique is thought to be used at some manufacturing plants for the production of the SiGe base of heterojunction bipolar transistors, and commercial UHV/CVD reactors up to the 200-mm standard are currently available on the market.

5. Rapid Thermal Chemical Vapor Deposition

Rapid thermal chemical vapor deposition (RT-CVD) is a very interesting technique because rapid switching of the temperature enables the thermal budget to be minimized, different films to be grown using different "adapted" temperatures, and some relaxing of the stringent epi/temperature constraints. Gibbons and coworkers [12] first reported this technique and labeled it limited reaction processing (LRP); a stable gas flow was established with the wafer at a low temperature and growth was initiated and terminated by rapidly heating and cooling the wafer. However, as the majority of investigators preferred to use gas flow switching rather than temperature ramps to control growth, this technique has been more often reported as RT-CVD. Within a few years, a considerable amount of research was devoted to the application of this technique to silicon and SiGe epitaxies, and various reactors were developed. Originally, the minimal common features of these reactors were

- single-wafer processing,
- no susceptor,
- tungsten–halogen lamp heating,
- cold walls, and
- low pressures (1–10 Torr).

Then some important refinements, such as an infrared (IR) pyrometer for temperature control and load-lock for moisture contamination, were introduced [13, 14]. This technique was rapidly applied to the growth of Si/SiGe structures for heterojunction bipolar transistor (HBT) structures [15], and various experiments were carried out to investigate CVD capabilities within these materials [14, 16, 17]. Thus, excitons were observed in Si/SiGe structures and short-period Si/SiGe superlattices, very abrupt boron profiles were demonstrated, etc. All reports led to the conclusion that RT-CVD, which is a relatively simple technique, was able to grow Si and SiGe epi films with a very high structural quality, often exceeding that of the films obtained using MBE. However, in the majority of experimental RT-CVD reactors, uniformity was not good because of temperature nonuniformity as well as a failure to adapt the operation pressure to the chamber design.

Around 1990, ASM Company introduced the Epsilon One, a new production epi reactor [18]. Although developed for conventional silicon epi, this CVD reactor has

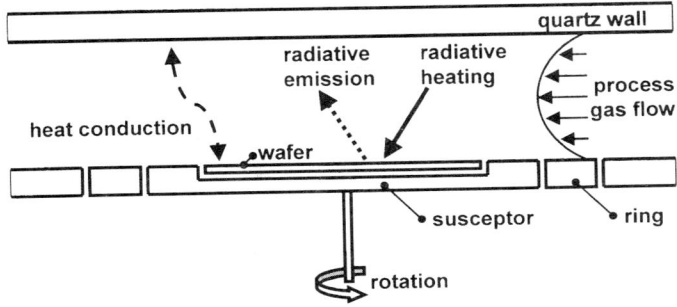

FIG. 1. Schematic cross section of a typical RT-CVD reactor. The process gases make a single pass over the rotating wafer and susceptor assembly, which are heated from top and bottom by tungsten halogen lamps. The temperature is measured by either a pyrometer or a thermocouple.

some of the above-mentioned features, single-wafer processing, lamp heating, and load-lock, and others, such as susceptor and substrate rotation for uniformity improvement. Thus, in the author's opinion, even if the temperature ramps were slowed down by the presence of a rotating susceptor, the Epsilon One could be considered to be an RT-CVD epi reactor. More recently, Applied Material has also introduced the AMAT HTF reactor, which is a similar system. The general schematic of these systems is similar to that of the CBE, except in the chamber design, as shown in Fig. 1. These systems process a single wafer at a time in a low-profile, horizontal deposition chamber. The wafer and susceptor assembly are radiantly heated from both sides by tungsten–halogen lamps. Process gases enter from the right in the figure, and laminar flow makes a single pass over the rotating wafer and susceptor. Typical processes and a number of process details are given in Section III. However, the major advantages of this important technique can be listed here,

- cold walls (at least significantly colder that the substrate),
- an external heater,
- cleanliness of gases,
- process flexibility (temperature and pressure range),
- chemistry flexibility (hydrides and chlorides),
- temperature agility,
- ability to work in both deposition regimes (surface and mass transport-limited kinetics),
- *in situ* chamber cleaning (usually with HCl at high temperatures), and
- mature and industrial technique,

as well as the main drawbacks,

- temperature uniformity and temperature control,
- complexity of CVD growth kinetics, and
- productivity of low-temperature/growth rate processes.

Comparing these pros and cons to those of other techniques, the balance is clearly in favor of RT-CVD. This certainly explains the extensive development and success of these systems. Today, they are sold in large numbers for conventional epi at manufacturing plants, as well as for advanced epi and R&D at major semiconductor suppliers and institutes.

III. Epitaxy Processes

This section presents and discusses some of the issues that are specific to the epi processes required by new applications in Si technology. All the well-established epi processes in production today are based on CVD and use a Si precursor gas diluted in H_2 as a carrier gas. Because they concern wafer fabrication or operations at the very front end of technology, they are run at a high temperature (1000–1150°C), in a domain where the Si surface is easily/naturally cleaned, using silicon chlorides. The epi quality is high, and high growth rates (GRs) are achieved. In addition, these processes are relatively simple in terms of chemistry (one gas used for deposition and one for doping), kinetics (GR and doping are proportional to gas flow), and structure (one individual blanket film deposited on a full-sheet silicon wafer).

Today epi depositions are highly desirable in the course of device fabrication to improve the electrical performances of advanced technologies. However, as soon as epi has to be run in the course of circuit fabrication, the process faces additional difficulties: a limited thermal budget, thin deposits (down to a few nanometers), abrupt dopant profiles, structured substrates, complex film stacks, etc. On the other hand, the advances in Si technology, with smaller geometries and larger wafers, also impose developments in epi.

In the following, the challenges of these new processes are detailed. We also discuss the chemistry and growth regimes chosen to manage these challenges and the points specific to these processes.

1. New Challenges

Today, epi technology is facing new challenges, which come from the continuous progress in silicon technology, namely, larger wafers and smaller geometries, and from new epi applications that will be required in future Si technologies. Due to progress in technology, the wafer size is switching from 200 to 300 mm, and epi films have to be made slightly thinner. These challenges are not addressed in the text that follows. On the other hand, as technology and devices, especially CMOS transistors, approach their theoretical limits (optical lithography limit, tunnel current through the gate oxide, etc.), great efforts are being made to improve or exchange device architecture. Consequently, new epi films are required. Most of these applications require a substantial amount of the device to be completed prior to epi running. Consequently, conventional epi processes, with their very high thermal budget, can no longer be used, and the epi film has to be deposited on various "structured" wafers. Moreover, these epitaxies (epis) can consist of complex layer stacks where many parameters have to be controlled. At present, these actual or potential epi applications are very numerous and disparate, each one having precise specifications. Although, it is not possible to examine all the details, the major issues to be addressed are listed here.

- The very limited thermal budget (H_2 bake)
- The low deposition temperature (complex kinetics and structural quality problem)
- Abrupt dopant profiles (at the nanometer scale)
- Strained epis: SiGe and SiGeC
- Different doping types and levels
- Cleaning of structured substrates
- Deposition on structured substrates (very different compared to prime Si wafers)
- Epitaxy in the presence of various materials: TEOS, Si_3N_4, SiO_2, poly/amorphous silicon, etc.
- Epitaxy in very complex structures
- Epitaxy after etching and implant steps
- Selective epi: Si, SiGe, SiGeC, etc.
- Control of facets or poly/epi interfaces
- Local/global loading effects
- Complex film stacks
- The differential epi/poly GR
- Control of poly roughness in nonselective epi

(To the author's knowledge, this is the only such list to have been published.)

While a variety of growth conditions has been developed in response to these points, a general trend toward low-temperature processes is visibly apparent. Low-temperature epi is known to give rise to extreme difficulties: surface cleaning, gas purity, complex deposition kinetics, etc. Some of these are addressed in Chapters 2, 3, 4, and 10, and others are discussed below.

2. GROWTH REGIMES

a. Background

Epitaxial silicon can be deposited using $SiCl_4$, $SiHCl_3$, SiH_2Cl_2 (DCS), SiH_4, or even Si_2H_6 as precursor. The GR depends on several parameters: temperature, gas source, gas flow rate, deposition pressure, and concentration. According to the gas source, it is admitted that the GR increases from $SiCl_4$ to SiH_4 [19] under similar

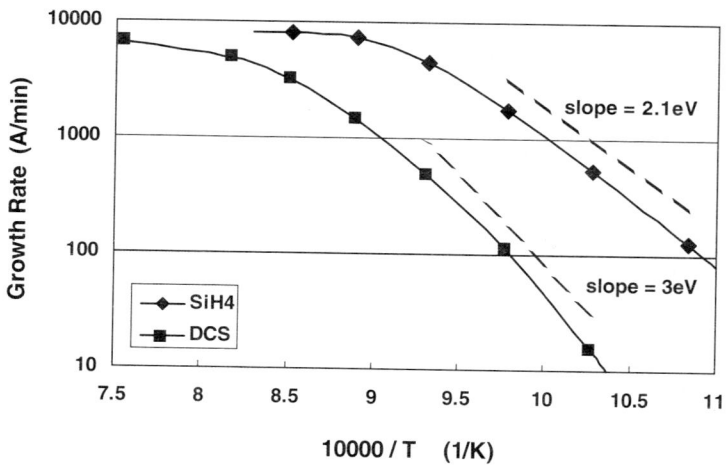

FIG. 2. Arrhenius plot for the silicon growth rate from silane and dichlorosilane under similar conditions. In the surface kinetics-limited regime (low temperature), silane chemistry presents a much higher kinetics compared to dichlorosilane.

conditions. Thus, the major advantage of the silane reaction is that deposition can occur at a lower temperature than any chlorosilane decomposition.

Figure 2 shows a comparison of Si GR versus temperature (Arrhenius plot) for dichlorosilane (DCS) and silane. In each case, two regimes are observed. At a high temperature, gas-phase transport is dominant and the GR is only weakly dependent on the temperature. At a low temperature, the GR depends on the surface-controlled reactions, which vary exponentially with the temperature. As the curves correspond to similar conditions of gas flow, we observe that the advantage of silane, compared to DCS, corresponds to more than one decade in terms of GR, or 100 K in terms of temperature. This significant variation will enable the thermal budget to be reduced in some applications.

Conventional epis are usually run in the high-temperature domain, and the processes are then gas mass transport limited. This means that the performance of the epi system will depend mainly on the gas flow pattern and the depletion of the precursor in this flow. Such processes, not being very sensitive to temperature, can generally tolerate temperature deviations of several degrees within wafers or wafer to wafer; temperature uniformity constraints often coming from thermal stress and possible crystal slip.

As soon as the low-temperature regime has to be used, the situation is very different. The GR is extremely sensitive to temperature. For example, 1 K will correspond to about 4 and 6% thickness variations with silane and DCS, respectively. In other words, any industrial process will have to be controlled to within 1 K: temperature uniformity and repeatability. While on the subject of temperature control, it should be noted that the nature of heat exchanges changes considerably between the low- and the high-temperature domains: temperature stabilization and balancing rely more and more on gas conduction. Indeed, as radiative exchanges vary much more rapidly with temperature (T^4) than gas conduction, the ratio conduction/radiation is increased by a factor of 4–5 between 1100 and 600°C. Thus, in the low-temperature domain, temperature is very influential, and the GRs are much lower than those achieved at a high temperature. This means that Si precursor impoverishment is negligible. However, what is the situation for other species also used in the process? Typically, dopant gases such as diborane, as well as germane and methylsilane, which are much more reactive than DCS or silane, will undergo depletion in the gas phase first. For example, this phenomenon is very important in the case of SiGe epitaxy; depending on the process, it can cause very large nonuniformities within the wafer and macro and micro loading effects.

In conclusion, the low-temperature domain presents difficulties in terms of process control (thickness, composition, doping, uniformity, etc.) because more experimental parameters have to be closely controlled.

b. Low-Temperature Silane-Based Epitaxy

In the low-temperature domain, the Arrhenius plot in Fig. 2 reveals an exponential dependence of GR on temperature, which indicates surface reaction rate-limited growth. The activation energy associated with this growth process, about 46 kcal mol^{-1}, is in accordance with the majority of values reported for various experimental conditions [14, 20–22]. All these values accord very well with the activation energy for hydrogen desorption from a Si $\langle 100 \rangle$ surface, 47 kcal mol^{-1} [23].

As shown in Fig. 3, growth kinetics is correlated with equilibrium hydrogen surface coverage in the following way: the reactive adsorption of silane molecules produces adsorbed hydrogen atoms whose desorption is not immediate; this hydrogen surface coverage regulates further SiH$_4$ adsorption [20, 24]. Consequently, hydrogen desorption limits the GR (via silane adsorption), which is ultimately independent of silane pressure. Initially developed in a domain of very low pressure, this model was then applied in a number of experiments. It was also further refined. For example, the molecular hydrogen adsorption was quantitatively taken into account [14]. Finally, these calculations were recently compared to experimental data

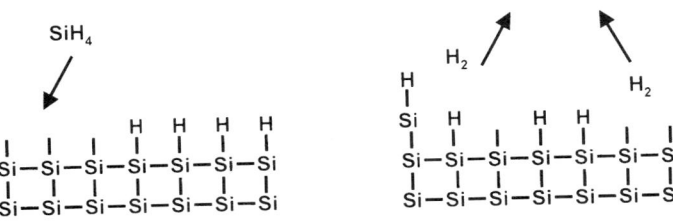

FIG. 3. Schematic of low-temperature silane-based epitaxy. Silane molecules are supposed to adsorb on two neighboring Si sites, leaving two adsorbed H atoms that passivate the surface. This hydrogen surface coverage impedes further silane adsorption, and growth is then limited by hydrogen desorption.

covering a wide range of pressures and temperatures [25]. The fact that they accorded extremely well suggests a high level of understanding of the growth kinetics of silicon epi using silane. A simplified form of the reaction pathway can be written

$$SiH_4 + 2\_ \rightarrow Si + 2\underline{H} + H_2 \quad (1)$$

$$\underline{H}^* + \underline{H} \rightarrow H_2 + 2\_ \quad (2)$$

$$H_2 + 2\_ \rightarrow 2\underline{H} \quad (3)$$

where $\underline{X}$ denotes a species which is adsorbed, $\underline{H}^*$ corresponds to an excited state of hydrogen (following Sinniah et al. [23], hydrogen desorption is a first-order process relative to the adsorbed atomic hydrogen), and no difference is made between Si on surface and Si in the bulk.

Since the first reports, this model has also been modified to fit the deposition of SiGe alloys using silane and germane [22, 25]. The pseudo-Arrhenius plot corresponding to the deposition of these alloys is, as shown in Chapter 10, very similar to the silicon plot, even when shifted to lower temperatures. In this case, germane molecules are supposed to behave like silane ones, but with a larger ($\times 5$) sticking coefficient. In addition, desorption of atomic hydrogen is considerably facilitated by the presence of Ge in the solid, so that the GR is increased by a large factor (more or less one decade, depending on the Ge content).

On the other hand, some precautions must be taken when using silane. First, the presence of impurities in the gas phase, even at very low concentrations, will not be purged in a compressed gas (such as silane) in the same way that they are in a liquid like DCS. Second, the low stability of silane molecules makes it susceptible to gas-phase nucleation of particles, which can "rain down" and become incorporated in the growing film. However, the thermal decomposition of silane (or eventually disilane), occurring at relatively low temperatures (500–900°C), is very advantageous for technologies that require a limited thermal budget. The epi layers are generally not too thick, and the processing temperature is kept low so that the dopant profile can be made very sharp.

In conclusion, it should be noted that this hydrogen coverage of the silicon surface during low-temperature growth is of considerable importance. It will control or directly influence a number of epi characteristics, such as dopant incorporation, film morphology, differential poly/mono growth, and the structural quality of epi. etc. The author also considers this phenomenon, which has an important surfactant-like effect, to be the main differentiation between CVD and MBE techniques (to the advantage of CVD).

c. *Low-Temperature DCS/HCl-Based Epitaxy*

Selective epi growth (SEG) is very desirable in terms of both process simplification and the development of novel device structures (this is illustrated in Section IV). The concept of SEG is based on the fact that silicon nucleation and growth are slower over a dielectric than over silicon. Various SEG processes have been published, however, the most common are based on DCS/HCl/H$_2$ chemistry. Today, selective processes are required at a moderate temperature (<900°C), and DCS is preferred to silane, as its decomposition produces chlorines, which are species known to favor SEG. Apart from the choice of DCS, the addition of a certain flow of HCl is generally required to inhibit totally any deposition on the dielectric, at least when utilizing an industrial reactor working at pressures exceeding 10 Torr. As such, DCS-based epi and DCS/HCl-based SEG are addressed below.

As DCS is already widely used in conventional epi, and has also been used for SiGe deposition at low temperatures, experimental kinetics has been reported in a number of papers. However, the modeling and understanding of this process are not as clear as for the silane process. Surface reactions are still uncertain, although they are supposedly predominant during low-temperature epi processes, more so when using modern single-wafer cold-wall reactors. Indeed, most of the models have been applied to the high-temperature domain, which is of little interest here. In the low-temperature regime, it is acknowledged that DCS molecules adsorb directly on the silicon surface and undergo a reactive dissociation into Si, H, and Cl, and that these species can desorb, depending on the temperature, as H$_2$, HCl, or SiCl$_2$. A simplified form of the reaction pathway proposed first by Coon et al. [26] may be written

$$SiH_2Cl_2 + 4\_ \rightarrow \underline{Si} + 2\underline{H} + 2\underline{Cl}, \qquad (4)$$

$$\underline{H}^* + \underline{H} \rightarrow H_2 + 2\_, \qquad (2)$$

$$\underline{H} + \underline{Cl} \rightarrow HCl + 2\_, \qquad (5)$$

$$2\underline{Cl} \rightarrow Cl_2 + 2\_, \qquad (6)$$

$$\underline{Si} + 2\underline{Cl} \rightarrow SiCl_2 + 2\_, \qquad (7)$$

where $\underline{X}$ denotes a species which is adsorbed, $\underline{H}^*$ corresponds to an excited state of hydrogen (following Sinniah et al. [23], hydrogen desorption is a first-order process relative to the adsorbed atomic hydrogen, and no difference is made between Si on the surface and Si in the bulk.

However, because epi from DCS takes place in a hydrogen ambient, and hydrogen is known to increase the GR of silicon compared to Ar, N$_2$, or He [27], it is probably preferable to include the influence of the carrier gas in the model. The following reaction is suggested in Ref. 28:

$$2\underline{Cl} + H_2 \rightarrow 2HCl + 2\_ \qquad (8)$$

however, a mechanism via atomic hydrogen, reactions (3) + (5); cannot be ruled out.

The Arrhenius plot corresponding to DCS/H$_2$ epi kinetics, also in Fig. 2, shows that deposition is kinetically limited for temperatures below 800°C. The apparent activation energy (the limiting step may be not unique) is about 70 kcal/mol. This value accords with those in a number of reports and is also very similar

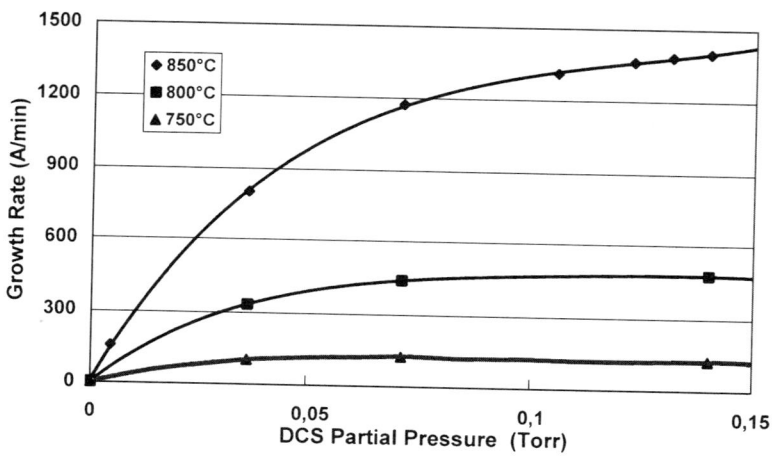

FIG. 4. Silicon growth rate as a function of dichlorosilane partial pressure. The clear saturation of kinetics, especially at the lowest temperatures, 800 and 750°C, indicates a very effective surface-limited reaction.

to the activation energies of both reaction (5) and reaction (8), which are two mechanisms of desorption working in parallel. These mechanisms undoubtedly correspond to the limiting step. Another interesting characteristic of deposition kinetics in the DCS/H_2 system is illustrated in Fig. 4, where GR is plotted as a function of DCS flow (or partial pressure) for different temperatures, all other parameters being constant. For low values, the kinetics is proportional to the gas flow. However, above a given DCS flow, the GR exhibits a very clear saturation, which indicates the presence of a very effective surface-limited reaction. At 750°C, for example, the GR is roughly constant for DCS pressures exceeding 0.05 Torr. Thus, the surface coverage would regulate DCS adsorption and GR. This situation is somewhat similar to that of SiH_4-based chemistry, but it is more complex due to the presence and interaction of two species on the surface. In addition, when referring to calculations carried out by Hierlemann et al. [27], one can speculate that chlorine coverage would be greater than hydrogen coverage under most experimental conditions. Compared to silane chemistry, this chlorine coverage brings about an important variation, which can explain different types of behavior, such as in dopant incorporation [29].

Silicon SEG using the $DCS/HCl/H_2$ system is also based on this chlorine chemistry. Since the first extensive work of Borland and Drowley in 1983 [30], it has given rise to a large amount of study, and huge progress has been made in terms of temperature reduction, process control, and epi characteristics. However, no (or almost no) attempts have been made to relate the SEG process to the underlying surface reaction physics. In addition, SEG is a complex process with a number of parameters, and as experiments are conducted using different conditions, any comparison is very difficult. The important parameters of SEG are the following.

- Temperature, pressure, and gas flow
- Reactor geometry
- DCS and HCl partial pressures
- Doping level and Ge (or C) content in the case of alloys
- Nature of the dielectric (including modifications due to previous steps: etch, implant, etc.)
- Silicon/dielectric ratio and size of silicon windows

Because of this complexity (to the author's knowledge), there is no precise physical model/understanding of SEG available in the literature to date. Basically, the reactions mentioned previously are still valid; but because of the significant HCl pressure, reaction (7) is certainly much more effective in depressing the GR and removing Si ad-atoms from the dielectric surface, compared to the DCS/H_2 chemistry.

Thus, within the DCS/HCl/H_2 system, perfect selectivity can be obtained by optimizing the HCl flow. This is illustrated in Fig. 5, where the nucleus density is plotted as a function of the HCl flow. These results were obtained on full-sheet Si_3N_4 CVD films on 200-mm wafers, using a commercial epi reactor. The nuclei were measured with a Surfscan, which enables the inspection of a full wafer. In this way, selectivity can be strictly tested. Whereas huge nucleus densities (over the Surfscan range) were observed at low HCl gas flows, selectivity was achieved above 130 sccm of HCl. However, it should be noted that this curve is somewhat

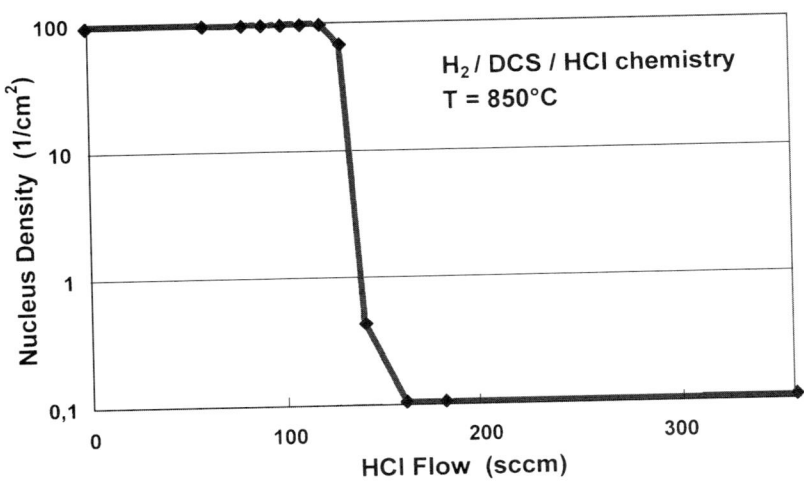

FIG. 5. Selectivity threshold of silicon deposition on CVD Si_3N_4. The nucleus density was measured as light point defects using a Surfscan, which allows the entire 200-mm wafer to be inspected. Under the present conditions, a perfect selectivity is obtained when using 140 sccm of HCl or more. Note that the points located on the bottom and top axes correspond to densities below 0.1 and above 100 LPD/cm^2.

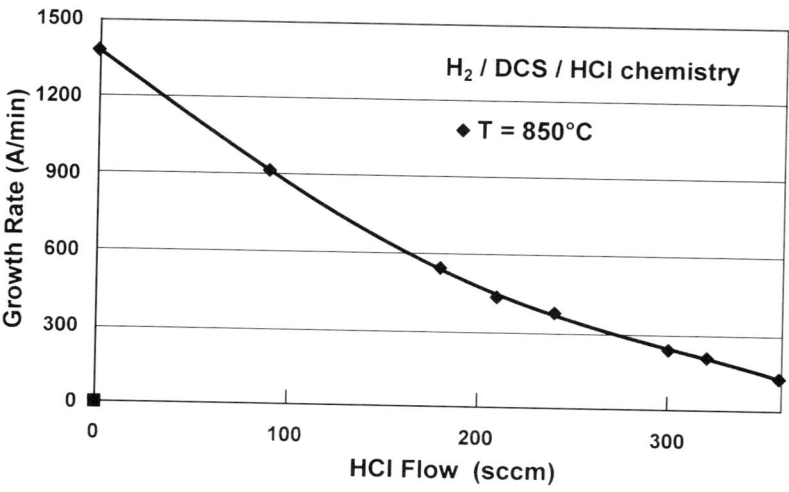

FIG. 6. Effect of HCl gas flow on Si growth rate. The deposition kinetics is measured on full-sheet silicon wafers and the conditions are similar to those in Fig. 5.

arbitrary because it does correspond to a particular choice of gas flow, duration, and substrate nature, including its preparation (clean, bake, etc.). We can observe that the selectivity threshold, about 140 sccm, is relatively abrupt since a 10% variation in HCl flow changes the level of defects by two to three decades. But the most interesting point is that selectivity can attain a very high degree of perfection (not published, to the author's knowledge), as nucleus densities or defectivities lower than $0.1/cm^2$ have been measured. The latter fact certainly proves that SEG can be utilized in very complex technologies. In addition to the selectivity, the deposition kinetics must also be known. The GRs corresponding to the conditions in Fig. 4 and obtained on full-sheet silicon wafers are given in Fig. 6. Not surprisingly, the GR is found to decrease severely with the HCl gas flow, and silicon etching would even occur if excessive amounts of HCl were added. Film thickness uniformity constitutes the third condition to be met to make the selective epi applicable. Indeed, a specific effect, called the local or micro "loading effect," must be controlled. This basically relates to a modification of the GR depending on the substrate pattern detail of masked and unmasked areas. Very often, the larger the silicon opening, the thinner the deposition, but the contrary can also be obtained under some conditions. In the case of silicon SEG, this micro loading effect is greatly suppressed when using low temperatures (<850°C) and reduced pressures (<50 Torr) [31]. The case of SiGe, which is detailed in Chapter 10, is more delicate. Since germane is a very reactive molecule and germanium is a growth catalyst, the micro loading suppression is more difficult and less perfect.

Apart from these three basic features of SEG, other important points, more closely related to substrates, are given in Section III.3. In conclusion, the following general rule can be made: The minimal HCl flow required to obtain full selectivity has to be increased with any of the following parameters—DCS flow (but the DCS/HCl ratio can be decreased), temperature, pressure, doping level (boron), and Si coverage.

3. PROCESS DETAILS

a. Surface Preparation for Low-Temperature Epitaxy

Epitaxial growth in general relies critically upon the silicon surface preparation, and all the following conditions must be met before deposition is initiated.

- No oxide on the surface
- A good crystal quality (no crystal defects due to etching or implant)
- No dopant precipitates (in the case of a very high doping level)
- No surface contamination or particles

Conventional epi processes use both a wet *ex situ* clean and an *in situ* hydrogen bake. The *ex situ* clean generally consists of an RCA cleaning procedure [32], which eliminates particles, native oxide, organics, metal, and carbon and repassivates the reactive silicon surface with a thin (0.8-nm) layer of suboxide, which is hydrophilic, stable, and relatively easily removed *in situ*. The epi process includes a high-temperature hydrogen bake (around 1100°C), which is used to remove the RCA-regrown oxide, to anneal eventual crystal imperfections, to dissolve/diffuse eventual high dopant concentrations, and to dissolve/evaporate most particles. All these effects are beneficial to the epi, and indeed this combination enables high-quality films to be deposited.

In the case of low-temperature epi, two situations exist. In the first situation, epi has to be run at a low temperature, because strained epi or sharp dopant profiles have to be created, but the substrate can undergo high temperatures. These conditions are met for processes placed at the very front end of the technology, or for the development of R&D structures, which are basically deposited on full-sheet wafers. In this case, a conventional surface preparation is recommended, and defect levels as low as a few defects per wafer are commonly achieved.

In the second situation, when the structure cannot support high temperatures, the strategy of the surface preparation must be changed. Indeed, as the etching rate of SiO_2 under hydrogen decreases very rapidly with temperature [11], it becomes impossible to remove the RCA oxide completely at bake temperatures below 1000°C. This situation corresponds to the majority of advanced applications, where the epi has to be run after a significant part of the devices has been built. In this case, the best solution would be to remove the chemical oxide *in situ*, using etch processes, which are efficient at low enough temperatures, directly in the epi chamber or in a clustered annex. In the literature, various treatments based on plasma, HF vapor, or other chemistries have been proposed, and encouraging results have been obtained. However, these "new chemistries" are generally carried out in prototype reactors and are not yet very common.

Consequently, most researchers today adopt a strategy based on the more conventional "HF-last" clean. The HF bath removes the oxide from the wafer surface and passivates the silicon surface with atomic hydrogen. Originally, the wafers were etched in diluted HF just prior to loading, without any water rinse. The UHV-CVD growth process reported by Meyerson *et al.* is an example of such a HF-last clean [33]. However, this procedure presents two major difficulties. First, transport and manipulation of the wafers coming from HF acid without any rinse are very critical, especially in the manufacturing environment. Second, removal of the residual HF is very difficult when hydrophilic patterns are present on the wafer surface. Because of these difficulties, a water rinse associated with drying is often used, even at the expense of some surface contamination. The small amount of oxygen present on the surface, due to water rinse and transportation between the cleaning and the epi equipment, is generally removed by a hydrogen bake at a moderate temperature. Thus, a possible and more widely used surface preparation procedure is as follows.

- RCA clean
- HF last
- Water rinse
- Isopropyl alcohol (IPA) dry
- Loading in the epi tool under an inert gas
- *In situ* hydrogen bake at 800–900°C for 1–5 min

Of course, a wet clean using an *in situ* rinse, made by displacing an extremely diluted HF solution with DI water in the same tank, is recommended because it eliminates the transfer of hydrophobic wafers from the HF to the DI-water bath. Clustering this precleaning with the epi tool is also preferable [34]. In this cleaning procedure, the quality of the chemicals is crucial. In particular, DI water with a total organic content and dissolved oxygen in the range of 1–2 ppb and hydrogen with less than 10 ppb of impurities are commonly used for the respective water rinse and the hydrogen bake processes. In the author's opinion, this type of surface preparation allows high-quality epi to be grown: typically, no oxygen or carbon is visible by secondary-ion mass spectroscopy (SIMS) at the interface (detection limit, about 1×10^{13} atoms/cm^2), and defect densities are in the range of 0.1–10 defects/cm^2 (light point defects, >0.2 μm), depending on the precise experimental conditions. Such low values are usually measured on silicon full sheets, but similar results are also expected on patterned wafers. On the other hand, some operations, such as dry-etch or implant, can cause the silicon at the surface and in the subsurface region to be highly defective and not compatible with high-quality epi growth. In these cases, the processes of such operations require modifications to respect the crystalline quality of the silicon (for example, by adding a soft-etch step to an etch recipe), and/or a sacrificial oxide, typically 10–20 nm thick, can be grown (and removed) for damage removal before epi.

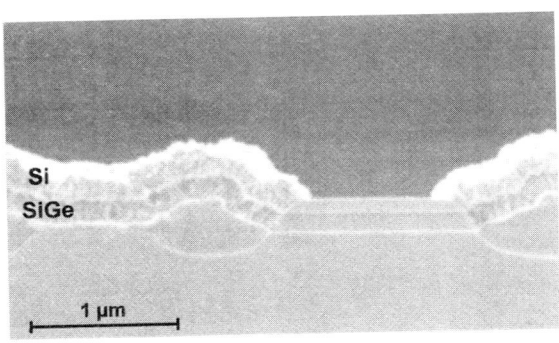

FIG. 7. Si/SiGe nonselective epitaxial growth on a LOCOS CMOS wafer. The cross section was SECCO-etched before SEM observation: both films are well decorated and SiGe is clearly delineated.

b. Epitaxy on Patterned Substrates

To exploit fully the capabilities of low-temperature epi in new applications, the epi has to be deposited on structured substrates. Here, a "structured substrate" means wafers where patterns made of different materials (generally dielectrics) together with monocrystalline silicon areas are present on the surface. In this section, some of the difficulties caused by these patterns are discussed for the case of nonselective deposition.

Figure 7 shows an example of a low-temperature Si/SiGe epi deposited on a substrate where the pattern consists of polycrystalline silicon on local oxidation of silicon (LOCOS). Polycrystalline material is deposited on top of the poly/LOCOS stack, and epi is grown on single-crystal regions. To highlight details such as dislocations, grain boundaries, and Si/SiGe interfaces, this cross section was submitted to chemical decoration before scanning electron microscopy (SEM) observation. Different characteristics of this sample may be of interest depending on the application's aims: epi perfection, poly/epi thickness ratio, poly/epi doping level ratio, poly roughness, etc. In the specific case of Fig. 7, we can note perfect epi growth, significant (not huge) poly roughness, and poly/epi thickness ratios of 1.3 for SiGe and 2 for Si. The majority of these film characteristics are known to depend significantly on the deposition conditions and, thus, can be modified purposefully. However, up to now, these characteristics and their variations have rarely been reported in publications, nor can they all be detailed here.

In the author's opinion, an important difficulty encountered when developing epi depositions on structured wafers concerns the precise control of growth kinetics. Indeed, a fixed process (chemistry, temperature target, pressure, etc.) does not give the same deposition on different substrates. Figure 8 shows the SiGe growth kinetics as a function of the germane gas flow, measured on full-sheet wafers and on the structures shown in Fig. 7. The comparison shows clearly that the GR is lower when using structured substrates compared to full-sheet wafers. Basically,

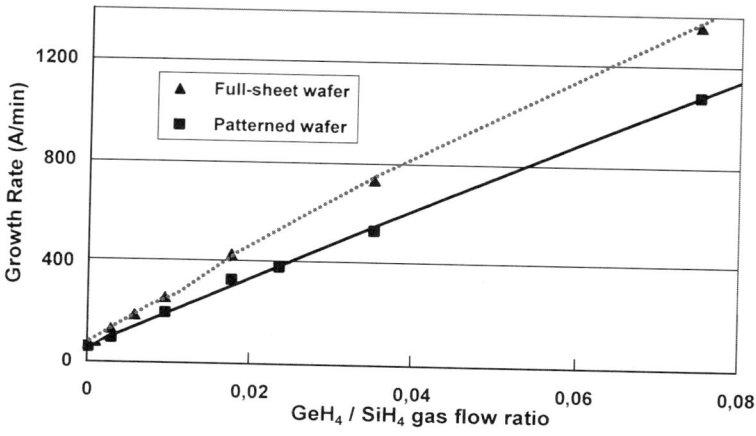

FIG. 8. SiGe growth kinetics as a function of the germane/silane flow ratio for silicon full-sheet and device wafers. In this case, it was established that the growth rate variation is due to both a thermal loading effect and a chemical loading effect.

this change can be attributed to two causes: a global "thermal" loading effect, which is detailed below, and a global "chemical" loading effect. The latter, usually referred to as the loading effect, is certainly due to the following mechanism. Since poly deposition is larger than epi deposition, it will induce a more significant impoverishment of the reactant gas and, therefore, a global reduction of kinetics. The actual causes of the GR variations presented in Fig. 8 have been studied in detail [35]. In this specific case, it was clearly established that

- silicon deposition did not present any significant chemical loading effect and the observed GR variation is due to the thermal effect, and
- SiGe growth did present both a thermal loading effect (to the same extent in degrees as silicon growth) and a chemical loading effect, both effects having comparable intensities.

The present results are specific to the conditions used for deposition. Nevertheless, they demonstrate that, in a general way, the deposition on device wafers is different from an eventual calibration using full-sheet wafers.

As no clear reports are found in the literature, it is of interest here to detail the thermal loading effect mechanisms. During deposition, the wafer rests on a SiC susceptor; both pieces are lamp heated inside a cold-wall chamber and under hydrogen. In such equipment, contrary to conventional furnaces, which follow the "blackbody emission" laws, the actual temperature of an object depends on its own optical properties. Referring to Fig. 1, the main incoming heat exchange contributions to the wafer can be easily listed and a simplified form given:

Radiative heating from lamps:

$$+\varepsilon^{\text{lamp}} P \qquad (9)$$

IR radiative emission of the wafer:

$$-\varepsilon \sigma T^4 \qquad (10)$$

Radiative heat exchange with the susceptor:

$$+\varepsilon \sigma T_{\text{susc}}^4 - \varepsilon \sigma T^4 \qquad (11)$$

Conductive heat exchange with the susceptor:

$$+K_1(T_{\text{susc}} - T) \qquad (12)$$

Conductive heat exchange with the chamber:

$$+K_2(298 - T). \qquad (13)$$

In Eqs. (9)–(13), $\varepsilon^{\text{lamp}}$ is the emissivity of the wafer corresponding to the lamp wavelengths, P the radiative power density from the lamps, ε the silicon emissivity corresponding to its own emission, T the wafer temperature, T_{susc} the susceptor temperature, K_1 the constant of heat conduction between the wafer and the susceptor backside, and K_2 the conduction constant between the wafer and the chamber; the other letters have their usual meaning. Note that the susceptor emissivity was taken as unity.

The stationary wafer temperature will correspond to the exact balance of all these contributions. Since T and T_{susc} are very close, T can be written $T_{\text{susc}} + \Delta T$; and therefore, the equilibrium temperature can be expressed as

$$\Delta T = [\varepsilon^{\text{lamp}} P - \varepsilon \sigma T_{\text{susc}}^4 - K_2(T_{\text{susc}} - 298)]/(8 \varepsilon \sigma T_{\text{susc}}^3 + K_1 + K_2) \qquad (14)$$

In practical cases, we have the relations $K_1 \gg K_2$ and $K_1 \gg 8 \varepsilon \sigma T_{\text{susc}}^3$. Equation (14) can thus be simplified as

$$\Delta T = [\varepsilon^{\text{lamp}} P - \varepsilon \sigma T_{\text{susc}}^4 - K_2(T_{\text{susc}} - 298)]/K_1 \qquad (15)$$

In the majority of epi reactors, T_{susc} is controlled during the process and can be considered as constant and repeatable. Note that in reactors that use thermocouple measurement, the susceptor temperature is not directly controlled, and K_1 has to be modified so that the temperature offset between the susceptor and the thermocouple may be taken into account. In any case, Eqs. (14) and (15) establish that in a general way, ΔT does not have a zero value and varies with wafer emissivity, top/bottom lamp power repartition (via P), and susceptor design as well as gas conductivity (via K_1). Despite severe simplifications being made, Eq. (15) explains perfectly how the actual process temperature varies with the wafer's optical properties, giving rise to the so-called thermal loading effect. In the same way, based on the same physics, one can easily demonstrate that this thermal loading effect also has a great influence on the deposition uniformity.

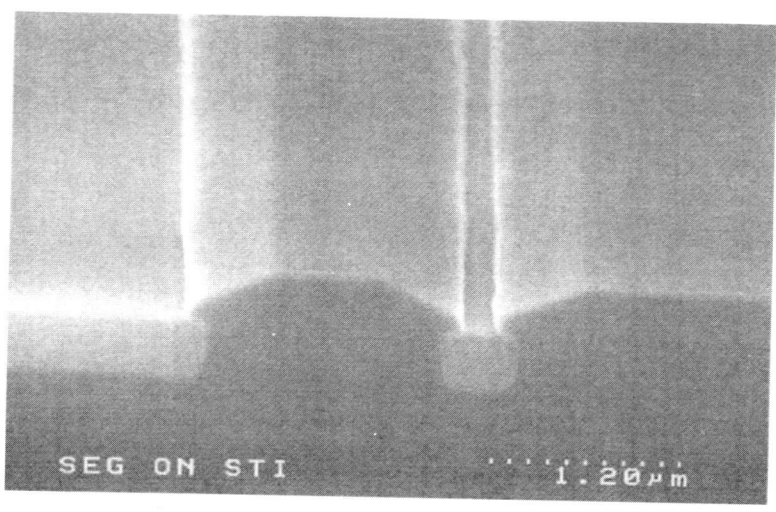

FIG. 9. Edge-view SEM photomicrograph of a selective silicon epitaxy deposited on a substrate with shallow trench isolation. On this test structure, the epitaxy was thick enough to make the {311} facets very visible.

In this section, RT-CVD is shown to exhibit both thermal and chemical loading effects. As these effects induce large deposition variations depending on the substrate, for example, between device wafers and full sheet wafers, they have to be taken into account in any accurate process calibration and control. This constitutes a major difficulty in low-temperature RT-CVD.

c. *Selective Epitaxial Growth*

Figure 9 shows an example of a selective silicon epi deposited on a substrate whose patterns consist of shallow trench isolation (STI) (SiO_2 filling). In such a process, all the following points may be of interest.

- Epitaxy perfection
- Global loading effect
- Thickness and doping in the different windows (micro loading effect)
- Facet nature and development
- Lateral overgrowth on dielectric

In the specific case in Fig. 9, we can note perfect epi growth, a significant overgrowth on the STI, and an important faceting. On this test structure, the epi was thick enough to make the {311} facets very visible as they extend along a large part of the epi. We have also verified that the STI structure was basically unchanged, without any noticeable SiO$_2$ etching. Again, despite the fact that the above-mentioned characteristics depend significantly on the deposition conditions, their variations have rarely been reported in the literature and not all of them can be detailed here.

Similarly to the case of nonselective epi, discussed in the preceding section, SEG presents loading effects with two possible components: a thermal component and a chemical one. The thermal loading effect works exactly as explained in the previous section. The situation is even a little simpler, as the emissivity of the wafer is constant during growth because of the absence of poly deposition. Conversely, as the chemical loading effect can have important and very visible local effects, it has been reported for some time in the literature. This point is discussed in Section III.2.c. However, as discussed below, this chemical component can also have an important global effect on the deposition.

Figure 10 shows the GR of silicon SEG plotted as a function of the silicon coverage, 100% corresponding to a full silicon sheet. These results were obtained using perfectly selective processes developed in an industrial reactor with the DCS/HCl/H$_2$ chemistry. It should also be mentioned that, in these results, any thermal contribution was eliminated, so that the curves give only the chemical contribution. As far as the author is aware, this is not the case in other data reported in the literature, where both contributions are mixed. One curve was obtained using relatively high HCl and DCS flows. In this case, the GR varies strongly with the silicon coverage, especially in the range of 0–20%; the larger the coverage,

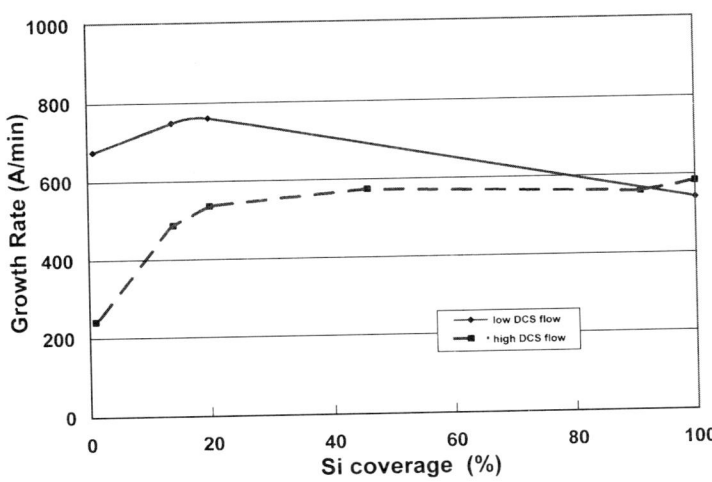

FIG. 10. Global loading effects in silicon SEG obtained with two DCS (and HCl) gas flows. As the thermal contribution has been eliminated, the variations correspond to the chemical effect.

the higher the GR [35]. Note that such a variation is counterintuitive. Indeed, the loading effect is very often attributed to gas-phase depletion, and as this depletion can be presumed to be more significant at high Si/SiO_2 ratios, a lower GR should be expected. Experimental values follow exactly the reverse variation. Our current interpretation is that the GR is dominated by HCl, and this gas is supposed to be depleted by the silicon regions. Therefore, the larger the silicon coverage, the larger the HCl impoverishment and the higher the GR. The other curve corresponds to another SEG process developed with lower DCS and HCl gas flows. In this case, a more "conventional" loading effect was obtained; the larger the coverage, the lower the GR. This variation is supposed to be due to DCS depletion, which certainly dominates the growth kinetics.

Finally, both thermal and chemical loading effects have a global effect on SEG deposition and must be taken into account. To simplify matters, we have chosen to discuss the case of silicon epi here. However, these loading effects also have a great impact on doping levels and alloy compositions. In the specific case of SiGe alloys, as the Ge is known to "catalyze" the growth kinetics, huge global (and local) loading effects can be observed in selective depositions.

IV. New Applications

In this section, we present and discuss some of the new or possible future applications of epi in silicon technology. We use the same approach to discuss each application in detail: we (i) present the motivation behind implementing the new process in technology, (ii) give the important film characteristics that are required and the process issues that need to be addressed, and, finally, (iii) present some process solutions or recommendations and some results that have been reported by crystal growth or device colleagues.

The list is not exhaustive, and the structures were selected by considering different parameters, such as technological importance, research activity, and difficulties in terms of epi growth. In this way, the reader will get a complete panorama of possible epi processes. For similar reasons, the deposition of the SiGe base of RF bipolar transistors is discussed in detail because it is certainly the leading application of "advanced epi." Indeed, it has attracted major technological and economical interest, and epi deposition also presents very interesting challenges.

1. Low-Cost Epitaxial Wafers

More than four decades of development in single-crystal growth have brought enormous progress in the production of high-quality ingots of increasing size. However, the growth is a long and complex process, as the temperature is very high (1410°C), and the furnace geometry as well as the melt constituents changes during pulling. In others respects, epi wafers were originally developed to create a different conductivity layer and to fix the latch-up problem in logic CMOS. Owing to their low oxygen content and the absence of crystal-originated particles (COPs), epi layers are also beneficial in terms of defect levels, and they have become a widely used standard in recent CMOS and DRAM technologies. Because of the history, the required epi was simply deposited on available polished wafers, which were high-quality and full-valued CMOS wafers.

Today, the situation is as follows. On one hand, the technical difficulties in Czochralski (Cz) growth increase considerably with the ingot size, and the pull rate has to be significantly reduced to avoid COPs in 300-mm ingots [36]. On the other hand, the advanced technologies have increasingly strict specifications and epi wafers will certainly be the only solution for an increasing number of markets. With these considerations in mind, a new strategy could emerge for the production of 300-mm wafers [37]. As active devices are designed within a few micrometers at the surface of the wafer, the specifications of this active region have to be tight but the specifications of the underlying substrate could be relaxed. As soon as the substrate is liberated from the specifications of active regions, Cz growth of ingots and wafer preparation can be greatly improved in terms of productivity and cost. High pull rates can be achieved as the resulting increase in COPs can be tolerated in the substrate, and a larger part of the ingot is usable because of the rather broad specifications. Additionally, the surface preparation of the substrate before epi can be simplified. Within this strategy, epi wafers would become the unique standard: the substrate fabrication would be cheaper and compensate for the epi process cost. Epitaxy would become the key growth process of these wafers, with the following advantages:

- high Si quality, low oxygen content, good surface quality, and no COPs;
- direct growth to the desired resistivity; and
- high radial and wafer-to-wafer uniformity.

For CMOS applications, a lightly doped p^- epi deposited on a highly doped p^+ substrate is required; and in this case, the major technical problem comes from the p-type autodoping due to boron evaporation from the backside of the substrate during deposition. It should be noted here that this problem may disappear in future CMOS technologies. Indeed, as the evolution of CMOS technology relies on a decrease in operating voltage, the most commonly used voltage will soon not be high enough to cause latch-up, eliminating the p^-/p^+ stack requirement. However, in the short term, this development cannot be considered certain because CMOS circuits designs currently use dual-voltage CMOS.

For this application, the thermal budget of epi is not critical, and the deposition, a few micrometers thick, is easily controlled. The requirements and specifications of the epi process are very similar to those for the standard fabrication of p^-/p^+ CMOS wafers, on which precise information is given in previous chapters. In short, an appropriate process could consist of a H_2 bake at about 1100°C and a silicon deposition in the temperature range of 1050–1100°C using DCS/H_2 chemistry. Both steps should be at atmospheric pressure to minimize boron autodoping and to keep the process cost low.

2. SOI Wafers

Silicon-on-insulator (SOI) substrates are very promising and may become a standard in future MOS field effect transistor (MOSFET) technologies. Because of the isolation provided by the underlying insulator, they have some major advantages compared to bulk silicon:

- a reduced junction capacitance,
- higher RF performances,
- a lower power consumption,
- a high soft error immunity, and
- simpler technology.

For these reasons, great efforts have been made in the last 20 years to fabricate SOI films. Among the techniques that have been developed to a certain point, some are available for commercial use. Today, most promising are the well-known "bond-and-etch back SOI" (BESOI) techniques with different refinements, and the various types of a more recent technique sometimes referred to as "smart-cut," namely, UNIBOND [38], Genesis [39], and ELTRAN [40].

This smart-cut technique can be considered an actual breakthrough. It allows the fabrication of SOI films by industrial means using an economical process with reusable donor wafers. The films are monocrystals of a very high quality (no precipitates or dislocations) on the entire surface of the wafer (200 mm today), without any design constraint. In short, this technique starts with a "donor wafer" and a "handle wafer" and combines a bonding method together with a very specific cleavage technique. Figure 11 shows the sequence of fabrication for LETI's UNIBOND and Singen's Genesis techniques. First, at least one of the two starting wafers is oxidized; the donor wafer is submitted to a hydrogen implantation to fragilize an interface buried under the surface. The implant peak location can be considered to correspond to stressed silicon. Second, the atomic bonds at the surface of both wafers are activated by wet chemistry or plasma, and the donor wafer is joined to the handle wafer by bonding. Owing to the surface bond activation, the strength of this bonding is relatively high, and it can be reinforced by an additional thermal treatment. Third, the seed wafer is split through the fragilized interface by means of a thermal or mechanical treatment, leaving a monocrystalline SOI film on the handle wafer. It should be noted that, in the majority of cases, the upper part of the donor wafer is epi silicon, which has all the advantages mentioned in Section IV.1.

As the SOI surface corresponds to the cleavage, surface finishing is used to improve its quality. This finishing can be done by chemical mechanical polishing

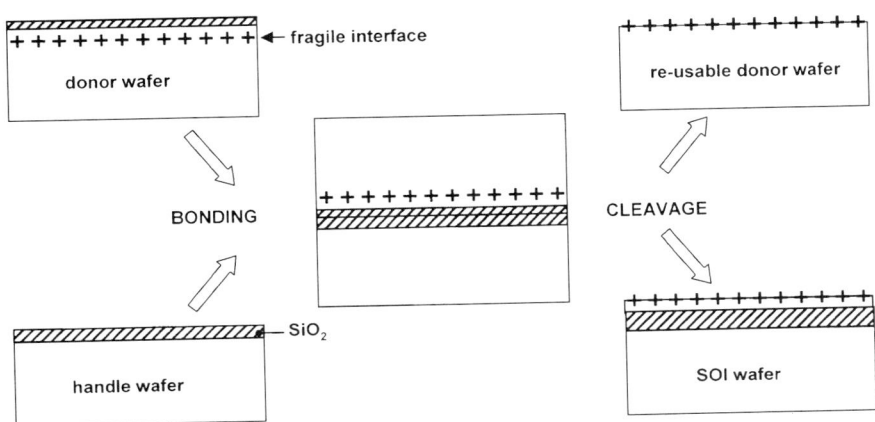

FIG. 11. Sequence of fabrication for LETI's UNIBOND and Singen's Genesis techniques. Bonding and "smart-cut" (cleavage along a fragilized interface) are combined to obtain high-quality SOI wafers. Epitaxy can be used for "donor" wafer preparation and for SOI film improvement or thickening.

(CMP), and very encouraging results have been reported in term of roughness, with the root mean square (RMS) reduced from 2 nm to 0.1 nm [39]. However, this result was obtained by removing a part of the SOI film thickness, and since the uniformity of such thin removals is not well controlled, the uniformity of the thin SOI film left on the handle wafer was degraded, from 8.5 nm (total range) before to 26 nm after CMP. For this reason, other surface finishing is desirable. Treatments at high temperatures using an epi reactor were experimented on with success. The roughness RMS was reduced from 2 to 0.2 nm, with only a very small uniformity degradation [39]. In this case, the uniformity of the removed thickness could be controlled within 10%: 4 nm, compared to 26 nm with CMP, for a 40-nm thickness removal. The author is not familiar with the details of the process, but a high-temperature hydrogen bake is known to smooth the SOI surface without reducing the film thickness [41], and this annealing may be associated with an etching action of gaseous HCl to improve the results further.

These smart-cut techniques also suffer from limitations in terms of SOI film thickness. Indeed, the cleavage corresponds to the implant depth range, which has to be maintained within certain limits for the smart-cut to be feasible with a good yield. Thus, the naturally fabricated SOI thickness lies within 100–200 nm. However, very different thicknesses are required by the potential applications. Bipolar technologies may require films up to a few micrometers thick, whereas MOSFET technologies may require them to be about 20 nm for fully depleted devices and about 100 nm for partially depleted ones; the choice between the two regimes is still being debated [42].

For ultrathin film applications, thinning is currently obtained by oxidation or CMP, which is not discussed here, as it is not related to epi. On the contrary,

SOI films can easily be made thicker using epi. Dealing with the preparation of full-sheet wafers before the fabrication of any device, this application presents no very specific difficulties. For the deposition of lightly doped silicon, a relatively standard epi process, described in the previous section, can be used. However, as the optical properties of SOI wafers are different from those of prime wafers, the reactor tuning must be adjusted, especially in accordance with the SiO_2 thickness. In addition, the hydrogen bake can be somewhat reduced to avoid any defectivity problem at the edge of the wafer, due to exaggerated undercutting at the Si/SiO_2 interfaces. In some applications, very specific stacks, using SiGe(C), abrupt dopant profiles, etc., may also be required. In this case, the process will be specific to the stack but similar to that used on prime wafers.

The ELTRAN process is similar to the two others, except for the nature of the cleavage interface, which is in double-layered porous silicon. In this case, a high-quality epi layer has to be deposited on a porous Si layer formed on the donor wafer. Silicon epi growth is possible, as porous silicon is perfectly monocrystalline and able to transmit the substrate orientation in certain porosity and thickness windows. However, to get a good yield and a relatively smooth surface at the cleavage operation, the porous silicon structure has to be respected during the epi process. So, whereas a very high quality is required, compared to a standard epi the temperature of the hydrogen bake has to be reduced significantly to avoid important restructuration and grain coarsening. On the other hand, the deposition temperature must not be set too low to get a process that is not too conformal and, thus, not able to fill the pores of porous silicon. Thus, an epi process with a hydrogen bake at a moderate temperature (800–900°C), possibly associated with an HF-last cleaning, together with a deposition using SiH_4 chemistry at an intermediate temperature (700–800°C depending on the gas pressure), seems to be suitable for this application. It should be noted here that the quality of the epi deposited over the porous silicon may constitute the main limitation of this technique today, as the best results report a defect density that is not excellent (about $6 \times 10^2/cm^2$) [40]. On the other hand, this SOI layer formed by epi growth features a high controllability and flexibility of film thickness: thick films of several micrometers to ultrathin films of 50 nm have been reported. This technique also requires surface finishing to remove the porous silicon and to decrease the roughness. For this technique, surface smoothing via a high-temperature bake under hydrogen seems to be preferential, perhaps because the initial roughness has a short wavelength.

3. Bases of RF Bipolar Transistors

a. Background

Bipolar transistors generally offer the best speed performances in silicon technology. For RF applications (apart from an acceptable device breakdown voltage), the most desired figures of merit are the cutoff frequency, f_T, and the maximum frequency of oscillation, f_{MAX}. F_T, which is the frequency at which the current gain of the transistor decreases to unity, can be approximated as

$$f_T = 1/(\tau_b + \tau_{bc} + \tau_{capa})/(2\pi), \qquad (16)$$

where τ_b is the transit time of minority carriers through the neutral base, τ_{bc} the delay associated with the transit through the base-collector depletion region, and τ_{capa} the delay associated with the charging/discharging of different capacitances [43].

A high collector doping can reduce the time constant τ_{bc}, which is related to the extension of the base-collector depletion region, but a maximal doping value is imposed by a minimal emitter–collector breakdown voltage. The base transit time is a key parameter in the f_T value, and epi bases, which can be fabricated as thin as desired, will represent an important breakthrough in high-speed bipolar transistors, compared to implanted bases.

On the other hand, f_{MAX}, which takes into account more parasitics of the transistor, is usually considered as a more realistic indication of transistor performances in a real circuit. This figure of merit is given by

$$f_{MAX} = f_T^{1/2}(8\pi R_b C_{jbc})^{-1/2}, \qquad (17)$$

where R_b is the base series resistance, and C_{jbc} the base/collector junction capacitance.

Low R_b values are necessary to achieve high frequencies. This resistance is usually considered as the sum of two components: the intrinsic and extrinsic resistances. The intrinsic base resistance, which corresponds to the active part of the base located beneath the emitter, has to be compromised with other electrical parameters of the device such as the current gain. On the other hand, the extrinsic base resistance, which links the edge of the active base to the base contact, has to be minimized through correct choice of the bipolar architecture or, eventually, by improved alignment of photolithography levels. Today, self-aligned emitter/base structures are preferred because they reduce the extrinsic part of R_b by decreasing the distance between the high-resistance active base and the low-resistance extrinsic base and because they also reduce the collector/base capacitance C_{jbc}.

When using a heterojunction bipolar transistor (HBT), i.e., the emitter material has a bandgap larger than that of the base material by ΔE_G, the current gain is increased by a factor $\exp(\Delta E_G/kT)$, compared to the equivalent homojunction bipolar. This property, which allows the base doping to be significantly increased while conserving an acceptable current gain, is the capital benefit of HBTs. The heterojunction can also be used to create an electric field, which will accelerate the minority carriers into the neutral region of the base. Today, there are two main approaches to the design of the vertical structure of HBTs. One approach is to use the doping structure of a standard bipolar transistor and to introduce a moderate, graded percentage of germanium into the base. This will improve the base transit time and f_T due to the electric field. A second approach is to use a base that is more heavily doped than the emitter along with a high germanium content ($>20\%$), which is necessary to obtain a standard current gain. This option results in a thin base with a very low resistance, which increases the f_T and f_{MAX} and also reduces the high-frequency noise.

The most straightforward way to utilize the HBT concept would have been to increase the bandgap of the emitter; however, despite intense research, no completely successful process has yet been reported. The other way is to fabricate the base of the bipolar with a material that has a bandgap smaller than that of silicon. Strained SiGe epis, which present a significant bandgap reduction (about 90 meV for 10% Ge), are an excellent candidate for Si-based HBTs. Indeed, in 1987 Iyer *et al.* [44] reported the first very promising results on Si/SiGe/Si HBTs. Even if the base current was not ideal, possibly because of some defects in the epi, the current gain was found to increase with the presence of Ge in the base, and different electrical characteristics clearly demonstrated HBT behavior. In this pioneering work, the structure was grown by MBE and the devices were simply formed by mesa etching. Immediately following this, considerable efforts were devoted to the development of SiGe epi via MBE as well as CVD techniques [45], as well as to the development of HBT structures. Consequently, huge and rapid progress was made in improving the static electrical characteristics, ideal Gummel plots were rapidly obtained, and very impressive dynamic performances were achieved. A double-mesa HBT [based on an all-epi structure grown by MBE, with a considerably more doped base than the emitter and with a high Ge content (above 20%)] was reported to have an $f_{MAX} = 160$ GHz [46]. More recently, an integrated HBT with a base deposited by CVD has demonstrated an f_T as high as 154 GHz [47]. Apart from these record values, discrete HBTs and the first HBT-based bipolar and bipolar CMOS (BiCMOS) technologies are commercially available today. More sophisticated functions and high-speed operation are required for various technologies, namely, wireless phones and local networks, millimeter-wave systems, satellite communications, and fiber-optic transmissions. Thus, major semiconductor manufacturers [48–50] are currently developing technologies which include high-performance (f_T and f_{MAX} between 60 and 200 GHz) HBTs and will operate from 1–2 to about 40 GHz. Frequency, noise, complexity, or cost must be favored, depending on the precise application.

In the literature, some distinction is usually made between "true" HBTs and drift-field bipolars. Presumably, this classification was done partly because the physics and electrical characteristics are somewhat different and partly because they were originally fabricated in very different ways: the mesa structure was grown by MBE and the integrated bipolar by CVD, respectively. However, with regard to epi, at least three classifications may be made:

- "true" HBTs versus drift-field bipolars,
- base doping higher or lower than emitter doping, and
- nonselective versus selective epi growth.

The third classification is preferred in the following discussion, because it corresponds to major changes in the deposition process.

b. *The Nonselective HBT Process*

In the literature, a number of HBT structures are based on nonselective epitaxial growth (NSEG). In mesa structures, the entire emitter/base/collector profile is grown by epi and a series of etching steps is used to isolate and contact the base and collector. As the starting substrate is a full-sheet wafer, the epi difficulties are not very important or specific. In addition, these structures are not suitable for complex circuit integration. They are not considered in the discussion that follows. Some self-aligned structures have been developed by IBM [51], but they are rather complex. Thus, NSEG is currently used to form relatively simple non-self-aligned HBT structures in bipolar and BiCMOS technologies.

Not all the structures can be presented here; therefore we will take an example now and discuss any relevant concerns for epi base deposition. The quasi-self-aligned double-poly SiGe HBT architecture shown in Fig. 12 is based on the original concept of Hong *et al.* [52] and has been integrated into an existing 0.35-μm CMOS process to form a BiCMOS technology suitable for the fabrication of highly integrated wireless communication circuits [53]. A basic feature is the use of nonselective Si/SiGe epi to form the intrinsic epi base and the polycrystalline base electrode. The process flow uses a modular approach where the CMOS and bipolar processing are largely uncoupled. The CMOS process steps are first carried out until the formation of source/drain extensions. The bipolar active regions are then opened in an a-Si/SiO$_2$ stack. The a-Si serves as a hard mask during the wet oxide etching from the base area prior to nonselective Si/SiGe epi growth. It also improves polycrystalline deposition over the field and CMOS regions and minimizes loading effects during epi. In such a BiCMOS technology, the HBT vertical structure has to be chosen in accordance with the CMOS flowchart. The base stack has to resist the remaining CMOS thermal budget: the strained SiGe/Si stack has to be stable enough against the implant and annealing so as not to relax, and the boron profile has to be confined in SiGe so as not to create parasitic barriers at the base/collector junction. Here, the typical drift-field HBT stack consists of three layers: 30 nm of undoped SiGe with 12% Ge, 30 nm of B-doped SiGe with

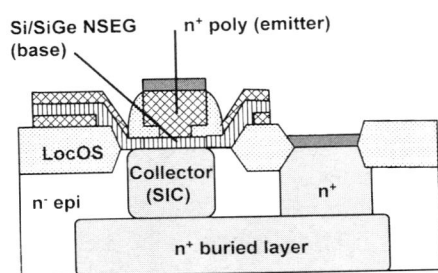

FIG. 12. Schematic cross section of a NSEG-based double-poly SiGe HBT. This architecture, proposed in Ref. 52, has recently been integrated in a 0.35-μm CMOS [53].

12–0% Ge grading, and a 35-nm B-doped Si cap. On the other hand, the epi process has to be designed in such a way that it will not degrade/shift the CMOS characteristics.

In the present application, the main challenges of the base epi are the following.

- Si/SiGe deposition on patterned wafers (SiO_2 and mono- and polycrystalline silicon are present)
- A very moderate thermal budget (to let the CMOS transistors remain unchanged)
- A high epi quality
- An acceptable level of defects
- No boron spike at the substrate/epi interface (this would create a parasitic energy barrier)
- Good process control and very good thickness control (especially when using a drift-field HBT concept with a poly-emitter)
- A low poly roughness (to avoid parasitic spacer formation around CMOS gates during poly-base patterning)
- A controlled poly/epi transition
- A poly/mono GR ratio above 1, which is favorable with regard to base resistance

Today, UHV-CVD as well as RT-CVD has been demonstrated to be efficient in growing nonselective Si/SiGe epi in such integrated HBT structures. In both cases, the surface preparation includes an HF-last treatment to avoid the need for a high-temperature preepi bake, and (H_2)/SiH_4/GeH_4/B_2H_6 chemistry is used to obtain "conforming" nonselective deposition. In the case of RT-CVD, a hydrogen bake is usually carried out at a moderate temperature (about 850–900°C for 1–5 min); in this way, the silicon surface is atomically smooth and very clean, without any oxygen or carbon being detected by SIMS analysis (below 5×10^{12} and 1×10^{12} cm^{-2}, respectively). The epi is deposited at 650–700°C, very often under reduced pressure. In the case of UHV-CVD (LP-CVD furnace configuration), the wet clean is much more critical because of the absence of a hydrogen bake, and even if unreported in publications, one can presume that some traces of oxygen and, more probably, of carbon could be present at the substrate/epi interface. On the other hand, due to the very low pressure and GR used in UHV-CVD, the polycrystalline deposit is usually thinner than the epi base.

Figure 13 shows a transmission electron micrograph (TEM) cross section of the HBT architecture, described previously. The epi base was grown by RT-CVD. In the photograph, we can note the absence of crystal defects in the base, smooth epi, conformal poly/epi interface, and low polycrystal roughness. We can also observe that the polycrystalline material grows faster than the epi layer, which is favorable for the extrinsic base resistance. In the present example, as the drift-field concept is used, the transistor characteristics are very sensitive to the precise emitter/base junction location. Because an As-doped poly-emitter is used, this junction is defined by the diffusion of arsenic from the n^+ poly-emitter through the

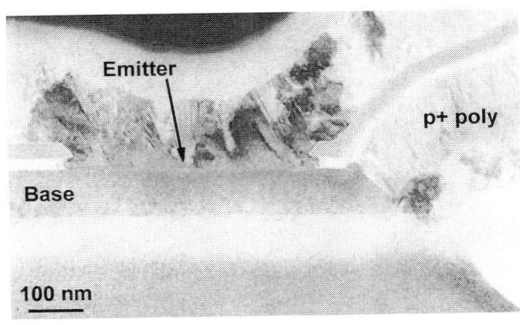

FIG. 13. TEM cross section of the NSEG-based HBT architecture presented in Fig. 12. Note the absence of crystal defects in the base, smooth epitaxy, and low polycrystal roughness.

Si cap and a part of the Ge ramping. Thus, the thickness of these two films, as well as the thermal annealing, is very critical. This fact is illustrated in Fig. 14, where the boxes give the influence of the Si cap thickness on the bipolar current gain: as the cap is changed from 35 to 30 nm, the current gain varies from 50 to 100. This means that a 10% variation in the Si cap thickness will correspond to a variation of 40% in the current gain (around 100). In addition, since the current gain as well as the majority of HBT characteristics depends on the neutral base, the base must also be well controlled in terms of Ge content, thickness, and boron doping (Gummel

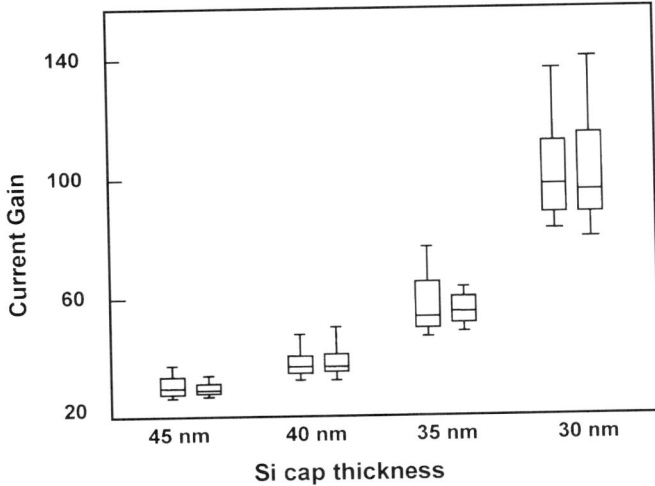

FIG. 14. Current gain box plots showing the influence of Si cap thickness. The cap thickness is varied from 45 to 30 nm in steps of 5 nm (left to right): at a current gain of about 100, small variations in thickness correspond to significant current gain variations.

number). In Fig. 14, we can observe that the different splits are well defined and separated. This proves that, with the present process, the uniformity of the epi base characteristics is good enough within the wafer and from wafer to wafer.

For industrial purposes, the HBT characteristics have to be repeatable in the long term and, if possible, not dependent on the HBT's size and circuit layout; the different loading effects which can occur in the epi process must then be minimized. In such a low-temperature NSEG using $(H_2)/SiH_4/GeH_4/B_2H_6$ chemistry, with a significant activation energy (about 2 eV), thermal loading effects are a major difficulty because of the wafer patterns and the nonselective nature of the deposit. Indeed, the optical properties of the wafer are different and vary continuously during the epi deposition (see Section III.3.b). On the other hand, as soon as the kinetics of polycrystalline and epi depositions are not identical, some chemical loading effects are expected. This is clearly observed in boron doping in RT-CVD: as boron incorporation is much more important in the polycrystal compared to epi, gas depletion is increased and boron doping in the epi region is reduced compared to those in Si full-sheet wafers [35]. Figure 15 illustrates the dependence of the current gain on the intrinsic base (or emitter) width. We can observe that the current gain ranges from 155 to 140 when the transistor surface is varied by a factor of 50. In this case, the devices do not use the drift-field concept, and consequently, the current gain depends mainly on the Ge content and the doping level of the base. Thus, the good current gain conservation demonstrates that the micro loading effect has been effectively minimized during the epi base deposition, presumably owing to the utilization of a reduced pressure. It should be noted here that the UHV-CVD

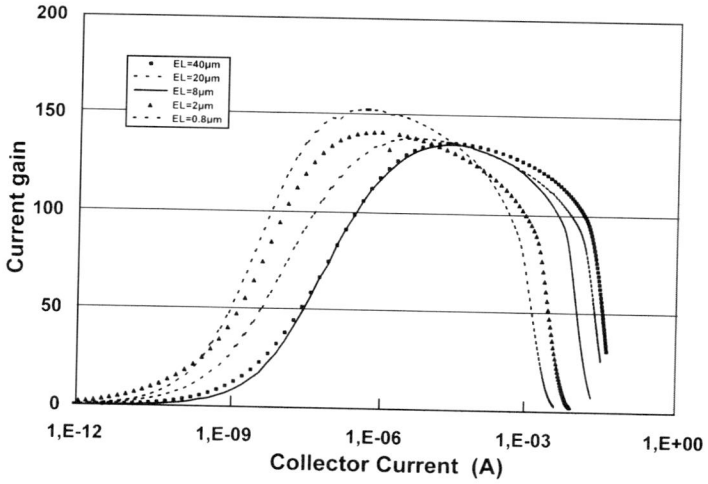

FIG. 15. Current gain characteristics for one series of SiGe HBTs with emitter lengths varying from 0.8 to 40 μm (the emitter width is 0.4 μm). The good conservation of the current gain demonstrates that the micro loading effect during epitaxy has been minimized.

technique with an LP-CVD configuration (furnace), which is not subject to thermal loading effects, is believed to exhibit global and local chemical loading effects, even if it is not reported in the literature.

c. *The Selective HBT Process*

Si/SiGe SEG is a promising base formation process because selective growth is well suited to a self-aligned emitter-base structure. Such structures, which produce HBTs with a low parasitic base resistance and collector capacitance, are very desirable for achieving ultrahigh-speed operation.

For example, Fig. 16 shows a self-aligned SEG-based SiGe HBT structure that is under consideration for new BiCMOS technologies at STMicroelectronics. The emitter-base structure is similar to that reported by Sato *et al.* [54], and this structure or, rather, closely related variants are currently being experimented on by different groups. For BiCMOS implementation, the HBT structure is usually fabricated after source/drain extensions. The bipolar active regions are opened in a nitride/TEOS stack on top of the p^+-doped base poly, which is on top of an oxide layer. Thin nitride spacers are formed to seal the poly sidewalls. Then the oxide is undercut with diluted HF, and the selective Si/SiGe epi is grown inside the structure. Note that the intrinsic base is linked to the extrinsic base (poly) during the simultaneous epi and poly deposition in the oxide undercut. The majority of the remarks made previous section concerning the choice of the HBT vertical structure and of the epi processes are still valid. Again, we can list the main requirements of the present epi.

- Selectivity (nitride, oxide, and mono- and polycrystalline silicon are present)
- Si/SiGe deposition on patterned wafers
- A very moderate thermal budget (to let the CMOS transistors remain unchanged)
- A high epi quality

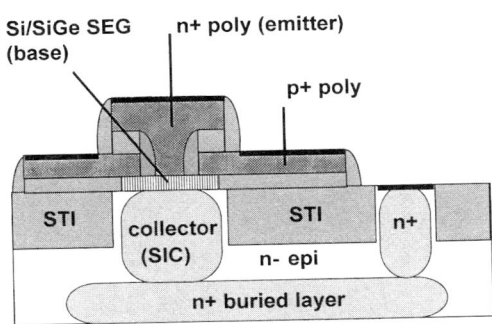

FIG. 16. Schematic cross section of a SEG-based double-poly SiGe HBT. This architecture, similar to that of Sato *et al.* [54], is self-aligned and is very desirable for achieving ultrahigh f_{MAX}.

- An acceptable level of defects
- No boron spike at the substrate/epi interface (autodoping from the p^+-poly is possible)
- Good process control (very good thickness control if the drift-field HBT concept and a poly-emitter are used) including the various loading effects
- A good poly/epi electrical link (important for dynamic performances)

Today, RT-CVD and GS-MBE (sometimes referred as single-wafer UHV-CVD) have proved to be efficient in growing selective Si/SiGe epi in similar HBT structures. In both techniques, the surface preparation includes an HF-last *ex situ* treatment and an *in situ* hydrogen bake, which is usually carried out at about 850–900°C for 1–5 min; in this way, high-quality epi are grown. In the case of RT-CVD, the epi is deposited using $(H_2)/SiH_2Cl_2/GeH_4/B_2H_6$ chemistry in the temperature range 650–800°C at a reduced pressure. In the case of GS-MBE, $Si_2H_6/GeH_4/B_2H_6$ chemistry is used between 550 and 600°C at very low pressures.

A TEM cross section of the HBT architecture described previously is shown in Fig. 17. The epi base was grown by RT-CVD. In the photograph, we can note the absence of crystal defects in the base, smooth epi, and good physical link between the intrinsic epi base and the extrinsic poly base. More generally, excellent static and dynamical electrical characteristics have been obtained in our pilot line and reported in the literature [47, 49, 50]. This proves that this structure can be managed in such a way that high-quality epi is grown. It also shows that the structure is effective in minimizing the parasitic resistance and capacitance. On the other hand, it seems that the different loading effects can be minimized or compensated for in such a way that the devices can be produced with acceptable characteristic dispersions [55].

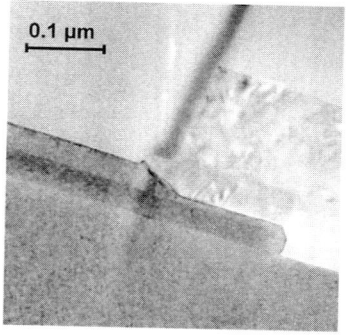

FIG. 17. TEM cross section of the SEG-based SiGe HBT presented in Fig. 16; a global view is given on the left, and a detail of the epi/poly connection is given on the right. Note the perfect epitaxial deposition and good physical link between the extrinsic epitaxial base and the extrinsic poly base.

d. SiGe:C Epitaxy for HBT

The last important development in the epitaxy of HBT bases concerns the utilization of SiGe:C alloys, i.e., SiGe alloys doped with carbon. In 1995, Lanzerotti *et al.* clearly demonstrated that carbon incorporation in the epi base of HBTs dramatically reduces boron diffusion as well as transient enhanced diffusion, due to the implant of an extrinsic base or selective implanted collector or emitter [56]. Very interesting results have been reported since that date.

Definitely, the presence of carbon at levels of about 10^{20} atoms/cm^3 in substitutional sites blocks boron movement during subsequent process operations. On one hand, this allows an increase in HBT performance by changing the trade-off between the resistance and the thickness of the base; on the other hand, it provides better flexibility in technologies in terms of the thermal budget. The latter point could be decisive in BiCMOS technologies, where HBT structures are very often submitted to high thermal budgets, necessary for CMOS gate/drain/source activation. On the other hand, the presence of carbon is found to have no deleterious effects on the electrical characteristics of HBTs, including yield, or their reliability. As such, HBT devices, compatible with CMOS technologies, were reported with excellent high-frequency and DC parameters [57].

Carbon incorporation in Si and SiGe films has been demonstrated using MBE, RT-CVD, and UHV-CVD techniques. The epi must be carried out at a temperature low enough to avoid SiC formation and place the carbon atoms in substitutional sites. In CVD techniques, methylsilane, SiH_3CH_3, is widely used as a carbon precursor without any noticeable difficulty in terms of the epi process. Apparently, the carbon dose does not have to be precisely controlled, and the doping level is low enough so that the SiGe kinetics is unchanged [58]. As such, epi issues are very similar to those for previous SiGe epis.

4. ELEVATED SOURCES AND DRAINS IN CMOS

Current silicon-based CMOS transistors with conventional architectures are encountering performance limits as the size (and operation voltage) is scaled down. One of the essential issues for high-performance advanced (0.1-μm) CMOS is the trade-off between the source/drain (S/D) junction depth and the silicide film thickness. Thick standard silicides ($TiSi_2$, $CoSi_2$, etc.) are mandatory for low-access electrical resistance, but they are not compatible with shallow S/D junctions, which are required to avoid short channel effects (SCEs).

Elevated S/Ds are a very attractive approach to both maintaining/increasing the salicide film thickness and to avoiding SCE and junction leakages. Typically, the self-aligned silicide (salicide) contact on the gates, sources, and drains is formed by depositing a metal (Ti or Co) and reacting the metal with the silicon to form the silicide ($TiSi_2$ or $CoSi_2$) with annealing. Silicide formation, which consumes approximately 50 nm of silicon, will be a limitation to the utilization of very shallow junctions. Elevated S/Ds, fabricated by SEG, will act as a source of "sacrificial" silicon for the formation of silicide and will allow the junction to be more elevated compared to the channel position. They usually consist of 20- to 60-nm-thick Si epi, but as dopant activation is easier in SiGe, SiGe alloys or SiGe/Si stacks may be preferred. In the literature, different strategies have been reported—with or without S/D extensions, before or after S/D implantations, etc.—but very often, elevated S/Ds are deposited just after extension and spacer formation. As this new process may be required in future generations (with $C_D = 0.1$ µm or below), it will probably be used in conjunction with thin (about 30-nm) composite (SiO_2/Si_3N_4) spacers. Elevated S/Ds will also concern very dense CMOS technologies and very complex circuits (about 10^8 transistors). As such, the level of defects, whatever the cause (particles or selectivity loss), will have to be controlled at very low values. With regard to the defectivity, it is usually argued that the quality of the epi is not a concern in this application. Indeed, the deposition will form a region insensitive to defects, and it will be consumed in a large part by silicide formation. However, the author believes that dislocation loops or other extended defects present may move during the activation annealing operation because of the STI stress, causing a problem of leakage and circuit yield. In addition, defects in epi are very often correlated with a certain morphology, which can be highly detrimental in terms of junction depth. For the same reason, facets along the spacer are undesirable.

Consequently, even if the SEG process for elevated S/Ds seems very simple, its industrialization, which has never been demonstrated until now, will probably present a certain level of difficulty. The main challenges or concerns are the following.

- Choice of the surface preparation strategy: whether or not to use a sacrificial oxide to restore the silicon surface, whether or not to elevate the transistor gate during the SEG
- Controlled undercutting of the sidewall oxide during surface desoxidation and preparation
- Si/(SiGe) deposition on patterned wafers (SiO_2, nitride, and silicon are present)
- A moderate thermal budget (to respect the S/D extensions and the preimplanted gates)
- A good selectivity (to avoid any S/D bridging)
- No gate–S/D bridging (no shortcut after salicide formation)
- A very low level of defects
- Control of the epi facet at the spacer edge
- Acceptable thickness control

In the literature, Si or SiGe elevated S/Ds have been deposited using various RT-CVD techniques with different chemistries: $H_2/SiH_2Cl_2/HCl$ [59], $H_2/SiH_4/HCl/$ (GeH_4) [60], and Si_2H_6 and Cl_2 (growth and etch steps) [61]. Before the epi process, the sacrificial oxide is removed from the S/D regions and, eventually, from the polysilicon gates. Then, the HF-treated wafers are submitted to a moderate (850–900°C for 1–5 min) H_2 bake and deposition is carried out at a temperature

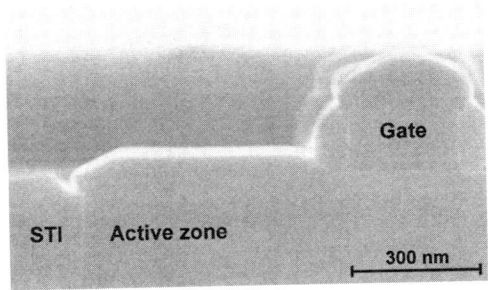

FIG. 18. SEM cross section of a silicon elevated source/drain/gate deposited by RT-CVD. Note that there is no significant facet at the spacer edge and the gate has been significantly thickened.

chosen according to the chemistry used (800–850°C with DCS/HCl, about 600°C with disilane). The facet formation along the spacer depends on a number of parameters, in particular, the spacer nature, spacer slope and presence of a pedestal oxide, and deposition process. It has been found that nitride spacers that are nearly vertical are favorable for facet elimination. If an oxide pedestal is present, it has to be significantly undercut, in such a way that the bottom corner of the Si_4N_3 spacer intercepts the (100) horizontal growing plan and not the (311) facet, which is done naturally during the wet desoxidation of the surface. The electrical results reported in the literature clearly demonstrate that elevated S/Ds are effective in eliminating junction leakages and in decreasing SCEs. Simultaneously, growth on uncapped polysilicon gate structures leads to a reduced salicide resistance.

In Fig. 18, a SEM micrograph shows an example of a CMOS structure with elevated S/D fabrication. In the present CMOS, n-type gates are preimplanted, and SiO_2/Si_4N_3 composite spacers are used. With our chosen strategy of uncapped gates during SEG, the desoxidation before epi has to be sufficient to remove completely the sacrificial oxide, which is thicker on the gates due to their polycrystalline nature and preimplantation. This leads to an increased undercutting of the composite spacers, which can help in the elimination of the facet. In the individual case of the sample presented in Fig. 18, the desoxidation was exaggerated and the undercutting severe. The elevated S/D and gate were deposited by RT-CVD: the HF-treated surface was submitted to a moderated H_2 bake, and Si selective deposition was carried out with H_2/DCS/HCl chemistry. In this cross section, we can observe that there are no defects in the epi, no deposition on the Si_3N_4 spacers, and no significant faceting at the spacer edge and that the poly gate has been significantly thickened and widened. Figure 19 shows a SEM micrograph top-view of an elevated S/D. This micrograph confirms the previous points and provides complementary information: the epi is smooth, faceting is visible at the STI border but barely visible at the spacer edge, poly grains are visible but are not too rough, and the spacers are well respected, without any nucleation points. In addition, thickness measurements showed that micro loading effects have been eliminated and macro loading effects

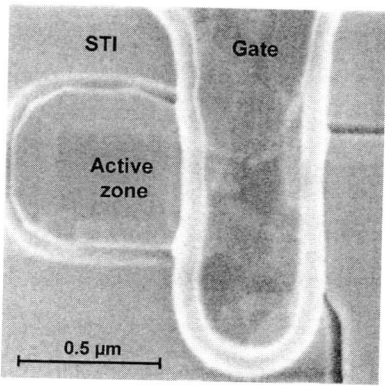

FIG. 19. SEM top-view of a silicon elevated source/drain/gate deposited by RT-CVD. Epitaxial deposition is very smooth; facets are clearly visible at the STI edge, and not along the spacer.

minimized, owing to a sufficiently low pressure and a correct DCS/HCl gas flow ratio, respectively. On the other hand, optical microscopy as well as electrical results confirmed a very good selectivity and demonstrated a very low level of defects.

In conclusion, such encouraging results, together with the various benefits reported in the literature (shallower junctions, improved silicidation and low contact resistance, minimized SCEs with shallow or no extension, etc.), lead to the belief that elevated S/Ds, which are classified as a difficult challenge on the SIA roadmap [62], could be a viable route in future CMOS technologies (with $C_D = 0.1$ μm or below).

5. Channel Engineering in CMOS

For advanced CMOS transistors (0.13 μm and beyond), conventional channels with a uniform dopant distribution would need doping concentrations of 2×10^{18} cm^{-3} or higher. Consequently, electrical parameters such as junction capacitance and carrier mobility would be deteriorated and SCEs would be overpronounced. Today, the retrograde profile, obtained by the implantation of heavy ions, is the usual answer to achieve a maximum bulk concentration that limits a punch-through phenomenon, together with a low surface concentration that maximizes the carrier mobility. However, the dopants have to be precisely placed, and the use of heavy ions, such as indium for n-MOS, suffers from poor activation and from carrier freeze-out.

Thus, significant improvements are still desirable in CMOS channel engineering. Epitaxial layers of silicon and/or SiGe:(C) alloys in the CMOS channels are possible breakthroughs for improving both the SCEs and the transistor transconductance. As such, various solutions with a number of variants have been suggested in the literature and are studied in the pilot lines of the main SC manufacturers. An exhaustive classification of these solutions would be very difficult, but the majority of them will fall into one of the following three categories.

- Epitaxial Si channels
- Si/SiGe channels
- Strained Si layers for CMOS

This classification has been selected for the following discussion.

a. *Epitaxial Si Channels for CMOS*

Compared to the retrograde profiles obtained by heavy-ion implantation, undoped silicon epi deposited on p- and n-type preimplanted channels is able to create steeper doping profiles. This also enables the use of more effective light dopant atoms such as boron [63, 64]. Such an ideal "box-like" channel dopant profile is shown schematically in Fig. 20. The highly doped part provides a ground-plane effect that enables suppression of SCEs. It also limits the lateral extension of the electric field and, therefore, is expected to decrease drain-induced barrier lowering (DIBL). The low doping level of the upper part of the silicon active zone will decrease the Coulomb interaction between carriers and ionized dopant impurities and will lead to improved carrier motilities and transconductances.

Therefore, the growth of an undoped SEG of silicon on p- and n-type preimplanted channels in a single process seems promising, because the process is not overly complex. The thickness of the epi has to be chosen so as to balance an increase in mobility and SCE and DIBL improvements. The thicker the layer, the larger the carrier mobility, but the poorer the short channel control. According to recent studies, epi thicknesses of about 30 nm seem to be suitable for 0.1-μm n- and p-CMOS optimization [64–66]. This operation must be carried out after well, anti-punch-through, and threshold voltage (V_{th}) implantations and before gate oxide growth. Consequently, the dopant atoms that have already been

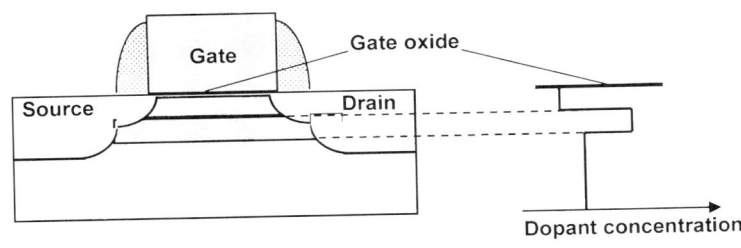

FIG. 20. Schematic cross section of an ideal-stepped doping profile for CMOS channels. The profile, given on the right, is expected to allow good electrical control of the channel (doping level box) and to optimize the carrier mobility (low doping level of the upper part).

implanted must be respected; i.e., the subsurface region, at a scale of tens of nanometers, must not be impoverished by dopant out-diffusion and desorption. This leads to severe limitation of the thermal budget, especially if boron is used in the n-MOS channel. Thus, despite the fact that some differences can be found, the challenges of this epi are somewhat similar to those of elevated S/Ds and are listed below.

- No undercutting at the shallow trench/silicon interface during desoxidation and H_2 bake
- Si deposition on patterned wafers (only SiO_2 is present besides silicon)
- A moderate thermal budget (to respect the initial surface doping)
- A good epi quality
- A very good selectivity and very low level of defects
- Control of the epi facet at the STI edge
- Acceptable thickness control

From a general point of view, the epi techniques that are suitable for elevated S/Ds can be used for this application. In the literature, RT-CVD techniques, with or without UHV capability, and using DCS or Si_2H_6 chemistries are the most often reported [63–66]. An example of a selective epi deposited on an active zone by RT-CVD in our pilot line is shown in Fig. 21. This structure was grown under conditions similar to those used for elevated S/Ds previous section. On the SEM cross section, we can observe that the STI level and morphology have been not changed by the preepi surface preparation and that the epi is smooth, without any crystalline or morphological defects. We can also note that the hydrogen bake, carried out at a moderate temperature, did not induce any significant etching of the near-vertical STI/substrate interface.

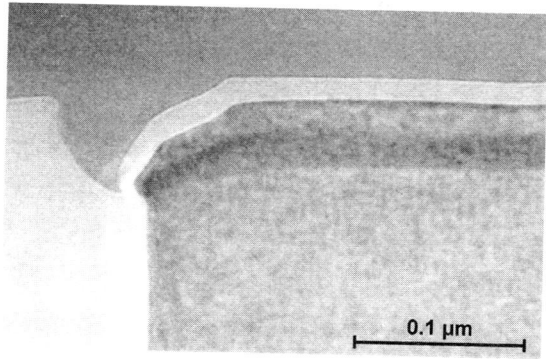

FIG. 21. TEM cross section of a selective epitaxy deposited by RT-CVD on a CMOS active zone. On this specific sample, a SiGe (dark color)/Si stack was deposited to reveal the epitaxy morphology precisely. The STI level and morphology were not changed by the preepitaxy surface preparation, and no defect is visible in the epitaxy.

In the literature, the use of an epi channel has been reported to improve the short channel control and to increase the carrier mobility in the channel. Some important transconductance increases have been reported in long transistors (10 to 50%), however, this gain is partially lost in short transistors, presumably due to the transverse electric field [65]. The carrier mobility is supposed to be increased mainly because of the low doping level in the vicinity of the carrier channel. However, one can suppose that epi can also improve the surface roughness and thus further increase the carrier mobility (by a reduction in scattering through roughness). This concept was recently studied in Ref. 66, where epi and hydrogen annealing were associated to optimize the surface roughness, but the improvements in carrier mobility and transistor transconductance were not very significant. In addition, we can note that the incorporation of carbon atoms in the epi channel would have beneficial effects. Indeed, moderate carbon concentrations (10^{19}–10^{20} atoms/cm^3) would suppress the transient enhanced diffusion of impurities in the channel, in the same way as boron diffusion in the SiGe:C base of a bipolar [56], and contents around 1% would create a tensile strain in the channel, leading to a significant increase in carrier mobility.

In conclusion, the use of epi channels could be an attractive and acceptable solution. This should enable both a moderate increase of carrier mobility in the channel and a better control of SCE in future CMOS technologies, at the expense of relatively low "apparent" difficulty in process integration. However, it must be noted that selective silicon epitaxy has still to be demonstrated to be viable in such industrial applications, i.e., able of high production yields for dense and complex circuits.

b. Strained SiGe Channels for CMOS

The much lower mobility of holes compared to electrons in Si is responsible for the gap between the performance of n-type and that of p-type devices in CMOS circuits. High-transconductance p-MOSFETs are thus desirable for the fabrication of high-speed and high-performance circuits. Strained SiGe epis on silicon provide an effective two-dimensional (2-D) hole confinement because of a large valence band offset; and strain effects lift the band degeneracy of the light and heavy holes and modify the effective mass in each band, resulting in an enhanced mobility. Indeed, a confined hole density and mobility as high as 2×10^{12}/cm^2 and 3000 cm^2 V^{-1} s^{-1} can be achieved at a low temperature [67]. Therefore, strained SiGe/Si structures are very attractive for boosting the performances of p-channel devices. While some Modulation-doped field effect transistors (MODFETs) have been reported in the literature [68], MOSFETs (or MOS-MODFETs) are preferred due to the much lower gate leakage currents and the ability to support positive and negative biases.

Figure 22 shows a schematic view of a p-MOSFET, which exploits the advantages of the strained SiGe channels in FETs. This device can be integrated

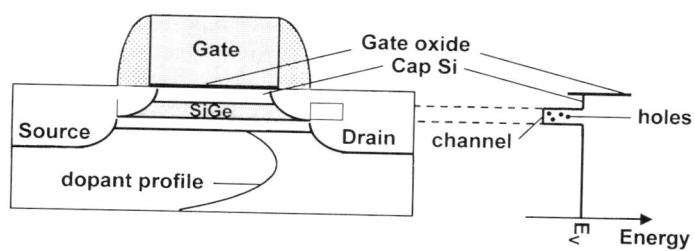

FIG. 22. Schematic cross section of the SiGe-channel p-MOSFET. The strained SiGe channel will confine holes and improve their mobilities.

into CMOS silicon technologies performing the epi operation on active zones of CMOS wafers after well, anti-punch-through, and V_{th} adjustment implants and before gate dielectric formation. In Fig. 22, the epi stack consists of a Si pedestal, a SiGe channel, and a Si cap, which are undoped. The SiGe channel has to be optimized to provide a well deep enough to confine a high hole density (the barrier of confinement and quantum confinement shift depend mainly on the Ge content and thickness, respectively) and not to relax during the subsequent operations in MOS fabrication. Channel thicknesses of about 10 nm and a germanium content of about 30% are usually used. In the present case (undoped epitaxial films deposited on previously implanted active zones), the silicon pedestal serves to separate the 2-D hole gas confined in the SiGe channel from the ionized dopant atoms and, thereby, enables an increase in hole mobility. However, this parameter must be compromised, as an overlarge distance between the doped region and the channel would result in poor SCE characteristics. The silicon cap is necessary for different reasons in MOSFETs. First, a silicon cap is used to form the gate oxide because thermal oxidation of SiGe alloys would result in a pileup of Ge at the oxide/SiGe interface, leading to a degraded interface, with high interface state densities, and, eventually, to a lower crystal quality. Second, as the relaxation of a SiGe epi that is buried in silicon would need two misfit dislocation segments (only one in the case of a simple superficial SiGe film), the Si cap makes the structure more resistant in terms of relaxation. Third, the silicon cap increases the hole mobility by reducing the scattering attributable to both the roughness of the semiconductor/oxide interface and the charges within the oxide. On the other hand, the cap layer must be thin enough to avoid the formation of a second channel at the Si/oxide interface, especially for high gate biases, which would degrade the global transport properties of the device. Consequently, cap thicknesses typically range from 4 to 10 nm, depending on gate dielectric formation and on device optimization.

This epi, which is also performed on active zones of CMOS wafers after V_{th} adjustment implantation, is somewhat similar to that of Si channels, and therefore it will share all the challenges listed in Section IV.5.a. Because of the film stack complexity and the 2-D hole gas presence, the following requirements are also important:

- a good crystalline quality of the strained SiGe channel (no misfit dislocations),
- low roughness at the SiGe channel/Si cap interface (presence of the 2-D hole gas), and
- acceptable SiGe thickness and composition control.

In the literature, mesa MODFETs or MOSFETs were first fabricated using MBE and UHV-CVD, and very soon afterward RT-CVD was also used with success. Thereafter, effective low field mobilities enhanced by 50% at room temperature and by over 100% at 90 K, compared to control Si devices, were demonstrated [69]. On the other hand, RT-CVD, with a large process flexibility and with the ability to grow perfectly selective epi (with chlorine chemistry), is certainly the easiest way to achieve the integration of such devices in a complex (CMOS) technology when selective epis of pure Si and SiGe alloys are required. Recently, different studies have reported encouraging results: a 150% mobility gain in long devices (low field mobility) and a 40% gain in 0.15-μm p-MOS using a $Si_{0.7}Ge_{0.3}$ channel on silicon [70] and a 2-D hole gas mobility of 930 cm^2/V s and a hole sheet density of 2.6×10^{12} cm^{-2} at room temperature, as well as an f_T and f_{MAX} as high as 62 and 68 GHz, respectively, using a more complex structure consisting of a strained 80–70% Ge-rich channel on a 30%-rich SiGe buffer [71].

However, the integration of strained SiGe channels in advanced CMOS technologies is not immediate because they have to be compatible with the subsequent operations. To facilitate this integration, the use of multiple SiGe quantum wells instead of a single SiGe film has been suggested [72]. First, this structure has been found to be more resistant to strain relaxation. Second, apart from the improvement in p-MOS, multiwells would also enhance the electron mobility. The latter effect may be due to tensile stress present in the upper Si layer due to a possible strain shared between the Si and the SiGe films. Figure 23 shows a TEM cross section of a MOS-MODFET, which includes a triple SiGe well in the channel. In this case, the epi was carried out by RT-CVD at a reduced pressure and using H_2/DCS/GeH$_4$/HCl chemistry at about 650°C. The TEM view shows very well-defined wells, and, if some defects are observed due to the S/D implant, we can note the absence of defects beneath the gate.

In conclusion, such structures, which are based on selective Si and SiGe epis, could be of interest, if they are shown to be fully compatible with the other technological steps and if they bring about significant improvements in circuits. Their fabrication would not necessarily be much more delicate than that of Si epi channels (see Section IV.5.a).

c. Strained Si Channels for CMOS

Biaxial tensile strain in thin Si layers grown pseudomorphically on a relaxed SiGe layer is known to provide both electrons and holes with low effective masses and high mobilities. Strained-Si MOSFET is therefore a promising device structure

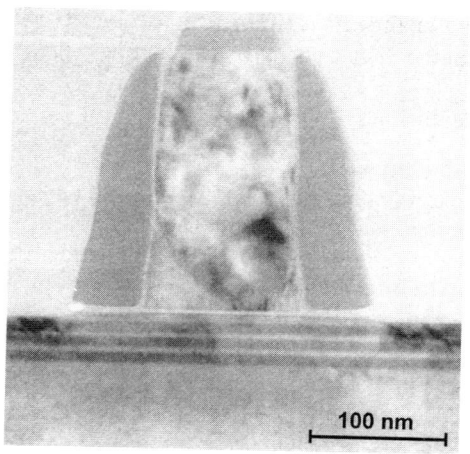

FIG. 23. TEM cross section of a MOS-MODFET with multiple SiGe well channels. Epitaxy was deposited by RT-CVD using H_2/DCS/HCl chemistry, and the three SiGe wells (dark color) are well defined. (Courtesy of T. Skotnicki, ST Microelectronics R&D.)

for sub-0.1-μm high-speed and low-power CMOS circuits, and indeed both n-channel and p-channel MOSFETs with higher electron and hole mobilities have been demonstrated [73, 74]. In addition, owing to the significant conduction band offset present in these structures, the possible spatial separation of electrons both from the ionized dopant atoms and from the gate interface can be used to increase the carrier mobility further. This effect has been exploited in MOSFETs with a buried Si channel [75] and in MODFETs [76], where 2-D-electron gases are formed at a Si/SiGe interface.

Figure 24 shows the schematic of a strained Si n-MOSFET, which is certainly the simplest structure to exploit the advantage of strained Si in FETs of Si technology.

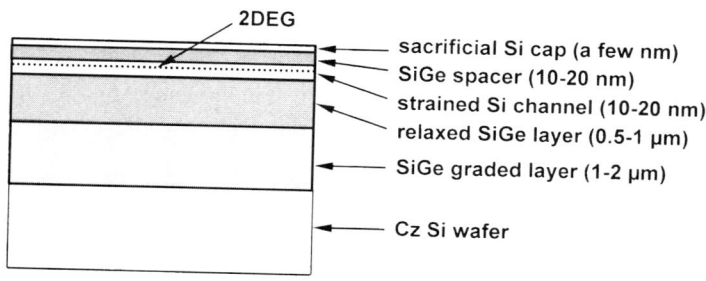

FIG. 24. Schematic cross section of the common heterostructure for the fabrication of strained Si channel MOSFETs. The strained Si channel will confine electrons and holes and will provide them with an increased mobility. Note that buffer layers are thick (micrometer range) and channel and spacers are very thin (10–20 nm).

First, a SiGe pseudo-substrate has to be fabricated. The SiGe epi is deposited in such a way that its initial strain is relieved and the SiGe surface has its own lattice parameter. A thin (10- to 20-nm) pseudomorphic Si epi is then deposited to form the channel material. As this silicon takes the in-plane SiGe lattice parameter, it is under tensile strain; the higher the Ge content in the pseudo-substrate, the higher the tensile strain in the Si channel. Finally, a MOS structure is created on this silicon material. It must be noted that the upper part of the strain-relieved SiGe layer may be boron-doped to create anti-punch-through profiles, but the Si channel will not, to optimize the mobility (see Section IV.6.a). For the fabrication of buried-Si channel MOSFETs, one additional SiGe spacer must be added onto the Si channel to confine the electrons, and a sacrificial silicon cap is generally grown for the gate oxide formation. For MODFET fabrication, the sequence of layers grown includes an additional n^+-doped SiGe layer inserted between the SiGe spacer and the Si cap of the previous structure. For example, Ismail et al. [76] give a complete sequence of layers grown for MODFET fabrication.

As this application is still in laboratories and is not mature enough yet to be used in industry, transistors have not been integrated in complex technologies, and epis are usually grown on full-sheet silicon wafers. Thus, below we list only the challenges that are very specific to the vertical structure of epi.

- A good crystalline quality of the SiGe pseudo-substrates (no threading dislocations)
- Low surface roughness of the SiGe pseudo-substrates
- A good crystalline quality of the strained-Si channel (no misfit dislocations)
- Low surface roughness of the strained-Si channel

Fabrication of SiGe pseudo-substrates, or almost completely relaxed SiGe layers, of a high quality is not easily achievable. Several methods have been developed, but the most common approach is to grow a compositionally graded layer plus a uniform SiGe capping layer [77]. The graded layer enables a dramatic reduction in the density of threading dislocations that are often found to propagate through the buffer. This is supposed to work as follows [78]. A low-slope grading allows a more progressive relaxation and reduces the surface strain. Nonplanar growth [due to the Stranski–Krastanov (SK) growth mode] and its pinning effect on threading dislocations are then avoided. In this way, relaxation takes place with long dislocation loops and only a few threading dislocations are generated. The capping layer has to be thick enough to relax the strain of the mismatch and to bury the resulting misfit dislocations. In such pseudo-substrates, during deposition the dislocation loops have to cross the whole wafer in such a way that they will leave a long (wafer-size) misfit segment and the threading segment will escape at the wafer edge. Thus, the dislocations have to be very mobile, and relatively high growth temperatures are desirable. Conversely, the strained Si channel has to be grown at a low temperature to avoid any dislocation generation and any SK growth-induced roughness or disruption of the surface [79].

In the literature, these structures are usually grown by RT-CVD and UHV-CVD. MBE is also reported, but unless very sophisticated techniques such as "atomic hydrogen irradiation" [80] are used, a lower film quality seems to be obtained. Owing to its natural versatility, RT-CVD is certainly the most efficient technique. The pseudo-substrate part can be grown at between 750° and 900°C with a relatively good quality. Threading dislocation densities of about 10^3 cm^{-2} were reported in a pioneer study [78]. On the other hand, a strained silicon film can be grown successfully using H_2/SiH_4, as well as H_2/Si_2H_6 chemistries at about 600°C.

Mobility and transconductance enhancement ratios of 2 have been reported for long-channel n-MOSFETS and n-MODFETs [75, 76]. Measurements on transistors with channel lengths of 0.1 μm have also demonstrated a significant improvement in mobility (75%) and transconductance (45%) [81]. Results have shown that the saturation velocity of electrons, which becomes a limiting factor, would also be increased by strain. They have also established that if DC measurements are used, the self-heating effect, due to the low thermal conductivity of SiGe, gives rise to transconductance degradation. This effect may limit the utilization of such structures in dense circuits. In addition, when using the 2-D confinement concept (MODFETs or buried-channel MOSFETs), the electron density in the Si channel (estimated to be about 1×10^{12} cm^{-2} with a strain corresponding to 20–30% Ge) could be a further limitation in terms of maximal current and performance. Finally, the integration difficulties, namely, the buffer thickness (more than 1 μm) and the sensitivity of the silicon channel to the thermal budget, are very large.

Because of these limitations and difficulties, these structures are not ready for industrial applications and/or may have no future in standard CMOS devices of future technologies. However, because of their very interesting characteristics, they may be useful in more specific applications such as high-speed low-power applications.

6. New CMOS Architectures

The ultimate limit of gate length in MOSFETs is estimated to about 20–25 nm. However, device downsizing in the deep 0.1-μm domain faces several obstacles, such as lithography limitation (What else after deep UV?), SCEs, trade-off between the diffusion and the activation of the dopant, the difficulties in down-scaling the equivalent oxide thickness (direct tunneling, boron penetration, etc., in thin-gate SiO_2 and high-K oxide integration), and the current drivability, which is no longer automatically scaled with the feature size. In addition, as circuit performances have become more and more limited by RC delays related to the connection capacitances, they will no longer be directly scaled with the feature size.

As such, new possible types of architectures, aimed at overcoming these difficulties, have been proposed in the literature as alternatives to planar CMOS. The solutions that have been explored have different variants, so that most of them will surely never be developed to an industrial level. Nevertheless, in the following, two types of architectures are briefly discussed as illustrations: the vertical CMOS architecture, which has been extensively studied; and the silicon-on-nothing (SON) structure, which is an innovative concept based mainly on epi deposition.

a. Vertical MOS

In a vertical MOSFET, carrier conduction occurs on the walls of a silicon pillar and is therefore perpendicular to the substrate surface. The surrounding gate transistor, first reported by Takato et al. in 1988 [82], was the first concept utilizing a silicon pillar surrounded by a polysilicon gate and self-aligned S/D formed by ion implantation. A very clear schematic drawing of this vertical CMOS is given in Ref. 82. Other concepts use epi layers, which can be grown with a very good thickness control and a doping profile in the sub-0.1 μm range.

Compared to conventional planar MOSFET, the vertical architecture has the potential to remove some important obstacles in CMOS development. The first, consisting in lithography limitations, can be eliminated, as the vertical channel may be defined by epi, typically with a $p^+/n/p^+$ stack for p-MOSFETs, down to the nanometer range. Also, when using epi for the fabrication of the pillar, the doping concentration can be varied on a very small scale along the channel; and from a more general point of view, concepts of channel engineering such as spacers, delta doping, and Si/SiGe(C) heterostructures can be utilized. Another advantage of vertical CMOS concerns the drive current per device unit width, which is significantly higher because each rectangular device pillar (at the minimal lithographic dimension) contains two MOSs driving in parallel. This characteristic could be used in high-performance logic applications to reduce the delays that are dominated by interconnect capacitances. Other advantages, such as a higher device density, decreased SCEs, and subthreshold slope improvement, have also been claimed, depending on the exact structure of the vertical CMOS. However, the realization of vertical MOSFETs also suffers from the inherent difficulties in combining all the essential characteristics of advanced MOSFETs, namely, a high-quality gate oxide, low parasitic capacitances, self-aligned S/Ds, and sufficient control of electrical parameters.

A typical vertical transistor is located at the circumference of a silicon mesa. The epi stack is first deposited and then structured by dry-etch to form an active pillar. Then the gate oxide is formed by thermal oxidation and a doped polysilicon gate is formed by a spacer technique. Starting from this architecture, much more sophisticated variants have been proposed in the literature. For example, Yang and Sturm [83] have introduced $Si_{0.796}Ge_{0.2}C_{0.004}$ films in the epi $p^+/n/p^+$ stack of p-channel vertical MOSFETs to suppress the boron diffusion induced by the gate oxide formation. Dopant diffusion from the S/D into the channel region during oxidation indeed is a severe limit to MOSFET scaling. They found that heavily doped SiGeC spacers or even SiGeC/Si stacks were preferable to undoped spacers. As the channel is totally constrained in silicon material, the subthreshold slope of devices is sharper, and possible deleterious effects from carbon are eliminated. In this configuration, the oxidation-enhanced boron diffusion from the S/D or even from the p^+ regions near the n channel is still suppressed. In this way, vertical p-channel MOSFETs with channels lengths of 80 nm have been fabricated with well-behaved characteristics, and working devices have been demonstrated with channel lengths down to 25 nm. On the other hand, Liu et al. [84] have fabricated a novel vertical-sidewall strained Si channel n-MOSFET without a relaxed SiGe layer. The structure consists of a strained SiGe pillar, on which a strained silicon epi is grown. As the SiGe pillar has a vertical lattice parameter larger than silicon, the silicon epi that is deposited on the sidewall is under tensile tetragonal strain. Thus, the MOSFET fabricated on this sidewall will also take advantage of the tensile-strained silicon channel. In the first reports, dislocation-free structures were demonstrated, and evidence of band offsets was given. In addition, transconductance measurements demonstrated the role of enhanced electron mobility as the tensile strain is increased. These two examples, chosen from various others, illustrate well the interest in epi layers to remove important roadblocks to CMOS development. In a third example, Hergenrother et al. [85] have proposed a structure called a vertical replacement gate (VRG) MOSFET, which combines a gate length controlled through the film thickness, a channel self-aligned on the gate, and a gate oxide grown on a single-crystal Si channel. With a 200-nm device, excellent current driving capabilities, $I_{on} = 1.1$ mA/μm and $I_{off} = 11$ pA/μm, were reported [85].

Up to now, most of the structures reported in publications have not been integrated into advanced technologies. In the case of integration, the fabrication of the pillar by epi may be carried out on CMOS substrates after completion of the isolation structure (STI). The epi process will then have most of the same requirements as Si (or SiGe) channels listed in Section IV.5.a (or IV.5.b). In addition, as the vertical epi will define the doping profile along the conduction path in the channel, other requirements will be necessary:

- heavy doping levels (n-type and/or p-type for the fabrication of S/D),
- very steep doping profiles (for both rising and falling slopes), and
- a relatively low thermal budget.

In the literature, most vertical MOSFET structures have been grown by MBE and RT-CVD (sometimes with UHV-CVD capability). MBE is generally limited to full-sheet depositions, and RT-CVD is the most appropriate if selective deposition is needed for integration. We also believe that RT-CVD is preferable to UHV-CVD if (abrupt) n^+ doping is required.

Today, vertical MOSs are not ready for industrial applications. In addition, they may have more successful competitors such as fully depleted SOI MOSFETs or even double-gated MOSFETs [86] in the future.

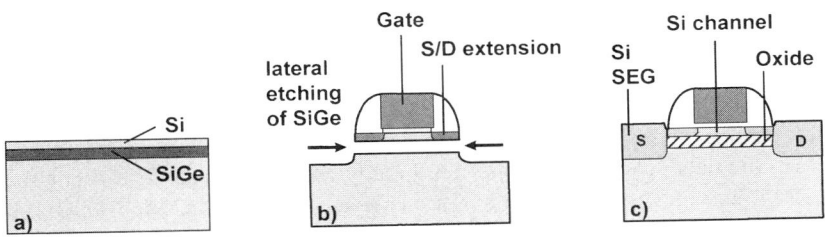

FIG. 25. Process sequence for the fabrication of SON MOSFETs: (a) SEG of SiGe and Si on active zones of a CMOS wafer; (b) air-tunnel formation by anisotropic Si etching and selective SiGe etching; (c) selective epitaxy of S/D regions.

b. *The Silicon-on-Nothing (SON) Concept*

The SON concept refers to an innovative process enabling the fabrication of SOI-like CMOS transistors, which are expected to exhibit excellent electrical performances almost without SCE and DIBL, thanks to the extremely thin silicon channel (5 to 20 nm) and the buried insulator layer (10 to 30 nm) [87].

The main operations in the formation of SON structures are depicted in Fig. 25. The fabrication starts with the successive depositions of a sacrificial SiGe SEG and a silicon SEG that will form the MOSFET channel on the active zones of an isolated CMOS wafer. It is worth noting that the epi films may be extremely thin and well controlled, as epi growth is capable of a nanometric thickness with excellent uniformity. The conventional CMOS process operations are then carried out as far as the spacer formation. Next, trenches in S/D regions are opened through the silicon cap by anisotropic plasma etching, and SiGe is selectively etched using either specific wet chemistry or isotropic plasma processing. The SEM given in Fig. 26 illustrates this situation. Note that etching the tunnel does not degrade the silicon cap and that excellent crystalline quality and thickness uniformity are

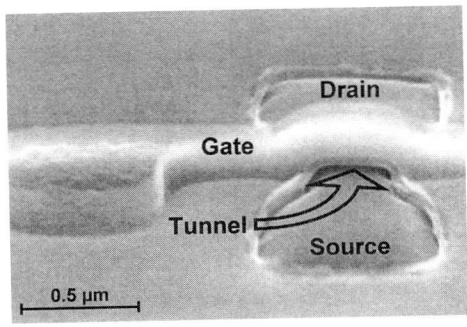

FIG. 26. SEM view of the SON structure after formation of the air tunnel.

maintained. At this point, an air tunnel is formed between the silicon channel and the bulk, thus giving the SON its name, and the silicon channel is supported by the gate poly resting on the STI zones at both sides. This air tunnel is then refilled with a dielectric, to form a thin embedded insulator/an SOI-like device. After this operation, the trenches in the S/D regions are filled with a second silicon SEG, which also connects the channel. Finally, S/D implants and RTA complete the front-end process flow. One major advantage of this structure over all other conventional SOI technologies is related to the fact that the thicknesses of silicon and buried dielectric are both defined by epi on the nanometer scale.

In this structure, the SiGe epi is a sacrificial deposition that will be replaced by a dielectric, and the silicon epi corresponds to the silicon channel. SiGe first has to transfer the crystalline information from the silicon substrate to the silicon channel and then must be removed by selective etching. The selectivity of both wet and dry etchings increases with the Ge content. However, the fabrication of high-quality Si channels also requires that there be no strain relaxation in this SiGe film. As a consequence, 20- to 30-nm thick films with 20–30% Ge, which are stable enough against misfit dislocation generation, and which enable the formation of long enough tunnels (0.1–1 μm), correspond to an appropriate trade-off. On the other hand, as very thin channels are highly desirable for obtaining excellent electrical performances, the thickness of the silicon cap typically ranges between 5 and 20 nm.

The first epi (Si/SiGe stack), which is performed on active zones of CMOS wafers, is somewhat similar to that of strained SiGe channels and shares most of the challenges listed in Section IV.5.b. In comparison, perhaps the thermal budget could be slightly increased, as no precise and abrupt V_{th} adjustment implantations are required in the present application. The second epi, used to contact the channel, is very similar to that for elevated S/Ds, and the challenges it presents are listed in Section IV.4. In addition, as this epi is used to connect the bulk Si to the Si channel, the deposition thickness has to be well controlled, and special attention must be paid to the facet on the channel side. In our pilot line, the two epis requested for the realization of SON structures are successfully carried out by RT-CVD using H_2/DCS/GeH_4/HCl chemistry. In the case of the SON structure presented in Fig. 26, the channel epi (30 nm of $Si_{0.75}Ge_{0.25}$ + 20 nm of Si) was developed at a relatively low temperature (600–650°C), which was chosen both to obtain a perfectly strained Si/SiGe structure without misfit dislocation and to control the small thicknesses of films. On the other hand, the SEG of silicon for the S/D elevation, which is thicker, is generally carried out at temperatures as high as 800–850°C.

Today, electrical characteristics of SON-MOSFETs are not available in the literature. However, extensive simulations have clearly shown all the potential advantages of this structure compared to conventional bulk transistors or to the state-of-the-art SOI transistors [87]. As a result of the very thin Si channel and buried dielectric, the SON structure should enable excellent electrical performances with an almost-ideal subthreshold slope and a quasi-total suppression of

SCEs. In addition, SON structures with S/D regions in contact with the substrate and with a thinner buried dielectric would provide much better heat dissipation than SOI films and would not present significant channel self-heating.

In conclusion, the SON structure, which is based on advanced Si/SiGe epis, can be considered as a first approach to very thin SOI films, and such films may become the dominating technology for CMOS below 70 nm.

7. OTHER APPLICATIONS

Beside the above-mentioned applications, new epi processes are also used in a variety of experimental structures or devices. Some of these applications, which give rise to a nonnegligible level of R&D activity, are listed below. The list is not exhaustive and not every item can be detailed here. However, this section illustrates how wide the field of epi applications can be, and recent bibliographic references have been included to assist the reader in finding further relevant details.

Firstly, epi, which is able to fabricate monocrystalline silicon with large modulations in boron doping or germanium content, is highly appreciated in the field of micromachining. Indeed, highly boron-doped and SiGe films, in combination with wet or dry etching techniques, are known as either "etch stop" layers or sacrificial selectively removed films [88]. Furthermore, because silicon is mechanically superior to compound semiconductor materials, and as the process technology for silicon has been highly developed for VLSI, silicon is the dominant material for micromachining. In this way, varied devices such as sensors, actuators, chemical reactors, and display devices, which are based on different physical structures (bridge, cantilever beam, grating, and membrane), can be designed for a wide range of applications using SiGe and highly-doped Si epis. It should be noted that etch stops are also used in BESOI techniques and that different SOI techniques, such as smart-cut [38, 39], could be improved or local SOI structures created using the same concept of selective Si or SiGe removal.

It is commonly accepted that integrated high-performance bipolar transistors have to incorporate a poly-Si emitter. This indeed applies to all-silicon bipolar technologies, where the high emitter efficiency of the poly-emitter is a nontradable parameter in the device design. However, the introduction of HBT with a SiGe base, which considerably relaxes the trade-off between the base resistance and the current gain, calls for a revision of this notion. Indeed, it has recently been demonstrated that the introduction of epitaxially aligned silicon emitters in an HBT technology can lead to devices which combine a $1/f$ noise figure of merit as low as 7.2×10^{-10} μm^2 and an f_{MAX} as high as 70 GHz [89]. The epi nature of the emitter and the absence of SiO_2 interface lead to excellent low $1/f$ noise performances, whereas the emitter efficiency loss is not an issue owing to the SiGe base properties. In this specific case, the epi, of possibly poor quality, was carried out after the emitter cut opening using RT-CVD with SiH_4 chemistry and a very low thermal budget.

Whereas the fabrication of optoelectronic devices relies on III–V and II–VI compounds, the goal of combining optoelectronics and silicon integrated circuit technology is highly desirable, and the epi of Si/SiGe structures has demonstrated the possibility of fabricating far-IR and 1.3- to 1.6-μm-wavelength photodetectors, as well as light-emitting devices.

SiGe/Si heterojunction internal photoemission (HIP) detectors have been fabricated with cutoff wavelengths ranging from 2 to 16-μm and nearly ideal thermionic emission dark-current characteristics [90, 91]. Such a detector consists of a heavily doped p^+ SiGe epi layer on a p^- Si substrate. The barrier height is determined by the valence offset, and therefore the detector cutoff wavelength can be tailored by adjusting the germanium content. There is a trade-off between the germanium content (cutoff wavelength) and the film thickness and doping (quantum efficiency). In this way, high-quality thermal imagery without uniformity correction has been demonstrated using 400×400 focal plane arrays of $Si_{0.56}Ge_{0.44}$/Si HIP photodiodes with a cutoff wavelength of 9.3 μm and monolithic CCD readout circuitry [91]. Alternatively, long-wavelength detectors can also be made using free carrier absorption in a SiGe quantum well or in Si/SiGe superlattices, and tunable long-wavelength detectors based on PtSi/SiGe/Si Schottky diodes have been also reported [92].

One of the most important optical applications for Si/SiGe structures is the fabrication of photodetectors near 1.3 and 1.55 μm for optical fiber communications. Early results using a SiGe strained-layer superlattice photodetector operating near 1.3 μm were reported by Temkin et al. [93]. More recently, SiGe/Si planar detectors with a similar waveguide-type structure have been demonstrated with an external quantum efficiency of 25–29% at 0.98 μm, a dark current of 0.5 pA/μm^2, and a 3-dB bandwidth of 10.5 GHz [94]. The absorption structure consisted of a 30-period superlattice of undoped 3-nm $Si_{0.9}Ge_{0.1}$/32-nm Si, grown selectively in a previously etched trench. The selective deposition was carried out in a cold-wall UHV reactor using cycles of deposition and etch with Si_2H_6/GeH_4 and Cl_2 chemistries, respectively. Besides, staircase bandgap SiGe/Si p–i–n photodiodes were also found to exhibit a high optical response, with a sensitivity peak of 27.8 A/W at 0.96 μm, and a relatively low dark-current density [95]. In this application, the staircase bandgap structure would increase the photocurrent by impact ionization for holes and enhanced multiplication effect.

Si has an indirect bandgap, which severely restricts its applications for emitters such as lasers and light-emitting diodes (LEDs). However, numerous studies have been devoted to the quasi-direct bandgap predicted in short-period Si/Ge superlattices due to zone folding and to the IR photoluminescence from SiGe alloys. Whereas the evidence for a direct bandgap remained inconclusive, the IR near-bandgap photoluminescence was found to be much more efficient than previously anticipated. Indeed, it has been demonstrated that a quantum efficiency as high as $11.5 \pm 2\%$ can be achieved in strained Si/SiGe superlattices by eliminating nonradiative channels [96]. This efficient photoluminescence has been observed in RT-CVD SiGe and has been identified to be due to excitons localized by random fluctuations in germanium composition. The random nature of the SiGe alloys is believed to relax the usual wave vector conservation rules and, therefore, to give strong no-phonon transitions despite the indirect bandgap. In 1995, by combining similar films with high-resolution lithography, Tang *et al.* fabricated quantum dot LEDs, and their 50-nm quantum dots based on $Si/Si_{0.7}Ge_{0.3}$ superlattices were reported to emit light efficiently at 1.3 μm with a threshold injection current of approximately 0.1 pA/dot at room temperature [97].

Another way toward light emission in silicon technology is the growth of III–V compounds on silicon wafers. For example, germanium and GaAs have lattice constants that are very close, and some efforts have been made to fabricate Ge pseudo-substrates on silicon and GaAs/Ge heteroepi. A possible complete epi stack would be Si wafer/gradual $Si_{1-x}Ge_x$/Ge pseudo-substrate/GaAs active substrate. However, there are still some difficulties to be overcome for industrial applications: first, the quality of the Ge pseudo-substrates is not ideal; and, second, the epi of a compound semiconductor such as GaAs on any elementary semiconductor such as Ge or Si is known to exhibit some antiphase domains with related crystal defects.

As we have reached the conclusion, it is not possible to present any new device that is or could be based on advanced Si epi. Let me merely mention the following as examples: resonant cavity photodiodes, SiGe fast-switching power diodes [98], velocity modulation transistors [99], and negative differential resistances based on different types of tunnel diodes [100].

V. Conclusion

Even though no new epi technique has emerged during the last 10–15 years, a new generation of RT-CVD systems has been developed, besides UHV-CVD, for Si and Si-based epis. The use of load-locking and high vacuum integrity together with high-purity gases reduces water vapor and the contaminant level to those achieved by UHV systems. In this way, the total thermal budget of epi can be dramatically reduced, and high-quality epis can be deposited at temperatures as low as 500–600°C. Thus, these load-locked single-wafer RT-CVD reactors, and, to a lesser extent, UHV-CVD reactors, have become worthy systems for a number of advanced epi processes.

As the thermal budget is reduced, epi deposition may be introduced in the course of device fabrication, and the number of possible device architectures and epi applications is dramatically extended. A low deposition temperature also enables thin films to be controlled in the nanometer range, and strained heteroepis, such as SiGe, to be grown. However, these "new" epi processes also face a number of new difficulties, such as silicon surface cleaning, structured substrates, complex stacks, selectivity, and complex deposition kinetics. At low temperatures, nonselective and selective epis are usually deposited using SiH_4 and DCS/HCl chemistries, respectively. As both systems are operated in the surface rate-limited regime, their kinetics is dominated by hydrogen (or chlorine) adsorption/desorption and varies rapidly with temperature. Thus, any industrial application will require very precise (<1 K) temperature control. The variability of these processes with the type of substrate, the so-called loading effect, is another important difficulty in their utilization. Loading effects for selective epi have been reported extensively in the literature, but as shown here, they are also significantly present in nonselective deposition. These effects consist of a thermal contribution, which is induced by the film stack present on the substrate in the case of cold-wall CVD, and a chemical contribution, which comes from a pattern-induced gas-phase depletion. Note that thermal effects in RT-CVD have been detailed here, because they do not appear in the literature.

Finally, the most important applications of epi in the new and future silicon technologies have been reviewed. The applications listed here are not guaranteed to be industrialized in the future, and others, not presented here, may well be, and very soon; this is a matter for speculation. However, from low-cost substrate fabrication to quantum devices, the prospect is rather wide and the list covers a broad field of "advanced epi."

Among all the applications, the deposition of the SiGe base of RF bipolar transistors is the most important and certainly one of the most challenging. The HBT application, which is sufficiently well developed to be incorporated into mainstream silicon integrated circuit technologies, is important because it presents considerable technological and economical interests. It is challenging because the thermal budget must be very limited (<1 min at 850–900°C in a number of BiCMOS technologies) and because the deposition temperature has to be low enough to control uniform thin films and abrupt doping profiles and to avoid any strain relaxation. Consequently, after having effectively pushed the research in "advanced epi" forward, this application is now establishing it as a standard operation in semiconductor manufacturing.

On the other hand, as very important roadblocks, such as lithography limitations, gate oxide, and tunnel current, are present on the roadmap of CMOS developments, the R&D in CMOS may be pushed toward new architectures that would get around these limitations of physics and techniques. In the author's opinion, advanced epi growth of silicon or Si-based materials will certainly be the basic brick for designing and building these new device architectures. In addition, new functionalities, such as optoelectronic ones, which are highly desirable in microelectronics, may also fuel the research in Si-based epi.

The next decade should be as exciting as the last one!

Acknowledgments

My work on SiGe epitaxy has been carried out in part at the CNET (France Telecom research) and in part at ST Microelectronics (central R&D). Present or former graduate students, J. C. Guerin, P. Warren, P. Jerier, C. Fellous, and P. Ribot, have also made essential contributions.

References

1. Sangster, R. C., E. F. Maverick, and M. L. Croutch, *J. Electrochem. Soc.* **104**, 317 (1957).
2. Kasper, E., H. Kibbel, and F. Schaffler, *J. Electrochem. Soc.* **136**, 1154 (1989).
3. Bean, J. C., in *"Germanium Silicon: Physics and Properties,* Semiconductors and Semimetals, Vol. 56, edited by R. Hull and J. C. Bean, p. 8, Academic Press, San Diego, CA, 1998.
4. Matteson, S., and R. A. Bowling, *J. Vac. Sci. Technol.* **A6**, 2504 (1988).
5. Hiramaya, H., M. Hiroi, K. K. Koyama, and T. Tatsumi, *Appl. Phys. Lett.* **56**, 1107 (1990).
6. Riber Company, Riber BP 231, Rueil-Malmaison, France.
7. Hirayama, H., T. Tatsumi, and N. Aizaki, *Appl. Phys. Lett.* **52**, 2242 (1988).
8. Donahue, T. J., and R. Reif, *J. Appl. Phys.* **57**, 2757 (1985).
9. Hsu, T., B. Anthony, R. Qian, J. Irby, D. Kinosky, A. Mahajan, S. Banerjee, C. Magee, and A. Tash, *J. Electr. Mater.* **21**, 65 (1992).
10. Meyerson, B. S., *Appl. Phys. Lett.* **48**, 797 (1985).
11. Ghidinni, T. G., and F. W. Smith, *J. Electrochem. Soc.* **131**, 2924 (1984).
12. Gibbons, J. F., C. M. Gronet, and K. E. Williams, *Appl. Phys. Lett.* **47**, 721 (1985).
13. Sturm, J. C., P. V. Schwartz, E. J. Prinz, and H. Manohara, *J. Vac. Sci. Technol.* **B9**, 2011 (1991).
14. Dutartre, D., P. Warren, I. Berbezier, and P. Perret, *Thin Solid Films* **222**, 52 (1992).
15. King, C. A., J. L. Hoyt, C. M. Gronet, J. F. Gibbons, M. P. Scott, and J. Turner, *IEEE Electron. Device Lett.* **10**, 52 (1989).
16. Dutartre, D., G. Bremond, A. Souifi, and T. Benyattou, *Phys. Rev. B* **44**, 11525 (1991).
17. Sturm, J. C., H. Manoharan, L. C. Lenchyshyn, M. L. W. Thewalt, N. L. Rowel, and D. C. Houghton, *Phys. Rev. Lett.* **66**, 1362 (1991).
18. Robinson, McD., and L. H. Lawrence, *Semiconductor Fabrication: Technology and Metrology,* ASTM STP 990, edited by D. C. Gupta, American Society for Testing and Materials, Philadelphia, 1989.
19. Bollen, L. J. M., *Acta Electron.* **21**(3), 185 (1978).
20. Lier, M., C. M. Greelief, S. R. Kasi, and M. Offenberg, *Appl. Phys. Lett.* **56**, 629 (1990).
21. Racanelli, M., and D. W. Greve, *Appl. Phys. Lett.* **56**, 2524 (1990).
22. Robbins, D. J., J. L. Glasper, A. G. Cullis, and W. Y. Leong, *J. Appl. Phys.* **69**, 3729 (1991).
23. Sinniah, K., M. Sherman, L. Lewis, W. Weinberg, J. Yates, Jr., and K. C. Janda, *Phys. Rev. Lett.* **58**, 2963 (1989).
24. Gates, S. M., and S. K. Kulgarni, *Appl. Phys. Lett.* **58**, 2963 (1991).
25. Greve, D. W., *Thin Films Vol.* 23, Academic Press, New York, 1998.
26. Coon, P. A., M. L. Wise, and S. M. George, *J. Cryst. Growth* **130**, 162 (1993).
27. Ohshita, Y., and N. Hosoi, *J. Cryst. Growth* **131**, 495 (1993).
28. Hierlemann, M., A. Kersch, C. Werner, and H. Schaefer, *J. Electrochem. Soc.* **142**(1), 259 (1995).
29. Ito, S., T. Nakamura, and S. Nishikawa, *J. Appl. Phys.* **78**, 2716 (1995).
30. Borland, J. O., and C. I. Drowley, *Solid State Technol.* **28**(8), 141 (1985).
31. Pagliaro, R., J. F. Corboy, L. Jastrzebski, and R. Soydan, *J. Electrochem. Soc.* **134**(5), 1237 (1987).
32. Kern, W., and D. A. Puotinen, *RCA Rev.,* p. 187 (1970).
33. Meyerson, B. S., F. J. Himsel, and K. J. Uram, *Appl. Phys. Lett.* **57**, 1034 (1990).
34. Frystak, D. C., R. Wise, P. Grothe, J. Barnett, B. Fowler, and R. Carpio, *Proceedings of the Electrochemical Society,* Fall meeting, Paris, France, Sept. 1–5, 1997.
35. Dutartre, D., C. Fellous, and P. Ribot, in press, *J. Mater. Sci. Mats. Electron.* (2001).
36. Simpson, T., SEH, personal communication (1999).
37. Russ, M., in *Proceedings of 1999 Epitaxial Technology Strategic Symposium,* hosted by Applied Materials, Unterschleissheim, Germany, 1999.
38. LETI's UNIBOND SOI wafers, available from SOI Tec, Bernin, 38000 France.
39. Malik, I. J., S. Kang, and A. L. Thilderkvist, in *Proceedings of 1999 Epitaxial Technology Strategic Symposium,* hosted by Applied Materials, Unterschleissheim, Germany, May 19, 1999.
40. Sagaguchi, K., and T. Yonehara, *Solid State Technol.* **43**(6), 88 (2000).
41. Sato, N., and T. Yonehara, *Appl. Phys. Lett.* **65**, 1924 (1994).

42. Anthoniadis, D. A., A. Wei, and A. Lochtefeld, in *Proceedings of 1999 ESSDERG*, p. 1924, 1999.
43. Ashburn, P., in Design and realization of bipolar transistors, edited by D. V. Morgan and H. R. Grubin, John Wiley & Sons, New York, 1988.
44. Iyer, S. S., G. L. Patton, S. S. Delage, S. Tiwari, and J. M. C. Stork, *1987 IEDM Tech. Digest*, p. 874 (1987).
45. Patton, G. L., D. L. Harame, J. M. C. Stork, B. S. Meyerson, G. Scilla, and E. Ganin, *IEEE Electron. Device Lett.* **EDL-10**, 534 (1989).
46. Schuppen, A., U. Erben, A. Gruhle, H. Kibbel, H. Schumacher, and U. Konig, *1995 IEDM Tech. Digest*, p. 743 (1995).
47. Oda, K., E. Ohue, M. Tanabe, H. Shimamoto, and K. Washio, *Thin Solid Films*, **369**(2), 358 (2000).
48. STMicroelectronics, *1998 ESSDERG Tech. Digest*, p. 448 (1998).
49. Infineon, *1999 ESSDERG Tech. Digest*, p. 88 (1999).
50. Hitachi, IBM, Lucent, and Motorola, *1999 IEDM Tech. Digest*, p. 557 (1999).
51. Harame, D. L., J. H. Comfort, J. D. Cressler, E. F. Crabbe, J. Y. C. Sun, B. S. Meyerson, and T. Tice, *Trans. Electron. Devices*, **42**(3), 469 (1995).
52. Hong, M., E. de Frezard, J. Steele, A. Zlotnicka, C. Stein, G. Tam, M. Racanelli, L. Knoch, Y. C. See, and K. Evans, *IEEE Electron. Device Lett.* **14**(10), 478 (1993).
53. Monroy, A., M. Laurens, M. Marty, D. Dutartre, D. Gloria, J. L. Carbonero, A. Perrotin, M. Roche, and A. Chantre, *1999 BCTM Tech. Digest*, p. 121 (1999).
54. Sato, F., T. Hashimoto, T. Tasumi, H. Kitahata, and T. Tashiro, *1992 IEDM Tech. Digest*, p. 397 (1992).
55. Kuhn, R., S. Decoutere, M. Caymax, F. Vleugels, E. Vershooten, R. Loo, and J. L. Loheac, *1999 ESSDERG Proceedings*, p. 437 (1999).
56. Lanzerotti, L. D., J. C. Sturm, E. Stach, R. Hull, T. Buyuklimanli, and C. Magee, *1996 IEDM Tech. Digest*, p. 249 (1996).
57. Knoll, D., B. Heinemann, H. J. Osten, K. E. Ehwald, B. Tillack, P. Schley, R. Barth, M. Matthes, K. S. Park, Y. Kim, and W. Winkler, *1998 IEDM Tech. Digest*, p. 703 (1998).
58. Ichikawa, A., Y. Hirose, T. Ikeda, T. Noda, M. Fujiu, T. Takatsuka, A. Moriya, M. Sakuraba, T. Matsuura, and J. Murota, *Thin Solid Films* **369**, 167 (2000).
59. Violette, K. E., C. P. Chao, R. Wise, and S. Unnikrishnan, *J. Electrochem. Soc.* **146**(5), 1895 (1999).
60. Uchino, T., T. Shiba, K. Ohnishi, A. Miyauchi, M. Nakata, Y. Inoue, and T. Suzuki, *1997 IEDM Tech. Digest*, p. 479 (1997).
61. Wakabayashi, H., T. Yamamoto, T. Tatsumi, K. Tokunaga, T. Tamura, T. Mogami, and T. Kunio, *1997 IEDM Tech. Digest*, p. 99 (1997).
62. SIA Technology Roadmap, (1999).
63. Oghuro, T., *et al.*, *1993 IEDM Tech. Digest*, p. 433 (1993).
63a. Hori, A., *1993 IEDM Tech. Digest*, p. 909 (1993).
64. Oghuro, T., H. Naruse, H. Sugaya, H. Kimijima, E. Morifuji, T. Yoshitomi, T. Morimoto, H. S. Momose, Y. Katsumata, and H. Iwai, *1998 IEDM Tech. Digest*, p. 927 (1998).
65. Alieu, J., Ph.D. thesis, Institut National des Sciences Appliquées, Lyon, France, July 1999.
66. Ha, J. M., personal communication (2000).
67. Warren, P., I. Sagnes, D. Dutartre, P. A. Badoz, J. M. Berroir, Y. Guldner, J. P. Vieren, and M. Voos, *Microelectron. Eng.* **25**, 171 (1994).
68. Venkataraman, V., P. V. Schwartz, and J. C. Sturm, *Appl. Phys. Lett.* **59**, 2871 (1991).
69. Garone, P. M., V. Venkataraman, and J. C. Sturm, *IEEE Electron. Devices. Lett.* **13**, 56 (1992).
70. Alieu, J., P. Bouillon, R. Gwoziecki, D. Moi, G. Bremond, and T. Skotnicki, *1998 ESDERG Tech. Digest*, p. 144 (1998).
71. Lu, W., R. Hammond, S. J. Koester, X. W. Wang, J. O. Chu, T. P. Ma, and I. Adesida, *1999 IEDM Tech. Digest*, p. 577 (1999).

72. Alieu, J., T. Skotnicki, E. Josse, J. L. Regolini, and G. Bremond, *2000 Symp. VLSI Tech. Digest*, p. 130 (2000).
73. Welser, J., J. L. Hoyt, S. Takagi, and J. F. Gibbons, *1994 IEDM Tech. Digest*, p. 373 (1994).
74. Rim, K. et al., *1995 IEDM Tech. Digest*, p. 517 (1995).
75. Jurczak, M., T. Skotnicki, G. Ricci, Y. Campidelli, C. Hernandez, and D. Bensahel, in *1999 ESSDERG Proc.* p. 304, 1999.
76. Ismail, K., S. Rishton, J. O. Chu, K. Chan, and B. S. Meyerson, *IEEE Electron. Devices Lett.* **14**, 3177 (1993).
77. Fitzgerald, E. A., Y. H. Xie, M. L. Green, D. Brasen, A. R. Kortan, J. Michel, Y. J. Mii, and B. E. Weir, *Appl. Phys. Lett.* **59**, 811 (1991).
78. Dutartre, D., P. Warren, F. Provenier, F. Chollet, and A. Perio, *J. Vac. Sci. Technol.* **A12**(4), 1009 (1994).
79. Chollet, F., P. Warren, D. Dutartre, and E. Andre, *Jpn. J. Appl. Phys.* **33**, 6437 (1994).
80. Sugii, N., K. Nakagawa, S. Yamaguchi, S. K. Park, and M. Miyao, *Thin Solid Films* **369**, 365 (1999).
81. Rim, K., J. L. Hoyt, and J. F. Gibbons, *1998 IEDM Tech. Digest*, p. 707 (1998).
82. Takato, H., K. Sunouchi, N. Okabe, A. Nitayama, K. Hieda, F. Horiguchi, and F. Masuoka, *1988 IEDM Tech. Digest*, p. 222 (1988).
83. Yang, M., and J. C. Sturm, *Thin Solid Films* **369**, 366 (2000).
84. Liu, K. C. et al., *1999 IEDM Tech. Digest*, p. 63 (1999).
85. Hergenrother, J. M. et al., *1999 IEDM Tech. Digest*, p. 75 (1999).
86. Wong, H. S. P., K. K. Chan, and Y. Taur, *1997 IEDM Tech. Digest*, p. 427 (1997).
87. Jurczak, M., T. Skotnicki, M. Paoli, B. Tormen, J. Martins, J. L. Regolini, D. Dutartre, P. Ribot, D. Lenoble, R. Pantel, and S. Monfray, *IEEE Trans. Electron. Devices* **47**, 2179 (2000).
88. Wu, K., E. Fitzgerald, and J. Borenstein, U.S. patent PCT/US/99/07849 (1999).
89. Jouan, S. et al., *IEEE Trans. Electron. Devices* **46**, 1525 (1999).
90. Lin, T. L., and J. Maserjian, *Appl. Phys. Lett.* **57**, 1422 (1990).
91. Tsaur, B. Y., C. K. Chen, and S. A. Marino, *IEEE Electron. Devices Lett.* **12**, 293 (1991).
92. Jimenez, J. R., X. Xiao, J. C. Sturm, and P. W. Pellegrini, *Appl. Phys. Lett.* **67**, 506 (1995).
93. Temkin, H., T. P. Pearsall, J. C. Bean, R. A. Logan, and S. Luryi, *Appl. Phys. Lett.* **48**, 963 (1986).
94. Sugiyama, M., T. Morikawa, T. Tatsumi, T. Hashimoto, and T. Tashiro, *IEDM Tech. Digest*, p. 583 (1995).
95. Lo, Z., R. Jiang, Y. Zheng, L. Zang, Z. Chen, S. Zhu, X. Cheng, and X. Liu, *Appl. Phys. Lett.* **77**, 1548 (2000).
96. Lenchyshyn, L. C., M. L. W. Thewaalt, J. C. Sturm, P. V. Schwartz, N. L. Rowell, J. P. Noël, and D. C. Houghton, *J. Electron. Mater.* **22**, 233 (1993).
97. Tang, Y. S., W. X. Ni, C. M. S. Torres, and G. V. Hansson, *Electron. Lett.* **31**, 1385 (1995).
98. Brown, A. R., G. A. M. Hurkx, H. G. A. Huizing, M. S. Peter, W. B. de Boer, J. G. M. van Berkum, P. C. Zalm, E. Huang, and N. Koper, *IEDM Tech. Digest*, p. 699 (1998).
99. Sakaki, H., *Jpn J. Appl. Phys.* **21**, L381 (1982).
100. Rommel, S. M., T. E. Dillon, P. R. Berger, R. Lake, P. E. Thompson, K. D. Hobart, A. C. Seabaugh, and D. S. Simons, *IEDM Tech. Digest*, p. 1035 (1998).

Index

A

Activation energy, negative, 86
Adatoms
  conservation, 215–216
  nucleation rate, 217
Adsorption
  kinetics, chlorosilane, 75–80
  Langmuir-type, 135–136
  precursor-mediated process, 86
Aerosol CVD
  inside deposition reactors, 43–46
  optical fibers, 41–43
    chemistry basics, 49
  outside deposition reactors, 47–49
  reactor principles, 43
Air-gap, CV, 245–247
Annealing, prior to epitaxial growth, 333–334
Architectures, CMOS, new, 446–451
Arsenic
  GaAs active substrate, 453
  GaAs solar cells, $Si_{1-x}Ge_x$ pseudosubstrates, 358–359
Arsenic mirror, 4
Atmospheric-pressure reactors
  barrel, 20
  cellular, 22
  continuous-type, 17–20
  coupled with transport-limited regimes, 11–12
  horizontal single-wafer, 24
  pancake, 21
  ultraclean, 129
  vertical-flow, 22–24
Atomic absorption spectroscopy, detection of surface defects, 269
Attenuation, light signal, 42

Autodoping
  buried layer parameters affecting, 322–324
  characterization, 321–322
  doping profile in presence of, 318–320
  pre-epitaxial bake effect, 324–325
  problem of, 316–318

B

Backsealing, for IGBTs, 289
Back-side defects, 291–292
Bake
  H_2
    absence in UHV-CVD before deposition, 403
    high-temperature, 425
    native oxide removal by, 363–364
    temperature of, 390–391
  pre-epitaxial
    effect on autodoping, 324–325
    hydrogen, 304
Band alignment, at Si/Ge heterointerfaces, 169–170
Barrel reactors
  gas injection design, 123
  simulations, 206–209
  susceptor insertion variations, 20
BiCMOS, 301–303, 429
Bipolar devices, small-signal, 285, 287
Bipolar transistors
  DFT: $Si_{1-x}Ge_x$, 352–353
  HBT: $Si_{1-x}Ge_x$, 350–352
  high-voltage, 283–285
  IGBT, 288–289
  medium-voltage mesa, 285
  and mixed technologies, 300–303
  RF, bases of, 426–435

Boron
  autodoping, prevention, 326
  doping
    HBT, 351
    in $Si_{1-x}Ge_x$ epitaxial growth, 138–143
Breakdown voltage
  discrete transistors, 301
  in epitaxial junctions, 280–281
Buffer layers
  optimized, 358
  SiGe, growth, 174–175
Buried layers
  parameters affecting autodoping, 322–324
  on patterned wafers, 319–320
Burn systems, epitaxial process chamber, 110

C

Capacitance–voltage
  air-gap, 245–247
  Hg probe, 241–243
Cap layer, effect on autodoping, 325–326
Carbon
  inclusion in $Si_{1-x}Ge_x$, 378–382
  role in $Si_{1-x}Ge_x$, 362
Carbon pileups, on Si substrate, 134–135
Carrier concentrations, B-doped films, 141–142
Carrier density, resistivity and, 283
Carriers, minority, recombination, 263–267
Cellular reactors, susceptor for, 22
Channels
  for CMOS
    epitaxial Si, 439–441
    strained Si, 443–446
    strained SiGe, 441–443
  enhanced mobility, $Si_{1-x}Ge_x$ for, 356–358
Chemical beam epitaxy
  control of SiGe alloys, 166–169
  doping, 163–165
  as example of surfactant-mediated growth, 173–174
  growth mechanisms, 158–160
  machines, 154–155
  selective growth, 175–177
  system schematics, 400–401
Chemical etching
  delineation of defects, 341, 343
  detection of subsurface defect, 268
  evaluation of pattern integrity, 314–315
  vapor-phase, of native oxide, 366–368

Chemical mechanical polishing, 424–425
Chemical mechanism, CVD reactions, 197–198
Chemical vapor deposition
  aerosol, 41–49
  bulk polycrystalline Si processes, 7–10
  classification paths, 5–6
  definition, 51
  film deposition
    products and chemistries, 37–41
    reactors, 10–36
  historical overview, 4–5
  kinetic scheme, 52–56
  machines for low-temperature epi growth, 128–133
  plasma-enhanced, 401–402
  rapid thermal, 404–406, 453–454
  reactor model conservation equations, 193–195
  UHV, 402–404, 430, 432–433, 453
Chlorine atoms, effect on autodoping, 325–327
Chlorosilanes
  adsorbed, structure and energetics, 74–75
  adsorption and desorption kinetics, 75–80
  effluents, 292
  gas-phase
    decomposition pathways, 64–69
    structure and energetics, 61–64
  molar ratio, 115–116, 119
  Si deposition from, 81–84
  thermodynamic equilibrium calculations, 69
Cladding, optical fiber, 42
Cleaning
  peroxide, 156
  pre-epitaxial, 306–308
Clean room, epitaxial growth facilities, 103–104
CMOS
  channel engineering in, 438–446
  elevated sources and drains in, 435–438
  new architectures, 446–451
  $Si_{1-x}Ge_x$
    elevated source/drain structure, 353–354
    enhanced mobility channels, 356–358
    gate electrodes, 354–356
  technology evolution, 423
Coatings, application of CVD, 4–5
Cold-wall reactors, 13, 16, 129
  AP reactors as, 17
  deposition process temperature, 188
Collision kinetic constant, 83
Condensation, liquid HCl, avoidance, 293

Conducting plates, for plasma generation, 30–31
Conservation, adatoms, 215–216
Conservation equations, CVD reactor model, 193–195
Contamination
  air, effect on Si deposition, 131
  in epitaxial growth facilities, 100–101
  general considerations, 255
  metal and molecular, 261–270
  O and C
    from exhaust, 132–133
    on Si substrate, 134–135
  on Si substrate, removal, 133–135
Continuous-type reactors
  key elements, 17
  resistance heating, 18–19
Critical thickness
  SiGe layers on Si (100), 170–172
  $Si_{1-x}Ge_x$, 347
Crystal morphology, control of, 215–220
CV, see Capacitance–voltage
CVD, see Chemical vapor deposition
Cylinder reactors, see Barrel reactors
Czochralsky pulling, 9–10

D

Dark-field signal, for microroughness measurement, 260
DCS, see Dichlorosilane
Decomposition pathways, gas-phase, $SiHCl_3$, 64–69
Deep-level capacitance transient spectroscopy, 267–268
Defects
  crystallographic layer, on patterned wafers, 327–343
  epitaxial
    crown, 290
    identification, 256–258
  subsurface, detection by chemical etching, 268
Density functional theory, study of SiHCl molecules, 64
Denuded zone treatment, prior to epitaxial growth, 331
Design
  CVD reactors, 10–17
  electron cyclotron resonance reactors, 35–36
  injectors, for Watkins Johnson belt reactor, 19–20
  reaction chamber, 121–122
Desorption kinetics, chlorosilane, 75–80
Dichlorosilane
  low-temperature epitaxy based on, 410–415
  for reduced pressure processing, 105
Diffusion treatment, prior to epitaxial growth, 329
Diodes, epitaxy process for, 290
Dislocations
  generated on epi films, 335–336
  threading, 347
Dissociation
  kinetic constant, $SiHCl_3$, 65–67
  silanes, 35–36
Distortion, of pattern, 312–313
Dopant
  dilution system, 122
  redistribution in epitaxial process, 282–283
  reincorporation, 318
  species, vapor pressure, 323
  unwanted, 316–317
Doping, see also Autodoping
  B and P, in $Si_{1-x}Ge_x$ epitaxial growth, 138–143
  gas-source MBE, 163–165
  profile, in presence of autodoping, 318–320
  solid-source MBE, 162–163
Doping control
  basic considerations, 235–238
  four-point probe, 239–241
  mercury probe CV, 241–243
  noncontact resistivity measurement, 245–247
  SIMS, 247–248
  SRP, 243–245
Drift field enhanced transistor, $Si_{1-x}Ge_x$, 352–353

E

Electrical power supply, CVD reactor, 15
Electrodes
  parallel-plate plasma reactors, 33–34
  $Si_{1-x}Ge_x$ gate, 354–356
Electron cyclotron resonance reactors, 35–36
Electron impact reactions, 55
Elevated source/drain structure, Si alloys in, 353–354
Ellipsometry, $Si_{1-x}Ge_x$ layers, 388

Emissive devices, superlattices for, 358
Energy balance equation, 194, 210
Energy source
  CVD, 6
  reaction activation, 13–14
Enhanced mobility channels, $Si_{1-x}Ge_x$ for, 356–358
Environmental considerations, effluents, 292
Epitaxial crown, 290
Epitaxial growth
  high-quality $Si/Si_{1-x}Ge_x/Si$ heterostructure, 143–147
  low-temperature
    CVD machines, 128–133
    surface treatment, 133–135
  mechanisms, 135–143
    terrace step kink, 192–193
  selective, 410–415
  stability, 218–220
  thermal treatments prior to, 329–331
Epitaxial growth facilities
  clean room, 103–104
  contamination problem, 100–101
  cooling water, 103
  general requirements, 102
  power supply, 102–103
Epitaxial junctions, breakdown voltage, 280–281
Epitaxial reactor
  computational issues, 196–197
  conservation equations, 193–195
  2D and 3D simulations
    barrel reactor, 206–209
    horizontal reactor, 198–203
    vertical reactor, 204–206
  estimation of model parameters, 197–198
  exhaust treatment, 108–111
  gas injection and flow patterns, 123
  gas panel, 122
  geometry, 91
  high-resistivity control, 119–121
  high-thickness deposition capability, 111–113
  induction heating, 95–96
  low-temperature applications, 123–124
  for patterned wafers, 306
  process gas and delivery, 104–108
  process pressure, 98
  quartzware, 98
  radiant heating, 94
  reaction chamber design, 121–122
  run-to-run repeatability, 113–119
  single-wafer, 96
  susceptors, 96–97
    design, 113
    temperature, 121
Epitaxial recipe
  basic steps, 304–306
  parameters affecting autodoping, 324–327
  for patterned wafers, 343
Epitaxial silicon
  atomic scale link with reactor scale, 217–218
  channels, for CMOS, 439–441
  deposition, 80–81
    reduced-order models, 209–214
  growth, 188
    modeling on terrace scale, 215–217
  growth rates, 407–409
Epitaxy
  at high voltage and at high currents, 279–280
  liquid-phase, 152
  low-temperature
    DCS/HCl-based, 410–415
    silane-based, 409–410
    surface preparation for, 415–416
  MBE, *see* Molecular beam epitaxy
  on patterned substrates, 417–420
  silicon
    multiscale modeling, 190–192
    physical and chemical aspects, 189–190
    selective growth, 412–413
  vapor-phase, 152
Epitaxy process
  for diodes and thyristors, 290
  dopant redistribution in, 282–283
  growth regimes, 407–415
  micromachining applications, 451
  new challenges, 406–407
  patterned substrates, 417–420
  for power and discrete devices, 283–290
  selective epi growth, 420–422
  surface preparation for low-temperature epitaxy, 415–417
Etching
  chemical, *see* Chemical etching
  oxide patterned by, 297–298
  wafer, by HCl, 108
Etch rate, thermal oxide, 367–368
Excess stress, critical thickness and, 171
Exhaust systems, gas, 15
Exhaust treatment, abatement equipment, 108–109

F

Faceting, 291
Film
   B- and P-doped, 141–143
   epitaxial, dislocations on, 335–336
   growth rate
      estimation, 195
      HCl effect, 81–82
      temperature-dependent, 189–190
   impurities in, 29
   new, 406–407
   silicon-on-insulator, 424–426
   $Si_{1-x}Ge_x$
      high-quality epitaxial growth, 143–147
      polycrystalline and amorphous, 354–356
   structure, as subdivision of CVD, 6
Film deposition
   atmospheric-pressure reactors, 17–24
   low-pressure reactors, 24–27
   plasma peculiarities, 41
   plasma reactors, 28–36
   properties of films, 37–38
   rapid thermal process reactors, 27–28
   reaction chemistries, 38–40
   reactor design principles, 10–17
Filtration, gas, 107
Fluidized bed reactor, 8–9
Fourier transform infrared spectroscopy, for thickness measurement, 248–252
Four-point probe
   layer thickness measurements, 239–241
   $Si_{1-x}Ge_x$ layers, 388–389
FTIR, see Fourier transform infrared spectroscopy
Furnace LP-CVD reactors
   horizontal, 24–26
   vertical, 26–27

G

Gallium
   GaAs active substrate, 453
   GaAs solar cells, $Si_{1-x}Ge_x$ pseudosubstrates, 358–359
Gas
   chemical species in, 53
   composition, effect on Si alloy deposition, 371–373
   delivery systems, 108
   epitaxial reactors, Si sources, 104–105
   exhaust systems, 15
   exhaust treatment, 108–111
   filtration, 107
   flow, HCl and DCS, 413–415
   injection and flow patterns, 123
   inject nozzles, in barrel reactors, 207
   motion, in vertical-flow reactors, 22–24
   purity, 105–107
   recycling, 8
Gas panels, design aspects, 122
Gas-phase autodoping region, 320
Gas-phase chemistry, subscheme of CVD, 53–54
Gas-phase reactivity
   chlorosilane, 61–69
   silane, 56–57
Germanium
   dots, 179, 181
   high fractions, $Si/Si_{1-x}Ge_x/Si$ heterostructure growth at, 143–147
Germanium oxide, application to optical fibers, 49
Getters, gas purifier based on, 106
Glasses, reaction chemistries, 38–39
Global loading effect
   on Si selective epi growth, 421–422
   thermal and chemical, 418
Graphite, sintered, for susceptors, 96–97

H

Haze test, for subsurface defects, 268
HBT, see Heterojunction bipolar transistor
Heat exchange, contributions to wafer, 418–419
Heat transport, and mass transport: boundary layer, 210–211
Heterojunction bipolar transistor
   nonselective process, 429–433
   selective process, 433–434
   SiGe base, 451
   SiGe:C epitaxy for, 435
   $Si_{1-x}Ge_x$, 350–352
   true and drift-field, 428
   vertical structure, 427–428
Heterojunction internal photoemission detectors, 452
Heterostructures
   Si/Ge
      band structures, 169–170
      critical thickness, 170–172
      surface segregation, 172
      surfactant-mediated growth, 173–174

Heterostructures (*Continued*)
    Si/Si$_{1-x}$Ge$_x$/Si, growth at high Ge fractions, 143–147
HF bath, in low-temperature epi, 416
HF dip, native oxide removal by, 364–365
High-temperature reactors, cold-wall, 12
High-thickness deposition, capability for power epitaxy, 111–113
High voltage
    bipolar transistors, 283–285
    epitaxy at, 279
Horizontal furnace LP-CVD reactors, 24–26
Horizontal reactors
    3D simulations, 198, 200, 203
    single-wafer, 24
Hot-wall reactors, 13, 16, 128–131
Hydrogen
    effect on Si growth rate, 84
    elimination from growing surface, 160
    point of use purification, 124
    pre-epitaxial bake, 304
    product of SiH$_2$Cl$_2$ dissociation, 67–68
Hydrogen bake
    absence in UHV-CVD before deposition, 403
    high-temperature, 425
    native oxide removal by, 363–364
    temperature of, 390–391
Hydrogen chloride
    effect on growing film, 81–82
    effluents, 292
    high-purity, 106–107
    in situ anhydrous etch, 307–308
    liquid, avoidance of condensation, 293
    low-temperature epitaxy based on, 410–415
    surface kinetics, 78
    wafer etch, 108

I

IGBT, *see* Insulated gate bipolar transistors
Implantation, lattice damage and recovery after, 332–334
Impurity pileups, 132–133
Induction heating
    epitaxial reactor, 95–96
    in power epitaxy, 112–113
Inductively coupled plasma-MS, 269–270
Injectors, design, for Watkins Johnson belt reactor, 19–20

Inside deposition reactors
    IVPO basics, 43–44
    modified CVD, 44–45
    plasma CVD, 45–46
    plasma modified CVD, 46
Insulated gate bipolar transistors, 288–289
Integrity
    geometrical pattern, 308–315
    pattern, 299
Interface smearing, 172
Ionization reactions, 55
Iron, in p-type Si, 265
Islands
    formation, critical thickness and, 171–172
    Ge, 179
Isodesmic reaction, chlorosilane species, 63
IV–IV semiconductors, 40

K

Kinetic constants, estimation, 52–53
Kinetics
    DCS/H$_2$ epi, 411–412
    growth, precise control of, 417–418
    surface
        chlorosilane, 74–80
        in horizontal reactors, 200
        silanes, 71–74
Kinetic scheme
    CVD, 52–56
    gas-phase and surface chemistry of silanes, 84–86
    gas-phase precursors to deposition and, 80–84
Komatsu reactor, 9

L

Lapping and staining method, 315
Lattice constant
    Si, 359
    Si$_{1-x}$Ge$_x$, 346–347
Lattices, *see also* Superlattices
    damage and recovery after implantation, 332–334
Leakage current, dependence on layer thickness, 146–147
Leaks, allowing oxygen to enter reactor, 385
Light signals, transmission via optical fibers, 41–42

Light sources, for reaction activation, 14
Liquid-phase epitaxy, 152
Liquid-phase HF process, native oxide removal by, 364–365
Load-lock, vacuum-capable, 389
Low-pressure reactors
  horizontal furnace, 24–26
  with kinetically limited regimes, 11–12
  Si deposition in, 57
  ultraclean hot-wall, 128–130
  ultrahigh-vacuum, 27
  vertical furnace, 26–27

M

Magnetron reactors, 34
Mapping, epitaxial layers, 234
Mass balance equations, 193, 209–210
Mass transport, and heat transport: boundary layer, 210–211
MBE, *see* Molecular beam epitaxy
Mercury probe CV, 241–243
Metal-oxide-semiconductor, *see also* BiCMOS; CMOS
  field effect transistor, 144, 146–147, 441–451
    modulation-doped, 441–446
    mixed bipolar/CMOS, 301–303
  power devices, 287
    low-voltage, 288
  vertical, 447–448
Metrology, $Si_{1-x}Ge_x$ layers, 386–393
Microautodoping, 317
Micromachining, epi process applications, 451
Microroughness, measurement, 259–261
Microwave photoconductive decay, 264–267
Migration, adsorbates on Si surface, 159–160
Misalignment, wafer, 310–311
Mixed deposition, nonselective, 376–378
Molecular beam epitaxy
  drawbacks, 399–400
  gas-source, *see* Chemical beam epitaxy
  SiGe buffer layers, growth, 174–175
  Si/Ge heterostructures
    band structures, 169–170
    critical thickness, 170–172
    surface segregation, 172
    surfactant-mediated growth, 173–174
  solid-source
    doping, 162–163

    growth mechanisms, 157–158
    growth of SiGe(C) alloys, 166
    machines, 153–154
    superlattices, 177–181
    surface treatment, 155–157
Momentum balance equation, 193
MOS, *see* Metal-oxide-semiconductor
MOSFET, *see* Metal-oxide-semiconductor, field effect transistor
Multiscale modeling, silicon epitaxy, 190–192
Mýller–Plesset correlation energy correction, 61–63

N

Neutral dissociation reactions, 55
Neutron activation analysis, 268
Nitrogen, purification of, 385–386
NMOS device, 355–357
Nomarski interference contrast, 299–300, 341
Nonconducting walls, plasma induced through, 31
Nucleation
  adatoms, rate of, 217
  epitaxial stacking faults, 337
Numerical solutions, predicted, 196–197

O

Optical fibers
  aerosol CVD, 41–43
  chemistry basics for, 49
  communications, Si/SiGe structures in, 452
Orange peel defect, 339–340
Organics, contamination by, 270
Out-diffusion region, component of autodoping, 320
Outside deposition reactors
  OVD, 47–48
  vapor axial deposition, 48–49
Oxide
  containing Ge, 357
  native, removal by
    H_2 bake, 363–364
    liquid-phase HF process, 364–365
    reactive ion etch, 368–369
    vapor-phase HF process, 365–368
  patterned by etching, 297–298
  poly seal on top of, 376–378

Oxygen
   control, in $Si_{1-x}Ge_x$ alloys, 382–386
   pileups, on Si substrate, 134–135

P

Pancake reactors, design aspects, 21
Parallel-plate plasma reactors, 32–34
Pattern
   distortion and washout, 312–313
   on epitaxial wafer, 296–300
   geometrical, integrity, 313–315
   integrity, 299
      doping problems, 316–327
   shift, 309–311
   V-groove, Si substrate, 178
Patterned wafer
   crystallographic epi layer defectivity on, 327–343
   epitaxy on, 417–420
   (111) oriented, shift, 310–311
   surface preparation: pre-epitaxial cleaning, 306–308
Peroxide cleaning, 156
Phosphorus doping, in $Si_{1-x}Ge_x$ epitaxial growth, 138–143
Photo-assisted CVD, 14
Photodetectors, superlattices for, 358
Plasma-enhanced CVD, 401–402
Plasma reactors
   aerosol CVD, 45–46
   electron cyclotron resonance, 35–36
   parallel-plate, 32–34
   plasma basics for use in CVD, 28–30
   plasma generation techniques, 30–31
   remote, 36
   for Si deposition, electronic reactions, 72–73
   silane precursor, 55
Plasma systems, susceptor for, 16
Plumbing, process chamber, 104
PMOS device, 355–357
Polishing, chemical mechanical, 424–425
Poly-seal layer, on top of oxide, 376–378
Power devices
   mixed epitaxy, 289–290, 300–303
   MOS
      low-voltage, 288
      medium- and high-voltage, 287
   RF, 287

Power spectral density function, evaluation of microroughness, 259–260
Power supply, epitaxial growth facilities, 102–103
Precursor
   depletion, 25–26
   gas-phase, 69, 80–84
   organic, 37
   Si, maximum flow rate, 391
   $SiCH_6$, 379
   Si:Ge, partial pressure, 371–372
   silane, 55
Pre-epitaxial bake
   effect on autodoping, 324–325
   hydrogen, 304
Pre-epitaxial cleaning, 306–308
Preform
   colloidal, in OVD reactors, 47–48
   optical fiber, 43
Pressure effects
   C and Ge contents of films, 381–382
   Si alloy deposition, 371–373
Pressure regimes, definitions, 12–13
Process chamber
   chemical beam epitaxy, 400–401
   CVD reactor, 16
   plumbing, 104
   wet scrubbers, 109–110
Process monitoring, through wafers, 271–272
Process pressure, epitaxial reactor, 98
Profiling, temperature, 121
Profilometric measurements, on alignment targets, 315
Pseudomorphic layer, $Si_{1-x}Ge_x$, 346–347
Pseudosubstrate
   for GaAs solar cells, 358–359
   SiGe, 445–446
Purity, gas, 105–107
Pyramid defect, 340
Pyrometers, reliability problems, 114

Q

Quantum wells, in MBE, 177–181
Quartz
   parts for epitaxial reactor, 98
   usage in cold- and hot-wall reactors, 16
Quartz-wall reactors, *see* Cold-wall reactors

R

Radiant heating
  epitaxial reactor, 94
  in power epitaxy, 112
Rapid thermal CVD, 404–406, 453–454
Rapid thermal process reactors, 27–28, 129
  $SiHCl_3$ as precursor, 82
Reaction chamber
  design, 121–122
  leaks in, 385
Reactive ion etch technology, 368–369
Recombination lifetime, 263–267
Recycling, gas, 8
Reduced-order models, epitaxial Si deposition, 209–214
Reduced pressure reactors, 98, 100
  facilities guidelines, 123–124
Remote plasma reactors, 36
Repeatability
  run-to-run, in power epitaxy, 113–119
  wafer measurement, 229
Resistance heating
  continuous-type reactors, 18–19
  in hot- and cold-wall systems, 13–14
Resistivity
  in B- and P-doped films, 141–143
  and carrier density, 283
  epitaxial layer, 119, 121, 303
  measured by SRP, 243–245
  on multilayer structures, 253–255
  noncontact, measurements, 245–247
  very high, control of, 279
Roughness, *see also* Microroughness
  orange peel defect, 339–340
  surface, of degraded samples, 146–147

S

Scanning auger, $Si_{1-x}Ge_x$ layers, 387
Scatterometry, 258–259
Scrubber treatment, pre-epitaxial, 308
Secondary-ion mass spectrometry
  background level of oxygen, 384–385
  detection of surface defects, 269–270
  limitations, 238
  profiling by, 247–248
  $Si_{1-x}Ge_x$ layers, 386–387
Selective deposition, $Si_{1-x}Ge_x$, 378
Selective epi growth
  DCS/HCl-based, 410–415

gas-source MBE advantage for, 175–177
  perfect, 420–421
  Si, global loading effect, 421–422
SEMI standards, 231
$SiCH_6$, partial pressure, 379–382
Sidebursts, FTIR spectrum, 248–250
Siemens reactors, 7–8
SiGe
  alloys
    formation and control by gas-source MBE, 166–169
    solid-source MBE, 166
  buffer layer growth, 174–175
  global loading effects, 422
  heterostructures, 40
  sacrificial deposition, 450
  SiGe:C epitaxy for HBT, 435
  Si/SiGe, selective epi growth, 433–434
  strained, channels for CMOS, 441–443
$Si_{1-x}Ge_x$
  bipolar applications, 349–353
  CMOS applications, 353–358
  deposition, process chemistry, 370–386
  epitaxial growth
    B- and P-doping in, 138–143
    on (100) surface, 135–138
  film, 143–147
  high-throughput epi process, 389–393
  layer deposition, 346–347
  layers
    four-point probe, 388–389
    scanning auger, 387
    SIMS, 386–387
  optoelectronics, 358
  psuedosubstrates for GaAs solar cells, 358–359
  role of carbon, 362
Si/Ge heterostructures
  band structures, 169–170
  critical thickness, 170–172
  surface segregation, 172
  surfactant-mediated growth, 173–174
Silane
  dissociation, 35–36
  gas-phase and surface chemistry, kinetic scheme, 84–86
  gas-phase reactivity, 56–57
  low-temperature epitaxy based on, 409–410
  precursor, 55
  surface reactivity, 71–74
Silicides, reaction chemistries, 40

Silicon
    adsorbed species, see Adatoms
    alloys, 360–362
        deposition process chemistry, 370–386
        native oxide removal, 363–369
    deposition step, 304–305
    deposits, cleaned out by HCl flow, 108
    doped and undoped, 38
    epitaxial, see Epitaxial silicon
    sources for epitaxial reactor gases, 104–105
    strained, channels for CMOS, 443–446
    surface treatment, 133–135
        for MBE, 155–157
Silicon dioxide, plasma processes for, 41
Silicon epitaxy
    multiscale modeling, 190–192
    physical and chemical aspects, 189–190
    selective growth, 412–413
Silicon nitride, reaction chemistries, 39
Silicon-on-insulator films, 424–426
Silicon-on-nothing concept, 449–451
Silicon oxide, reaction chemistries, 38–39
Silicon oxinitrides, reaction chemistries, 39–40
Silicon polycrystal reactors
    Komatsu and Union Carbide, 9
    Siemens, 7–8
SIMS, see Secondary-ion mass spectrometry
Single-wafer process, high-throughput, 389–393
Single-wafer reactors
    horizontal, 24
    with 300-mm wafers, 270–271
    thickness uniformity, 96
Slip-lines
    detection, 258
    frequency, 95–96
    production of, 330–331
    shapes, 336
Slip problem, and susceptor design, 113
Solar cells, GaAs, $Si_{1-x}Ge_x$ pseudosubstrates, 358–359
Solid-source MBE
    doping, 162–163
    growth mechanisms, 157–158
    machines, 153–154
    surface segregation, 172
Source/drain junction, elevated, in CMOS, 435–438
Spreading resistance profilometry, 230, 243–245
    two-point probe, 321–322

Sputtering
    magnetron reactors for, 34
    plasma-exposed surfaces, 36
    in SIMS, 247
SRP, see Spreading resistance profilometry
Stability, epitaxial growth, 218–220
Stacking faults, epitaxial, 336–339
Step fluctuation, evolution around straight shape, 218–219
Sticking coefficient
    HCl, 78–79
    $SiHCl_3$, 81
Substrate, see also Pseudosubstrate
    contamination removal, 133–135
    and epitaxial layer: interface, 319–320
    patterned, epitaxy on, 417–420
    Si, V-groove pattern, 178
    temperature, in plasma reactors, 29
Superlattices
    in MBE, 177–181
    for photodetectors, 358
    short-period Si/Ge, 452–453
Surface charge profiler, 238
Surface plasma CVD reactors, 45–46
Surface preparation
    low-temperature, 362–369
    for low-temperature epitaxy, 415–416
    pre-epitaxial cleaning, 306–308
Surface quality
    metal and molecular contamination, 261–270
    microroughness, 259–261
    scatterometry, 258–259
    visual inspection, 256–258
Surface reaction
    and gas-phase species, 54
    in low-temperature DCS/HCl system, 411–413
Surface reactivity
    chlorosilane, 74–80
    silane, 71–74
Surface segregation, in solid-source MBE, 172
Surface treatment
    for MBE, 155–157
    silicon, 133–135
    wafer, 243
Surfactant-mediated growth, gas-source MBE as example, 173–174
Susceptor
    barrel reactors, 20
    cellular reactors, 22
    classes, 16

design, for power epitaxy, 113
disk, geometry, 91
epitaxial reactor, 96–97
pancake reactors, 21
rotating, 200

T

TCS, *see* Trichlorosilane
Temperature
  clean room, 103–104
  control, 408–409
  effect on Si alloy deposition, 371–373
  film growth rate dependent on, 189–190
  H_2 bake, 390–391
  Komatsu process, 10
  ramp-up in H_2, 304
  repeatability of reading, 114–115
  Siemens process, 9
  substrate, in plasma reactors, 29
  uniformity and profiling, 121
Temperature-programmed desorption, chlorosilane species, 76–80
Terrace step kink
  mechanism of epitaxial growth, 192–193
  scale, modeling epitaxial Si growth on, 215–217
Thermal decomposition, including water scrubber, 110–111
Thermal treatments, prior to epitaxial growth, 329–331
Thermodynamic equilibrium, chlorosilanes, 69
Thermophoresis, in aerosol CVD reactors, 43
Thickness
  critical
    SiGe layers on Si (100), 170–172
    $Si_{1-x}Ge_x$, 347
  epitaxial, defects and problems, 290–292
  layer
    four-point probe, 239–241
    leakage current dependence on, 146–147
    measurement
      model-based FTIR, 251–252
      multilayer structures, 253–255
      on n/N and p/P structures, 252–253
      standard FTIR, 248–250
  silicon-on-insulator films, 425–426
  uniformity, single-wafer reactors, 96
  within-wafer variations, 230–234
Thyristors, epitaxy process for, 290
Transfer chamber, N_2-purged, 131

Transistors
  bipolar, *see* Bipolar transistors
  IGBT, 288–289
  PNP, 285, 287
Transition-state theory, loose, 66
Trichlorosilane
  molar ratio, 115–116, 119
  reduction in Siemens reactor, 7–8
Troe's formalism, 57, 73

U

Ultrahigh-vacuum CVD, 402–404, 430, 432–433, 453
Ultrahigh-vacuum reactors
  design aspects, 27
  hot-wall, 128
Union Carbide reactor, 9

V

Vapor axial deposition reactors, 48–49
Vapor-phase epitaxy, 152
Vapor-phase HF process, native oxide removal by, 365–368
Vertical-flow reactors
  gas motion, 22–24
  simulations, 204–206
Vertical furnace LP-CVD reactors, 26–27
V-groove pattern, Si substrate, 178
Vibrational frequency, chlorosilane species, 62–63

W

Wafer
  bare and patterned, 303–306
  constant temperature, 200
  contamination control, 261–270
  diameters, 15–16
  edge shrinking, 113
  etching by HCl, 108
  heat exchange contributions to, 418–419
  in horizontal furnace LP-CVD reactors, 25
  important parameters, 227
  low-cost epitaxial, 422–424
  measurement
    pattern, mapping, and variation, 230–235
    precision and accuracy, 229–230

Wafer (*Continued*)
  microroughness, 259–261
  misalignment, 310–311
  300-mm, 270–271
  patterned, *see* Patterned wafer
  placed in load-lock, 389
  sampling and cost, 229
  silicon-on-insulator, 424–426
  size, 392–393
Washout, pattern, 312–313
Water-cooling system, epitaxial growth facilities, 103
Watkins Johnson belt reactor, 17–18
  injector design, 19–20

Wet scrubbers, epitaxial process chamber, 109–110

X

X-ray diffraction, $Si_{1-x}Ge_x$ layers, 388
X-rays, detection of surface defects, 268–270

Z

Zero mask, 299

Contents of Volumes in This Series

Volume 1 **Physics of III–V Compounds**

C. Hilsum, Some Key Features of III–V Compounds
F. Bassani, Methods of Band Calculations Applicable to III–V Compounds
E. O. Kane, The k-p Method
V. L. Bonch-Bruevich, Effect of Heavy Doping on the Semiconductor Band Structure
D. Long, Energy Band Structures of Mixed Crystals of III–V Compounds
L. M. Roth and P. N. Argyres, Magnetic Quantum Effects
S. M. Puri and T. H. Geballe, Thermomagnetic Effects in the Quantum Region
W. M. Becker, Band Characteristics near Principal Minima from Magnetoresistance
E. H. Putley, Freeze-Out Effects, Hot Electron Effects, and Submillimeter Photoconductivity in InSb
H. Weiss, Magnetoresistance
B. Ancker-Johnson, Plasma in Semiconductors and Semimetals

Volume 2 **Physics of III–V Compounds**

M. G. Holland, Thermal Conductivity
S. I. Novkova, Thermal Expansion
U. Piesbergen, Heat Capacity and Debye Temperatures
G. Giesecke, Lattice Constants
J. R. Drabble, Elastic Properties
A. U. Mac Rae and G. W. Gobeli, Low Energy Electron Diffraction Studies
R. Lee Mieher, Nuclear Magnetic Resonance
B. Goldstein, Electron Paramagnetic Resonance
T. S. Moss, Photoconduction in III–V Compounds
E. Antoncik and J. Tauc, Quantum Efficiency of the Internal Photoelectric Effect in InSb
G. W. Gobeli and I. G. Allen, Photoelectric Threshold and Work Function
P. S. Pershan, Nonlinear Optics in III–V Compounds
M. Gershenzon, Radiative Recombination in the III–V Compounds
F. Stern, Stimulated Emission in Semiconductors

Volume 3 Optical of Properties III–V Compounds

M. Hass, Lattice Reflection
W. G. Spitzer, Multiphonon Lattice Absorption
D. L. Stierwalt and R. F. Potter, Emittance Studies
H. R. Philipp and H. Ehrenveich, Ultraviolet Optical Properties
M. Cardona, Optical Absorption above the Fundamental Edge
E. J. Johnson, Absorption near the Fundamental Edge
J. O. Dimmock, Introduction to the Theory of Exciton States in Semiconductors
B. Lax and J. G. Mavroides, Interband Magnetooptical Effects
H. Y. Fan, Effects of Free Carries on Optical Properties
E. D. Palik and G. B. Wright, Free-Carrier Magnetooptical Effects
R. H. Bube, Photoelectronic Analysis
B. O. Seraphin and H. E. Bennett, Optical Constants

Volume 4 Physics of III–V Compounds

N. A. Goryunova, A. S. Borschevskii, and D. N. Tretiakov, Hardness
N. N. Sirota, Heats of Formation and Temperatures and Heats of Fusion of Compounds $A^{III}B^{V}$
D. L. Kendall, Diffusion
A. G. Chynoweth, Charge Multiplication Phenomena
R. W. Keyes, The Effects of Hydrostatic Pressure on the Properties of III–V Semiconductors
L. W. Aukerman, Radiation Effects
N. A. Goryunova, F. P. Kesamanly, and D. N. Nasledov, Phenomena in Solid Solutions
R. T. Bate, Electrical Properties of Nonuniform Crystals

Volume 5 Infrared Detectors

H. Levinstein, Characterization of Infrared Detectors
P. W. Kruse, Indium Antimonide Photoconductive and Photoelectromagnetic Detectors
M. B. Prince, Narrowband Self-Filtering Detectors
I. Melngalis and T. C. Harman, Single-Crystal Lead-Tin Chalcogenides
D. Long and J. L. Schmidt, Mercury-Cadmium Telluride and Closely Related Alloys
E. H. Putley, The Pyroelectric Detector
N. B. Stevens, Radiation Thermopiles
R. J. Keyes and T. M. Quist, Low Level Coherent and Incoherent Detection in the Infrared
M. C. Teich, Coherent Detection in the Infrared
F. R. Arams, E. W. Sard, B. J. Peyton, and F. P. Pace, Infrared Heterodyne Detection with Gigahertz IF Response
H. S. Sommers, Jr., Macrowave-Based Photoconductive Detector
R. Sehr and R. Zuleeg, Imaging and Display

Volume 6 Injection Phenomena

M. A. Lampert and R. B. Schilling, Current Injection in Solids: The Regional Approximation Method
R. Williams, Injection by Internal Photoemission
A. M. Barnett, Current Filament Formation

R. Baron and J. W. Mayer, Double Injection in Semiconductors
W. Ruppel, The Photoconductor-Metal Contact

Volume 7   Application and Devices
Part A

J. A. Copeland and S. Knight, Applications Utilizing Bulk Negative Resistance
F. A. Padovani, The Voltage-Current Characteristics of Metal-Semiconductor Contacts
P. L. Hower, W. W. Hooper, B. R. Cairns, R. D. Fairman, and D. A. Tremere, The GaAs Field-Effect Transistor
M. H. White, MOS Transistors
G. R. Antell, Gallium Arsenide Transistors
T. L. Tansley, Heterojunction Properties

Part B

T. Misawa, IMPATT Diodes
H. C. Okean, Tunnel Diodes
R. B. Campbell and Hung-Chi Chang, Silicon Junction Carbide Devices
R. E. Enstrom, H. Kressel, and L. Krassner, High-Temperature Power Rectifiers of $GaAs_{1-x}P_x$

Volume 8   Transport and Optical Phenomena

R. J. Stirn, Band Structure and Galvanomagnetic Effects in III–V Compounds with Indirect Band Gaps
R. W. Ure, Jr., Thermoelectric Effects in III–V Compounds
H. Piller, Faraday Rotation
H. Barry Bebb and E. W. Williams, Photoluminescence I: Theory
E. W. Williams and H. Barry Bebb, Photoluminescence II: Gallium Arsenide

Volume 9   Modulation Techniques

B. O. Seraphin, Electroreflectance
R. L. Aggarwal, Modulated Interband Magnetooptics
D. F. Blossey and Paul Handler, Electroabsorption
B. Batz, Thermal and Wavelength Modulation Spectroscopy
I. Balslev, Piezopptical Effects
D. E. Aspnes and N. Bottka, Electric-Field Effects on the Dielectric Function of Semiconductors and Insulators

Volume 10   Transport Phenomena

R. L. Rhode, Low-Field Electron Transport
J. D. Wiley, Mobility of Holes in III–V Compounds
C. M. Wolfe and G. E. Stillman, Apparent Mobility Enhancement in Inhomogeneous Crystals
R. L. Petersen, The Magnetophonon Effect

Volume 11 Solar Cells

H. J. Hovel, Introduction; Carrier Collection, Spectral Response, and Photocurrent; Solar Cell Electrical Characteristics; Efficiency; Thickness; Other Solar Cell Devices; Radiation Effects; Temperature and Intensity; Solar Cell Technology

Volume 12 Infrared Detectors (II)

W. L. Eiseman, J. D. Merriam, and R. F. Potter, Operational Characteristics of Infrared Photodetectors
P. R. Bratt, Impurity Germanium and Silicon Infrared Detectors
E. H. Putley, InSb Submillimeter Photoconductive Detectors
G. E. Stillman, C. M. Wolfe, and J. O. Dimmock, Far-Infrared Photoconductivity in High Purity GaAs
G. E. Stillman and C. M. Wolfe, Avalanche Photodiodes
P. L. Richards, The Josephson Junction as a Detector of Microwave and Far-Infrared Radiation
E. H. Putley, The Pyroelectric Detector—An Update

Volume 13 Cadmium Telluride

K. Zanio, Materials Preparations; Physics; Defects; Applications

Volume 14 Lasers, Junctions, Transport

N. Holonyak, Jr. and M. H. Lee, Photopumped III–V Semiconductor Lasers
H. Kressel and J. K. Butler, Heterojunction Laser Diodes
A Van der Ziel, Space-Charge-Limited Solid-State Diodes
P. J. Price, Monte Carlo Calculation of Electron Transport in Solids

Volume 15 Contacts, Junctions, Emitters

B. L. Sharma, Ohmic Contacts to III–V Compounds Semiconductors
A. Nussbaum, The Theory of Semiconducting Junctions
J. S. Escher, NEA Semiconductor Photoemitters

Volume 16 Defects, (HgCd)Se, (HgCd)Te

H. Kressel, The Effect of Crystal Defects on Optoelectronic Devices
C. R. Whitsett, J. G. Broerman, and C. J. Summers, Crystal Growth and Properties of $Hg_{1-x}Cd_xSe$ alloys
M. H. Weiler, Magnetooptical Properties of $Hg_{1-x}Cd_xTe$ Alloys
P. W. Kruse and J. G. Ready, Nonlinear Optical Effects in $Hg_{1-x}Cd_xTe$

Volume 17 CW Processing of Silicon and Other Semiconductors

J. F. Gibbons, Beam Processing of Silicon
A. Lietoila, R. B. Gold, J. F. Gibbons, and L. A. Christel, Temperature Distributions and Solid Phase Reaction Rates Produced by Scanning CW Beams

A. Leitoila and J. F. Gibbons, Applications of CW Beam Processing to Ion Implanted Crystalline Silicon
N. M. Johnson, Electronic Defects in CW Transient Thermal Processed Silicon
K. F. Lee, T. J. Stultz, and J. F. Gibbons, Beam Recrystallized Polycrystalline Silicon: Properties, Applications, and Techniques
T. Shibata, A. Wakita, T. W. Sigmon, and J. F. Gibbons, Metal-Silicon Reactions and Silicide
Y. I. Nissim and J. F. Gibbons, CW Beam Processing of Gallium Arsenide

Volume 18 Mercury Cadmium Telluride

P. W. Kruse, The Emergence of $(Hg_{1-x}Cd_x)Te$ as a Modern Infrared Sensitive Material
H. E. Hirsch, S. C. Liang, and A. G. White, Preparation of High-Purity Cadmium, Mercury, and Tellurium
W. F. H. Micklethwaite, The Crystal Growth of Cadmium Mercury Telluride
P. E. Petersen, Auger Recombination in Mercury Cadmium Telluride
R. M. Broudy and V. J. Mazurczyck, (HgCd)Te Photoconductive Detectors
M. B. Reine, A. K. Soad, and T. J. Tredwell, Photovoltaic Infrared Detectors
M. A. Kinch, Metal-Insulator-Semiconductor Infrared Detectors

Volume 19 Deep Levels, GaAs, Alloys, Photochemistry

G. F. Neumark and K. Kosai, Deep Levels in Wide Band-Gap III–V Semiconductors
D. C. Look, The Electrical and Photoelectronic Properties of Semi-Insulating GaAs
R. F. Brebrick, Ching-Hua Su, and Pok-Kai Liao, Associated Solution Model for Ga-In-Sb and Hg-Cd-Te
Y. Ya. Gurevich and Y. V. Pleskon, Photoelectrochemistry of Semiconductors

Volume 20 Semi-Insulating GaAs

R. N. Thomas, H. M. Hobgood, G. W. Eldridge, D. L. Barrett, T. T. Braggins, L. B. Ta, and S. K. Wang, High-Purity LEC Growth and Direct Implantation of GaAs for Monolithic Microwave Circuits
C. A. Stolte, Ion Implantation and Materials for GaAs Integrated Circuits
C. G. Kirkpatrick, R. T. Chen, D. E. Holmes, P. M. Asbeck, K. R. Elliott, R. D. Fairman, and J. R. Oliver, LEC GaAs for Integrated Circuit Applications
J. S. Blakemore and S. Rahimi, Models for Mid-Gap Centers in Gallium Arsenide

Volume 21 Hydrogenated Amorphous Silicon
Part A

J. I. Pankove, Introduction
M. Hirose, Glow Discharge; Chemical Vapor Deposition
Y. Uchida, di Glow Discharge
T. D. Moustakas, Sputtering
I. Yamada, Ionized-Cluster Beam Deposition
B. A. Scott, Homogeneous Chemical Vapor Deposition
F. J. Kampas, Chemical Reactions in Plasma Deposition
P. A. Longeway, Plasma Kinetics

H. A. Weakliem, Diagnostics of Silane Glow Discharges Using Probes and Mass Spectroscopy
L. Gluttman, Relation between the Atomic and the Electronic Structures
A. Chenevas-Paule, Experiment Determination of Structure
S. Minomura, Pressure Effects on the Local Atomic Structure
D. Adler, Defects and Density of Localized States

Part B

J. I. Pankove, Introduction
G. D. Cody, The Optical Absorption Edge of a-Si: H
N. M. Amer and W. B. Jackson, Optical Properties of Defect States in a-Si: H
P. J. Zanzucchi, The Vibrational Spectra of a-Si: H
Y. Hamakawa, Electroreflectance and Electroabsorption
J. S. Lannin, Raman Scattering of Amorphous Si, Ge, and Their Alloys
R. A. Street, Luminescence in a-Si: H
R. S. Crandall, Photoconductivity
J. Tauc, Time-Resolved Spectroscopy of Electronic Relaxation Processes
P. E. Vanier, IR-Induced Quenching and Enhancement of Photoconductivity and Photo luminescence
H. Schade, Irradiation-Induced Metastable Effects
L. Ley, Photoelectron Emission Studies

Part C

J. I. Pankove, Introduction
J. D. Cohen, Density of States from Junction Measurements in Hydrogenated Amorphous Silicon
P. C. Taylor, Magnetic Resonance Measurements in a-Si: H
K. Morigaki, Optically Detected Magnetic Resonance
J. Dresner, Carrier Mobility in a-Si: H
T. Tiedje, Information about band-Tail States from Time-of-Flight Experiments
A. R. Moore, Diffusion Length in Undoped a-Si: H
W. Beyer and J. Overhof, Doping Effects in a-Si: H
H. Fritzche, Electronic Properties of Surfaces in a-Si: H
C. R. Wronski, The Staebler-Wronski Effect
R. J. Nemanich, Schottky Barriers on a-Si: H
B. Abeles and T. Tiedje, Amorphous Semiconductor Superlattices

Part D

J. I. Pankove, Introduction
D. E. Carlson, Solar Cells
G. A. Swartz, Closed-Form Solution of I–V Characteristic for a a-Si: H Solar Cells
I. Shimizu, Electrophotography
S. Ishioka, Image Pickup Tubes

P. G. LeComber and W. E. Spear, The Development of the a-Si: H Field-Effect Transistor and Its Possible Applications
D. G. Ast, a-Si: H FET-Addressed LCD Panel
S. Kaneko, Solid-State Image Sensor
M. Matsumura, Charge-Coupled Devices
M. A. Bosch, Optical Recording
A. D'Amico and G. Fortunato, Ambient Sensors
H. Kukimoto, Amorphous Light-Emitting Devices
R. J. Phelan, Jr., Fast Detectors and Modulators
J. I. Pankove, Hybrid Structures
P. G. LeComber, A. E. Owen, W. E. Spear, J. Hajto, and W. K. Choi, Electronic Switching in Amorphous Silicon Junction Devices

Volume 22 Lightwave Communications Technology
Part A

K. Nakajima, The Liquid-Phase Epitaxial Growth of InGaAsP
W. T. Tsang, Molecular Beam Epitaxy for III–V Compound Semiconductors
G. B. Stringfellow, Organometallic Vapor-Phase Epitaxial Growth of III–V Semiconductors
G. Beuchet, Halide and Chloride Transport Vapor-Phase Deposition of InGaAsP and GaAs
M. Razeghi, Low-Pressure Metallo-Organic Chemical Vapor Deposition of $Ga_x In_{1-x} AsP_{1-y}$ Alloys
P. M. Petroff, Defects in III–V Compound Semiconductors

Part B

J. P. van der Ziel, Mode Locking of Semiconductor Lasers
K. Y. Lau and A. Yariv, High-Frequency Current Modulation of Semiconductor Injection Lasers
C. H. Henry, Special Properties of Semiconductor Lasers
Y. Suematsu, K. Kishino, S. Arai, and F. Koyama, Dynamic Single-Mode Semiconductor Lasers with a Distributed Reflector
W. T. Tsang, The Cleaved-Coupled-Cavity (C^3) Laser

Part C

R. J. Nelson and N. K. Dutta, Review of InGaAsP InP Laser Structures and Comparison of Their Performance
N. Chinone and M. Nakamura, Mode-Stabilized Semiconductor Lasers for 0.7–0.8- and 1.1–1.6-μm Regions
Y. Horikoshi, Semiconductor Lasers with Wavelengths Exceeding 2 μm
B. A. Dean and M. Dixon, The Functional Reliability of Semiconductor Lasers as Optical Transmitters
R. H. Saul, T. P. Lee, and C. A. Burus, Light-Emitting Device Design
C. L. Zipfel, Light-Emitting Diode-Reliability
T. P. Lee and T. Li, LED-Based Multimode Lightwave Systems
K. Ogawa, Semiconductor Noise-Mode Partition Noise

Part D

F. Capasso, The Physics of Avalanche Photodiodes
T. P. Pearsall and M. A. Pollack, Compound Semiconductor Photodiodes
T. Kaneda, Silicon and Germanium Avalanche Photodiodes
S. R. Forrest, Sensitivity of Avalanche Photodetector Receivers for High-Bit-Rate Long-Wavelength Optical Communication Systems
J. C. Campbell, Phototransistors for Lightwave Communications

Part E

S. Wang, Principles and Characteristics of Integrable Active and Passive Optical Devices
S. Margalit and A. Yariv, Integrated Electronic and Photonic Devices
T. Mukai, Y. Yamamoto, and T. Kimura, Optical Amplification by Semiconductor Lasers

Volume 23 Pulsed Laser Processing of Semiconductors

R. F. Wood, C. W. White, and R. T. Young, Laser Processing of Semiconductors: An Overview
C. W. White, Segregation, Solute Trapping, and Supersaturated Alloys
G. E. Jellison, Jr., Optical and Electrical Properties of Pulsed Laser-Annealed Silicon
R. F. Wood and G. E. Jellison, Jr., Melting Model of Pulsed Laser Processing
R. F. Wood and F. W. Young, Jr., Nonequilibrium Solidification Following Pulsed Laser Melting
D. H. Lowndes and G. E. Jellison, Jr., Time-Resolved Measurement During Pulsed Laser Irradiation of Silicon
D. M. Zebner, Surface Studies of Pulsed Laser Irradiated Semiconductors
D. H. Lowndes, Pulsed Beam Processing of Gallium Arsenide
R. B. James, Pulsed CO_2 Laser Annealing of Semiconductors
R. T. Young and R. F. Wood, Applications of Pulsed Laser Processing

Volume 24 Applications of Multiquantum Wells, Selective Doping, and Superlattices

C. Weisbuch, Fundamental Properties of III–V Semiconductor Two-Dimensional Quantized Structures: The Basis for Optical and Electronic Device Applications
H. Morkoc and H. Unlu, Factors Affecting the Performance of (Al, Ga)As/GaAs and (Al, Ga)As/InGaAs Modulation-Doped Field-Effect Transistors: Microwave and Digital Applications
N. T. Linh, Two-Dimensional Electron Gas FETs: Microwave Applications
M. Abe et al., Ultra-High-Speed HEMT Integrated Circuits
D. S. Chemla, D. A. B. Miller, and P. W. Smith, Nonlinear Optical Properties of Multiple Quantum Well Structures for Optical Signal Processing
F. Capasso, Graded-Gap and Superlattice Devices by Band-Gap Engineering
W. T. Tsang, Quantum Confinement Heterostructure Semiconductor Lasers
G. C. Osbourn et al., Principles and Applications of Semiconductor Strained-Layer Superlattices

Volume 25 Diluted Magnetic Semiconductors

W. Giriat and J. K. Furdyna, Crystal Structure, Composition, and Materials Preparation of Diluted Magnetic Semiconductors
W. M. Becker, Band Structure and Optical Properties of Wide-Gap $A_{1-x}^{II}Mn_xB_{IV}$ Alloys at Zero Magnetic Field
S. Oseroff and P. H. Keesom, Magnetic Properties: Macroscopic Studies
T. Giebultowicz and T. M. Holden, Neutron Scattering Studies of the Magnetic Structure and Dynamics of Diluted Magnetic Semiconductors
J. Kossut, Band Structure and Quantum Transport Phenomena in Narrow-Gap Diluted Magnetic Semiconductors
C. Riquaux, Magnetooptical Properties of Large-Gap Diluted Magnetic Semiconductors
J. A. Gaj, Magnetooptical Properties of Large-Gap Diluted Magnetic Semiconductors
J. Mycielski, Shallow Acceptors in Diluted Magnetic Semiconductors: Splitting, Boil-off, Giant Negative Magnetoresistance
A. K. Ramadas and R. Rodriquez, Raman Scattering in Diluted Magnetic Semiconductors
P. A. Wolff, Theory of Bound Magnetic Polarons in Semimagnetic Semiconductors

Volume 26 III–V Compound Semiconductors and Semiconductor Properties of Superionic Materials

Z. Yuanxi, III–V Compounds
H. V. Winston, A. T. Hunter, H. Kimura, and R. E. Lee, InAs-Alloyed GaAs Substrates for Direct Implantation
P. K. Bhattacharya and S. Dhar, Deep Levels in III–V Compound Semiconductors Grown by MBE
Y. Ya. Gurevich and A. K. Ivanov-Shits, Semiconductor Properties of Supersonic Materials

Volume 27 High Conducting Quasi-One-Dimensional Organic Crystals

E. M. Conwell, Introduction to Highly Conducting Quasi-One-Dimensional Organic Crystals
I. A. Howard, A Reference Guide to the Conducting Quasi-One-Dimensional Organic Molecular Crystals
J. P. Pouquet, Structural Instabilities
E. M. Conwell, Transport Properties
C. S. Jacobsen, Optical Properties
J. C. Scott, Magnetic Properties
L. Zuppiroli, Irradiation Effects: Perfect Crystals and Real Crystals

Volume 28 Measurement of High-Speed Signals in Solid State Devices

J. Frey and D. Ioannou, Materials and Devices for High-Speed and Optoelectronic Applications
H. Schumacher and E. Strid, Electronic Wafer Probing Techniques
D. H. Auston, Picosecond Photoconductivity: High-Speed Measurements of Devices and Materials
J. A. Valdmanis, Electro-Optic Measurement Techniques for Picosecond Materials, Devices, and Integrated Circuits.
J. M. Wiesenfeld and R. K. Jain, Direct Optical Probing of Integrated Circuits and High-Speed Devices
G. Plows, Electron-Beam Probing
A. M. Weiner and R. B. Marcus, Photoemissive Probing

Volume 29 Very High Speed Integrated Circuits: Gallium Arsenide LSI

M. Kuzuhara and T. Nazaki, Active Layer Formation by Ion Implantation
H. Hasimoto, Focused Ion Beam Implantation Technology
T. Nozaki and A. Higashisaka, Device Fabrication Process Technology
M. Ino and T. Takada, GaAs LSI Circuit Design
M. Hirayama, M. Ohmori, and K. Yamasaki, GaAs LSI Fabrication and Performance

Volume 30 Very High Speed Integrated Circuits: Heterostructure

H. Watanabe, T. Mizutani, and A. Usui, Fundamentals of Epitaxial Growth and Atomic Layer Epitaxy
S. Hiyamizu, Characteristics of Two-Dimensional Electron Gas in III–V Compound Heterostructures Grown by MBE
T. Nakanisi, Metalorganic Vapor Phase Epitaxy for High-Quality Active Layers
T. Nimura, High Electron Mobility Transistor and LSI Applications
T. Sugeta and T. Ishibashi, Hetero-Bipolar Transistor and LSI Application
H. Matsueda, T. Tanaka, and M. Nakamura, Optoelectronic Integrated Circuits

Volume 31 Indium Phosphide: Crystal Growth and Characterization

J. P. Farges, Growth of Discoloration-free InP
M. J. McCollum and G. E. Stillman, High Purity InP Grown by Hydride Vapor Phase Epitaxy
T. Inada and T. Fukuda, Direct Synthesis and Growth of Indium Phosphide by the Liquid Phosphorous Encapsulated Czochralski Method
O. Oda, K. Katagiri, K. Shinohara, S. Katsura, Y. Takahashi, K. Kainosho, K. Kohiro, and R. Hirano, InP Crystal Growth, Substrate Preparation and Evaluation
K. Tada, M. Tatsumi, M. Morioka, T. Araki, and T. Kawase, InP Substrates: Production and Quality Control
M. Razeghi, LP-MOCVD Growth, Characterization, and Application of InP Material
T. A. Kennedy and P. J. Lin-Chung, Stoichiometric Defects in InP

Volume 32 Strained-Layer Superlattices: Physics

T. P. Pearsall, Strained-Layer Superlattices
F. H. Pollack, Effects of Homogeneous Strain on the Electronic and Vibrational Levels in Semiconductors
J. Y. Marzin, J. M. Gerard, P. Voisin, and J. A. Brum, Optical Studies of Strained III–V Heterolayers
R. People and S. A. Jackson, Structurally Induced States from Strain and Confinement
M. Jaros, Microscopic Phenomena in Ordered Superlattices

Volume 33 Strained-Layer Superlattices: Materials Science and Technology

R. Hull and J. C. Bean, Principles and Concepts of Strained-Layer Epitaxy
W. J. Schaff, P. J. Tasker, M. C. Foisy, and L. F. Eastman, Device Applications of Strained-Layer Epitaxy

S. T. Picraux, B. L. Doyle, and J. Y. Tsao, Structure and Characterization of Strained-Layer Superlattices
E. Kasper and F. Schaffer, Group IV Compounds
D. L. Martin, Molecular Beam Epitaxy of IV–VI Compounds Heterojunction
R. L. Gunshor, L. A. Kolodziejski, A. V. Nurmikko, and N. Otsuka, Molecular Beam Epitaxy of II–VI Semiconductor Microstructures

Volume 34 Hydrogen in Semiconductors

J. I. Pankove and N. M. Johnson, Introduction to Hydrogen in Semiconductors
C. H. Seager, Hydrogenation Methods
J. I. Pankove, Hydrogenation of Defects in Crystalline Silicon
J. W. Corbett, P. Deák, U. V. Desnica, and S. J. Pearton, Hydrogen Passivation of Damage Centers in Semiconductors
S. J. Pearton, Neutralization of Deep Levels in Silicon
J. I. Pankove, Neutralization of Shallow Acceptors in Silicon
N. M. Johnson, Neutralization of Donor Dopants and Formation of Hydrogen-Induced Defects in n-Type Silicon
M. Stavola and S. J. Pearton, Vibrational Spectroscopy of Hydrogen-Related Defects in Silicon
A. D. Marwick, Hydrogen in Semiconductors: Ion Beam Techniques
C. Herring and N. M. Johnson, Hydrogen Migration and Solubility in Silicon
E. E. Haller, Hydrogen-Related Phenomena in Crystalline Germanium
J. Kakalios, Hydrogen Diffusion in Amorphous Silicon
J. Chevalier, B. Clerjaud, and B. Pajot, Neutralization of Defects and Dopants in III–V Semiconductors
G. G. DeLeo and W. B. Fowler, Computational Studies of Hydrogen-Containing Complexes in Semiconductors
R. F. Kiefl and T. L. Estle, Muonium in Semiconductors
C. G. Van de Walle, Theory of Isolated Interstitial Hydrogen and Muonium in Crystalline Semiconductors

Volume 35 Nanostructured Systems

M. Reed, Introduction
H. van Houten, C. W. J. Beenakker, and B. J. van Wees, Quantum Point Contacts
G. Timp, When Does a Wire Become an Electron Waveguide?
M. Büttiker, The Quantum Hall Effects in Open Conductors
W. Hansen, J. P. Kotthaus, and U. Merkt, Electrons in Laterally Periodic Nanostructures

Volume 36 The Spectroscopy of Semiconductors

D. Heiman, Spectroscopy of Semiconductors at Low Temperatures and High Magnetic Fields
A. V. Nurmikko, Transient Spectroscopy by Ultrashort Laser Pulse Techniques
A. K. Ramdas and S. Rodriguez, Piezospectroscopy of Semiconductors
O. J. Glembocki and B. V. Shanabrook, Photoreflectance Spectroscopy of Microstructures
D. G. Seiler, C. L. Littler, and M. H. Wiler, One- and Two-Photon Magneto-Optical Spectroscopy of InSb and $Hg_{1-x}Cd_x$Te

Volume 37 The Mechanical Properties of Semiconductors

A.-B. Chen, A. Sher and W. T. Yost, Elastic Constants and Related Properties of Semiconductor Compounds and Their Alloys
D. R. Clarke, Fracture of Silicon and Other Semiconductors
H. Siethoff, The Plasticity of Elemental and Compound Semiconductors
S. Guruswamy, K. T. Faber and J. P. Hirth, Mechanical Behavior of Compound Semiconductors
S. Mahajan, Deformation Behavior of Compound Semiconductors
J. P. Hirth, Injection of Dislocations into Strained Multilayer Structures
D. Kendall, C. B. Fleddermann, and K. J. Malloy, Critical Technologies for the Micromachining of Silicon
I. Matsuba and K. Mokuya, Processing and Semiconductor Thermoelastic Behavior

Volume 38 Imperfections in III/V Materials

U. Scherz and M. Scheffler, Density-Functional Theory of sp-Bonded Defects in III/V Semiconductors
M. Kaminska and E. R. Weber, E12 Defect in GaAs
D. C. Look, Defects Relevant for Compensation in Semi-Insulating GaAs
R. C. Newman, Local Vibrational Mode Spectroscopy of Defects in III/V Compounds
A. M. Hennel, Transition Metals in III/V Compounds
K. J. Malloy and K. Khachaturyan, DX and Related Defects in Semiconductors
V. Swaminathan and A. S. Jordan, Dislocations in III/V Compounds
K. W. Nauka, Deep Level Defects in the Epitaxial III/V Materials

Volume 39 Minority Carriers in III–V Semiconductors: Physics and Applications

N. K. Dutta, Radiative Transitions in GaAs and Other III–V Compounds
R. K. Ahrenkiel, Minority-Carrier Lifetime in III–V Semiconductors
T. Furuta, High Field Minority Electron Transport in p-GaAs
M. S. Lundstrom, Minority-Carrier Transport in III–V Semiconductors
R. A. Abram, Effects of Heavy Doping and High Excitation on the Band Structure of GaAs
D. Yevick and W. Bardyszewski, An Introduction to Non-Equilibrium Many-Body Analyses of Optical Processes in III–V Semiconductors

Volume 40 Epitaxial Microstructures

E. F. Schubert, Delta-Doping of Semiconductors: Electronic, Optical, and Structural Properties of Materials and Devices
A. Gossard, M. Sundaram, and P. Hopkins, Wide Graded Potential Wells
P. Petroff, Direct Growth of Nanometer-Size Quantum Wire Superlattices
E. Kapon, Lateral Patterning of Quantum Well Heterostructures by Growth of Nonplanar Substrates
H. Temkin, D. Gershoni, and M. Panish, Optical Properties of $Ga_{1-x}In_xAs/InP$ Quantum Wells

Volume 41   High Speed Heterostructure Devices

F. Capasso, F. Beltram, S. Sen, A. Pahlevi, and A. Y. Cho, Quantum Electron Devices: Physics and Applications
P. Solomon, D. J. Frank, S. L. Wright, and F. Canora, GaAs-Gate Semiconductor-Insulator-Semiconductor FET
M. H. Hashemi and U. K. Mishra, Unipolar InP-Based Transistors
R. Kiehl, Complementary Heterostructure FET Integrated Circuits
T. Ishibashi, GaAs-Based and InP-Based Heterostructure Bipolar Transistors
H. C. Liu and T. C. L. G. Sollner, High-Frequency-Tunneling Devices
H. Ohnishi, T. More, M. Takatsu, K. Imamura, and N. Yokoyama, Resonant-Tunneling Hot-Electron Transistors and Circuits

Volume 42   Oxygen in Silicon

F. Shimura, Introduction to Oxygen in Silicon
W. Lin, The Incorporation of Oxygen into Silicon Crystals
T. J. Schaffner and D. K. Schroder, Characterization Techniques for Oxygen in Silicon
W. M. Bullis, Oxygen Concentration Measurement
S. M. Hu, Intrinsic Point Defects in Silicon
B. Pajot, Some Atomic Configurations of Oxygen
J. Michel and L. C. Kimerling, Electical Properties of Oxygen in Silicon
R. C. Newman and R. Jones, Diffusion of Oxygen in Silicon
T. Y. Tan and W. J. Taylor, Mechanisms of Oxygen Precipitation: Some Quantitative Aspects
M. Schrems, Simulation of Oxygen Precipitation
K. Simino and I. Yonenaga, Oxygen Effect on Mechanical Properties
W. Bergholz, Grown-in and Process-Induced Effects
F. Shimura, Intrinsic/Internal Gettering
H. Tsuya, Oxygen Effect on Electronic Device Performance

Volume 43   Semiconductors for Room Temperature Nuclear Detector Applications

R. B. James and T. E. Schlesinger, Introduction and Overview
L. S. Darken and C. E. Cox, High-Purity Germanium Detectors
A. Burger, D. Nason, L. Van den Berg, and M. Schieber, Growth of Mercuric Iodide
X. J. Bao, T. E. Schlesinger, and R. B. James, Electrical Properties of Mercuric Iodide
X. J. Bao, R. B. James, and T. E. Schlesinger, Optical Properties of Red Mercuric Iodide
M. Hage-Ali and P. Siffert, Growth Methods of CdTe Nuclear Detector Materials
M. Hage-Ali and P. Siffert, Characterization of CdTe Nuclear Detector Materials
M. Hage-Ali and P. Siffert, CdTe Nuclear Detectors and Applications
R. B. James, T. E. Schlesinger, J. Lund, and M. Schieber, $Cd_{1-x}Zn_xTe$ Spectrometers for Gamma and X-Ray Applications
D. S. McGregor, J. E. Kammeraad, Gallium Arsenide Radiation Detectors and Spectrometers
J. C. Lund, F. Olschner, and A. Burger, Lead Iodide
M. R. Squillante, and K. S. Shah, Other Materials: Status and Prospects
V. M. Gerrish, Characterization and Quantification of Detector Performance
J. S. Iwanczyk and B. E. Patt, Electronics for X-ray and Gamma Ray Spectrometers
M. Schieber, R. B. James, and T. E. Schlesinger, Summary and Remaining Issues for Room Temperature Radiation Spectrometers

Volume 44 II–IV Blue/Green Light Emitters: Device Physics and Epitaxial Growth

J. Han and R. L. Gunshor, MBE Growth and Electrical Properties of Wide Bandgap ZnSe-based II–VI Semiconductors
S. Fujita and S. Fujita, Growth and Characterization of ZnSe-based II–VI Semiconductors by MOVPE
E. Ho and L. A. Kolodziejski, Gaseous Source UHV Epitaxy Technologies for Wide Bandgap II–VI Semiconductors
C. G. Van de Walle, Doping of Wide-Band-Gap II–VI Compounds—Theory
R. Cingolani, Optical Properties of Excitons in ZnSe-Based Quantum Well Heterostructures
A. Ishibashi and A. V. Nurmikko, II–VI Diode Lasers: A Current View of Device Performance and Issues
S. Guha and J. Petruzello, Defects and Degradation in Wide-Gap II–VI-based Structures and Light Emitting Devices

Volume 45 Effect of Disorder and Defects in Ion-Implanted Semiconductors: Electrical and Physiochemical Characterization

H. Ryssel, Ion Implantation into Semiconductors: Historical Perspectives
You-Nian Wang and Teng-Cai Ma, Electronic Stopping Power for Energetic Ions in Solids
S. T. Nakagawa, Solid Effect on the Electronic Stopping of Crystalline Target and Application to Range Estimation
G. Müller, S. Kalbitzer and G. N. Greaves, Ion Beams in Amorphous Semiconductor Research
J. Boussey-Said, Sheet and Spreading Resistance Analysis of Ion Implanted and Annealed Semiconductors
M. L. Polignano and G. Queirolo, Studies of the Stripping Hall Effect in Ion-Implanted Silicon
J. Stoemenos, Transmission Electron Microscopy Analyses
R. Nipoti and M. Servidori, Rutherford Backscattering Studies of Ion Implanted Semiconductors
P. Zaumseil, X-ray Diffraction Techniques

Volume 46 Effect of Disorder and Defects in Ion-Implanted Semiconductors: Optical and Photothermal Characterization

M. Fried, T. Lohner and J. Gyulai, Ellipsometric Analysis
A. Seas and C. Christofides, Transmission and Reflection Spectroscopy on Ion Implanted Semiconductors
A. Othonos and C. Christofides, Photoluminescence and Raman Scattering of Ion Implanted Semiconductors. Influence of Annealing
C. Christofides, Photomodulated Thermoreflectance Investigation of Implanted Wafers. Annealing Kinetics of Defects
U. Zammit, Photothermal Deflection Spectroscopy Characterization of Ion-Implanted and Annealed Silicon Films
A. Mandelis, A. Budiman and M. Vargas, Photothermal Deep-Level Transient Spectroscopy of Impurities and Defects in Semiconductors
R. Kalish and S. Charbonneau, Ion Implantation into Quantum-Well Structures
A. M. Myasnikov and N. N. Gerasimenko, Ion Implantation and Thermal Annealing of III–V Compound Semiconducting Systems: Some Problems of III–V Narrow Gap Semiconductors

Volume 47 Uncooled Infrared Imaging Arrays and Systems

R. G. Buser and M. P. Tompsett, Historical Overview
P. W. Kruse, Principles of Uncooled Infrared Focal Plane Arrays
R. A. Wood, Monolithic Silicon Microbolometer Arrays
C. M. Hanson, Hybrid Pyroelectric-Ferroelectric Bolometer Arrays
D. L. Polla and J. R. Choi, Monolithic Pyroelectric Bolometer Arrays
N. Teranishi, Thermoelectric Uncooled Infrared Focal Plane Arrays
M. F. Tompsett, Pyroelectric Vidicon
T. W. Kenny, Tunneling Infrared Sensors
J. R. Vig, R. L. Filler and Y. Kim, Application of Quartz Microresonators to Uncooled Infrared Imaging Arrays
P. W. Kruse, Application of Uncooled Monolithic Thermoelectric Linear Arrays to Imaging Radiometers

Volume 48 High Brightness Light Emitting Diodes

G. B. Stringfellow, Materials Issues in High-Brightness Light-Emitting Diodes
M. G. Craford, Overview of Device issues in High-Brightness Light-Emitting Diodes
F. M. Steranka, AIGaAs Red Light Emitting Diodes
C. H. Chen, S. A. Stockman, M. J. Peanasky, and C. P. Kuo, OMVPE Growth of AlGaInP for High Efficiency Visible Light-Emitting Diodes
F. A. Kish and R. M. Fletcher, AlGaInP Light-Emitting Diodes
M. W. Hodapp, Applications for High Brightness Light-Emitting Diodes
I. Akasaki and H. Amano, Organometallic Vapor Epitaxy of GaN for High Brightness Blue Light Emitting Diodes
S. Nakamura, Group III–V Nitride Based Ultraviolet-Blue-Green-Yellow Light-Emitting Diodes and Laser Diodes

Volume 49 Light Emission in Silicon: from Physics to Devices

D. J. Lockwood, Light Emission in Silicon
G. Abstreiter, Band Gaps and Light Emission in Si/SiGe Atomic Layer Structures
T. G. Brown and D. G. Hall, Radiative Isoelectronic Impurities in Silicon and Silicon-Germanium Alloys and Superlattices
J. Michel, L. V. C. Assali, M. T. Morse, and L. C. Kimerling, Erbium in Silicon
Y. Kanemitsu, Silicon and Germanium Nanoparticles
P. M. Fauchet, Porous Silicon: Photoluminescence and Electroluminescent Devices
C. Delerue, G. Allan, and M. Lannoo, Theory of Radiative and Nonradiative Processes in Silicon Nanocrystallites
L. Brus, Silicon Polymers and Nanocrystals

Volume 50 Gallium Nitride (GaN)

J. I. Pankove and T. D. Moustakas, Introduction
S. P. DenBaars and S. Keller, Metalorganic Chemical Vapor Deposition (MOCVD) of Group III Nitrides
W. A. Bryden and T. J. Kistenmacher, Growth of Group III–A Nitrides by Reactive Sputtering
N. Newman, Thermochemistry of III–N Semiconductors
S. J. Pearton and R. J. Shul, Etching of III Nitrides

S. M. Bedair, Indium-based Nitride Compounds
A. Trampert, O. Brandt, and K. H. Ploog, Crystal Structure of Group III Nitrides
H. Morkoc, F. Hamdani, and A. Salvador, Electronic and Optical Properties of III–V Nitride based Quantum Wells and Superlattices
K. Doverspike and J. I. Pankove, Doping in the III-Nitrides
T. Suski and P. Perlin, High Pressure Studies of Defects and Impurities in Gallium Nitride
B. Monemar, Optical Properties of GaN
W. R. L. Lambrecht, Band Structure of the Group III Nitrides
N. E. Christensen and P. Perlin, Phonons and Phase Transitions in GaN
S. Nakamura, Applications of LEDs and LDs
I. Akasaki and H. Amano, Lasers
J. A. Cooper, Jr., Nonvolatile Random Access Memories in Wide Bandgap Semiconductors

Volume 51A Identification of Defects in Semiconductors

G. D. Watkins, EPR and ENDOR Studies of Defects in Semiconductors
J.-M. Spaeth, Magneto-Optical and Electrical Detection of Paramagnetic Resonance in Semiconductors
T. A. Kennedy and E. R. Glaser, Magnetic Resonance of Epitaxial Layers Detected by Photoluminescence
K. H. Chow, B. Hitti, and R. F. Kiefl, μSR on Muonium in Semiconductors and Its Relation to Hydrogen
K. Saarinen, P. Hautojärvi, and C. Corbel, Positron Annihilation Spectroscopy of Defects in Semiconductors
R. Jones and P. R. Briddon, The Ab Initio Cluster Method and the Dynamics of Defects in Semiconductors

Volume 51B Identification of Defects in Semiconductors

G. Davies, Optical Measurements of Point Defects
P. M. Mooney, Defect Identification Using Capacitance Spectroscopy
M. Stavola, Vibrational Spectroscopy of Light Element Impurities in Semiconductors
P. Schwander, W. D. Rau, C. Kisielowski, M. Gribelyuk, and A. Ourmazd, Defect Processes in Semiconductors Studied at the Atomic Level by Transmission Electron Microscopy
N. D. Jager and E. R. Weber, Scanning Tunneling Microscopy of Defects in Semiconductors

Volume 52 SiC Materials and Devices

K. Järrendahl and R. F. Davis, Materials Properties and Characterization of SiC
V. A. Dmitriev and M. G. Spencer, SiC Fabrication Technology: Growth and Doping
V. Saxena and A. J. Steckl, Building Blocks for SiC Devices: Ohmic Contacts, Schottky Contacts, and p-n Junctions
M. S. Shur, SiC Transistors
C. D. Brandt, R. C. Clarke, R. R. Siergiej, J. B. Casady, A. W. Morse, S. Sriram, and A. K. Agarwal, SiC for Applications in High-Power Electronics
R. J. Trew, SiC Microwave Devices

J. Edmond, H. Kong, G. Negley, M. Leonard, K. Doverspike, W. Weeks, A. Suvorov, D. Waltz, and C. Carter, Jr., SiC-Based UV Photodiodes and Light-Emitting Diodes
H. Morkoç., Beyond Silicon Carbide! III–V Nitride-Based Heterostructures and Devices

Volume 53 Cumulative Subject and Author Index Including Tables of Contents for Volume 1–50

Volume 54 High Pressure in Semiconductor Physics I

W. Paul, High Pressure in Semiconductor Physics: A Historical Overview
N. E. Christensen, Electronic Structure Calculations for Semiconductors under Pressure
R. J. Neimes and M. I. McMahon, Structural Transitions in the Group IV, III–V and II–VI Semiconductors Under Pressure
A. R. Goni and K. Syassen, Optical Properties of Semiconductors Under Pressure
P. Trautman, M. Baj, and J. M. Baranowski, Hydrostatic Pressure and Uniaxial Stress in Investigations of the EL2 Defect in GaAs
M. Li and P. Y. Yu, High-Pressure Study of DX Centers Using Capacitance Techniques
T. Suski, Spatial Correlations of Impurity Charges in Doped Semiconductors
N. Kuroda, Pressure Effects on the Electronic Properties of Diluted Magnetic Semiconductors

Volume 55 High Pressure in Semiconductor Physics II

D. K. Maude and J. C. Portal, Parallel Transport in Low-Dimensional Semiconductor Structures
P. C. Klipstein, Tunneling Under Pressure: High-Pressure Studies of Vertical Transport in Semiconductor Heterostructures
E. Anastassakis and M. Cardona, Phonons, Strains, and Pressure in Semiconductors
F. H. Pollak, Effects of External Uniaxial Stress on the Optical Properties of Semiconductors and Semiconductor Microstructures
A. R. Adams, M. Silver, and J. Allam, Semiconductor Optoelectronic Devices
S. Porowski and I. Grzegory, The Application of High Nitrogen Pressure in the Physics and Technology of III–N Compounds
M. Yousuf, Diamond Anvil Cells in High Pressure Studies of Semiconductors

Volume 56 Germanium Silicon: Physics and Materials

J. C. Bean, Growth Techniques and Procedures
D. E. Savage, F. Liu, V. Zielasek, and M. G. Lagally, Fundamental Crystal Growth Mechanisms
R. Hull, Misfit Strain Accommodation in SiGe Heterostructures
M. J. Shaw and M. Jaros, Fundamental Physics of Strained Layer GeSi: Quo Vadis?
F. Cerdeira, Optical Properties
S. A. Ringel and P. N. Grillot, Electronic Properties and Deep Levels in Germanium-Silicon
J. C. Campbell, Optoelectronics in Silicon and Germanium Silicon
K. Eberl, K. Brunner, and O. G. Schmidt, $Si_{1-y}C_y$ and $Si_{1-x-y}Ge_xC_y$ Alloy Layers

Volume 57 Gallium Nitride (GaN) II

R. J. Molnar, Hydride Vapor Phase Epitaxial Growth of III–V Nitrides
T. D. Moustakas, Growth of III–V Nitrides by Molecular Beam Epitaxy
Z. Liliental-Weber, Defects in Bulk GaN and Homoepitaxial Layers
C. G. Van de Walle and N. M. Johnson, Hydrogen in III–V Nitrides
W. Götz and N. M. Johnson, Characterization of Dopants and Deep Level Defects in Gallium Nitride
B. Gil, Stress Effects on Optical Properties
C. Kisielowski, Strain in GaN Thin Films and Heterostructures
J. A. Miragliotta and D. K. Wickenden, Nonlinear Optical Properties of Gallium Nitride
B. K. Meyer, Magnetic Resonance Investigations on Group III-Nitrides
M. S. Shur and M. Asif Khan, GaN and AlGaN Ultraviolet Detectors
C. H. Qiu, J. I. Pankove, and C. Rossington, III–V Nitride-Based X-ray Detectors

Volume 58 Nonlinear Optics in Semiconductors I

A. Kost, Resonant Optical Nonlinearities in Semiconductors
E. Garmire, Optical Nonlinearities in Semiconductors Enhanced by Carrier Transport
D. S. Chemla, Ultrafast Transient Nonlinear Optical Processes in Semiconductors
M. Sheik-Bahae and E. W. Van Stryland, Optical Nonlinearities in the Transparency Region of Bulk Semiconductors
J. E. Millerd, M. Ziari, and A. Partovi, Photorefractivity in Semiconductors

Volume 59 Nonlinear Optics in Semiconductors II

J. B. Khurgin, Second Order Nonlinearities and Optical Rectification
K. L. Hall, E. R. Thoen, and E. P. Ippen, Nonlinearities in Active Media
E. Hanamura, Optical Responses of Quantum Wires/Dots and Microcavities
U. Keller, Semiconductor Nonlinearities for Solid-State Laser Modelocking and Q-Switching
A. Miller, Transient Grating Studies of Carrier Diffusion and Mobility in Semiconductors

Volume 60 Self-Assembled InGaAs/GaAs Quantum Dots

Mitsuru Sugawara, Theoretical Bases of the Optical Properties of Semiconductor Quantum Nano-Structures
Yoshiaki Nakata, Yoshihiro Sugiyama, and Mitsuru Sugawara, Molecular Beam Epitaxial Growth of Self-Assembled InAs/GaAs Quantum Dots
Kohki Mukai, Mitsuru Sugawara, Mitsuru Egawa, and Nobuyuki Ohtsuka, Metalorganic Vapor Phase Epitaxial Growth of Self-Assembled InGaAs/GaAs Quantum Dots Emitting at 1.3 μm
Kohki Mukai and Mitsuru Sugawara, Optical Characterization of Quantum Dots
Kohki Mukai and Mitsuru Sugawara, The Photon Bottleneck Effect in Quantum Dots
Hajime Shoji, Self-Assembled Quantum Dot Lasers
Hiroshi Ishikawa, Applications of Quantum Dot to Optical Devices
Mitsuru Sugawara, Kohki Mukai, Hiroshi Ishikawa, Koji Otsubo, and Yoshiaki Nakata, The Latest News

Volume 61 Hydrogen in Semiconductors II

Norbert H. Nickel, Introduction to Hydrogen in Semiconductors II
Noble M. Johnson and Chris G. Van de Walle, Isolated Monatomic Hydrogen in Silicon
Yurij V. Gorelkinskii, Electron Paramagnetic Resonance Studies of Hydrogen and Hydrogen-Related Defects in Crystalline Silicon
Norbert H. Nickel, Hydrogen in Polycrystalline Silicon
Wolfhard Beyer, Hydrogen Phenomena in Hydrogenated Amorphous Silicon
Chris G. Van de Walle, Hydrogen Interactions with Polycrystalline and Amorphous Silicon—Theory
Karen M. McNamara Rutledge, Hydrogen in Polycrystalline CVD Diamond
Roger L. Lichti, Dynamics of Muonium Diffusion, Site Changes and Charge-State Transitions
Matthew D. McCluskey and Eugene E. Haller, Hydrogen in III–V and II–VI Semiconductors
S. J. Pearton and J. W. Lee, The Properties of Hydrogen in GaN and Related Alloys
Jörg Neugebauer and Chris G. Van de Walle, Theory of Hydrogen in GaN

Volume 62 Intersubband Transitions in Quantum Wells: Physics and Device Applications I

Manfred Helm, The Basic Physics of Intersubband Transitions
Jerome Faist, Carlo Sirtori, Federico Capasso, Loren N. Pfeiffer, Ken W. West, Deborah L. Sivco, and Alfred Y. Cho, Quantum Interference Effects in Intersubband Transitions
H. C. Liu, Quantum Well Infrared Photodetector Physics and Novel Devices
S. D. Gunapala and S. V. Bandara, Quantum Well Infrared Photodetector (QWIP) Focal Plane Arrays

Volume 63 Chemical Mechanical Polishing in Si Processing

Frank B. Kaufman, Introduction
Thomas Bibby and Karey Holland, Equipment
John P. Bare, Facilitization
Duane S. Boning and Okumu Ouma, Modeling and Simulation
Shin Hwa Li, Bruce Tredinnick, and Mel Hoffman, Consumables I: Slurry
Lee M. Cook, CMP Consumables II: Pad
François Tardif, Post-CMP Clean
Shin Hwa Li, Tara Chhatpar, and Frederic Robert, CMP Metrology
Shin Hwa Li, Visun Bucha, and Kyle Wooldridge, Applications and CMP-Related Process Problems

Volume 64 Electroluminescence I

M. G. Craford, S. A. Stockman, M. J. Peanasky, and F. A. Kish, Visible Light-Emitting Diodes
H. Chui, N. F. Gardner, P. N. Grillot, J. W. Huang, M. R. Krames, and S. A. Maranowski, High-Efficiency AlGaInP Light-Emitting Diodes
R. S. Kern, W. Götz, C. H. Chen, H. Liu, R. M. Fletcher, and C. P. Kuo, High-Brightness Nitride-Based Visible-Light-Emitting Diodes
Yoshiharu Sato, Organic LED System Considerations
V. Bulović, P. E. Burrows, and S. R. Forrest, Molecular Organic Light-Emitting Devices

Volume 65 Electroluminescence II

V. Bulović and S. R. Forrest, Polymeric and Molecular Organic Light Emitting Devices: A Comparison
Regina Mueller-Mach and Gerd O. Mueller, Thin Film Electroluminescence
Markku Leskelä, Wei-Min Li, and Mikko Ritala, Materials in Thin Film Electroluminescent Devices
Kristiaan Neyts, Microcavities for Electroluminescent Devices

Volume 66 Intersubband Transitions in Quantum Wells: Physics and Device Applications II

Jerome Faist, Federico Capasso, Carlo Sirtori, Deborah L. Sivco, and Alfred Y. Cho, Quantum Cascade Lasers
Federico Capasso, Carlo Sirtori, D. L. Sivco, and A. Y. Cho, Nonlinear Optics in Coupled-Quantum-Well Quasi-Molecules
Karl Unterrainer, Photon-Assisted Tunneling in Semiconductor Quantum Structures
P. Haring Bolivar, T. Dekorsy, and H. Kurz, Optically Excited Bloch Oscillations—Fundamentals and Application Perspectives

Volume 67 Ultrafast Physical Processes in Semiconductors

Alfred Leitenstorfer and Alfred Laubereau, Ultrafast Electron–Phonon Interactions in Semiconductors: Quantum Kinetic Memory Effects
Christoph Lienau and Thomas Elsaesser, Spatially and Temporally Resolved Near-Field Scanning Optical Microscopy Studies of Semiconductor Quantum Wires
K. T. Tsen, Ultrafast Dynamics in Wide Bandgap Wurtzite GaN
J. Paul Callan, Albert M.-T. Kim, Christopher A. D. Roeser, and Eriz Mazur, Ultrafast Dynamics and Phase Changes in Highly Excited GaAs
Hartmut Haug, Quantum Kinetics for Femtosecond Spectroscopy in Semiconductors
T. Meier and S. W. Koch, Coulomb Correlation Signatures in the Excitonic Optical Nonlinearities of Semiconductors
Roland E. Allen, Traian Dumitrică, and Ben Torralva, Electronic and Structural Response of Materials to Fast, Intense Laser Pulses
E. Gornik and R. Kersting, Coherent THz Emission in Semiconductors

Volume 68 Isotope Effects in Solid State Physics

Vladimir G. Plekhanov:, Elastic Properties; Thermal Properties; Vibrational Properties; Raman Spectra of Isotopically Mixed Crystals; Excitons in LiH Crystals; Exciton–Phonon Interaction; Isotopic Effect in the Emission Spectrum of Polaritons; Isotopic Disordering of Crystal Lattices; Future Developments and Applications; Conclusions

Volume 69 Recent Trends in Thermoelectric Materials Research I

H. Julian Goldsmid, Introduction
Terry M. Tritt and Valerie M. Browning, Overview of Measurement and Characterization Techniques for Thermoelectric Materials

Mercouri G. Kanatzidis, The Role of Solid-State Chemistry in the Discovery of New Thermoelectric Materials

B. Lenoir, H. Scherrer, and T. Caillat, An Overview of Recent Developments for BiSb Alloys

Ctirad Uher, Skutterudites: Prospective Novel Thermoelectrics

George S. Nolas, Glen A. Slack, and Sandra B. Schujman, Semiconductor Clathrates: A Phonon Glass Electron Crystal Material with Potential for Thermoelectric Applications

Volume 70 Recent Trends in Thermoelectric Materials Research II

Brian C. Sales, David G. Mandrus, and Bryan C. Chakoumakos, Use of Atomic Displacement Parameters in Thermoelectric Materials Research

S. Joseph Poon, Electronic and Thermoelectric Properties of Half-Heusler Alloys

Terry M. Tritt, A. L. Pope, and J. W. Kolis, Overview of the Thermoelectric Properties of Quasicrystalline Materials and Their Potential for Thermoelectric Applications

Alexander C. Ehrlich and Stuart A. Wolf, Military Applications of Enhanced Thermoelectrics

David J. Singh, Theoretical and Computational Approaches for Identifying and Optimizing Novel Thermoelectric Materials

Terry M. Tritt and R. T. Littleton, IV, Thermoelectric Properties of the Transition Metal Pentatellurides: Potential Low-Temperature Thermoelectric Materials

Franz Freibert, Timothy W. Darling, Albert Migliori, and Stuart A. Trugman, Thermomagnetic Effects and Measurements

M. Bartkowiak and G. D. Mahan, Heat and Electricity Transport through Interfaces

Volume 71 Recent Trends in Thermoelectric Materials Research III

M. S. Dresselhaus, Y.-M. Lin, T. Koga, S. B. Cronin, O. Rabin, M. R. Black, and G. Dresselhaus, Quantum Wells and Quantum Wires for Potential Thermoelectric Applications

D. A. Broido and T. L. Reinecke, Thermoelectric Transport in Quantum Well and Quantum Wire Superlattices

G. D. Mahan, Thermionic Refrigeration

Rama Venkatasubramanian, Phonon Blocking Electron Transmitting Superlattice Structures as Advanced Thin Film Thermoelectric Materials

G. Chen, Phonon Transport in Low-Dimensional Structures

ISBN 0-12-752181-X